Fishes Out of Water

Biology and Ecology of Mudskippers

Marine Science Series

The CRC Marine Science Series is dedicated to providing state-of-the-art coverage of important topics in marine biology, marine chemistry, marine geology, and physical oceanography. The series includes volumes that focus on the synthesis of recent advances in marine science.

CRC MARINE SCIENCE SERIES

SERIES EDITORS

Michael J. Kennish, Ph.D. and Judith S. Weis, Ph.D.

PUBLISHED TITLES

Acoustic Fish Reconnaissance, I.L. Kalikhman and K.I. Yudanov
Artificial Reef Evaluation with Application to Natural Marine Habitats, William Seaman, Jr.
The Biology of Sea Turtles, Volume I, Peter L. Lutz and John A. Musick
Chemical Oceanography, Third Edition, Frank J. Millero
Coastal Ecosystem Processes, Daniel M. Alongi
Coastal Lagoons: Critical Habitats of Environmental Change, Michael J. Kennish and Hans W. Paerl
Coastal Pollution: Effects on Living Resources and Humans, Carl J. Sindermann
Climate Change and Coastal Ecosystems: Long-Term Effects of Climate and Nutrient Loading on Trophic Organization, Robert J. Livingston
Ecology of Estuaries: Anthropogenic Effects, Michael J. Kennish
Ecology of Marine Bivalves: An Ecosystem Approach, Second Edition, Richard F. Dame
Ecology of Marine Invertebrate Larvae, Larry McEdward
Ecology of Seashores, George A. Knox
Environmental Oceanography, Second Edition, Tom Beer
Estuarine Indicators, Stephen A. Bortone
Estuarine Research, Monitoring, and Resource Protection, Michael J. Kennish
Estuary Restoration and Maintenance: The National Estuary Program, Michael J. Kennish
Eutrophication Processes in Coastal Systems: Origin and Succession of Plankton Blooms and Effects on Secondary Production in Gulf Coast Estuaries, Robert J. Livingston
Fishes Out of Water: Biology and Ecology of Mudskippers, Zeehan Jaafar and Edward O. Murdy
Habitat, Population Dynamics, and Metal Levels in Colonial Waterbirds: A Food Chain Approach, Joanna Burger, Michael Gochfeld
Handbook of Marine Mineral Deposits, David S. Cronan
Handbook for Restoring Tidal Wetlands, Joy B. Zedler
Intertidal Deposits: River Mouths, Tidal Flats, and Coastal Lagoons, Doeke Eisma
Living Shorelines: The Science and Management of Nature-Based Coastal Protection, Donna Marie Bilkovic, Molly M. Mitchell, Megan K. La Peyre, and Jason D. Toft
Marine Chemical Ecology, James B. McClintock and Bill J. Baker
Ocean Pollution: Effects on Living Resources and Humans, Carl J. Sindermann
Physical Oceanographic Processes of the Great Barrier Reef, Eric Wolanski
Pollution Impacts on Marine Biotic Communities, Michael J. Kennish
Practical Handbook of Estuarine and Marine Pollution, Michael J. Kennish
Practical Handbook of Marine Science, Third Edition, Michael J. Kennish
Restoration of Aquatic Systems, Robert J. Livingston
Seagrasses: Monitoring, Ecology, Physiology, and Management, Stephen A. Bortone
Trophic Organization in Coastal Systems, Robert J. Livingston

Fishes Out of Water

Biology and Ecology of Mudskippers

EDITED BY

Zeehan Jaafar • Edward O. Murdy

CRC Press
Taylor & Francis Group
Boca Raton London New York

CRC Press is an imprint of the
Taylor & Francis Group, an **informa** business

CRC Press
Taylor & Francis Group
6000 Broken Sound Parkway NW, Suite 300
Boca Raton, FL 33487-2742

First issued in paperback 2020

ISBN 13: 978-0-367-57343-0 (pbk)
ISBN 13: 978-1-4987-1787-8 (hbk)

Library of Congress Cataloging-in-Publication Data

Names: Zeehan Jaafar, author. | Murdy, Edward O., author.
Title: Fishes out of water : biology and ecology of mudskippers / Zeehan Jaafar and Edward O. Murdy.
Description: Boca Raton : Taylor & Francis, 2017. | Series: CRC marine science series | Includes index.
Identifiers: LCCN 2016055509 | ISBN 9781498717878 (hardback : alk. paper)
Subjects: LCSH: Mudskippers.
Classification: LCC QL638.G7 Z44 2017 | DDC 597/.7—dc23
LC record available at https://lccn.loc.gov/2016055509

Visit the Taylor & Francis Web site at
http://www.taylorandfrancis.com

and the CRC Press Web site at
http://www.crcpress.com

Dedication

Dr. Toru Takita on an expedition to study mudskippers in Ca Mau, Vietnam, June 2013. (Courtesy of Atsushi Ishimatsu.)

IN MEMORY OF DR. TORU TAKITA

SEPTEMBER 18, 1936–AUGUST 2, 2014

Dr. Toru Takita was born in 1936 in Busan, Korea. He began his academic career at Nagasaki University, Japan, in March 1963 where he remained until his retirement in March 2002. Post retirement, Dr. Takita was appointed as Professor Emeritus in the Faculty of Fisheries at Nagasaki University. As a recognized international expert on the natural history of mudskippers, we originally envisioned him being involved in this edited volume. Regretfully, terminal illness prevented him from participating. Having studied mudskippers in many countries within the Indo-Pacific, Dr. Takita befriended and inspired many mudskipper enthusiasts, several of whom are authors in this volume. His contribution to our understanding of the ecology and natural history of mudskippers is immense, exemplified by the accounts of friends and colleagues below. We thus dedicate this book in his memory.

My first encounter with Takita-san was via an e-mail in 2001, when he contacted me about the work he had been doing on the mudskippers and mudflats in Broome, Western Australia, as he thought he had found at least one new mudskipper species and several new records for Australia. Takita-san sent me specimens, and a number of carefully prepared data sheets and observations on color and sexual variation. Finally, I got to meet him when he visited Darwin with several students and associates to carry out ecological studies on mudskippers and their habitats, on their way back from another Broome trip. He came again in 2004 with a different cohort of students and associates. I remember giving them all dinner on our veranda—as it was monsoon season with high humidity and temperature, everyone was very hot and sticky but still enjoyed their barramundi and kangaroo meal. I was greatly impressed by the way Takita-san worked. I have spent a lot of time in mangroves and mud and know how much hard work it is (wading or crawling through mud is not easy, especially when pursued by biting midges and 33°C heat). The length of time that 70-something-year-old Takita-san and his much younger group spent out in the mangroves was impressive; he came in only when the tide began to engulf them. Then he would write up notes, take photographs, record data, and always leave me with specimens, notes and data, and the odd mud-map when he flew home. We collaborated on several publications, taxonomic and ecological—and only misidentified a couple of mudskippers. I am very glad that Zeehan Jaafar and I named Periophthalmus takita *for him. One of my great regrets is that I did not photograph him. He still lives in my heart as one of the most generous scientists I have ever met.*

Helen K. Larson

I never had the privilege of meeting Takita-san, but we were in written correspondence since the 1990s. I remember the one time I could have met him when Dr. Helen Larson hosted me in Darwin, Australia in 2007, but I arrived after he had left. On Helen's table was a Japanese culinary delicacy, a tin of canned Japanese mudskippers, a gift from Takita-san. Professor Takita was an uncommon scholar in several ways. He was one of the few field ecologists who studied mudskippers, was an enthusiast of tropical tidal mudflats, and was a kind and resourceful mentor. Takita-san helped, encouraged, and inspired me for many years when I was a student from a remote "no-mudskippers" country with no institutional backup, and kept assisting me with ideas, scientific material, contacts, photos, and scientific references in Japanese and English. He would always reply to an e-mail. Takita-san was one of the last scientists (the only one I know of) who used the Japanese traditional technique called mutsukake, *to collect* Boleophthalmus *mudskippers. I found his specimens in several museum collections, marked by the unmistakable tiny holes of the needle-like harpoon. He was one of the very few scientists who would walk in deep thixotropic intertidal mud, chasing or observing mudskippers. Takita-san did that until late in life, and once he wrote me he was sorry his sore back prevented him from walking in the mud anymore. He was one of the few old-school ecologists, still extremely careful about the taxonomic discrimination of his specimens, and often collaborated with fish taxonomists, striving to provide keys to identify mudskippers in the field. Among his numerous contributions, he described the species* Periophthalmus spilotus *Murdy and Takita, 1999 and* Periophthalmus darwini *Larson and Takita, 2004, and provided descriptions of the spatial distribution of mudskipper communities of the Strait of Malacca and northern Australia. I was honored to be among the peer reviewers of the latter study. The mudskipper* Periophthalmus takita *Jaafar and Larson 2008, was named after him, in recognition of his many scientific contributions on mudskippers.*

Gianluca Polgar

I had the pleasure of corresponding with Toru Takita over several years discussing all things mudskipper. We corresponded via e-mails in which he freely shared his thoughts and ideas. I regret that we never met and that our communication started only later in my career; there was so much more to learn from this generous mentor. In 2008, Helen Larson and I named a mudskipper from Australia after him, a befitting tribute to a man devoted to pursuing odd critters in the mangroves and mudflats of the Indo-Pacific.

Zeehan Jaafar

I first met Toru Takita at the annual meeting of the Ichthyological Society of Japan (ISJ), in 1996. We had been in e-mail communication previously as Takita-san contacted me about what he thought was a new species of Periophthalmus *from Malaysia.*

I was dismissive of his notion as the pictures he sent me indicated that the pelvic fins were completely fused, which is atypical for the genus and was (at that time) only known in a single species of Periophthalmus *(*Ps. chrysospilos*). As his pictured mudskipper was not* Ps. chrysospilos, *I told him that I thought his "new" species was probably a juvenile* Periophthalmodon schosseri *(which also has united pelvic fins). How wrong I was as Professor Takita proceeded to tell me in thorough detail when we met at ISJ. After our first meeting, I never doubted him again. Subsequently, he and I described the new species as* Ps. spilotus *(Murdy and Takita 1999). I initially proposed naming this new species after him, but he demurred saying that he was not worthy of such an honor. In October 1997, Professor Takita invited me to visit him in Nagasaki. One of the places he wanted to show me was Isahaya Bay where the Ministry of Construction had undertaken a massive project to reclaim farmland. The resulting 7-kilometer seawall (dike) was completed earlier that year depriving Isahaya Bay of tidal waters from the Ariake Sea. What were once tidal wetlands had been converted into 1500 hectares of farmland. One tragic consequence of this construction project was the loss of marine life including all the resident mudskippers. Takita-san was distraught. He claimed this was a project to "spend money" and that the newly reclaimed farmland was not needed. The loss of mudskippers and their habitat was clearly difficult for him to bear. It was evident to me that his passion for mudskippers was sincere and abiding. He told me at that time that once he retired, he would spend all of his available time studying mudskippers. And … he did.*

Edward O. Murdy

Takita-sensei (teacher in Japanese) invited me to join his mudskipper research team in Penang in 1994. Our mudskipper study then expanded to other places in Malaysia, Japan, Korea, and Australia. And since 2011, we worked on another mudskipper study in the Mekong Delta, Vietnam. Nine years earlier, he had retired from Nagasaki University, but Takita-sensei's enthusiasm and energy in the field never subsided until his very last trip to Vietnam. We usually visited our study site early in the morning and worked until early afternoon. Under the tropical sunlight, working in mudflats or mangroves was exhausting after only half a day. So we would return to the hotel to rest. One afternoon at our hotel in Bac Lieu Province, I knocked on Takita-sensei's door.

*When I entered the room, I saw him busily working, sorting out larval and juvenile mudskippers from the samples we collected in the morning. While Takita-sensei was working, I along with the other members of our group and I took showers and then a nap, and maybe drank one can of beer. At that time, Takita-sensei was 75 years old. In March 2014, I received 11 manuscript files from Takita-sensei. His wish was to publish a book on the biology of mudskippers for the general public. He told me that we were responsible for disseminating the knowledge gained from scientific research into plain words for the general public because it was these people who supported science by paying taxes. He showed me a book on sparrows and said that he wanted to publish a book like that about mudskippers. At that time, he had cancer and probably knew that it would be difficult for him to finish the book alone. It took more than one year to realize his last dream, but it was too late for him. I can only hope that Takita-sensei would have liked our book, "*水から出た魚たち—Fish Emerging from Water—*" by Toru Takita and Atsushi Ishimatsu (July 2015).*

Atsushi Ishimatsu

REFERENCES

Jaafar Z. and Larson H.K. 2008. A new species of mudskipper, *Periophthalmus takita* (Teleostei: Gobiidae: Oxudercinae), from Australia, with a Key to the Genus. *Zoological Science* 25:946–952.

Larson H.K. and Takita T. 2004. Two new species of *Periophthalmus* (Teleostei: Gobiidae: Oxudercinae) from Northern Australia, and a re-diagnosis of *Periophthalmus novaeguineaensis*. The Beagle: *Records of the Museums and Art Galleries of the Northern Territory* 20:175–1855.

Murdy E.O. and Takita T. 1999. *Periophthalmus spilotus*, a new species of mudskipper from Sumatra (Gobiidae: Oxudercinae). *Ichthyological Research* 46:367–370.

Contents

Foreword

This book treats in detail, the biology of mudskippers, a group of fishes poorly known even to the scientific community. Mudskippers belong to one of the largest groups of fishes, commonly known as "gobies." Together with their close relatives, gobies number over 2000 species and many new species continue to be discovered every year. Almost a quarter of the gobioid (or gobies and relatives) species now known have only been described in scientific literature in the last 15–20 years. This highly diverse group of fishes are ubiquitous—they are found in freshwaters, estuaries, and marine habitats; in cold temperate, subtropical, and tropical environments; from shallow intertidal pools to depths of over 1000 meters; and from seashores to caves. Despite their high diversity and abundance, the knowledge of gobies as a group is largely in a discovery phase with relatively little information published on their biology and ecology.

Within the gobies, the mudskippers are unique in their ability to spend considerable time out of water. The habitats of the mudskippers, typically mosquito- and crocodile-infested mudflats and mangrove forest environments, often discourage scientists from studying these fishes. The few that prevailed have come to enjoy the challenges presented when working in areas such as these. Because of the dedicated efforts of these individuals, many of whom are contributing authors in this volume, mudskippers are studied more extensively than many other groups of gobies. It is not difficult to imagine why these authors were drawn to the unusual and fascinating behaviors of mudskippers. Aside from foraging, mudskipper activities while out of water include constructing nests and burrows, defending territories, and courtship displays. As you might imagine, such terrestrial doings are atypical of fishes and require physical adaptations for the aerial environment. Consequently, aspects such as mudskipper anatomy, ecology, and physiology have been studied and published in hundreds of scientific articles. However, until now, there has not been any major synthesis of the overall knowledge of mudskippers.

I believe this book will have wide appeal to scientists and amateur naturalists alike. The authors and editors have collaborated to produce an accurate and exceptional review of this diverse group of fishes. The broad range of expertise of these authors has ensured that aspects of mudskipper biology and ecology are treated comprehensively. Undoubtedly, the contents of this book will provide a framework that will help advance future studies on mudskippers and other organisms found in mangrove environments. The book is particularly timely as we experience continued massive destruction of mangrove environments in the Indo-Pacific region, effectively compromising the rendered ecosystem services and destroying the habitats of mudskippers.

In short, this book will allow the reader an understanding of the biology, ecology, and unique adaptations of these amphibious fishes. It is with great pleasure that I commend this book to you.

Doug Hoese
Senior Fellow
Australian Museum
Sydney, Australia

Preface

Mudskipper in mangrove. A mudskipper, *Periophthalmus* sp. rests on a pneumatophore of a mangrove plant, *Bruguiera* sp., during ebb tide at Paya Mangrove Forest, Tioman Island, Peninsular Malaysia. (Courtesy of Law Ing Sng.)

Mudskippers have long captured the fascination of scientists, naturalists, and aquarists because of their amphibious nature. Many species tolerate extended periods out of water, emerging during low tide to forage and seek mates. These fishes occur in mangrove forests and on mudflats throughout the Indo-West Pacific and tropical western coast of Africa. Their ability to move with speed and agility over the muddy substrate earned them the common name "mudskippers". The close association of mudskippers to mangrove forests and mudflats make these fishes excellent candidates as bioindicators for these habitats. Mudskippers are widely consumed and form part of the subsistence catches of coastal communities in Sierra Leone, Bangladesh, India, Japan, Vietnam, and Papua New Guinea. In China, Taiwan, and Thailand, mudskippers are reared in aquaculture facilities, although fry are still obtained from wild stocks. Raw mudskippers are considered an aphrodisiac in Peninsular Malaysia and juveniles are sometimes eaten whole for this purpose. So ubiquitous are these fishes that they feature in a Malay folklore about a mother abandoning her children for her craving of mudskipper eggs.

Greenshank and mudskipper. The common greenshank, *Tringa nebularia*, preys on a mudskipper, *Boleophthalmus pectinirostris*, on an exposed mudflat during ebb tide at Mai Po Nature Reserve, Hong Kong, China. (Courtesy of Wilson Dring.)

The aquatic and terrestrial habitats which mudskippers inhabit, pose drastically different challenges to these fishes. Yet, many mudskipper species display mechanical, physiological, and behavioral adaptations that allow unfettered use of both habitat types. Modifications to the skeleton and associated musculature compensate for the loss of buoyancy afforded by the aquatic medium while these fishes are on land. These are exemplified in the tail-standing behavior of *Scartelaos histophorus* and the climbing behavior of many species of *Periophthalmus*. Modifications to the gill structure and hyper-vascularization of epithelial tissues facilitate aerial respiration while out of water. Fundamental sensory organs and receptors are also adapted to simultaneously accommodate both terrestrial and aquatic habitats. Mudskippers also build burrows in the mud substrate into which some species retreat during the flood tide. In egg chambers within these burrows, mudskippers maintain developing eggs within air pockets. Air within the burrow is periodically replenished, typically by male mudskippers that transport air from the surface in their buccal cavities, to ensure enough oxygen for the developing eggs. These eggs come into contact with water only when they are ready to hatch, at which time, the egg chambers are flooded and the larvae disperse upon hatching.

These mechanical, physiological, and behavioral departures from those of typical fishes make mudskippers the subject of many scientific studies. The wealth of available published information, gray literature, and experiences of mudskipper researchers and enthusiasts galvanized us to gather and synthesize the disparate information. When approached, 24 colleagues from 9 countries readily agreed to contribute chapters on topics of their expertise. Together, we aim to fill the current void in available literature for a comprehensive and authoritative text on mudskippers. In the 15 chapters within this resultant volume, specialist contributors detail various aspects of mudskipper diversity, evolution, distribution, biology, ecology, and conservation. Offering novel insights, new data, and a fresh perspective, we hope this book is a primer for budding mudskipper enthusiasts in their quest to further the knowledge on this group of enigmatic fishes.

Zeehan Jaafar
Edward O. Murdy

Species List List of recognized mudskipper species and abbreviations used in this book. References for authority names appear at the end of Chapter 1.

Genera	Genus Abbreviation	Species	Authority
Apocryptes	*As.*	*bato*	(Hamilton 1822)
Apocryptodon	*Ap.*	*madurensis*	(Bleeker 1849)
		punctatus	Tomiyama 1934
		wirzi	Koumans 1937
Boleophthalmus	*B.*	*birdsongi*	Murdy 1989
		boddarti	(Pallas 1770)
		caeruleomaculatus	McCulloch and Waite 1918
		dussumieri	Valenciennes 1837
		pectinirostris	(Linnaeus 1758)
		poti	Polgar, Jaafar, and Konstantinidis 2013
Oxuderces	*O.*	*dentatus*	Eydoux and Souleyet 1837
		nexipinnis	(Cantor 1849)
Parapocryptes	*Pa.*	*rictuosus*	(Valenciennes 1837)
		serperaster	(Richardson 1846)
Periophthalmodon	*Pn.*	*freycineti*	Valenciennes 1837
		schlosseri	(Pallas 1770)
		septemradiatus	(Hamilton 1822)
Periophthalmus	*Ps.*	*argentilineatus*	Valenciennes 1837
		barbarus	(Linnaeus 1766)
		chrysospilos	Bleeker 1853
		darwini	Larson and Takita 2004
		gracilis	Eggert 1935
		kalolo	Lesson 1831
		magnuspinnatus	Lee, Choi, and Ryu 1995
		malaccensis	Eggert 1935
		minutus	Eggert 1935
		modestus	Cantor 1842
		novaeguineaensis	Eggert 1935
		novemradiatus	Hamilton 1822
		pusing	Jaafar, Polgar, and Zamroni 2016
		spilotus	Murdy and Takita 1999
		takita	Jaafar and Larson 2008
		variabilis	Eggert 1935
		walailakae	Darumas and Tantichodok 2002
		waltoni	Koumans 1941
		weberi	Eggert 1935
Pseudapocryptes	*Pd.*	*borneensis*	(Bleeker 1855)
		elongatus	(Cuvier 1816)
Scartelaos	*S.*	*cantoris*	(Day 1871)
		gigas	Chu and Wu 1963
		histophorus	(Valenciennes 1837)
		tenuis	(Day 1876)
Zappa	*Z.*	*confluentus*	(Roberts 1978)

Contributors

Chew, Shit F.
Natural Sciences and Science Education,
National Institute of Education
Nanyang Technological University
1 Nanyang Walk
Singapore 637616
E-mail: sfun.chew@nie.edu.sg

Clayton, David
Department of Biology
Faculty of Science
P.O. Box 36
Sultan Qaboos University
Al-Khoud 123, Sultanate of Oman
E-mail: david@squ.edu.om

Ip, Yuen K.
Department of Biological Sciences
National University of Singapore
10 Kent Ridge Road
Singapore 119260
E-mail: dbsipyk@nus.edu.sg

Ishimatsu, Atsushi
Institute for East China Sea Research
Nagasaki University
1551-7 Tairamachi
Nagasaki 851-2213, Japan
E-mail: a-ishima@nagasaki-u.ac.jp

Jaafar, Zeehan
Division of Fishes
National Museum of Natural History
Smithsonian Institution
P.O. Box 37012 MRC 159
Washington, DC 20013
E-mail: jaafarz@si.edu; jaafarz@nus.edu.sg

Kuciel, Michał
Poison Information Centre, Department of
Toxicology and Environmental Disease
Jagiellonian University Medical College
ul. Kopernika 15
31-501 Kraków, Poland
E-mail: michalkuciel@gmail.com

Larson, Helen K.
Museum and Art Gallery of the Northern
Territory
P.O. Box 4646
Darwin, Northern Territory 0801, Australia
E-mail: helen.larson@nt.gov.au

Lauriano, Eugenia R.
Department of Chemical, Biological,
Pharmaceutical and Environmental Sciences
University of Messina
Viale F. Stagno d'Alcontres 31
98166 Messina, Italy
E-mail: elauriano@unime.it

Maeda, Ken
Marine Genomics Unit
Okinawa Institute of Science and Technology
Graduate University
1919-1 Tancha, Onna
Okinawa 904-0495, Japan
E-mail: goby@live.jp

Malavasi, Stefano
Dipartimento di Scienze Ambientali,
Informatica e Statistica
Università Ca'Eoscari Venezia
Via Torino 155
30172 Venezia Mestre, Italy
E-mail: mala@unive.it

Martin, Karen L.M.
Department of Biology
Pepperdine University
Malibu, California 90263
E-mail: karen.martin@pepperdine.edu

Mleczko, Richard
c/o NOPTA
GPO Box 4634
Melbourne Victoria 3001, Australia
E-mail: richard.mleczko@bigpond.com

Murdy, Edward O.
Department of Biological Sciences
George Washington University
Washington, DC 20052
E-mail: eomurdy@gwu.edu

Nanami, Atsushi
Research Center for Sub-tropical Fisheries
Seikai National Fisheries Research Institute
Japan Fisheries Research and Education Agency
Fukai-Ota 148
Ishigaki, Okinawa 907-0451, Japan
E-mail: nanami@fra.affrc.go.jp

Pace, Cinnamon
Department of Biological and Environmental Sciences
Le Moyne College
Syracuse, New York 13214
E-mail: pacecm@lemoyne.edu

Parenti, Lynne R.
Division of Fishes, National Museum of Natural History
Smithsonian Institution
P.O. Box 37012 MRC 159
Washington, DC 20013-7012
E-mail: parentil@si.edu

Polgar, Gianluca
Environmental and Life Sciences Programme
Faculty of Science
Universiti Brunei Darussalam
Jalan Tungku Link, BE 1410
Brunei, Darussalam
E-mail: gianluca.polgar@gmail.com, gianluca.polgar@ubd.edu.bn

Polidoro, Beth
School of Mathematics and Natural Sciences
Arizona State University
4701 W Thunderbird Rd.
Glendale, Arizona 85306
E-mail: beth.polidoro@asu.edu

Rupp, Hans-Georg
Holunderweg 5
65760 Niederhöchstadt, Germany
E-mail: hagerupp@aol.com

Tachihara, Katsunori
Laboratory of Fisheries Biology and Coral Reef Studies
Faculty of Science
University of the Ryukyus
1 Senbaru
Nishihara, Okinawa 903-0213, Japan
E-mail: ktachiha@sci.u-ryukyu.ac.jp

Townsend, Kathy
School of Biological Sciences
University of Queensland
St. Lucia, Queensland 4072
Australia
E-mail: kathy.townsend@uq.edu.au

Tsuhako, Yuko
Laboratory of Fisheries Biology and Coral Reef Studies
Faculty of Science
University of the Ryukyus
1 Senbaru
Nishihara, Okinawa 903-0213, Japan
E-mail: ytsuhako@mail.goo.ne.jp

Zaccone, Giacomo
Department of Biomedical and Dental Sciences and Morphofunctional Imaging
Faculty of Science, Messina University
Viale Stagno d'Alcontres 31
I-98166 Messina, Italy
E-mail: gzaccone@unime.it

Żuwała, Krystyna
Department of Comparative Anatomy, Institute of Zoology and Biomedical Research
Faculty of Biology and Environmental Science
Jagiellonian University
ul. Gronostajowa 9
30-387 Kraków, Poland
E-mail: krystyna.zuwala@uj.edu.pl

CHAPTER 1

Taxonomy and Systematics Review

Edward O. Murdy and Zeehan Jaafar

CONTENTS

1.1 INTRODUCTION

Oxudercine gobies, commonly known as mudskippers, are tropical and subtropical fishes naturally occurring in shallow sublittoral, littoral, and supralittoral zones of the Indo-Pacific and western Africa. Ten genera—*Apocryptes*, *Apocryptodon*, *Boleophthalmus*, *Oxuderces*, *Parapocryptes*, *Periophthalmodon*, *Periophthalmus*, *Pseudapocryptes*, *Scartelaos*, and *Zappa*—comprising 43 species, are presently recognized (Jaafar and Parenti 2016; Polgar et al. 2013; Jaafar and Larson 2008; Murdy 1989). These fishes commonly associate with soft-bottom habitats, especially mangrove forests and exposed mudflats. Species such as *Periophthalmodon septemradiatus* and *Periophthalmus weberi* also occur further upriver, in areas where salinity is negligible (Larson 2008). Within the Indo-Pacific, mudskippers are distributed longitudinally from the Red Sea/East Africa (40° E) to Samoa/Tonga (165° W) and latitudinally from Japan/South Korea (35° N) to Australia (20° S) (Polgar et al. 2014; Murdy 1989). Along the western African coast, a single species of mudskipper, *Periophthalmus barbarus*, is recognized and its natural distribution is from Morocco south to northern Angola (Murdy 2016).

1.2 HISTORICAL ACCOUNTS OF OXUDERCINE GOBIES

The first published account of an oxudercine goby can be traced to a Dutch explorer, de Vlamingh (1701), who reported on his voyage to the Dutch East Indies (specifically Batavia or modern day

Figure 1.1 One of the first, if not the first, mudskipper illustration from the 1690s. (From Holthuis L.B. and Pietsch T.W. *Les planches inédites de poissons et autres animaux marins de l'Indo-Ouest Pacifique d'Isaac Johannes Lamotius*, Publications Scientifiques du Muséum, Muséum National d'Histoire Naturelle, Paris, 2006. Courtesy of National Museum of Natural History (Paris), Directorate of Libraries and Documentation.)

Jakarta, Indonesia) more than 400 years ago. Subsequent inclusion of oxudercine gobies in early published scientific literature also resulted from European voyages to tropical areas during which flora and fauna native to those locales were collected (Figure 1.1). These organisms, including the mudskippers, were unfamiliar to both voyagers and early European scholars; they were considered novel or at times mystical, and their accounts often featured crude figures and descriptions (Prevost 1747; Valentijn 1726; Ruysch 1718).

The earliest oxudercine gobies were described by Carl Linnaeus: *Gobius pectinirostris* (=*Boleophthalmus pectinirostris*) in 1758 and *Gobius barbarus* (=*Periophthalmus barbarus*) in 1766. The type specimen of *Gobius pectinirostris* is still extant in the Linnaean collection of the Uppsala University, Zoological Museum, Sweden. Important post-Linnaean studies on the systematics of oxudercine gobies include works from prominent ichthyologists, among them, Pallas (1770), Hamilton (1822), and Eggert (1935). Bloch and Schneider (1801) coined the name *Periophthalmus* and designated a new species they described, *Periophthalmus papilio (=Periophthalmus barbarus),* as the type species for this genus.

1.3 HIGHER LEVEL RELATIONSHIPS

Ichthyologists Albert Günther, Pieter Bleeker, and David Starr Jordan were among the first to propose classification systems based on the interrelationships of goby-like fishes (see Van Tassell et al. 2011, for details). Günther (1861) erected Oxudercidae for a single species, *Oxuderces dentatus*, whereas he placed all the other mudskipper genera within the Gobiidae together with some other non-mudskipper taxa. Bleeker (1874) classified all then-known mudskipper genera into two groups within the subfamily Gobiiformes: the Apocrypteini and Periophthalmini. In his extensive treatment of the suborder Gobioidei, Jordan (1923) placed mudskipper genera into two families, the Gobiidae and Periophthalmidae.

These seminal works formed the basis in the treatment of gobioid fishes by Koumans (1953; 1931) who essentially followed Bleeker (1874) in his classification of mudskipper genera. Unlike Bleeker, Koumans elevated the groupings of mudskippers to the subfamilial levels: the Apocrypteinae and Periophthalminae. Using osteological characters, Miller (1973) divided gobioid fishes in the suborder Gobioidei into two families: the Rhyacichthyidae and Gobiidae. The latter was further subdivided into seven subfamilies, one of which, the Gobionellinae, comprised all mudskipper genera and some non-mudskipper genera. Although the name Oxudercidae was a junior synonym of the name Gobiidae, Springer (1978) demonstrated that the name had priority over more recent subfamilial group names such as Apocrypteinae, Gobionellinae, and Periophthalminae. Hence, Oxudercinae came into use as the group name encompassing all mudskippers.

The Oxudercinae was one of four subfamilies within the family Gobiidae recognized by Hoese (1984); the other three subfamilies were the Amblyopinae, Gobiinae, and Sicydiinae. Seven characters defined the Oxudercinae *sensu* Hoese (1984): (1) a single pterygiophore preceding the first hemal spine; (2) the lateral process of sphenotic large and not in contact with the eye; (3) eyes displaced forward and upward; (4) tongue fused to floor of mouth; (5) teeth flattened; (6) second dorsal fin usually long-based; and (7) two epurals present. However, only the first three of the above characters are presently considered synapomorphies of the Oxudercinae as the rest are shared with at least one other group of gobiid fishes (Murdy 1989).

Murdy (1989) redefined the characters uniting the Oxudercinae in the following five synapomorphies: (1) a complex character involving displacement of the eyes anterodorsally, elongation of the frontals, shift in location of the sphenotics such that they are no longer in contact with the eye, and rearrangement of adductor mandibulae complex such that it also originates from the frontal as well as the sphenotic and preopercular bones; (2) anterior naris at tip of pendulous flap that overlaps the upper jaw; (3) the ventroposteriorly directed process of the palatine bone is greatly reduced such that there is little or no overlap with the ectopterygoid bone; (4) ascending processes of the premaxillae bones greatly reduced and oriented vertically; and (5) a single pterygiophore anterior to first hemal spine. Characters (1), (2), and (4) above are unique to the Oxudercinae, whereas character (5) is consistently found in two other gobiid genera, *Discordipinna* and *Kelloggella*, as well as in the gobiid subfamily Ptereleotrinae *sensu* Hoese (1984) (Murdy 1989). These non-oxudercine gobies possess character distributions that make a close relationship with the Oxudercinae extremely unlikely (Murdy 1989). Character state (3) is approximated in several amblyopine genera (Harrison 1989; Murdy 1989) and is further discussed in Section 1.4.

Using characters associated with the palatopterygoquadrate complex, Harrison (1989) considered mudskippers to be allied with the "*Oxyurichthys* lineage" within Gobionellinae *sensu* Miller (1973). The "*Oxyurichthys* lineage" of Harrison (1989) included *Oxyurichthys* as well as all oxudercine (mudskippers) and amblyopine (eel gobies) taxa. Pezold (1993) recognized a monophyletic Oxudercinae as one of five subfamilies within Gobiidae, the others being Amblyopinae, Gobiinae, Gobionellinae, and Sicydiinae. The classification and subfamilial names proposed by Pezold (1993) matched those of Hoese (1984), with the addition of the Gobionellinae. We consider the mudskippers to comprise a phylogenetically discrete group, the subfamily Oxudercinae, within the family Gobiidae.

The advent of phylogenetic studies using molecular data over the past two decades introduced yet another approach in elucidating intertaxa relationships. Hypotheses of higher level relationships of the Oxudercinae derived from molecular datasets differ from those derived from morphological data. The first molecular analysis of gobioid fishes used a single mitochondrial DNA gene, cytochrome *b* (*cytb*) sequences, from 28 gobioid species to assess relationships within this speciose group (Akihito et al. 2000). The oxudercine representative in this study was *Periophthalmus argentilineatus*, which was sister to the only amblyopine goby representative, *Taenioides limicola* (=*Trypauchenopsis intermedia*). Using 12S rRNA (1003 base pairs) and $tRNA^{Val}$ (75 base pairs) gene sequences from 43 gobioid species, Wang et al. (2001) obtained a clade comprising the Amblyopinae, Gobionellinae, Oxudercinae, and Sicydiinae. In her study using complete sequences from three mitochondrial DNA genes, *ND1*, *ND2*, and *COI* (3573 base pairs) from 67 gobioid species, Thacker (2003) sampled three oxudercine species from three genera (*Periophthalmus barbarus*, *Pseudapocryptes elongatus*, and *Scartelaos histophorus*) and one amblyopine species (*Odontamblyopus rubicundus*) and recovered *Scartelaos histophorus* as sister to *Odontamblyopus rubicundus*, and together, this clade was sister to a clade comprising *Pseudapocryptes elongatus* and *Periophthalmus barbarus*. These four species formed a clade indicating that the Oxudercinae was paraphyletic with respect to the Amblyopinae. The oxudercine/amblyopine taxa were nested within a monophyletic clade referred to as the "expanded Gobionellidae," which also included the

Gobionellinae and Sicydiinae (Thacker 2003). In a follow-up study, Thacker (2009) analyzed mitochondrial DNA sequence data from four genes *ND1*, *ND2*, *COI*, and *cytb*, for 107 gobioid species, but the same oxudercine and amblyopine taxa, as in the previous study (i.e., Thacker 2003) were used. Conclusions pertaining to the paraphyly of the oxudercine and amblyopine gobies in Thacker (2003) remained in Thacker (2009). Despite these taxa forming a monophyletic clade, the in-group relationships of the sampled oxudercine genera in Thacker (2009) were different from that recovered in Thacker (2003). Thacker (2009) recovered *Pseudapocryptes elongatus* as sister to *Odontamblyopus rubicundus*, which was in turn sister to *Scartelaos histophorus*. These three species were sister to *Periophthalmus barbarus*.

Based on the clades recovered from analyses of the above four mitochondrial genes, Thacker (2009) proposed a classification system and provided names to monophyletic clades but was later critiqued for not including any synapomorphic characters to diagnose the monophyletic clades. As stated by Pezold (2011), "the phylogenies constructed (by Thacker 2009) offer much insight, but demand greater corroboration through the addition of more taxa, additional independent genes and complementary morphological analyses". In a reanalysis of several mitochondrial DNA datasets including those of Thacker (2009; 2003), Agorreta and Rüber (2012) also recovered a clade that included amblyopine, gobionelline, oxudercine, and sicydiine gobies.

The largest molecular phylogenetic analysis of the Gobioidei (222 species) to date included seven oxudercine and three amblyopine genera (Agorreta et al. 2013). This study was the first to include both mitochondrial and nuclear genes (6000 base pairs from 12S, $tRNA^{VAL}$, partial 16S, *cytb*, *rag1*, *zic1*, and *gpr85*) and concluded *inter alia* that neither the oxudercine nor amblyopine gobies are monophyletic. Rather, both groups comprise the *Periophthalmus* lineage of the "gobionelline-like" Gobiidae (Agorreta et al. 2013). The *Periophthalmus* lineage of Agorreta et al. (2013) included the following genera: *Odontamblyopus*, *Taenioides*, and *Trypauchen* (considered by Murdy and Shibukawa (2001) as part of the monophyletic gobiid subfamily Amblyopinae) and *Apocryptes*, *Apocryptodon*, *Boleophthalmus*, *Oxuderces*, *Periophthalmus*, *Pseudapocryptes*, and *Scartelaos* (considered by Murdy (1989) as part of the monophyletic gobiid subfamily Oxudercinae). The results of Agorreta et al. (2013), Agorreta and Rüber (2012), and Thacker (2009) differ from that of Tornabene et al. (2013) who used two nuclear genes (1665 base pairs from rhodopsin and *RAG1*) and recovered the amblyopine gobiids as the sister group of the oxudercine gobiids. Tornabene et al. (2013) sampled two species of a single oxudercine genus (*Periophthalmus*) and two amblyopine genera (*Odontamblyopus* and *Taenioides*); the exclusion of other oxudercine genera in their study precludes them from testing the monophyly of the Oxudercinae.

In a phylogenetic study that utilized molecular (3258 base pairs from mitochondrial *ND1*, *ND2*, and *COI* genes) and morphological (i.e., bones involved with jaw suspension) data, Thacker (2013) reaffirmed her earlier findings that mudskippers and eel gobies comprised a monophyletic clade (i.e., *Periophthalmus* lineage) within the Gobionellidae. The *Periophthalmus* lineage (Thacker 2013: tab. 1) included the amblyopine genera of Murdy (2011a) and the oxudercine genera of Murdy (2011b). With the exception of *Oxyurichthys*, the *Periophthalmus* lineage of Thacker (2013) was consistent with the "*Oxyurichthys* lineage" of Harrison (1989) whose groupings were based solely on the comparative morphology of the jaw suspensorium (i.e., bones of the palatopterygoquadrate complex). Thacker (2013) included *Oxyurichthys* in her *Stenogobius* lineage, whereas Harrison (1989) had *Stenogobius* as a stand-alone group. The *Periophthalmus* lineage of Thacker (2013) expanded the constituent genera of the *Periophthalmus* lineage of Agorreta et al. (2013) and essentially agreed with the conclusions of Agorreta and Rüber (2012) with respect to amblyopine and oxudercine gobies.

Using only morphological characters, Gill and Mooi (2012) diagnosed a monophyletic Gobiidae that included subfamilies *sensu* Pezold (1993): Amblyopinae, Gobiinae, Gobionellinae, Oxudercinae, and Sicydiinae. Gill and Mooi (2012) further demonstrated that the family-group name "Gobionellidae Bleeker 1874," utilized by Pezold (2011) and Thacker (2009, 2013), should not be used, and cited the following reasons: (1) Gobionellidae is not defined by any synapomorphy and

(2) the names Oxudercidae Günther 1861, and Amblyopina Günther 1861, predate Gobionellidae. Gill and Mooi (2012), and later followed by Tornabene et al. (2013), did not utilize subfamilial-group names for gobiid fishes, but instead, employed vernacular names for collective genera such as amblyopines, oxudercines, gobionellines, and gobiines to facilitate discussion. In their comprehensive listing of family-group names for fishes, Van Der Laan et al. (2014) followed Pezold (1993) in their recognition of subfamilial-group names for the Gobiidae: Amblyopinae, Gobionellidae, Gobiinae, Oxudercinae, and Sicydiinae. The classification of the Gobiidae *sensu* Van Der Laan et al. (2014) is followed here. We believe that the monophyly of both the Amblyopinae and Oxudercinae (provided by Murdy and Shibukawa (2001) and Murdy (1989), respectively) using synapomorphies based on morphological characters will eventually be recovered by molecular studies with robust gene and taxon sampling.

1.4 SISTER-GROUP RELATIONSHIPS

Both molecular and morphological studies demonstrated a close relationship between the amblyopine and oxudercine gobies (see Section 1.3). Several studies using osteological characters support the close relationship between these two groups. Harrison (1989) and Murdy (1989) noted the similarity in size, shape, and articulation of the palatine and ectopterygoid bones in oxudercines and amblyopines. Members of the "*Oxyurichthys* lineage" of Harrison (1989) share a short, stubby palatine and included six oxudercine genera (e.g., *Apocryptodon*, *Boleophthalmus*, *Parapocryptes*, *Periophthalmus*, *Pseudapocryptes*, and *Scartelaos*), two amblyopine genera (e.g., *Taenioides* and *Trypauchen*) and *Oxyurichthys*. However, Harrison (1989: 342) noted that the condition of the palatopterygoquadrate in *Trypauchen*, "... appears incongruent with those of other members of the '*Oxyurichthys* lineage'... " and speculated that this condition may be a further modification of the derived palatopterygoquadrate morphotype of *Trypauchen*. In addition to all 10 oxudercine genera, Murdy (1989) also included two amblyopine genera and species, *Brachyamblyopus sp.* (=*Caragobius urolepis*) and *Trypauchen vagina* in his analysis. Murdy (1989) found that while the ventroposteriorly directed process of the palatine was reduced in size, the extent of the reach of the palatine is greater in oxudercine gobies. In addition, the palatine and ectopterygoid were closely applied to one another in the sampled amblyopines, whereas there was little or no overlap between the palatine and ectopterygoid in oxudercines.

Murdy (1989) compared the Oxudercinae to other gobiids such as the *Taenioides* Group, the *Trypauchen* Group, the *Gobionellus* Group, and the *Sicydium* Group (all groups *sensu* Birdsong et al. 1988) in search of a sister group. Murdy (1989) concluded that the genus *Evorthodus* of the *Sicydium* Group most closely approximated the character states of many oxudercine genera, as they are similar especially in characters involving the jaw suspensorium, dentition, and branchial arch structure. Consequently, he proposed *Evorthodus* as a probable sister taxon to the Oxudercine but cautioned on the possibility that these similarities could be the result of convergences given these are all mud-dwelling, omnivorous fishes (Murdy 1989). This observation is corroborated by both morphological and molecular data; *Evorthodus* is indeed closely related to, but is unlikely sister to, the oxudercine gobies. The *Oxyurichthys* lineage of Harrison (1989), for example, was recovered as sister to the *Ctenogobius* lineage that included *Evorthodus*, *Ctenogobius*, *Gnatholepis*, *Gobionellus*, *Gobioides*, and *Oligolepis*. The *Ctenogobius* and *Oxyurichthys* lineages together formed the sister group to *Stenogobius*; these three taxa compose the Gobionellinae *sensu* Harrison (1989). Thacker (2003) recovered a monophyletic clade comprising oxudercine and amblyopine genera, which was sister to a clade that included *Evorthodus*, *Awaous*, *Ctenogobius*, *Gnatholepis*, *Sicyopterus*, *Stenogobius*, and *Stiphodon*. *Evorthodus* was recovered in the *Stenogobius* lineage of Thacker (2013) that was sister to the *Periophthalmus* lineage and in the *Stenogobius* lineage of Agorreta et al. (2013) that was sister to the *Periophthalmus* lineage. Tornabene et al. (2013), however, had the oxudercines and amblyopines in a sister-group relationship; their *Stenogobius* Group, which included *Evorthodus*, was sister to the oxudercine/amblyopine clade.

A sister-group relationship between the oxudercines and amblyopines gobiids appears likely considering current literature based on morphological and molecular data. Diagnostic characters that unite the Oxudercinae are listed above, while those that define the Amblyopinae are the ultimate dorsal- and anal-fin pterygiophores support a single ray (Murdy and Shibukawa 2002, 2001), and the fin element to vertebra ratio is approximately 2:1 (Birdsong et al. 1988). Both oxudercine and amblyopine gobies include elongate fishes that are fossorial in soft-bottom habitats or in muddy substrate. A number of oxudercine genera are amphibious and several are amphibious air breathers (Graham 1997), whereas no amblyopine genera are amphibious. However, unlike all oxudercine and many amblyopine gobies that are common in shallow coastal waters and exposed littoral surfaces, some amblyopine gobies, such as *Karsten totoyensis* (Garman 1903), can be found at depths up to 1122 m (Murdy 2002). Oxudercine gobies are carnivorous, herbivorous, or omnivorous (see Chapter 10), whereas amblyopine gobies appear to be strictly carnivorous (Fukagawa 1994). Oxudercine gobies differ morphologically from amblyopine gobies in having separate dorsal fins or fins united only at the base of the dorsal-fin elements (vs. a single, continuous dorsal fin in amblyopines or united throughout the length of the dorsal-fin elements); eyes not reduced and some amphibious species with excellent visual acuity (vs. eyes reduced or absent, with limited function); 26 total vertebrae (vs. 26–51 total vertebrae); first two pterygiophores of the second dorsal fin in separate interneural spaces (vs. first two pterygiophores of the second dorsal fin in the same interneural space, except for *Biendongella*, *Pseudotrypauchen*, and *Sovvityazius*); and one anal-fin pterygiophore preceding the first hemal spine (vs. 1–7 anal-fin pterygiophores preceding the first hemal spine) (Prokofiev 2015; Murdy 2011a; Murdy and Shibukawa 2002; Hoese 1984). The relationship between the Oxudercinae and Amblyopinae is not fully elucidated at this time and further study is warranted.

1.5 RELATIONSHIPS AMONG GENERA

Murdy (1989) hypothesized the relationships between oxudercine genera based on morphological characters in which the Oxudercinae were recovered in two distinct tribes: the Oxudercini and Periophthalmini (Figure 1.2). The former comprised three genera (*Apocryptodon*, *Oxuderces*, and *Parapocryptes*), whereas the latter comprised seven genera (*Apocryptes*, *Boleophthalmus*, *Periophthalmodon*, *Periophthalmus*, *Pseudapocryptes*, *Scartelaos*, and *Zappa*). All members of the Oxudercini typically have six first dorsal-fin spines, whereas the Periophthalmini (except for *Periophthalmodon* and *Periophthalmus*) typically have five first dorsal-fin spines; *Periophthalmodon* and *Periophthalmus* have variable numbers of first dorsal-fin spines ranging from 4 to 17 depending on the species (Murdy 1989). Members of the Oxudercini are united by the presence of finger-like projections in the maxillodentary ligament in the lip of the lower jaw (Murdy 1989). Two of the three Oxudercini, *Apocryptodon* and *Oxuderces*, share the condition of having the epaxialis muscle attaching anteriorly at the frontal and epioccipital junction and thus were considered sister genera (Murdy 1989).

Members of the Periophthalmini are united in having the origin of the retractor dorsalis on some portion of the third (rather than the fourth) vertebra (Murdy 1989). All members of the Periophthalmini are amphibious (Polgar et al. 2010; Graham 1997) and *Boleophthalmus*, *Periophthalmodon*, and *Periophthalmus* are considered amphibious air breathers (Graham 1997). These three genera are also united by derived muscle and osteology characters, whereas several osteological features link *Periophthalmus* and *Periophthalmodon* as sister genera (Murdy 1989). The phylogeny presented in Murdy (1989) indicates that the more highly derived genera of Oxudercinae have increasingly developed adaptations to the terrestrial environment (Polgar et al. 2010).

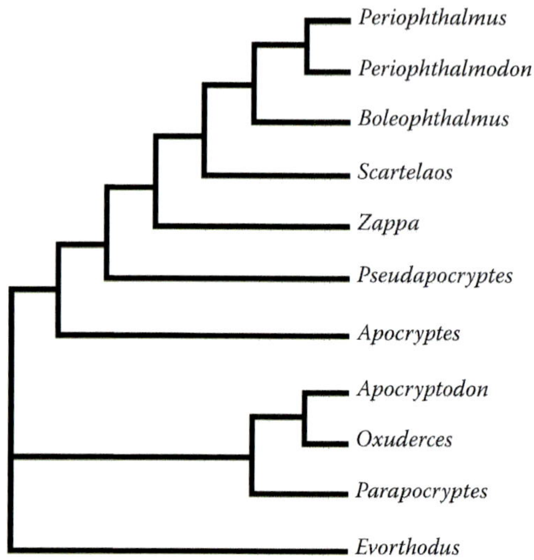

Figure 1.2 Hypothesized relationships of oxudercine genera based on morphological features. (Modified from Murdy E.O., *Records of the Australian Museum*, 11, 1–93, 1989. Courtesy of Gianluca Polgar.)

1.6 SPECIES ACCOUNTS

We consider mudskippers to comprise a monophyletic unit of goby-like fishes, belonging to the subfamily Oxudercinae within which, we recognize 10 genera and 43 species. We provide below, diagnostic characters for each genus and recognized species, listed in alphabetic order.

APOCRYPTES VALENCIENNES IN CUVIER AND VALENCIENNES 1837

Diagnosis. *Apocryptes* is unique among oxudercine gobies in possessing small, cycloid scales on the snout. Two species of *Periophthalmodon* also possess snout scales, but these scales are approximately five times as large as those in *Apocryptes* (Murdy 1989).

Included species. A single species, *As. bato* (Figure 1.3).

Apocryptes bato (Hamilton 1822)

Diagnosis. Total elements in second dorsal fin 21–23 (mean = 22.0); total elements in anal fin 20–24 (mean = 22.1); head length 22–24% of standard length (SL) (mean = 22.7%); in some preserved specimens, 6–7 vertical narrow brown bars along sides, anteriormost coursing from dorsum through pectoral base; caudal-fin length 23–33% SL (mean = 28.2%) (Murdy 1989).

Distribution. From Kuwait southward to India and east to Bangladesh and Myanmar (Global Biodiversity Information Facility 2016; Clayton and Wells 1994).

APOCRYPTODON BLEEKER 1874

Diagnosis. This genus is unique among oxudercines in possessing a posteriorly directed lamina on the parapophyses of the fourth vertebra (Murdy 1989).

Included species. Three species: *An. madurensis* (Figure 1.4), *An. punctatus* (Figure 1.5), and *An. wirzi* (Figure 1.6).

Figure 1.3 Lateral view of *Apocryptes bato* (~84 mm TL) from Calcutta, India. (Courtesy of Gianluca Polgar.)

Figure 1.4 Lateral view of *Apocryptodon madurensis* (~67 mm TL) from Vietnam. (Courtesy of Đac Đinh Trân.)

Figure 1.5 Lateral view of *Apocryptodon punctatus* (length unknown) from Japan. (Courtesy of Koji Nakamura.)

Figure 1.6 Lateral view of *Apocryptodon wirzi* (~82 mm TL) from Purutu Island, Fly River Delta, Papua New Guinea. (Courtesy of Gianluca Polgar.)

Apocryptodon madurensis (Bleeker 1849)

Diagnosis. This species is distinguished from other species of *Apocryptodon* in having a prominent spot on the first dorsal fin between the third and fifth dorsal-fin spines, and in having prepelvic scales extending to the tip of the lower jaw (Ni 2008).

Distribution. The distributional limits of this species are unclear. Valid records exist for Java (Indonesia) (Bleeker 1849), the Yellow Sea, East China Sea, Taiwan Strait, and South China Sea (Ni 2008).

Apocryptodon punctatus Tomiyama 1934

Diagnosis. Three to six vertical brown bars or blotches along the sides; predorsal scales 22–33 (mean = 27.9) (Murdy 1989).

Distribution. Records for this species exist from southwestern Japan (Global Biodiversity Information Facility 2016), Korea (Park et al. 2008), and Taiwan (Chen et al. 2013).

Apocryptodon wirzi Koumans 1937

Diagnosis. Total second dorsal-fin (D2) elements 29–31 (mean = 29.9); total anal-fin elements 27–30 (mean = 28.9); second dorsal and anal fins basally connected with caudal fin; pore situated posteriorly in interorbital region (Murdy 1989).

Distribution. Northern Australia and Papua New Guinea (Jaafar and Parenti 2016).

BOLEOPHTHALMUS VALENCIENNES IN CUVIER AND VALENCIENNES 1837

Diagnosis. Distinguishable from other oxudercine genera by the greatly thickened epidermis of the head and nape and in possessing a rectangular piece of cartilage spanning the width of the pelvic girdle (Murdy 1989).

Included species. Six species: *B. birdsongi* (Figure 1.7), *B. boddarti* (Figure 1.8), *B. caeruleomaculatus* (Figure 1.9), *B. dussumieri* (Figure 1.10), *B. pectinirostris* (Figure 1.11), and *B. poti* (Figure 1.12).

Boleophthalmus birdsongi Murdy 1989

Diagnosis. In combination, flattened horizontal dentary teeth lacking notches; length of first dorsal-fin (D1) base equal to or less than 13% SL; tip of adpressed pectoral fin reaching a vertical point slightly posterior to terminus of D1; D1 membrane proximally black and distally blue *in vivo*, without spots, completely black in preserved material; D2 with a black, broad medial band, bordered by a narrow whitish basal band and a broad yellowish to whitish marginal band, hyaline after preservation (Polgar et al. 2013).

Distribution. Northern Australia and Papua New Guinea (Global Biodiversity Information Facility 2016).

Boleophthalmus boddarti (Pallas 1770)

Diagnosis. In combination, flattened horizontal dentary teeth notched; longitudinal scales fewer than 80; dorsal margin of pectoral fin black, retained after preservation; *in vivo*, D1 bright yellow in smaller specimens; dark saddle-like bars often extending below the midline in the posterior third of the body, visible in preserved material (Polgar et al. 2013).

Distribution. Eastern and western coasts of India, Sunda Shelf, eastward to Sulawesi and the Moluccas (Polgar et al. 2013).

Figure 1.7 Lateral view of *Boleophthalmus birdsongi* (~84 mm TL) from Darwin, Northern Territory, Australia. (Courtesy of Gianluca Polgar.)

Figure 1.8 Lateral view of *Boleophthalmus boddarti* (~130 mm TL) from Tanjung Piai, Johor, Peninsular Malaysia. (Courtesy of Gianluca Polgar.)

Figure 1.9 Lateral view of *Boleophthalmus caeruleomaculatus* (~140 mm TL) from Sisikura Island, Fly River Delta, Papua New Guinea. (Courtesy of Gianluca Polgar.)

Figure 1.10 Lateral view of *Boleophthalmus dussumieri* (~89 mm SL) from Kuwait. (Courtesy of John E. Randall.)

Figure 1.11 Lateral view of *Boleophthalmus pectinirostris* (~196 mm TL) from Tanjung Piai, Johor, Peninsular Malaysia. (Courtesy of Gianluca Polgar.)

Figure 1.12 Lateral view of *Boleophthalmus poti* (~109 mm TL) from Sisikura Island, Fly River Delta, Papua New Guinea. (Courtesy of Gianluca Polgar.)

Boleophthalmus caeruleomaculatus McCulloch and Waite 1918

Diagnosis. In combination, flattened horizontal dentary teeth lacking notches; pelvic-fin length is equal to, or greater than, 16% SL; lateral scale rows more than 125; D1, D2 with no distinct marginal band; D1, D2, and caudal fin with series of sky-blue spots, becoming whitish or semi-transparent on preservation, along the whole length of rays; sky-blue skin fold below eyes, not visible after preservation (Polgar et al. 2013).

Distribution. Northern Australia and Gulf of Papua (Polgar et al. 2013).

Boleophthalmus dussumieri Valenciennes in Cuvier and Valenciennes 1837

Diagnosis. In combination, flattened horizontal dentary teeth notched; length of caudal fin equal to or greater than 22% SL; dark brown spots on D1, body and head (darker than background), visible in preservation (Polgar et al. 2013).

Distribution. Arabian Sea and Persian Gulf (Polgar et al. 2013) and India, Indonesia, Mauritius, and Pakistan (Global Biodiversity Information Facility 2016).

Boleophthalmus pectinirostris (Linnaeus 1758)

Diagnosis. In combination, flattened horizontal dentary teeth notched; length of caudal fin equal to or less than 22% SL; *in vivo*, scattered sky-blue spots on D1, body and head (paler than background) often not visible in preservation; black elongate blotches may be present in the distal portion of D1, visible after preservation (Polgar et al. 2013).

Distribution. Peninsular and eastern Malaysia, Sumatra (Indonesia), China, Taiwan, Korea, and southern Japan (Global Biodiversity Information Facility 2016, Chen et al. 2014, Polgar et al. 2013).

Boleophthalmus poti Polgar, Jaafar and Konstantinidis 2013

Diagnosis. In combination, flattened horizontal dentary teeth lacking notches; pelvic-fin length is equal to, or less than, 12% SL; lateral scale rows fewer than 120; D1 membrane with numerous tiny yellowish speckles (whitish after preservation) and a distinct whitish to sky-blue marginal band (whitish after preservation); D2 with a few whitish spots, proximally; caudal fin with no spots or speckles (Polgar et al. 2013).

Distribution. Only known from the Fly River Delta in the Gulf of Papua (Polgar et al. 2013).

OXUDERCES EYDOUX AND SOULEYET 1850

Diagnosis. The genus is unique among the subfamily in possessing a single, anterior interorbital pore that sits in a fleshy interorbital trough; highly thickened epidermis over the eye; a broad and spatulate neural spine of the fourth vertebra; having the anterior ceratohyal elongate and notched posterior to insertion of the fourth branchiostegal; and an acute angle between the metapterygoid-symplectic-quadrate strut and the anguloarticular (Jaafar and Parenti 2016).

Included species. Two species: *O. dentatus* (Figure 1.13) and *O. nexipinnis* (Figure 1.14).

Oxuderces dentatus Eydoux and Souleyet 1850

Diagnosis. Mouth terminal; dermal invagination posterior to the point of attachment of each pelvic-fin base absent; head relatively shorter than its congener (25.1–27.7% SL) (Jaafar and Parenti 2016).

Distribution. Macao and China (Jaafar and Parenti 2016).

Figure 1.13 Lateral view of *Oxuderces dentatus* (~80 mm SL) from Fengsien, Nanking, China. (Courtesy of Sandra Raredon.)

Figure 1.14 Lateral view of *Oxuderces nexipinnis* (~82 mm TL) from Tanjung Piai, Johor, Peninsular Malaysia. (Courtesy of Gianluca Polgar.)

Oxuderces nexipinnis (Cantor 1849)

Diagnosis. Mouth superior; dermal invagination posterior to the point of attachment of each pelvic-fin base present; head relatively longer than its congener (27.3–30.3% SL) (Jaafar and Parenti 2016).
Distribution. India, Malaysia, Thailand, and Indonesia (Jaafar and Parenti 2016).

PARAPOCRYPTES BLEEKER 1874

Diagnosis. The genus is unique among oxudercines in possessing a greatly expanded distal head of the fourth epibranchial (Murdy 1989).
Included species. Two species: *Pa. rictuosus* (Figure 1.15) and *Pa. serperaster* (Figure 1.16).

Parapocryptes rictuosus (Valenciennes in Cuvier and Valenciennes 1837)

Diagnosis. In most specimens, floor of mouth and tongue with many black flecks; two small pores anteriorly on nape; caudal fin extremely long, caudal-fin length 27–33% SL (mean = 30.1%); longitudinal scale count 81–94 (mean = 87.7); body depth 8–12% SL (mean = 9.8%); head depth 9–12% SL (mean = 9.9%); head width 8–11% SL (mean = 9.5%); transverse scale count from the anal-fin origin dorsoanteriorly to the first dorsal-fin base (TRF) 16–19 (mean = 17.4); transverse scale count from the anal-fin origin dorsoposteriorly to the second dorsal-fin base (TRB) 16–20 (mean = 17.8); predorsal scales 32–40 (mean = 35.9) (Murdy 1989).
Distribution. Only known from the east coast of India (Global Biodiversity Information Facility 2016).

Figure 1.15 Lateral view of *Parapocryptes rictuosus* (~139 mm TL) from Chollagi, Andhra Pradesh, India. (Courtesy of Gianluca Polgar.)

Figure 1.16 Lateral view of *Parapocryptes serperaster* (~135 mm TL) from lower Mekong River, Vietnam. (Courtesy of Đac Đinh Trân.)

Parapocryptes serperaster (Richardson 1846)

Diagnosis. No black pigment inside mouth; a single large pore posteriorly in interorbital region; caudal fin not greatly lengthened, caudal-fin length 19–27% SL (mean = 23.0%); longitudinal scale count 62–81 (mean = 71.9); body depth 11–17% SL (mean = 13.9%); head depth 10–14% SL (mean = 11.8%); head width 10–14% SL (mean = 11.9%); TRF 20–26 (mean = 22.2); TRB 15–26 (mean = 20.5); predorsal scales 23–38 (mean = 31.4) (Murdy 1989).

Distribution. Northeast India, Myanmar, Peninsular Malaysia, Singapore, Java (Indonesia), Vietnam, Macao, and southern China (Global Biodiversity Information Facility 2016).

PERIOPHTHALMODON BLEEKER 1874

Diagnosis. This genus is unique among oxudercines in possessing two rows of teeth in the upper jaw and in having a black stripe (brown in preservative) coursing from the eye to the caudal peduncle (Murdy 1989).

Included species. Three species: *Pn. freycineti* (Figure 1.17), *Pn. schlosseri* (Figure 1.18), and *Pn. septemradiatus* (Figure 1.19).

Periophthalmodon freycineti (Quoy and Gaimard 1824)

Diagnosis. Pectoral-fin rays 15–17 (mean = 16.0); spinous dorsal fin IV–V (mean = 4.1); length of D1 base 5–8% SL (mean = 6.3%); pelvic fins with frenum and completely united; isthmus lacking scales; snout completely scaled; dorsal fins not contiguous in adult males (Murdy 1989).

Figure 1.17 Lateral view of *Periophthalmodon freycineti* (~162 mm TL) from Purutu Island, Fly River Delta, Papua New Guinea. (Courtesy of Gianluca Polgar.)

Figure 1.18 Lateral view of *Periophthalmodon schlosseri* (~266 mm TL) from Tanjung Piai, Johor, Peninsular Malaysia. (Courtesy of Gianluca Polgar.)

Figure 1.19 Lateral view of an adult male *Periophthalmodon septemradiatus* (~115 mm TL) from Vietnam; note air bubble at pectoral-fin tip. (Courtesy of Đac Đinh Trân.)

Distribution. Central Philippines, eastern Indonesia, southern Papua New Guinea, and northern Australia (Global Biodiversity Information Facility 2016).

Periophthalmodon schlosseri (Pallas 1770)

Diagnosis. Pectoral-fin rays 16–19 (mean = 17.0); spinous dorsal fin VI–IX (mean = 8.0); length of D1 base 10–17% SL (mean = 13.9%); pelvic fins with frenum and completely united; isthmus lacking scales; snout completely scaled; dorsal fins not contiguous in adult males (Murdy 1989).

Distribution. Eastern India, Thailand, Peninsular Malaysia, Sarawak (Malaysia), Kalimantan (Indonesia), Singapore, Java (Indonesia) (Global Biodiversity Information Facility 2016), and Vietnam (Trân et al. 2013). Records exist for northern Australia and Papua New Guinea (Global Biodiversity Information Facility 2016); we consider these records as misidentifications.

Periophthalmodon septemradiatus (Hamilton 1822)

Diagnosis. Pectoral-fin rays 12–15 (mean = 13.6); spinous dorsal fin IV–XV (mean = 10.4); length of D1 base 3–18% SL (mean = 12.3%); pelvic fins lack frenum and are separate; isthmus completely scaled; snout lacking scales; dorsal fins contiguous in adult males (Murdy 1989).

Distribution. Eastern India, Bangladesh, Myanmar, Thailand, Peninsular Malaysia, Singapore, Sarawak (Malaysia), Kalimantan (Indonesia) (Global Biodiversity Information Facility 2016), and Vietnam (Trân et al. 2013).

PERIOPHTHALMUS BLOCH AND SCHNEIDER 1801

Diagnosis. Members of the genus are as yet undefined by a synapomorphy. However, species of this genus are characterized by the following features (which may or may not be shared with non-congeners: possession of compressed, convex posterior extensions on the neural, and hemal spines of caudal vertebrae 8–13 and in possessing 16 or fewer (typically 12–14) pectoral-fin rays (Murdy 1989).

Included species. Nineteen species: *Ps. argentilineatus* (Figure 1.20), *Ps. barbarus* (Figure 1.21), *Ps. chrysospilos* (Figure 1.22), *Ps. darwini* (Figure 1.23), *Ps. gracilis* (Figure 1.24), *Ps. kalolo* (Figure 1.25), *Ps. magnuspinnatus* (Figure 1.26), *Ps. malaccensis* (Figure 1.27), *Ps. minutus* (Figure 1.28), *Ps. modestus* (Figure 1.29), *Ps. novaeguineaensis* (Figure 1.30), *Ps. novemradiatus* (Figure 1.31), *Ps. pusing* (Figure 1.32), *Ps. spilotus* (Figure 1.33), *Ps. takita* (Figure 1.34), *Ps. variabilis* (Figure 1.35), *Ps. walailakae* (Figure 1.36), *Ps. waltoni* (Figure 1.37), and *Ps. weberi* (Figure 1.38).

Periophthalmus argentilineatus Valenciennes in Cuvier and Valenciennes 1837

Diagnosis. Pelvic fins lacking frenum; little or no membrane uniting medialmost pelvic rays; D1 height moderate, its margin usually convex, occasionally straight, with a black stripe inframarginally and numerous small, white spots posteriorly on fin, no elongate spines; D2 with single, dusky stripe mesially; dorsal fins not connected by membrane; D1 with 11–16 spines (mean = 13.9); longitudinal scale count 64–100 (mean = 81.1); head width 14–23% SL (mean = 18.7%); pelvic-fin length 11–15% SL (mean = 13.2%); length of anal-fin base 14–19% SL (mean = 16.8%); length of D2 base 18–24% SL (mean = 20.8%); total D2 elements 10–13 (mean = 12.0) total anal-fin elements 9–12 (mean = 11.0); transverse scale count from the D2 origin ventroposteriorly to the anal-fin base (TRDB) 18–26 (mean = 20.8) (Murdy 1989).

Figure 1.20 Lateral view of *Periophthalmus argentilineatus* (~51 mm SL) from Fiji. (Courtesy of John E. Randall.)

Figure 1.21 Lateral view of *Periophthalmus barbarus* (length unknown) from Cameroon. (Courtesy of Randall E. Brummett.)

Figure 1.22 Lateral view of a male *Periophthalmus chrysospilos* (~102 mm TL) from Vietnam. (Courtesy of Đac Đinh Trân.)

Figure 1.23 Lateral view of *Periophthalmus darwini* (~51 mm TL) from Purutu Island, Fly River Delta, Papua New Guinea. (Courtesy of Gianluca Polgar.)

Figure 1.24 Lateral view of a male *Periophthalmus gracilis* (~48 mm TL) from the Mekong Delta, Vietnam. (Courtesy of Đac Đinh Trân.)

Figure 1.25 Lateral view of *Periophthalmus kalolo* (~95 mm SL) from Djibouti. (Courtesy of John E. Randall.)

Distribution. One of the most widely distributed mudskippers found longitudinally from East Africa/Red Sea (30°–40° E) to Fiji/Samoa/Tonga (180°–175° W) and latitudinally from Japan/South Korea (35° N) to Australia (20° S) (Murdy 1989).

Note. Polgar et al. (2014) reported that *Ps. argentilineatus* is not monophyletic and comprises three distinct clades. As no species names were assigned to these clades, further taxonomic work is needed.

Figure 1.26 Lateral view of *Periophthalmus magnuspinnatus* (~65 mm TL) from Guandong, China. (Courtesy of Ray Cui.)

Figure 1.27 Lateral view of *Periophthalmus malaccensis* (~97 mm TL) from Wawontulap, North Sulawesi, Indonesia. (Courtesy of Gianluca Polgar.)

Figure 1.28 Lateral view of *Periophthalmus minutus* (~68 mm TL) from Darwin, Northern Territory, Australia. (Courtesy of Gianluca Polgar.)

Figure 1.29 Lateral view of *Periophthalmus modestus* (length unknown) from Kanagawa Prefecture, Japan. (Courtesy of Hiroshi Senou.)

Figure 1.30 Lateral view of *Periophthalmus novaeguineaensis* (~68 mm TL) from Purutu Island, Fly River Delta, Papua New Guinea. (Courtesy of Gianluca Polgar.)

Figure 1.31 Lateral view of *Periophthalmus novemradiatus* (~67 mm SL) from India. (Courtesy of Jim Van Tassell and Frank Greco.)

Periophthalmus barbarus (Linnaeus 1766)

Diagnosis. No frenum uniting pelvic fins; large adult size, approaching 150 mm SL; D1 height moderate, its margin straight with a blackish stripe inframarginally and, occasionally, with a few white spots posteriorly (spots often lacking), no elongate spines; D2 with single dusky stripe mesially; dorsal fins not connected by membrane; D1 with 10–14 spines (mean = 11.5);

Figure 1.32 Lateral view of *Periophthalmus pusing* (~35 mm TL) from Sumba Island, Indonesia. (Courtesy of Zeehan Jaafar.)

Figure 1.33 Lateral view of *Periophthalmus spilotus* (~85 mm TL) from Kukup Island, Peninsular Malaysia. (Courtesy of Gianluca Polgar.)

Figure 1.34 Lateral view of *Periophthalmus takita* (~58 mm SL) from Northern Territory, Australia. (Courtesy of Gianluca Polgar.)

longitudinal scale count 86–107 (mean = 96.6); head width 15–22% SL (mean = 18.1%); pelvic-fin length 14–16% SL (mean = 14.9%); length of anal-fin base 14–18% SL (mean = 15.5%); length of D2 base 21–24% SL (mean = 22.7%); total D2 elements 11–14 (mean = 12.9); total anal-fin elements 9–11 (mean = 9.9); TRDB 20–34 (mean = 25.5) (Murdy 1989).

Distribution. Morocco south to northern Angola (Murdy 2016).

Figure 1.35 Lateral view of *Periophthalmus variabilis* (~70 mm TL) from Mekong Delta, Vietnam. (Courtesy of Đac Đinh Trân.)

Figure 1.36 Lateral view of *Periophthalmus walailakae* (~67 mm TL) from Kukup Island, Peninsular Malaysia. (Courtesy of Gianluca Polgar.)

Figure 1.37 Lateral view of *Periophthalmus waltoni* (~98 mm SL) from Kuwait. (Courtesy of John E. Randall.)

Periophthalmus chrysospilos Bleeker 1853

Diagnosis. Pelvic fins totally united into a rounded disk; D1 high, first two spines elongate in males, only first spine elongate in females, first spine in males relatively longer than first spine in females; D2 with dusky stripe mesially; dorsal fins not connected by membrane; D1 with 7–10 spines (mean = 8.7); longitudinal scale count 64–77 (mean = 70.7); head width 13–22% SL (mean = 17.3%); pelvic-fin length 12–17% SL (mean = 14.3%); length of anal-fin base 17–21% SL (mean = 18.5%); length of D2 base 16–24% SL (mean = 21.2%); total D2 elements 12–13 (mean = 12.5); total anal-fin elements 11–13 (mean = 11.7); TRDB 15–19 (mean = 17.0) (Murdy 1989).

Figure 1.38 Lateral view of *Periophthalmus weberi* (~60 mm SL) from lower Fly River, Papua New Guinea. (Courtesy of Gianluca Polgar.)

Distribution. East coast of India to Gulf of Thailand, Java Sea (Murdy 1989) and Vietnam (Trân et al. 2013).

Periophthalmus darwini Larson and Takita 2004

Diagnosis. A slender-bodied *Periophthalmus* (body depth at anal-fin origin 11–15% of SL, mean = 12.7%), unique among *Periophthalmus* species in having a greatly reduced first dorsal fin (depressed dorsal 5–10% of SL) in both sexes, and typically with only five spines, fin widely separated from second dorsal-fin origin; pelvic fins united for half their length, distinct frenum present, posterior tips of fins pointed to bluntly pointed, fifth ray bifurcating close to base and branching about nine times; D1 plain blackish except for slightly paler base and narrow whitish margin; second dorsal fin translucent with broad submarginal black band and a row of black spots or blotches along fin base; no melanophores on anal fin; pelvic fins whitish or with some fine speckling; orange spots may be present on lower side of head and body when live; second dorsal-fin rays modally I,11; anal-fin rays modally I,12; pectoral-fin rays 11–13; caudal-fin rays typically with ventralmost 6–7 rays branched, dorsalmost 8–9 rays rarely branched; lateral scales 58–78; predorsal scales typically 24–25 (Larson and Takita 2004).

Distribution. Western Australia and Northern Territory, Australia (Larson and Takita 2004) and the Fly River Delta (Polgar 2005).

Periophthalmus gracilis Eggert 1935

Diagnosis. No pelvic frenum or if present, only visible with magnification; pelvic fins totally separate; D1 short and rounded, prominent black spot posteriorly, no elongate spines; D1 shorter than or equal in height to D2; interdorsal fin distance approximates the length of the first D1 spine; D2 with single dusky stripe mesially; dorsal fins not connected by membrane; D1 with 9–11 spines (mean = 9.4); longitudinal scale count 52–70 (mean = 59.2); head width 14–19% SL (mean = 17.1%); pelvic-fin length 12–15% SL (mean = 12.8%); length of anal-fin base 18–21% SL (mean = 19.3%); length of D2 base 19–21% SL (mean = 20.0%); total D2 elements 12–13 (mean = 12.1); total anal-fin elements 11–12 (mean = 11.9); TRDB 14–20 (mean = 17.1) (Murdy 1989).

Distribution. Indonesia, Malaysia, Australia, and the Philippines (Murdy 1989).

Periophthalmus kalolo Lesson 1831

Diagnosis. Frenum of pelvic fins vestigial, but visible macroscopically; pelvic fins united by membrane for about one half their length; D1 height moderate, its margin rounded, black stripe inframarginally with numerous white spots proximally, no elongate spines; D2 with single dusky stripe inframarginally; dorsal fins not connected by membrane; D1 with 11–15 spines (mean = 12.9); longitudinal scale count 66–86 (mean = 74.6); head width 17–23% SL (mean = 19.7%); pelvic-fin length 13–15% SL (mean = 14.5%); length of anal-fin base 16–19% SL (mean = 17.6%); length of D2 base 19–24% SL (mean = 20.9%); total D2 elements 12–13 (mean = 12.3); total anal-fin elements 11–12 (mean = 11.6); TRDB 18–22 (mean = 20.2) (Murdy 1989).

Distribution. Widely distributed; longitudinally from East Africa/Red Sea (30°–40° E) to Fiji/Samoa/Tonga (180°–175° W) and latitudinally from the Philippines (15° N) to Australia (20° S) (Global Biodiversity Information Facility 2016).

Note. Murdy (1989) designated MNHN A-1499 as lectotype of *Ps. argentilineatus*. Eschmeyer et al. (1998: 824) mistakenly reported that Murdy also designated this specimen as lectotype of *Ps. kalolo*. Kottelat (2013: 419) surmised that *Ps. kalolo* becomes the objective senior synonym of *Ps. argentilineatus*. This then results in *Ps. kalolo* of Murdy (1989) being referred to as *Ps. kallopterus* Bleeker 1853 and *Ps. argentilineatus* being referred to as *Ps. kalolo* (see Kottelat 2013 for further discussion). Given the nomenclatural confusion surrounding *Ps. argentilineatus* and *Ps. kalolo*, additional study is necessary to resolve this conundrum.

Periophthalmus magnuspinnatus Lee, Choi, and Ryu 1995

Diagnosis. D1 11–13; total D2 elements 13; total anal-fin elements 12–13; longitudinal scale count 86–90; predorsal scale count 31–32; length of first dorsal-fin base 25–30% SL; height of D1 greater than body depth; first two spines of D1 slightly elongate; pelvic fins with a prominent frenum and only partially united by membrane forming a deeply concave distal margin; D1 with no spots and a broad dusky submarginal band; dorsal fins not connected by membrane; caudal fin broad and rounded; sky-blue speckles on head and flanks (Polgar 2005).

Distribution. South Korea (Lee et al. 1995) and China (Wang et al. 2006).

Periophthalmus malaccensis Eggert 1935

Diagnosis. Pelvic frenum prominent; pelvic fins partially united by basal membrane; D1 height moderate, its margin slightly rounded, a wide brown stripe inframarginally with numerous white spots proximally, first spine elongate; D2 with single dusky stripe mesially; dorsal fins not connected by membrane; D1 with 9–10 spines (mean = 9.8); longitudinal scale count 47–61 (mean = 54); head width 14–20% SL (mean = 16.6%); pelvic-fin length 14–16% SL (mean = 14.5%); length of anal-fin base 15–21% SL (mean = 17.7%); length of D2 base 17–23% SL (mean = 19.8%); total D2 elements 11–12 (mean = 11.6); total anal-fin elements 11–12 (mean = 11.3); TRDB 14–17 (mean = 15.7) (Murdy 1989).

Distribution. Northern Vietnam, the Philippines, and eastern Indonesia (Global Biodiversity Information Facility 2016). Records exist for India and Vanuatu (Global Biodiversity Information Facility 2016); we consider these records as misidentifications.

Periophthalmus minutus Eggert 1935

Diagnosis. Pelvic frenum lacking or only visible with magnification; membrane uniting medial rays of pelvic fin slight to moderate; D1 height moderate, its margin straight to slightly convex, a brown stripe near the margin and many white spots proximally, no elongate spines; D2 with single dusky stripe mesially; dorsal fins not connected by membrane; D1 with 10–17 spines (mean = 12.5); longitudinal scale count 62–78 (mean = 69.8); head width 14–19% SL (mean = 17.1%); pelvic-fin length 12–15% SL (mean = 12.8%); length of anal-fin base 16–19% SL (mean = 17.7%); length of D2 base 20–23% SL (mean = 21.3%); total D2 elements 11–13 (mean = 11.9); total anal-fin elements 11–12 (mean = 11.4); TRDB 16–19 (mean = 17.5) (Murdy 1989).

Distribution. The Andaman Islands, Thailand, Peninsular Malaysia, Indonesia, the Philippines, and northern Australia (Global Biodiversity Information Facility 2016).

Periophthalmus modestus Cantor 1842

Diagnosis. Pelvic fins united anteriorly by a moderate to strong frenum; medial rays united by a membrane for about half their length; D1 height moderate, its margin rounded, a dusky band inframarginally and no spots on fin, no elongate spines; D2 with single dusky stripe inframarginally; dorsal fins not connected by membrane; D1 with 10–17 spines (mean = 13.1); longitudinal scale count 75–100 (mean = 85.3); head width 14–20% SL (mean = 17.1%); pelvic-fin length 12–15% SL (mean = 13.0%); length of anal-fin base 16–22% SL (mean = 19.8%); length of D2 base 20–24% SL (mean = 22.7%); total D2 elements 12–14 (mean = 12.6); total anal-fin elements 11–13 (mean = 12.3); TRDB 19–29 (mean = 24.1) (Murdy 1989).

Distribution. Southern Taiwan and Fuzhou (China) northward to South Korea and northeast to Tokyo Bay, Japan (Global Biodiversity Information Facility 2016). Records exist for Tanzania, Mozambique, and South Africa (Global Biodiversity Information Facility 2016); we consider these records as misidentifications.

Periophthalmus novaeguineaensis Eggert 1935

Diagnosis. Pelvic fins with a strong frenum; pelvic fins partially united by membrane for more than three-quarters of their length; D1 height moderate, its margin triangular, entirely plain gray to blackish in preservative, no spots or stripes on fin, with one broad, submarginal band, no elongate spines; D2 with one broad, submarginal black band, and a basal row of black spots or blotches; dorsal fins not connected by membrane; D1 with 9–14 spines; longitudinal scale count 71–101; when alive, small, bluish-white spots on side of head and body (Jaafar and Larson 2008).

Distribution. Known from Australia, Indonesia (Jaafar and Larson 2008), and Papua New Guinea (Polgar et al. 2010).

Periophthalmus novemradiatus (Hamilton 1822)

Diagnosis. Males possess greatly elongate first spine of first dorsal fin, reaching eighth element of D2 when appressed. In preservative, first dorsal fin brown or black with white distal edge; spots absent. Margin of first dorsal fin concave in males and concave to straight in females. Second dorsal fin with two black bands, broadest one located inframarginally and the other located medially; black spots basally on membrane between dorsal-fin elements. Pelvic fin with prominent frenum. Inner rays of pelvic fins fused for more than half of the length. First dorsal fin spines modally IX; D2 rays modally I,12; anal-fin rays modally I,13; caudal-fin rays modally 15 segmented rays, with first and last two rays unbranched; lateral scale rows 62–67; predorsal scales 18–24 (Jaafar et al. 2009).

Distribution. Eastern Indian Ocean, Andaman Sea, and western Pacific Ocean (Shibukawa and Ghaffar 2011).

Periophthalmus pusing Jaafar, Polgar, and Zamroni 2016

Diagnosis. No pelvic frenum, pelvic fins totally separate; D1 short and rounded, prominent black spot posteriorly, no elongate spines; D1 taller than D2; interdorsal fin distance less than half length of the first D1 spine; D2 with single dusky stripe mesially; dorsal fins not connected by membrane; D1 with 11–15 (mean = 12.6) spines; longitudinal scale count 42–56; total D2 elements 11–13 (mean = 11.5); total anal-fin elements 11–13 (mean = 11.5) (Jaafar et al. 2016).
Distribution. Sumba (Indonesia) and Fly River Delta, Papua New Guinea (Jaafar et al. 2016).

Periophthalmus spilotus Murdy and Takita 1999

Diagnosis. Pelvic fins totally united into rounded disc; D1 high, no spots nor elongate spines, first two spines slightly longer than others spines in males, appressed D1 reaching D2 origin; D2 with dusky stripe mesially, whitish spots medial to this stripe; dorsal fins not connected by membrane; caudal fin blackish medially; anal fin translucent, no spots nor stripes; D2 with 14 (rarely 15) total elements (mean = 14.1); anal fin with 14 or 15 (rarely 13) total elements (mean = 14.7); length of anal-fin base 22–29% SL (mean = 24.2%); length of D2 base (23–27% SL (mean = 24.4%) (Murdy and Takita 1999).
Distribution. Known only from Peninsular Malaysia and Sumatra (Indonesia) (Global Biodiversity Information Facility 2016).

Periophthalmus takita Jaafar and Larson 2008

Diagnosis. First dorsal-fin spines modally VIII; D2 rays modally I,12; anal-fin rays modally I, 11; caudal fin modally with 17 segmented rays; lateral scales 58–73; predorsal scales 22–31. Pelvic fin with prominent frenum; sparse, small brown spots on dorsal surface. Inner rays of pelvic fin fused for half but never more than three-quarters of their length. Head and body pale gray, whitish ventrally; small, rounded, blackish spots on cheeks, pectoral-fin base, and anterior part of body; round orange spots along lower half of head and body (less numerous posteriorly). First dorsal fin broadly rounded, gray to black, with very thin, white margin restricted to tips of spines. Second dorsal fin whitish gray with narrow pale marginal stripe, two broad black bands, one submarginal and one medial, and a basal row of black spots. Anal fin pale orange to creamy. Pectoral fins pale with fine dusky markings mostly following fin rays. Caudal fin dusky to pinkish gray (Jaafar and Larson 2008).
Distribution. Known only from the Northern Territory and Western Australia (Australia) (Jaafar and Larson 2008).

Periophthalmus variabilis Eggert 1935

Diagnosis. Typically, D1 of moderate height with white margin, and one grayish, brownish or blackish stripe inframarginally. First spine of D1 always longest, rarely elongate; rounded to elliptical red-orange spots present ventral to inframarginal stripe, turning black in preservative. Second dorsal fin with yellow-orange margin and one black band inframarginally, rounded red-orange spots present on membrane between elements at base, turning black in preservative. When alive, caudal and pectoral fins red-orange. Prominent pelvic frenum, inner rays fused for less than half their length. Ventroposterior branchiostegal membrane dusky. First dorsal fin spines modally X; D2 rays modally I,12; anal-fin rays modally I,11; caudal fin modally 14 segmented rays; lateral scales 48–60; predorsal scales 17–22 (Jaafar et al. 2009).

Distribution. Thailand, Malaysia, Singapore, Indonesia (Jaafar et al. 2009), and Vietnam (Trân et al. 2013).

Periophthalmus walailakae Darumas and Tantichodok 2002

Diagnosis. Innermost pelvic-fin rays joined by a basal membrane for their entire length; pelvic fins form a rounded disk. Strong pelvic frenum present. Series of dark brown speckles on caudal and pectoral-fin rays present in live and preserved specimens. No sexual dimorphism of the dorsal fins. Length of anal-fin base 15–18% SL (mean = 17.0); length of TRB 17–23% SL (mean = 19.0%); head depth 20–23% SL (mean = 21.1%) (Polgar and Khaironizam 2008).

Distribution. Southern Thailand (Darumas and Tantichodok 2002), Peninsular Malaysia (Polgar and Khaironizam 2008), and Singapore (Jaafar et al. 2006).

Periophthalmus waltoni Koumans 1941

Diagnosis. Moderate to strong frenum uniting pelvic fins; medially, pelvic fins united for about one half their length; D1 height moderate, its margin slightly rounded, no stripes on fin with only a few white spots posteriorly, no elongate spines; D2 with single dusky stripe inframarginally; dorsal fins not connected by membrane; D1 with 10–13 spines (mean = 12.0); longitudinal scale count 91–121 (mean = 103.4); head width 14–22% SL (mean = 17.0%); pelvic-fin length 12–14% SL (mean = 12.8%); length of anal-fin base 16–21% SL (mean = 19.0%); length of D2 base 23–27% SL (mean = 24.8%); total D2 elements 13–14 (mean = 13.8); total anal-fin elements 11–12 (mean = 11.8); TRDB 21–28 (mean = 25.1) (Murdy 1989).

Distribution. Kuwait, Bahrain, United Arab Emirates, Iran, and Pakistan (Global Biodiversity Information Facility 2016).

Periophthalmus weberi Eggert 1935

Diagnosis. Pelvic frenum lacking or only visible with magnification; in males, D1 height moderate, its margin almost straight, no stripe on fin only pale blue spots in life, white spots in preservative, all spines extend beyond membrane with the first two spines elongate, dorsal fins contiguous, connected by membrane; in females, D1 represented by only a few short spines; D2 in both sexes lacking stripes; D1 with 4–16 spines (mean = 9.9); longitudinal scale count 46–52 (mean = 50.1); head width 15–19% SL (mean = 17.4%); pelvic-fin length 15–17% SL (mean = 15.7%); length of anal-fin base 14–19% SL (mean = 16.5%); length of D2 base 21–27% SL (mean = 24.5%); total D2 elements 11–14 (mean = 13.1); total anal-fin elements 9–12 (mean = 10.9); TRDB 12–17 (mean = 14.0) (Murdy 1989).

Distribution. Irian Jaya (Indonesia), Papua New Guinea, and northern Australia (Global Biodiversity Information Facility 2016). Records also exist for the Ganges Delta (Global Biodiversity Information Facility 2016); we consider these records as misidentifications.

PSEUDAPOCRYPTES BLEEKER 1874

Diagnosis. *Pseudapocryptes* is unique among oxudercines in possessing 150 or more scales in a longitudinal series (Murdy 1989).

Included species. Two species: *Pd. borneensis* (Figure 1.39) and *Pd. elongatus* (Figure 1.40).

Figure 1.39 *Pseudapocryptes borneensis* as figured in the *Atlas of Fishes* (From Bleeker P., 1983, Atlas Ichthyologique, Smithsonian Institution Press, Washington, DC, pl. 440, fig. 3).

Figure 1.40 Lateral view of *Pseudapocryptes elongatus* (~125 mm TL) from Mekong Delta, Vietnam. (Courtesy of Đac Đinh Trân.)

Pseudapocryptes borneensis (Bleeker 1855)

Diagnosis. Anal fin with 27–29 total elements (mean = 28.4); body depth 14–16% SL (mean = 15.1%); head length 20–22% SL (mean = 20.6%); head depth 11–13% SL (mean = 12.1%); pectoral-fin length 13–17% SL (mean = 15.0%); pelvic-fin length 12–14% SL (mean = 13.2%); many small, brown spots on head and trunk (Murdy 1989).

Distribution. Singapore and Kalimantan (Indonesia) (type locality) (Murdy 1989).

Pseudapocryptes elongatus (Cuvier 1816)

Diagnosis. Anal fin with 28–31 total elements (mean = 29.3); body depth 10–14% SL (mean = 11.4%); head length 15–22% SL (mean = 17.7%); head depth 8–11% SL (mean = 9.3%); pectoral-fin length 10–13% SL (mean = 11.9); pelvic-fin length 9–12% SL (mean = 10.6%); few brown spots, if any, on body (Murdy 1989).

Distribution. East coast of India eastward to Myanmar, Thailand, Malaysia, Indonesia, and Vietnam (Global Biodiversity Information Facility 2016). One specimen each from Kuwait and Taiwan are also listed in the Global Biodiversity Information Facility online database; these records need to be verified as they would expand the range of this species.

SCARTELAOS SWAINSON 1839

Diagnosis. The only oxudercine genus with barbels on the ventral surface of the head (Murdy 1989).

Included species. Four species: *S. cantoris* (Figure 1.41), *S. gigas* (Figure 1.42), *S. histophorus* (Figure 1.43), and *S. tenuis* (Figure 1.44).

Figure 1.41 Lateral view of *Scartelaos cantoris* from the Andaman Islands (RMNH 2009, ~70 mm SL). (Courtesy of Gianluca Polgar.)

Figure 1.42 Lateral view of *Scartelaos gigas* (length unknown) from western coast of Taiwan. (Courtesy of Chen I-Shiung.)

Figure 1.43 Lateral view of *Scartelaos histophorus* (~ 118 mm TL) from Vietnam. (Courtesy of Đac Đinh Trân.)

Scartelaos cantoris (Day 1871)

Diagnosis. No barbel on ventral midline near symphysis of lower jaw; upper jaw teeth 25–37 (mean = 30.7); head depth 10–13% SL (mean = 11.3%); caudal-fin length 23–27% SL (mean = 24.3%); length of D1 base 9–11% SL (mean = 9.6%); 5–10 narrow, pale, vertical bars on body anterior to D2 origin; no large, black spots but three horizontal, brown-black lines basally on D2, most distinct posteriorly; D2 and anal fins connected by membrane to caudal fin (Murdy 1989).

Figure 1.44 Lateral view of *Scartelaos tenuis* (~150 mm TL) from Bandar Khamir, Hormozgan, Iran. (Courtesy of Gianluca Polgar.)

Distribution. The Andaman Islands (India) (Global Biodiversity Information Facility 2016). One record also exists from Phuket (Thailand) (Global Biodiversity Information Facility 2016) but we have since confirmed that this specimen, deposited in Royal Ontario Museum in Toronto, Canada, is *S. histophorus* (pers. comm. R. Winterbottom).

Scartelaos gigas Chu and Wu in Chu, Chan, and Chen 1963

Diagnosis. No barbel on ventral midline near symphysis of lower jaw; cheek and opercle with yellow transverse band; first dorsal fin yellow in midline, blackish anteriorly and posteriorly; caudal fin without dark band; no vertical bars on body; no large, black spots or stripes basally on posterior half of D2; D2 and anal fins not connected by membrane to caudal fin (Ni 2008; Murdy 1989).
Distribution. Korea, China, and Taiwan (Global Biodiversity Information Facility 2016).

Scartelaos histophorus (Valenciennes in Cuvier and Valenciennes 1837)

Diagnosis. Prominent barbel on ventral midline near symphysis of lower jaw; upper jaw teeth 19–31 (mean = 24.7); head depth 10–12% SL (mean = 11.2%); caudal-fin length 19–25% SL (mean = 22.2%); length of D1 base 6–8% SL (mean = 6.4%); 4–7 vertical, narrow, bluish gray bars on trunk; no large, black spots or stripes basally on posterior half of D2; D2 and anal fins connected by membrane to caudal fin (Murdy 1989).
Distribution. Pakistan to Japan and Australia (Global Biodiversity Information Facility 2016).

Scartelaos tenuis (Day 1876)

Diagnosis. No barbel on ventral midline near symphysis of lower jaw; upper jaw teeth 31–37 (mean = 32.8); head depth 10–11% SL (mean = 10.2%); caudal-fin length 22–26% SL (mean = 23.0%); length of Dl base 6–8% SL (mean = 7.1%); 15–18 vertical, brown bars on trunk; large, black spots basally on posterior half of second dorsal fin; D2 and anal fins not connected by membrane to caudal fin (Murdy 1989).
Distribution. Kuwait, Iran, and Pakistan (Global Biodiversity Information Facility 2016).

ZAPPA MURDY 1989

Diagnosis. Distinguishable from all other members of the subfamily in having the first spinous dorsal-fin pterygiophore bent posteriorly at a point three-quarters along its length to extend horizontally over tip of fourth neural spine (Murdy 1989).
Included species. A single species, *Z. confluentus* (Figure 1.45).

Figure 1.45 Lateral view of *Zappa confluentus* (~65 mm TL) from Fly River, Papua New Guinea. (Courtesy of Gianluca Polgar.)

Zappa confluentus (Roberts 1978)

Diagnosis. Total D2 elements 27–30 (mean = 28.1); caudal-fin length 21–28% SL (mean = 25.0); head length 22–26% SL (mean = 23.9%) (Murdy 1989).

Distribution. Known only from several river systems in Irian Jaya (Indonesia) and Papua New Guinea (Global Biodiversity Information Facility 2016).

ACKNOWLEDGMENTS

We thank Keiichi Matsuura, Lisa Palmer, Ted Pietsch, and Nalani Schnell for their help with procuring obscure images for this chapter. We are very grateful for the generosity of Randall E. Brummett, Chen I-Shiung, Ray Cui, Frank Greco, Koji Nakamura, Gianluca Polgar, Jack Randall, Sandra Raredon, Hiroshi Senou, Đac Đinh Trân, and Jim Van Tassell who graciously allowed us to use their images.

REFERENCES

Agorreta A. and Rüber L. 2012. A standardized reanalysis of molecular phylogenetic hypotheses of Gobioidei. *Systematics and Biodiversity* 10:375–390.

Agorreta A., San Mauro D., Schliewen U. et al. 2013. Molecular phylogenetics of Gobioidei and phylogenetic placement of European gobies. *Molecular Phylogenetics and Evolution* 69:619–633.

Akihito P., Iwata A., Kobayashi A. et al. 2000. Evolutionary aspects of gobioid fishes based upon a phylogenetic analysis of mitochondrial cytochrome *b* genes. *Gene* 259:5–15.

Birdsong R.S., Murdy E.O. and Pezold F. 1988. A study of the vertebral column and median fin osteology in gobioid fishes with comments on gobioid relationships. *Bulletin of Marine Science* 42:174–214.

Bleeker P. 1849. Bijdrage tot de kennis der Blennioiden en Gobioiden van den Soenda-Molukschen Archipel, met beschrijving van 42 nieuwe soorten. *Verhandelingen van het Bataviaasch Genootschap* 22:1–40.

Bleeker P. 1853. Nieuwe bijdrage tot de kennis der ichthyologische fauna van het eiland Banka. *Natuurkundig Tijdschrift voor Nederlandsch Indië* 3:715–738.

Bleeker P. 1855. Negende bijdrage tot de kennis der ichthijologische fauna van Borneo. Zoetwatervisschen van Pontianak en Bandjermasin. *Natuurkundig Tijdschrift voor Nederlandsch Indië* 9:415–430.

Bleeker P. 1874. Esquisse d'un systeme naturel des Gobioides. *Archives Nerlandaises des Sciences Exactes et Naturelles* 9:289–331.

Bleeker P. 1983. *Atlas Ichthyologique*. Washington, DC: Smithsonian Institution Press 11: pp. 1–22, plates 421–447. [Posthumous publication].

Bloch M.E. and Schneider J.G. 1801. *M. E. Blochii, Systema Ichthyologiae Iconibus cx Ilustratum*. Post obitum auctoris opus inchoatum absolvit, correxit, interpolavit Jo. Gottlob Schneider, Saxo. Berolini: Sumtibus Auctoris Impressum et Bibliopolio Sanderiano Commissum, 584 pp.

Cantor T.E. 1842. General features of Chusan, with remarks on the flora and fauna of that island. *Annals and Magazine of Natural History (New Series)* 9:481–493.

Chen H., Polgar G., Yin W. and Fu C-Z. 2014. Cryptic species and evolutionary history of the *Boleophthalmus pectinirostris* complex, along the Northwestern Pacific coast. *Acta Hydrobiologica Sinica* 38:75–86.

Chen I-S., Weng C-J., Chen Y-R. et al. 2013. The checklist of inland water and mangrove fish fauna of Kinmen Island, Fujian Province, Taiwan with comments on ecological conservation of native fishes. *Journal of Marine Science and Technology* 21:316–319.

Chu Y-T., Chan C-L. and Chen C-T. 1963. *Ichthyofauna of the East China Sea*. Beijing, 642 pp. [In Chinese.]

Clayton D. and Wells K. 1994. *Discovering Kuwait's wildlife*. Kuwait: Fahad Al-Marzouk Printing and Publishing, 253 pp.

Cuvier G. 1816. *Le Règne Animal Distribué d'après son Organisation Pour Servir de Base à l'histoire Naturelle des Animaux et d'introduction à l'anatomie Comparée. Les Reptiles, es Poissons, les Mollusques et les Annélides*. Edition 1, 532 pp.

Cuvier G. and Valenciennes A. 1837. *Histoire Naturelle des Poissons*. Vol. 12. Paris: F.G. Levrault, 507 pp.

Darumas U. and Tantichodok P. 2002. A new species of mudskipper from southern Thailand (Gobiidae: Oxudercinae). *Phuket Marine Biological Center. Research Bulletin* 64:101–107.

Day F. 1871. On the fishes of the Andaman Islands. *Proceedings of the Zoological Society of London* 1870 (3):677–705.

Day F. 1876. *The Fishes of India; Being a Natural History of the Fishes Known to Inhabit the Seas and Fresh Waters of India, Burma, and Ceylon*. Part 2: pp. 169–368. London: William Dawson & Sons.

de Vlamingh W. 1701. *Journaal wegens een voyagie, gedaan op order der Hollandsche Oost-Indische maatschappy in de jaaren 1696 en 1697*. Amsterdam: De Coup et al., booksellers, 24 pp.

Eggert B. 1935. Beitrag zur systematik, biologie und geographischen verbreitung der Periophthalminae. *Zoologische lahrbuecher Abteilung fiir Systematik Oekologie und Geographie der Tiere* 67:29–116.

Eschmeyer W.N. 1998. *Catalog of Fishes*. Center for Biodiversity Research and Information, Special Publication 1. San Francisco, CA: California Academy of Sciences, vols. 1–3:1–2905.

Eydoux J.F.T. and Souleyet F.A. 1850. Poissons. In *Voyage autour du monde exécuté pendant les années 1836 et 1837 sur la corvette La Bonite, commandée par M. Vaillant*, pp. 155–216. Zoologie, Vol. 1 (pt 2). Paris.

Fukagawa G. 1994. Life history of the eel goby in Ariake Sound, Japan *in Japanese*. Master's thesis, Nagasaki University, Nagasaki, Japan.

Garman S. 1903. Some fishes from Australasia. *Bulletin of the Museum of Comparative Zoology* 39(8):229–241.

Gill A. and Mooi R. 2012. Thalasseleotrididae, new family of marine gobioid fishes from New Zealand and temperate Australia, with a revised definition of its sister taxon, the Gobiidae (Teleostei: Acanthomorpha). *Zootaxa* 3266:41–52.

Global Biodiversity Information Facility (GBIF). *Free and Open Access to Biodiversity Data*. http://www.gbif.org (accessed October 31, 2016).

Graham J.B. 1997. *Air-Breathing Fishes*. San Diego, CA: Academic Press, 299 pp.

Günther A.C.L.G. 1861. *Catalogue of the Acanthopterygian Fishes in the Collection of the British Museum*. Vol. 3, 586 pp.

Hamilton F. 1822. *An account of the fishes in the River Ganges and its branches*. Edinburgh: Archibald Constable and Company, 405 pp.

Hoese D.F. 1984. Gobioidei: Relationships. In *Ontogeny and systematic of fishes* (eds. H.G. Moser, W.J. Richards, D.M. Cohen, M.P. Fahay, A.W. Kendall Jr. and S.L. Richardson) pp. 588–591. Lawrence: Allen Press.

Holthuis L.B. and Pietsch T.W. 2006. *Les Planches inédites de Poissons et autres Animaux marins de l'Indo-Ouest Pacifique d'Isaac Johannes Lamotius*. Paris: Publications Scientifiques du Muséum, Muséum National d'Histoire Naturelle, 292 pp.

Harrison I.J. 1989. Specialization of the gobioid palatopterygoquadrate complex and its relevance to gobioid systematics. *Journal of Natural History* 23:325–353.

Jaafar Z. and Larson H.K. 2008. A new species of mudskipper, *Periophthalmus takita* (Teleostei: Gobiidae: Oxudercinae), from Australia, with a Key to the Genus. *Zoological Science* 25:946–952.

Jaafar Z., Lim K.K.P. and Chou L.M. 2006. Taxonomical and morphological notes on two species of mudskippers, *Periophthalmus walailakae* and *Periophthalmodon schlosseri* (Teleostei: Gobiidae) from Singapore. *Zoological Science* 23:1043–1047.

Jaafar Z. and Parenti L.R. 2016. Systematics of the mudskipper genus *Oxuderces* Eydoux & Souleyet 1948 (Teleostei: Gobiidae: Oxudercinae) with resurrection from synonymy of *O. nexipinnis* (Cantor 1849). *Zoological Journal of the Linnean Society* 1–21. doi:10.1111/zoj.12482.

Jaafar Z., Perrig M. and Chou L.M. 2009. *Periophthalmus variabilis* (Teleostei: Gobiidae: Oxudercinae), a valid species of mudskipper, and a re-diagnosis of *Periophthalmus novemradiatus*. *Zoological Science* 26:309–314.

Jaafar Z., Polgar G. and Zamroni Y. 2016. Description of a new species of *Periophthalmus* (Teleostei: Gobiidae) from the Lesser Sunda Islands. *Raffles Bulletin of Zoology* 64:278–283.

Jordan D.S. 1923. A classification of fishes including families and genera as far as known. *Stanford University Publications, University Series, Biological Sciences* 3(2):77–243.

Kottelat M. The fishes of the inland waters of Southeast Asia: A catalogue and core bibliography of the fishes known to occur in freshwaters, mangroves and estuaries. *Raffles Bulletin of Zoology* Supplement No. 27:1–663.

Koumans F.P. 1931. *A preliminary revision of gobioid fishes with united ventral fins*. Leiden: Rijks-Universiteit, 174 pp.

Koumans F.P. 1937. Notes on Gobioid Fishes 10. On a collection of the Museum of Basle, with description of a new species of *Apocryptodon*. *Zoologisches Mededeelingen, Leiden Museum* 20:24–26.

Koumans F.P. 1941. Gobioid fishes of India. *Memoires of the Indian Museum* 13:205–329.

Koumans F.P. 1953. Gobioidea. In *Fishes of the Indo-Australian Archipelago*, vol.10 (eds. M. Weber and L.F. de Beaufort). Leiden: E.J. Brill, 423 pp.

Larson H.K. 2008. Weber's mudskipper *Periophthalmus weberi*: New record for the Daly River. *Northern Territory Naturalist* 20:19–21.

Larson H.K. and Takita T. 2004. Two new species of *Periophthalmus* (Teleostei: Gobiidae: Oxudercinae) from Northern Australia, and a re-diagnosis of *Periophthalmus novaeguineaensis*. *The Beagle: Records of the Museums and Art Galleries of the Northern Territory* 20:175–185.

Lee Y-J. Choi Y. and Ryu B-S. 1995. A taxonomic revision of the genus *Periophthalmus* (Pisces: Gobiidae) from Korea with description of a new species. *Korean Journal of Ichthyology* 7:120–127.

Lesson R.P. 1831. Poissons. In *Voyage Autour du Monde-sur la Corvette de la Majest, "La Coquille", Pendant 1822–1825*, pp. 66–238. Paris: L.I. Duperrey.

Linnæus C. 1758. *Systema Naturæ Per Regna Tria Naturæ, Secundum Classes, Ordines, Genera, Species, Cum Characteribus, Differentiis, Synonymis, Locis. Tomus I.* Editio decima, reformata. Stockholm: Holmiæ (Salvius), 824 pp.

Linnæus C. 1766. *Systema Naturæ Sive Regna Tria Naturæ, Secundum Classes, Ordines, Genera, Species, Cum Characteribus, Differentiis, Synonymis, Locis. Tomus I.* Editio duodecimo. Stockholm: Holmiæ (Salvius), 532 pp.

McCulloch A. R. and Waite E.R. 1918. Descriptions of two new Australian gobies. *Records of the South Australian Museum* 1:79–82.

Miller P. 1973. The osteology and adaptive features of *Rhyacichthys aspro* (Teleostei: Gobioidei) and the classification of gobioid fishes. *Journal of Zoology* 171: 397–434.

Murdy E.O. 1989. A taxonomic revision and cladistic analysis of the oxudercine gobies (Gobiidae: Oxudercinae). *Records of the Australian Museum*, Supplement 11:1–93.

Murdy E.O. 2002. *Karsten*: A new genus of eel goby (Gobiidae: Amblyopinae) with a key to "*Trypauchen*" group genera. *Copeia* 2002(3):787–791.

Murdy E.O. 2011a. Systematics of Amblyopinae. In *The biology of gobies* (eds. R.A. Patzner, J.L. Van Tassell, M. Kovacic and B. Kapoor) pp. 107–118. Enfield, NH: Science Publishers Inc.

Murdy E.O. 2011b. Systematics of Oxudercinae. In *The biology of gobies* (eds. R.A. Patzner, J.L. Van Tassell, M. Kovacic, and B. Kapoor) pp. 99–106. Enfield, NH: Science Publishers Inc.

Murdy E.O. 2016. Subfamily Oxudercinae. In *The living marine resources of the Eastern Central Atlantic. Volume 4: Bony fishes part 2 (Perciformes to Tetradontiformes) and Sea turtles* (eds. K.E. Carpenter and N. De Angelis) p. 2843. Rome: FAO Species Identification Guide for Fishery Purposes, FAO.

Murdy E.O. and Shibukawa K. 2001. A revision of the gobioid fish genus *Odontamblyopus* (Gobiidae: Amblyopinae). *Ichthyological Research* 48:31–43.

Murdy E.O. and Shibukawa K. 2002. Redescription of the gobiid fish genus *Pseudotrypauchen* (Gobiidae: Amblyopinae) and its significance in amblyopine phylogeny. *Marine and Freshwater Research* 53:253–258.

Murdy E.O. and Takita T. 1999. *Periophthalmus spilotus*, a new species of mudskipper from Sumatra (Gobiidae: Oxudercinae). *Ichthyological Research* 46:367–370.

Ni Y. 2008. Gobiidae: *Apocryptodon, Cristatogobius, Cryptocentroides, Cryptocentrus, Boleophthalmus, Oxuderces, Parapocryptes, Periophthalmus, Pseudapocryptes, Pseudogobius, Scartelaos.* In *Fauna Sinica. Osteichthys: Perciformes (V) Gobioidei*, (ed. H-L Wu) pp. 687–693, 898–901. Beijing: Science Press. (In Chinese with English language key to species.)

Pallas P. S., 1770. *Spicilegia Zoologica Quibus Novae Imprimis et Obscurae Animalium Species Iconibus, Descriptionibus Atque Commentariis Illustrantur. Tomus I.* Prostant apud Gottl. August. Lange, Berolini, 558 pp.

Park K.D., Kim J.K., Chang D.S., et al. 2008. Age and growth of the mudskipper, *Scartelaos gigas* (Perciformes, Gobiidae) from Korea. *Animal Cells and Systems* 12:305–311.

Pezold F. 1993. Evidence for a monophyletic Gobiinae. *Copeia* 1993(3):634–643.

Pezold F. 2011. Systematics of the family Gobionellidae. In *The biology of gobies* (eds. R.A. Patzner, J.L. Van Tassell, M. Kovacic and B. Kapoor) pp. 87–98. Enfield, NH: Science Publishers Inc.

Polgar, G. 2005. *The Mudskipper.* http://www.themudskipper.org (accessed October 31, 2016).

Polgar G., Jaafar Z. and Konstantinidis P. 2013. A new species of mudskipper, *Boleophthalmus poti* (Teleostei: Gobiidae: Oxudercinae) from the Gulf of Papua, Papua New Guinea, and a key to the genus. *The Raffles Bulletin of Zoology* 61(1):311–321.

Polgar G. and Khaironizam M.Z. 2008. First record of *Periophthalmus walailakae* (Gobiidae: Oxudercinae) from Peninsular Malaysia. *Cybium* 32:349–351.

Polgar G., Sacchetti A. and Galli P. 2010. Differentiation and adaptive radiation of amphibious gobies (Gobiidae: Oxudercinae) in semi-terrestrial habitats. *Journal of Fish Biology* 77:1645–1664.

Polgar G., Zane L., Babbucci M. et al. 2014. Phylogeography and demographic history of two widespread Indo-Pacific mudskippers (Gobiidae: *Periophthalmus*). *Molecular Phylogenetics and Evolution* 73:161–176.

Prevost A.F. 1747. Poissons extraordinaires d'Amboine. In *History Generale de Voyages, ou Nouvelle Collection de Touts Les Relation de Voyages par Mer et par Terre.* pp. 84–254. La Haye 17.

Prokofiev A.M. 2015. An overview of gobies of the subfamily Amblyopinae (Gobiidae) from the Western South China Sea (Vietnam and Indonesia) and Pacific waters of the Philippines. *Journal of Ichthyology* 55:655–663.

Quoy J.R.C. and Gaimard J.P. 1824–25. Description des Poissons. Chapter IX. In Freycinet, L. de, *Voyage autour du Monde...exécuté sur les corvettes de L. M. "L'Uranie" et "La Physicienne," pendant les années 1817, 1818, 1819 et 1820.* Paris. 192–401 [1–328 in 1824; 329–616 in 1825].

Richardson J. 1846. Report on the ichthyology of the seas of China and Japan. *Report of the British Association for the Advancement of Science 15th meeting* [1845]:187–320.

Roberts T.R. 1978. An ichthyological survey of the Fly River in Papua New Guinea with descriptions of new species. *Smithsonian Contributions to Zoology* 281:1–72.

Ruysch H. 1718. Collectivo nova piscium amboinensium partim ibi ad vivum delineatorum partim Ex Museuo Henrici Ruysch M.D. XX tabulis comprehensa. In *Theatrum universal omnium animalium ... ex scriptoribus tam antiquis quam recentioribus..a J. Jonstonio collectum, ac plus quam trecentis piscibus nuperrime ex Indiis orientalibus allatis... cumm enumeration morborum...ac notiti anomalium...cura H. Ruysch, etc.* vol. 1: pp. 1–40. Wetstein, Amsterdam.

Shibukawa K. and Ghaffar M.A. 2011. Family Gobiidae—Gobies. In *Fishes of Terengganu—East Coast of Malay Peninsula, Malaysia* (eds. M. Matsunuma, H. Motomura, K. Matsuura, N. A. M. Shazili and N. A. Ambak) pp. 193–209. Kagoshima: Kagoshima University Museum.

Springer V.G. 1978. Synonymization of the Oxudercidae, with comments on the identity of *Apocryptes cantoris* Day (Pisces: Gobiidae). *Smithsonian Contributions to Zoology* 270:1–14.

Swainson W. 1839. *On the natural history and classification of fishes, amphibians, & reptiles, or monocardian animals*, Vol. 2. London: Spottiswoode & Co., 448 pp.

Thacker C.E. 2003. Molecular phylogeny of the gobioid fishes (Teleostei: Perciformes: Gobioidei). *Molecular Phylogenetics and Evolution* 26:354–368.

Thacker C.E. 2009. Phylogeny of Gobioidei and placement within Acanthomorpha, with a new classification and investigation of diversification and character evolution. *Copeia* 1:93–104.

Thacker C.E. 2013. Phylogenetic placement of the European sand gobies in Gobionellidae and characterization of gobionellid lineages (Gobiiformes: Gobioidei). *Zootaxa* 3619:369–382.
Tomiyama I. 1934. Four species of gobies of Japan. *Journal of the Tokyo Imperial University, Faculty of Science Section IV* 3(3):325–334.
Tornabene L., Chen Y. and Pezold F. 2013. Gobies are deeply divided: Phylogenetic evidence from nuclear DNA (Teleostei: Gobioidei: Gobiidae). *Systematics and Biodiversity* 11(3):1–17.
Trân Đ. Đ., Shibukawa K., Nguyen P.T et al. 2013. *Fishes of the Mekong Delta, Vietnam.* Can Tho: Can Tho University Publishing, 174 pp.
Valentijn F. 1726. Omstandig verhaal van de geschiedenissen en zaakan het kerkelyke oft den godsdienst betreffunde, zoo in Amboinia...Band...,Tonkin, Cambodia, en Siam... In: *Oud en nieuw Oost-Indiën, vervattende een naaukerurige en uitvoerige verhandelinge van Ndederlands Mogentheyd in die Gewesten...* vol. 3: pp. 336–515. Dodrecht & Amsterdam: van Braam & Gerard onder de Linde.
Van Der Laan R., Eschmeyer W.N. and Fricke R. 2014. Family-group names of recent fishes. *Zootaxa* 3882(2):1–230.
Van Tassell J. L., Tornabene L. and Taylor M.S. 2011. A history of gobioid morphological systematics. In *The biology of gobies* (eds. R.A. Patzner, J.L. Van Tassell, M. Kovacic, and B. Kapoor) pp. 3–22. Enfield, NH: Science Publishers Inc.
Wang H-Y., Tsai M-P, Dean J. and Lee S-C. 2001. Molecular phylogeny of gobioid fishes (Perciformes: Gobioidei) based on mitochondrial 12s rRNA sequences. *Molecular Phylogenetics and Evolution* 20:390–408.
Wang Z-Q., Yang J-Q. and Tang W-Q. 2006. A long-term misidentified new record species of Gobiidae from China—*Periophthalmus magnuspinnatus. Acta Zootaxon Sinica* 31:906–910.

CHAPTER 2

The Natural Distribution of Mudskippers

Lynne R. Parenti and Zeehan Jaafar

CONTENTS

2.1 INTRODUCTION

Oxudercine gobies, the mudskippers, are distributed broadly throughout tropical and subtropical coastal habitats of the Indo-West Pacific (IWP) and the eastern Atlantic (Murdy 1989; Clayton 1993; Figure 2.1). These fishes live in close association with, or proximity to, mangrove forests and adjacent mudflats that distinguish global tropical and subtropical coastlines (Hogarth 2001). The distribution of oxudercine gobies is largely coincident with that of modern mangrove habitats except that mudskippers are absent from the western tropical Atlantic and eastern tropical Pacific (Figure 2.1).

We first summarize the biogeographic regions proposed for mangroves (Duke 1992; Duke et al. 1998; Figure 2.2) and mudskippers (Murdy 1989; Figure 2.3). We then present a revised classification of mudskipper areas based on updated distributional and taxonomic data (Figure 2.4). For example, Murdy (1989) had no scientific collections of mudskippers from Sulawesi; today, we have ample data on species of mudskippers collected during the past two decades from that island complex (e.g., Polgar et al. 2014; Parenti et al. 2015) and elsewhere throughout their range.

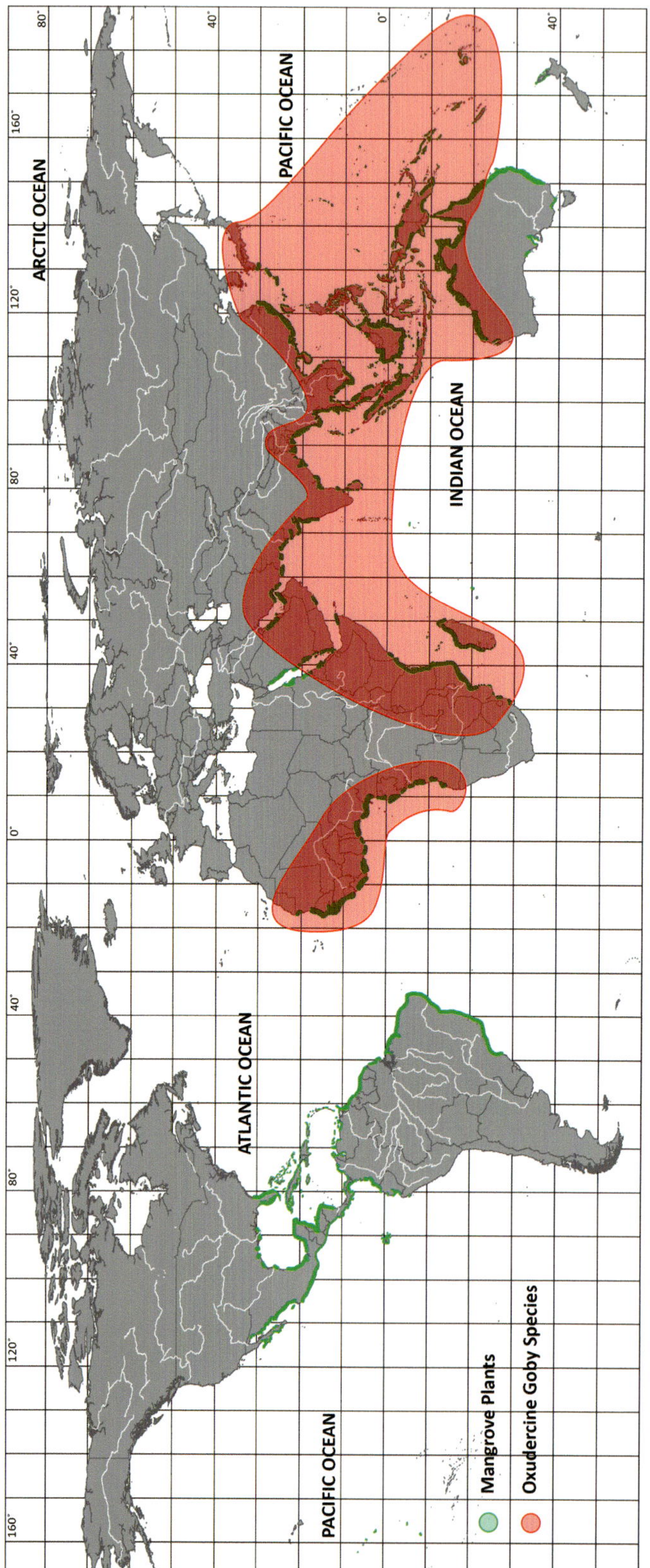

Figure 2.1 Green bands along coastlines indicate natural distribution of mangrove plants. Distribution of extant oxudercine gobiid species, within red solid lines, is largely coincident with the natural distribution of mangrove forests except in the western Atlantic and eastern Pacific, where oxudercine gobies are absent.

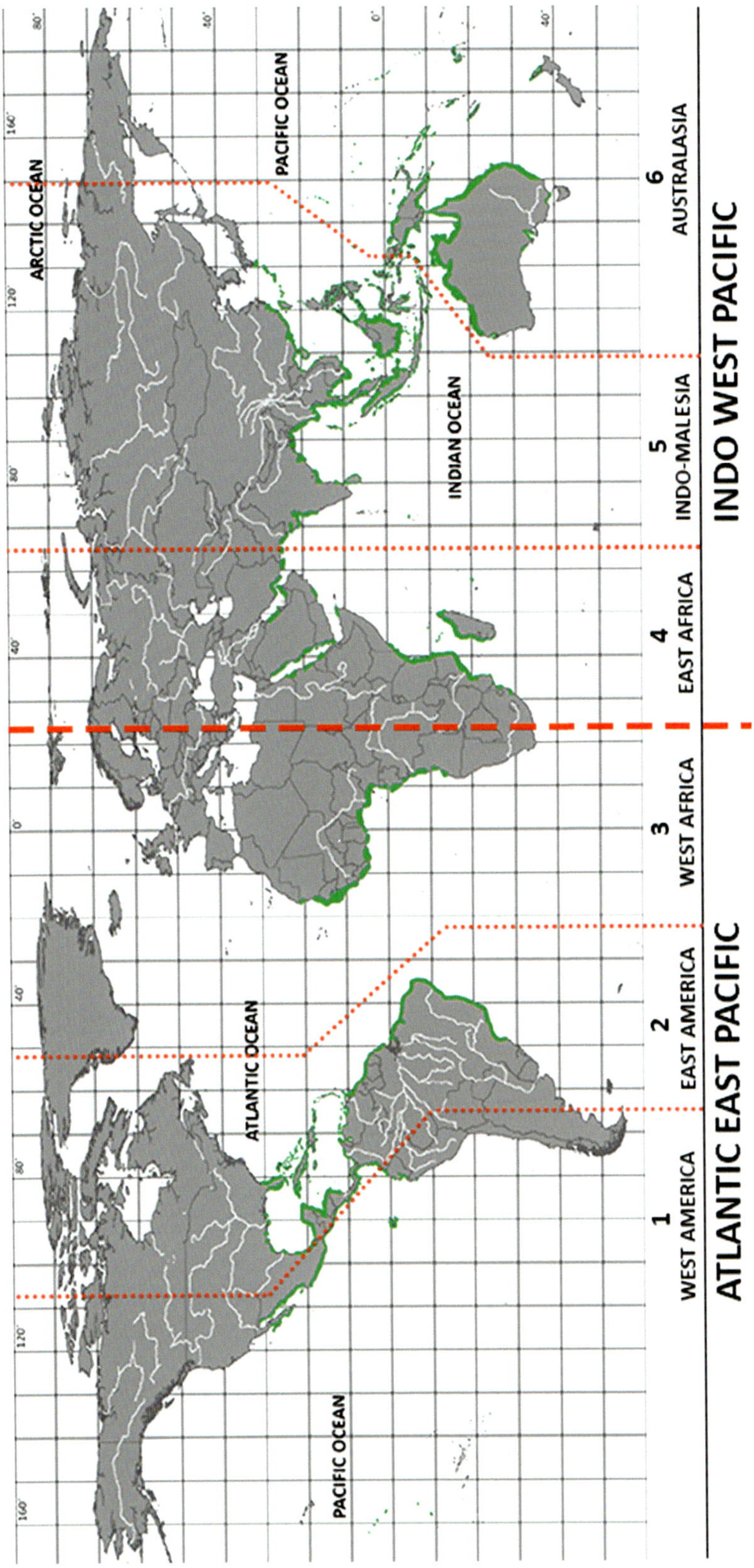

Figure 2.2 Six global bioregions in two hemispheres proposed by Duke (1992) based on mangrove flora. Green bands along coastlines indicate the natural distribution of mangrove plants. Dotted or dashed red lines divide mangrove regions. Duke (1992) did not indicate the eastern limit of the Australasia region or the western limit of the west America region. (Redrawn from Duke, N.C, *Tropical Marine Ecosystems*, (eds. A. I. Robertson and D. M. ALongi), 63–100, 1992.)

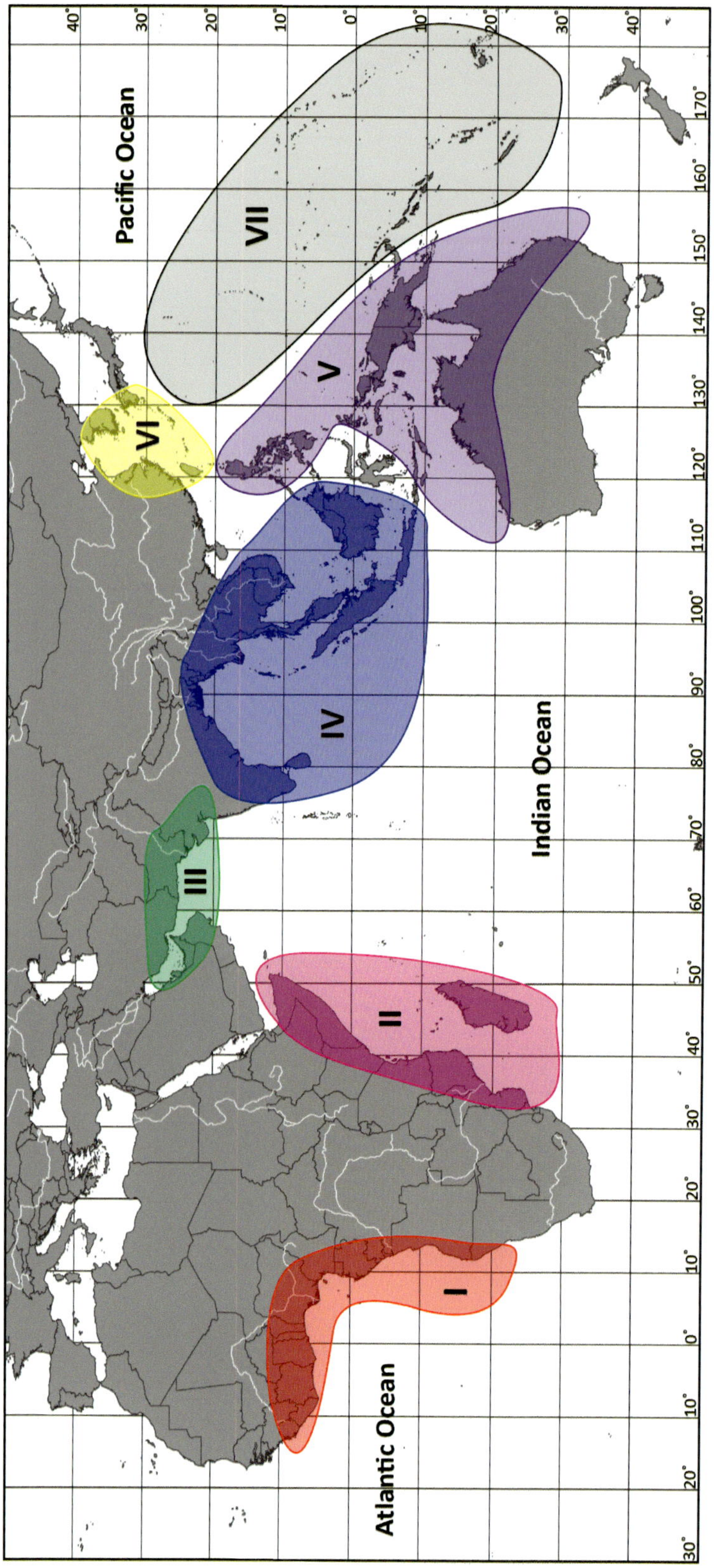

Figure 2.3 Seven biogeographic divisions proposed by Murdy (1989, Roman numerals were used in the original description of areas) based on distribution of oxudercine gobies: (I) West Africa; (II) East Africa; (III) Arabian Gulf; (IV) Indo-Malaya; (V) Australia/New Guinea; (VI) China Sea; and (VII) Oceania. Sulawesi was not included in any region, as no oxudercine was then known from that island complex. (Modified from Murdy E.O., *Records of the Australian Museum Supplement*, 11, 1–93, 1989.)

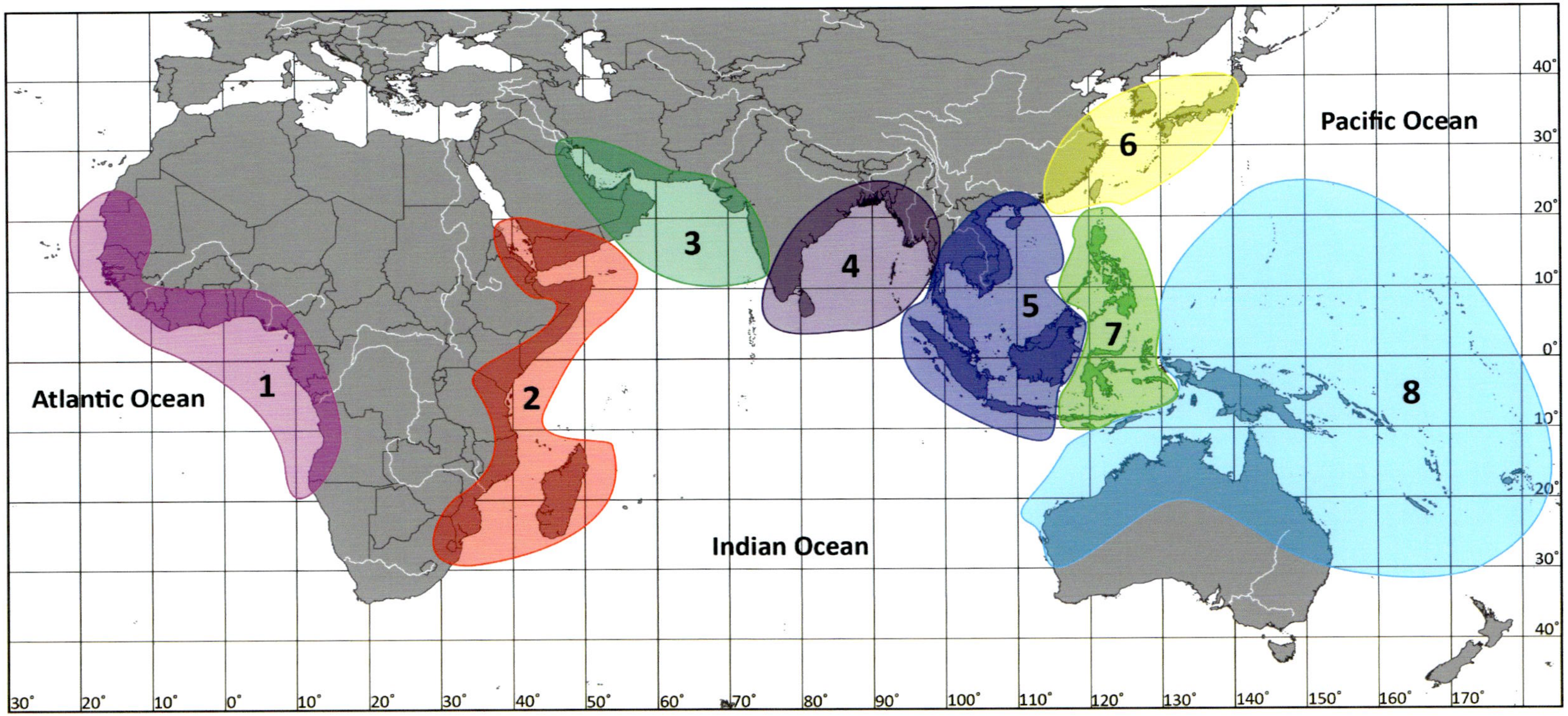

Figure 2.4 Eight proposed areas of endemism for oxudercine gobiid fishes based on current collections and taxonomy: (1) Tropical eastern Atlantic; (2) Southwestern Indian Ocean; (3) Arabian Sea; (4) Bay of Bengal; (5) South China Sea; (6) East China Sea; (7) Banda Sea; and (8) Oceania.

An understanding of the biogeography of the mudskippers requires a detailed description of species distributions and a preliminary proposal of their areas of endemism. One of our future goals is a comprehensive comparative biogeographic analysis of mudskippers, mangrove plants, and the associated biota of the marine intertidal zone, such as gastropods (e.g., Ozawa et al. 2015). In that analysis, we will use phylogenetic and distributional data of the mangrove biota to infer the historical, hierarchical relationships among areas of endemism; a general classification of those areas is expected to identify repeated phylogenetic patterns of the endemic taxa and, therefore, reflect the shared history of components of the mangrove biota (Parenti and Ebach 2009). We defer that analysis until completion of a comprehensive phylogenetic analysis of oxudercine gobies that is currently underway. Thus, the areas of endemism for mudskippers that we propose, below, are preliminary and await rigorous testing.

2.2 BIOGEOGRAPHIC REGIONS OF MANGROVES AND MUDSKIPPERS

Modern mangrove forests are distributed in the littoral zone of tropical and subtropical regions between 30° N and 30° S latitude and are restricted largely to climates with an average monthly minimum air temperature of 20°C (Giri et al. 2010; Ellison et al. 1999; Duke 1992). Mangrove forests extend farther south in eastern South America, Australia, and the North Island of New Zealand, likely as relict populations of once broader distributions (Duke 1992).

Modern mangrove floristic communities are classified in six biogeographic regions: (1) western America and eastern Pacific; (2) eastern America, including the Caribbean and parts of the Gulf of Mexico; (3) West Africa; (4) East Africa and Madagascar; (5) Indo-Malesia, including Asia; and (6) Australasia and the western Pacific (Duke 1992: fig. 1; Figure 2.2). These divisions are based on presence/absence data for 69 species of mangrove plants classified in 27 genera and 21 families (Duke 1992: tab. 3). Just one species, the widespread fern *Acrostichum aureum*, lives in all six biogeographic regions. There is otherwise no overlap of native species between what Duke (1992) recognized as two global biogeographic hemispheres: the Atlantic East Pacific (AEP) comprising the first three biogeographic regions including 11 native mangrove plant species in seven genera and the IWP comprising the latter three biogeographic regions with 58 native mangrove plant species in 23 genera. Duke (1992) was not explicit about how he delimited the six regions; those numbered 1, 2, 5, and 6 have at least one endemic species each, whereas regions 3 and 4 have no endemics and share species with the other areas in their own global biogeographic hemisphere.

Mudskippers are distributed in seven, disjunct divisions: (1) West Africa; (2) East Africa; (3) Arabian Gulf; (4) Indo-Malaya; (5) Australia/New Guinea; (6) China Sea; and (7) Oceania, according to Murdy (1989: fig. 98; Figure 2.3). These mudskipper divisions differ from those proposed for mangrove forests in several ways. First, of course, they do not include the modern mangrove regions of the western Americas and eastern Pacific or eastern Americas and the Caribbean in which there are no oxudercine gobies. Also the "East Africa" and "Arabian Gulf" divisions of Murdy (1989) together equal the "East Africa" region of Duke (1992) and Duke et al. (1998). Although Duke (1992) and Duke et al. (1998) included much of Asia in their "Indo-Malesia" region, Murdy (1989) treated these as two separate areas: "Indo-Malaya" and "China Sea."

Further, Duke (1992) and Duke et al. (1998) included the western Pacific, as well as the island of Maluku, in their "Australasia." In contrast, Murdy (1989) considered these to be two separate areas "Australia/New Guinea" and "Oceania"—and included the Moluccas (presumably the islands of Maluku and northern Maluku) in the former. The Philippine Islands and the Indonesian island of Timor are part of the "Indo-Malesia and Asia" region of Duke et al. (1998) but in the "Australia/New Guinea" division of Murdy (1989).

Many species of mudskippers and mangroves have restricted distributions and are endemic to the proposed biogeographic regions. The presence of at least one endemic taxon is needed to recognize an area as biogeographically distinct (Humphries and Parenti 1986) and at least two endemic taxa to make the area potentially informative in a comparative biogeographic analysis (see Harold and Mooi 1994; Parenti and Ebach 2009). Murdy (1989: 74) claimed that he recognized his "seven divisional associations … based on maximal endemism …" yet, as with the regions for mangroves, two of his mudskipper areas, "East Africa" and "Oceania" have no endemic species. We summarize the distribution of mudskippers to support our preliminary, revised proposal of areas of endemism in which each area has at least one endemic taxon and the areas are allopatric (Figure 2.4).

2.3 DISTRIBUTION OF OXUDERCINE GOBY GENERA AND SPECIES

We currently recognize 43 valid species of oxudercine gobies classified in 10 genera (Jaafar et al. 2016; Polgar et al. 2013; Jaafar and Larson 2008; Murdy 1989; Table 2.1). We reconfirmed identification of most of the specimens examined by Murdy (1989) and added data on oxudercine species described since 1989 (Jaafar et al. 2009, 2016; Polgar et al. 2013, 2014; Jaafar and Larson 2008; Larson and Takita 2004; Darumas and Tantichodok 2002; Murdy and Takita 1999; Lee et al. 1995; Figure 2.5).

Periophthalmus is the most speciose genus, with 19 species, and is also the most widely distributed, ranging from the Gulf of Guinea eastward to Samoa (Jaafar and Larson 2008). For clarity, we illustrate the distribution limits of the species of *Periophthalmus* in a series of six maps (Figures 2.6 through 2.11). A single species, *Periophthalmus barbarus*, lives on the tropical west coast of Africa and volcanic islands in the Gulf of Guinea (Figure 2.6). The remaining 18 species in the genus live throughout the IWP from East Africa to the western margin of the Pacific Plate (Springer 1982; Figures 2.6 through 2.11). Their distribution is coincident with that of *Ps. argentilineatus*, the most widely distributed species of mudskippers (Figure 2.7). A second widespread species, *Ps. kalolo*, is distributed from eastern Africa to Tonga, but does not live in the Arabian or East China seas (Figure 2.6). These are the only two species of mudskippers that live in the southwestern Indian Ocean.

Widespread species are uninformative for cladistic biogeographic analysis (Nelson and Platnick 1981). But Gill and Kemp (2002) noted that many presumed widespread species of fishes throughout the IWP may simply represent an artifact of poor taxonomic understanding or practice. To investigate the phylogeography of two widespread species, Polgar et al. (2014) collected samples of *Ps. argentilineatus* and *Ps. kalolo* from 12 localities in the South China Sea, northern Sulawesi, and the southwestern Indian Ocean. Based on an analysis of molecular sequence data from two mitochondrial genes (D-loop and 16*S* rDNA) and one nuclear gene (*rag1*), they recognized three distinct molecular lineages of *Ps. argentilineatus*—one from the southwestern Indian Ocean which was sister to a second from northern Sulawesi and a third from northern Sulawesi which was sister to samples of *Ps. kalolo* from the southwestern Indian Ocean and northern Sulawesi; also, they found only shallow genetic differentiation between samples of *Ps. kalolo* from the southwestern Indian Ocean and northern Sulawesi (Figure 2.5). Although preliminary, this study is significant for our proposal of areas of endemism because it provides the only evidence for an endemic oxudercine lineage, which could be recognized at the species level, from the southwestern Indian Ocean.

A single species of *Periophthalmus, Ps. waltoni*, lives in "… mudflat areas from the Arabian Gulf to Pakistan" (Murdy 1989: 44; Figure 2.9). The remaining 15 species of *Periophthalmus* live east of the southern tip of India. They have restricted, yet highly overlapping distributions (Figures 2.6 through 2.11).

The nine remaining oxudercine genera each have fewer species and smaller ranges, yet exhibit a high degree of endemism and allopatry that makes them ideal for a cladistic biogeographic analysis (Figures 2.12 through 2.20). Two of these genera, *Scartelaos* (Figure 2.12) and *Boleophthalmus* (Figure 2.13), each have a single species ranging from the Arabian Gulf to Pakistan or India (Table 2.1), respectively. The remaining oxudercine species live east of the southern tip of India. Two genera,

Table 2.1 Distribution of Oxudercine Gobies

		Areas of Endemism								
Genus	Species	Tropical Eastern Atlantic Ocean	Southwestern Indian Ocean	Arabian Sea	Bay of Bengal	South China Sea	East China Sea	Banda Sea	Oceania	No. of Areas
Apocryptes (*As.*)	*bato*	–	–	–	×	–	–	–	–	**1**
Apocryptodon (*An.*)	*madurensis*	–	–	–	×	×	–	×	×	**4**
	punctatus	–	–	–	–	–	×	–	–	**1**
	wirzi	–	–	–	–	–	–	–	×	**1**
Boleophthalmus (*B.*)	*birdsongi*	–	–	–	–	–	–	–	×	**1**
	boddarti	–	–	×	×	×	–	×	–	**4**
	caeruleomaculatus	–	–	–	–	–	–	–	×	**1**
	dussumieri	–	–	×	–	–	–	–	–	**1**
	pectinirostris	–	–	–	–	–	×	–	–	**1**
	poti	–	–	–	–	–	–	–	×	**1**
	cf. *pectinirostris*	–	–	–	–	×	–	–	–	**1**
Oxuderces (*O.*)	*dentatus*	–	–	–	–	–	×	–	–	**1**
	nexipinnis	–	–	–	×	×	–	–	–	**2**
Parapocryptes (*Pa.*)	*rictuosus*	–	–	–	×	–	–	–	–	**1**
	serperaster	–	–	–	–	×	×	–	–	**2**

(*Continued*)

Table 2.1 (*Continued*) **Distribution of Oxudercine Gobies**

		Areas of Endemism								
Genus	**Species**	**Tropical Eastern Atlantic Ocean**	**Southwestern Indian Ocean**	**Arabian Sea**	**Bay of Bengal**	**South China Sea**	**East China Sea**	**Banda Sea**	**Oceania**	**No. of Areas**
Periophthalmodon (*Pn.*)	*freycineti*	–	–	–	–	–	–	×	×	**2**
	schlosseri	–	–	–	×	×	–	–	–	**2**
	septemradiatus	–	–	–	×	×	–	–	–	**2**
Periophthalmus (*Ps.*)	*argentilineatus*	–	×	×	×	×	×	×	×	**7**
	barbarus	×	–	–	–	–	–	–	–	**1**
	chrysospilos	–	–	–	–	×	–	–	–	**1**
	darwini	–	–	–	–	–	–	–	×	**1**
	gracilis	–	–	–	–	×	–	×	×	**3**
	kalolo	–	×	–	×	×	–	×	×	**5**
	magnuspinnatus	–	–	–	–	–	×	–	–	**1**
	malaccensis	–	–	–	–	–	–	×	–	**1**
	minutus	–	–	–	–	×	–	×	×	**3**
	modestus	–	–	–	–	–	×	–	–	**1**
	novaeguineaensis	–	–	–	–	–	–	–	×	**1**
	novemradiatus	–	–	–	×	–	–	–	–	**1**
	pusing	–	–	–	–	–	–	×	–	**1**
	spilotus	–	–	–	–	×	–	–	–	**1**
	takita	–	–	–	–	–	–	–	×	**1**
	variabilis	–	–	–	–	×	–	–	–	**1**
	walailakae	–	–	–	×	×	–	–	–	**2**
	waltoni	–	–	×	–	–	–	–	–	**1**
	weberi	–	–	–	–	–	–	–	×	**1**

(*Continued*)

Table 2.1 (*Continued*) Distribution of Oxudercine Gobies

		Areas of Endemism								
Genus	**Species**	**Tropical Eastern Atlantic Ocean**	**Southwestern Indian Ocean**	**Arabian Sea**	**Bay of Bengal**	**South China Sea**	**East China Sea**	**Banda Sea**	**Oceania**	**No. of Areas**
Pseudapocryptes (*Pd.*)	*borneensis*	–	–	–	–	×	–	–	–	**1**
	elongatus	–	–	–	×	×	–	–	–	**2**
Scartelaos (*S.*)	*cantoris*	–	–	–	×	–	–	–	–	**1**
	gigas	–	–	–	–	–	×	–	–	**1**
	histophorus	–	–	×	×	×	×	×	×	**6**
	tenuis	–	–	×	–	–	–	–	–	**1**
Zappa (*Z.*)	*confluentus*	–	–	–	–	–	–	–	×	**1**
Total number of species		1	2	6	14	18	9	10	16	–
Number of endemic species		1	Possibly 1	3	3	5	6	2	9	–

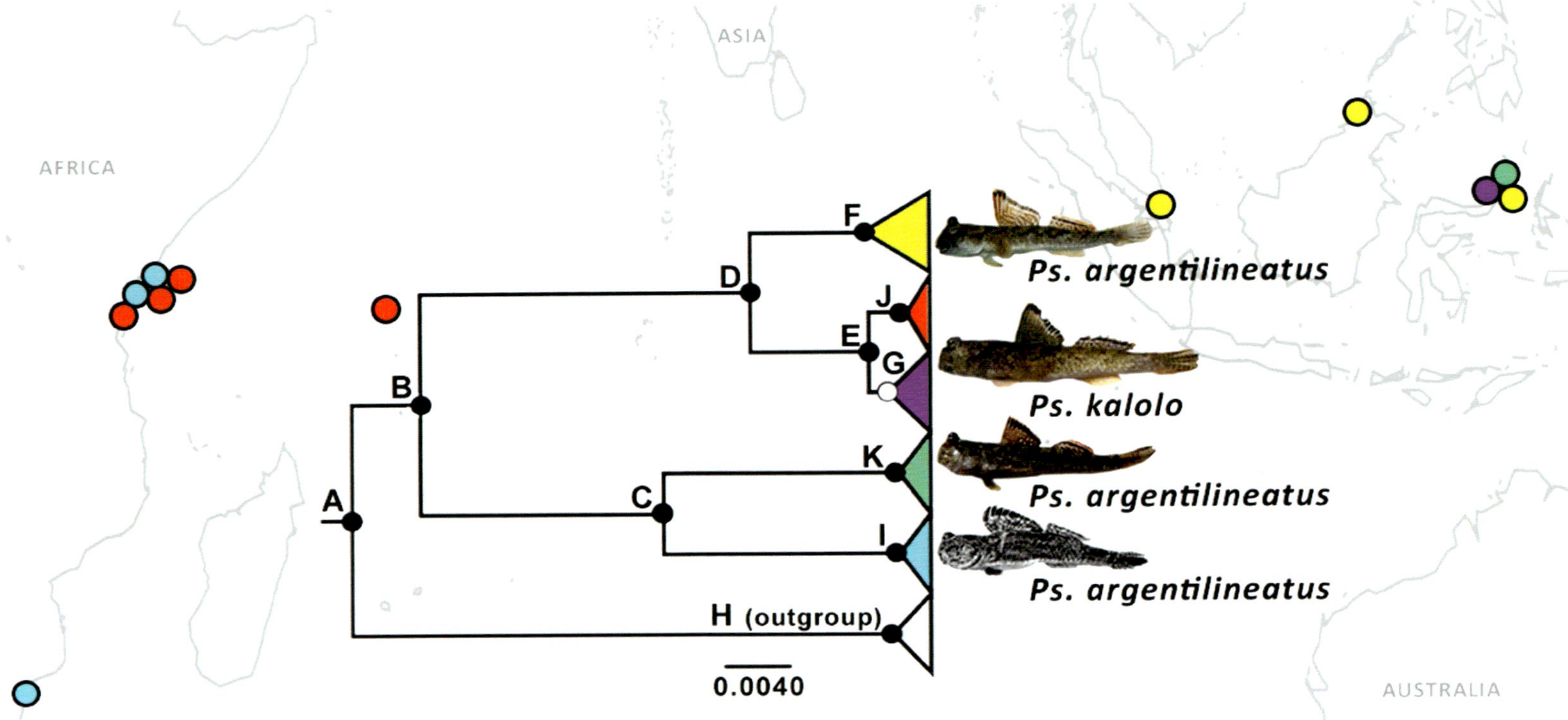

Figure 2.5 Distribution and phylogeny of populations of *Periophthalmus argentilineatus* (yellow, green, and light blue circles) and *Periophthalmus kalolo* (red and purple circles). (Modified from Polgar G. et al., *Molecular Phylogenetics and Evolution*, 73, 161–176, 2014. With permission.)

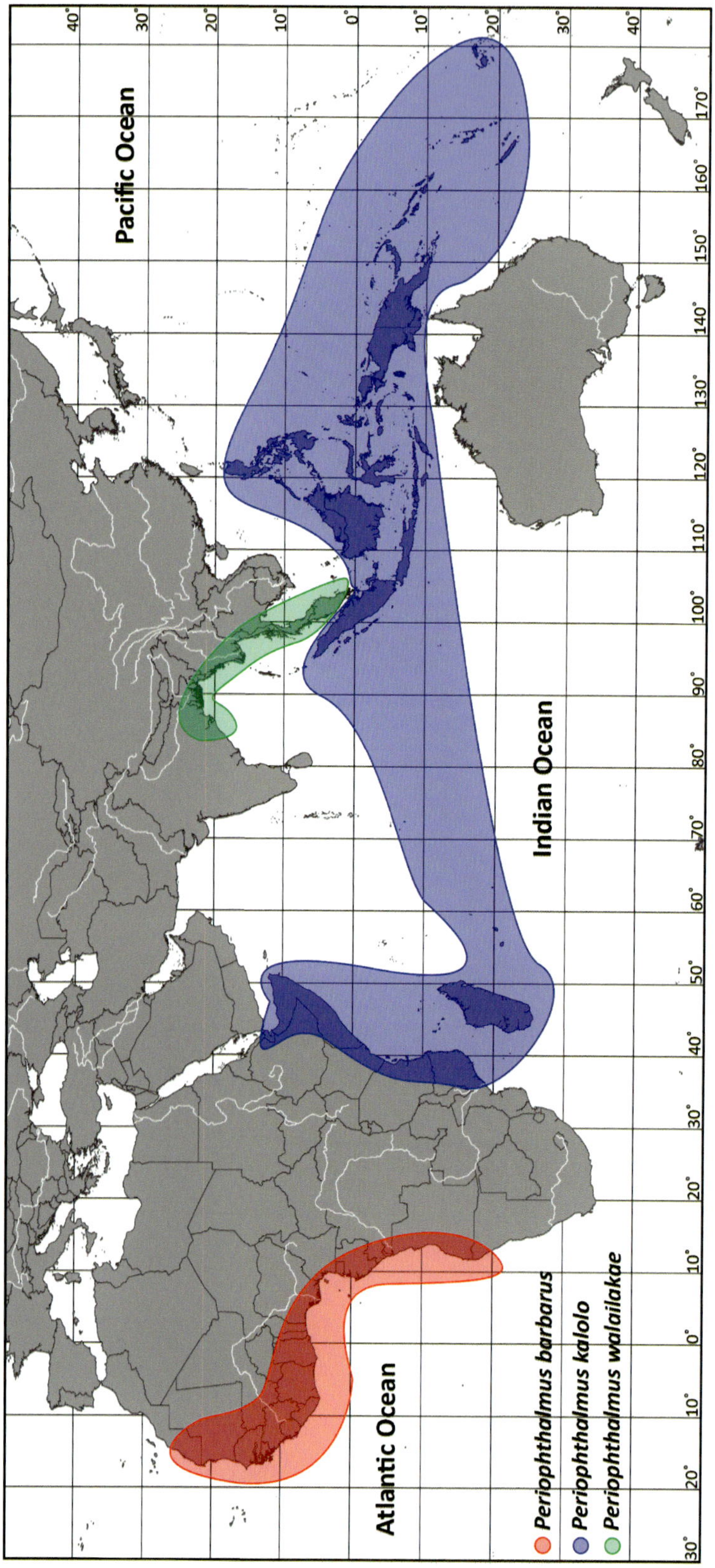

Figure 2.6 Distribution of the genus *Periophthalmus* (1): *Ps. barbarus*, *Ps. kalolo*, and *Ps. walailakae*.

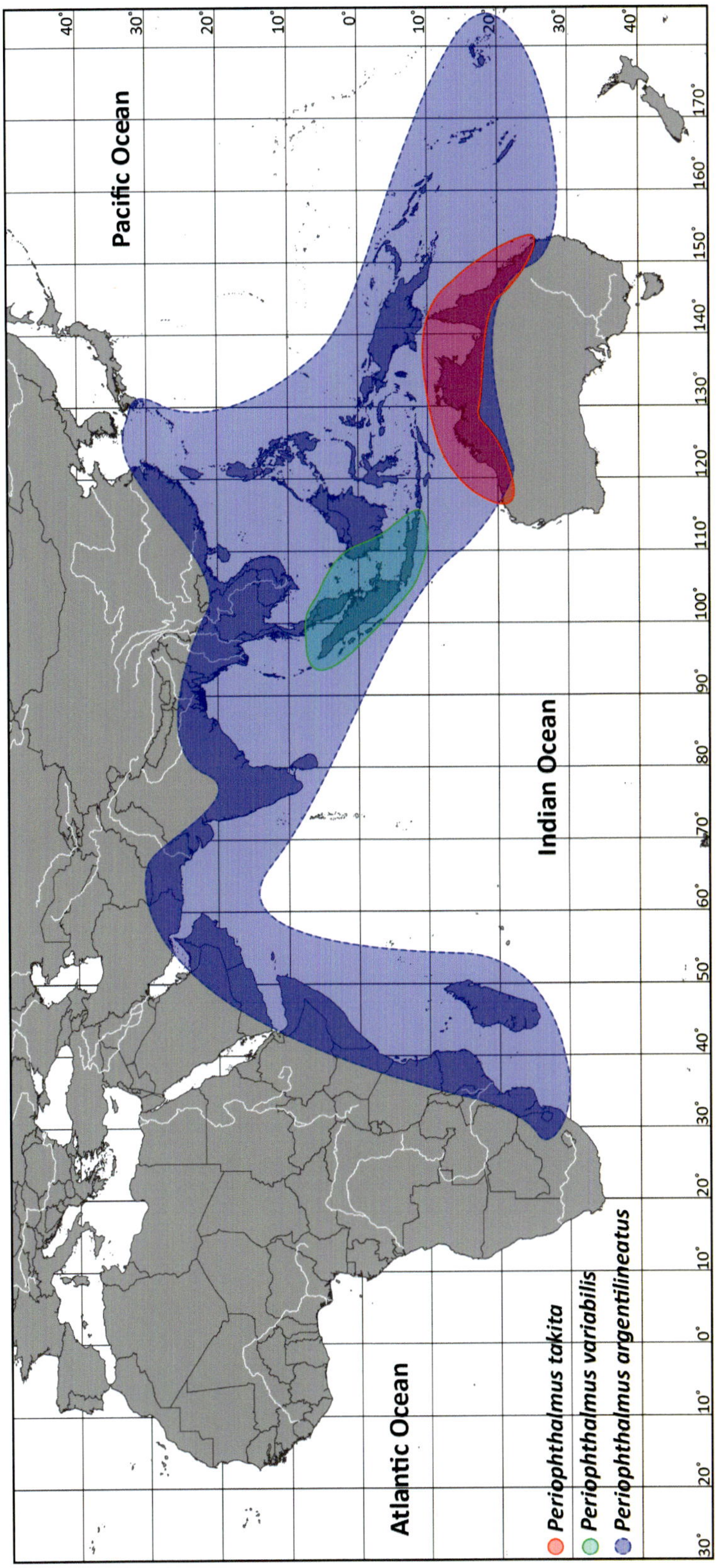

Figure 2.7 Distribution of the genus *Periophthalmus* (2): *Ps. argentilineatus*, *Ps. takita*, and *Ps. variabilis*.

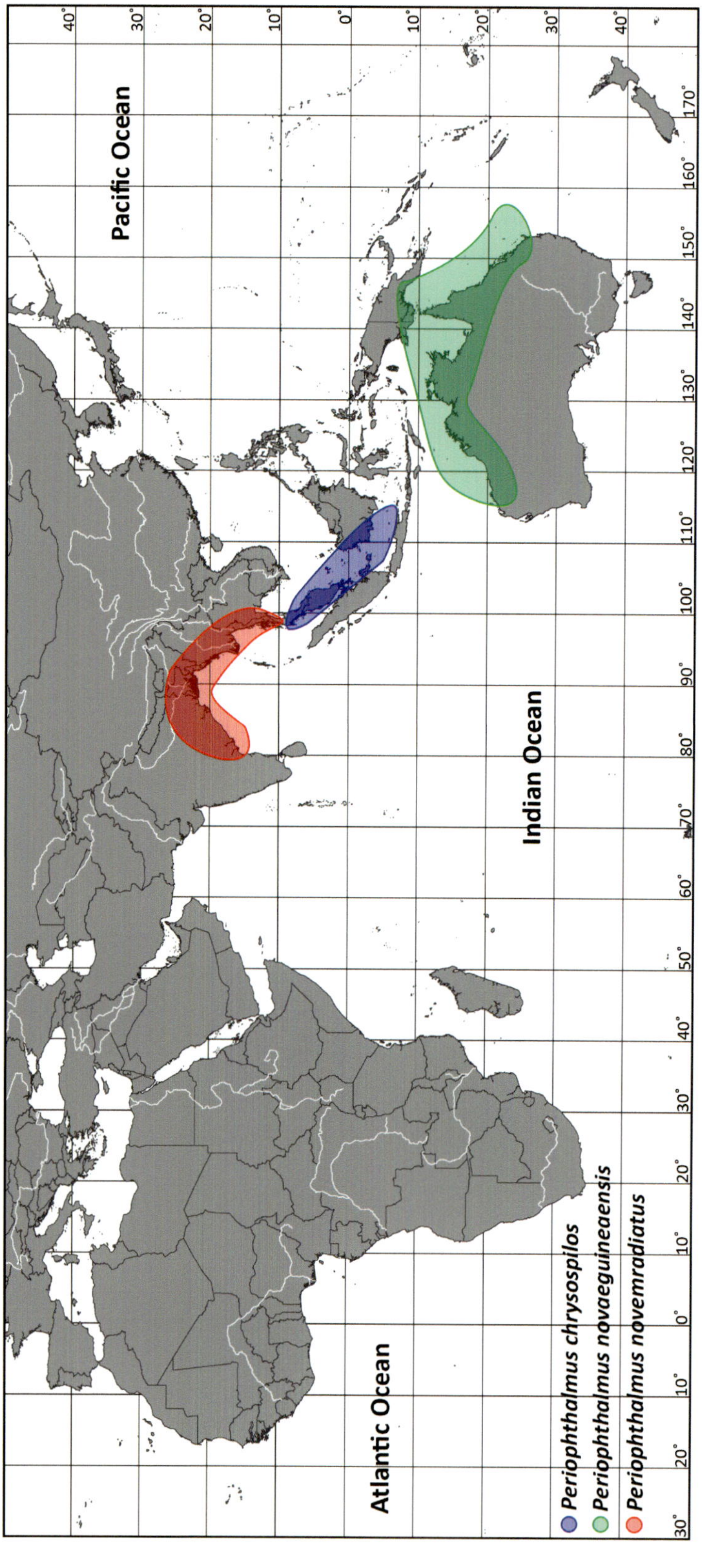

Figure 2.8 Distribution of the genus *Periophthalmus* (3): *Ps. chrysospilos*, *Ps. novaeguineaensis*, and *Ps. novemradiatus*.

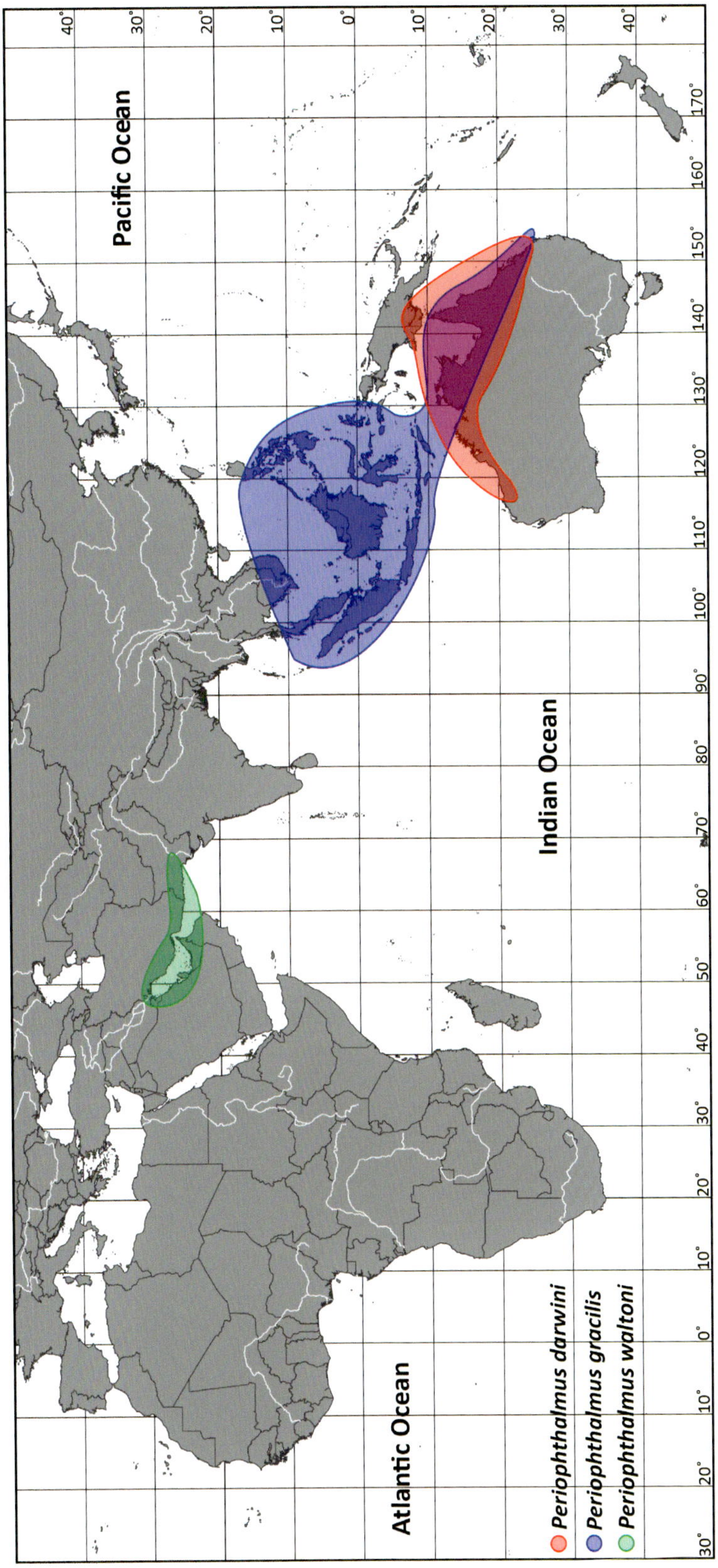

Figure 2.9 Distribution of the genus *Periophthalmus* (4): *Ps. darwini*, *Ps. gracilis*, and *Ps. waltoni*.

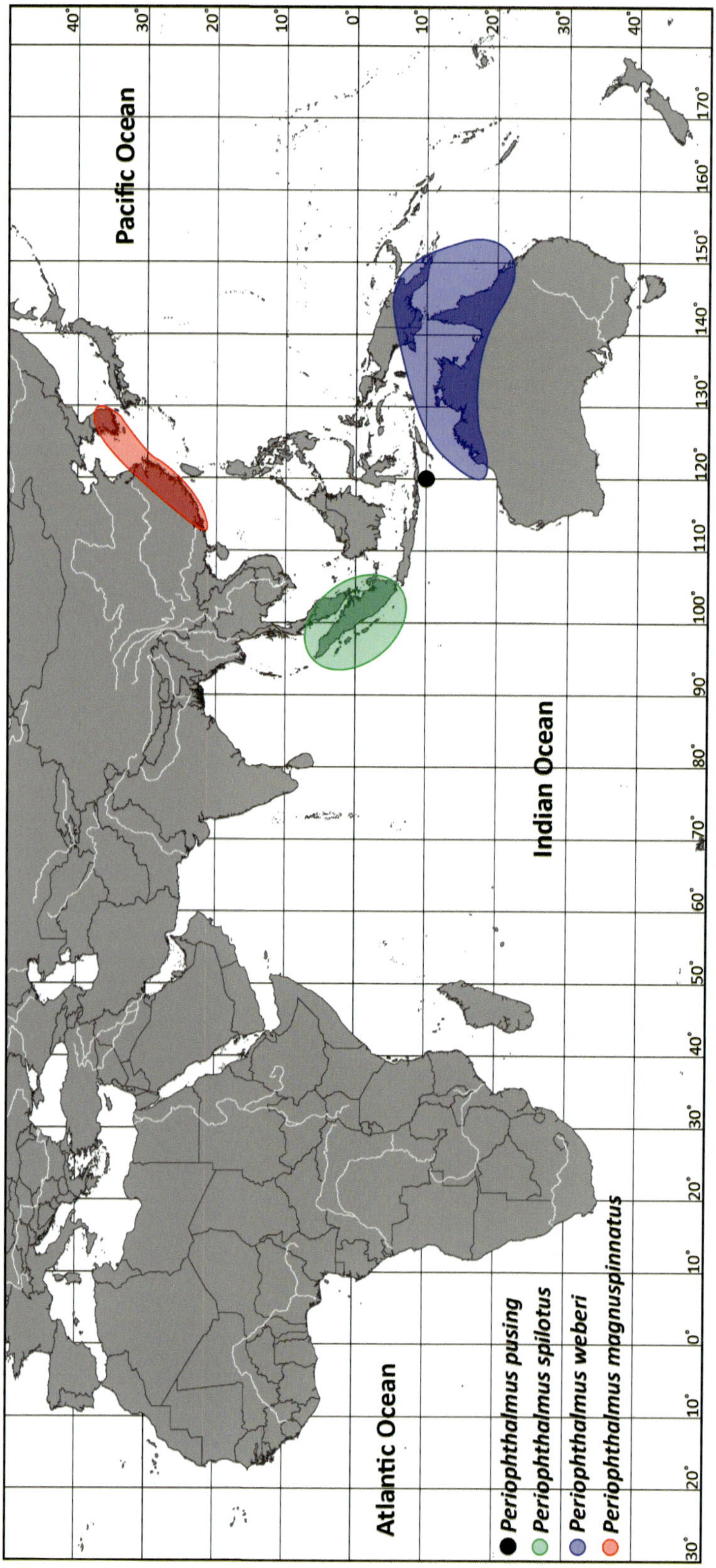

Figure 2.10 Distribution of the genus *Periophthalmus* (5): *Ps. pusing*, *Ps. magnuspinnatus*, *Ps. spilotus*, and *Ps. weberi.*

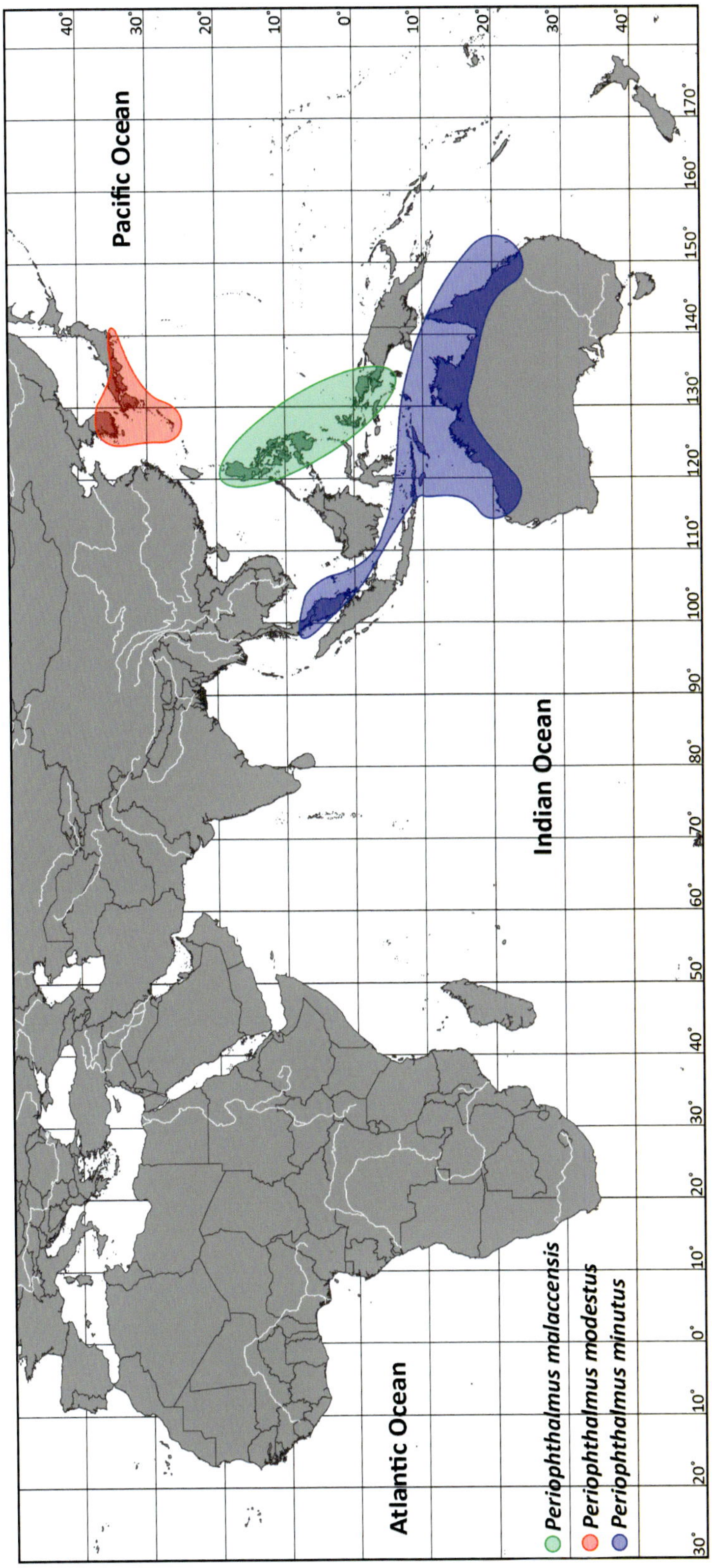

Figure 2.11 Distribution of the genus *Periophthalmus* (6): *Ps. malaccensis*, *Ps. modestus*, and *Ps. minutus*.

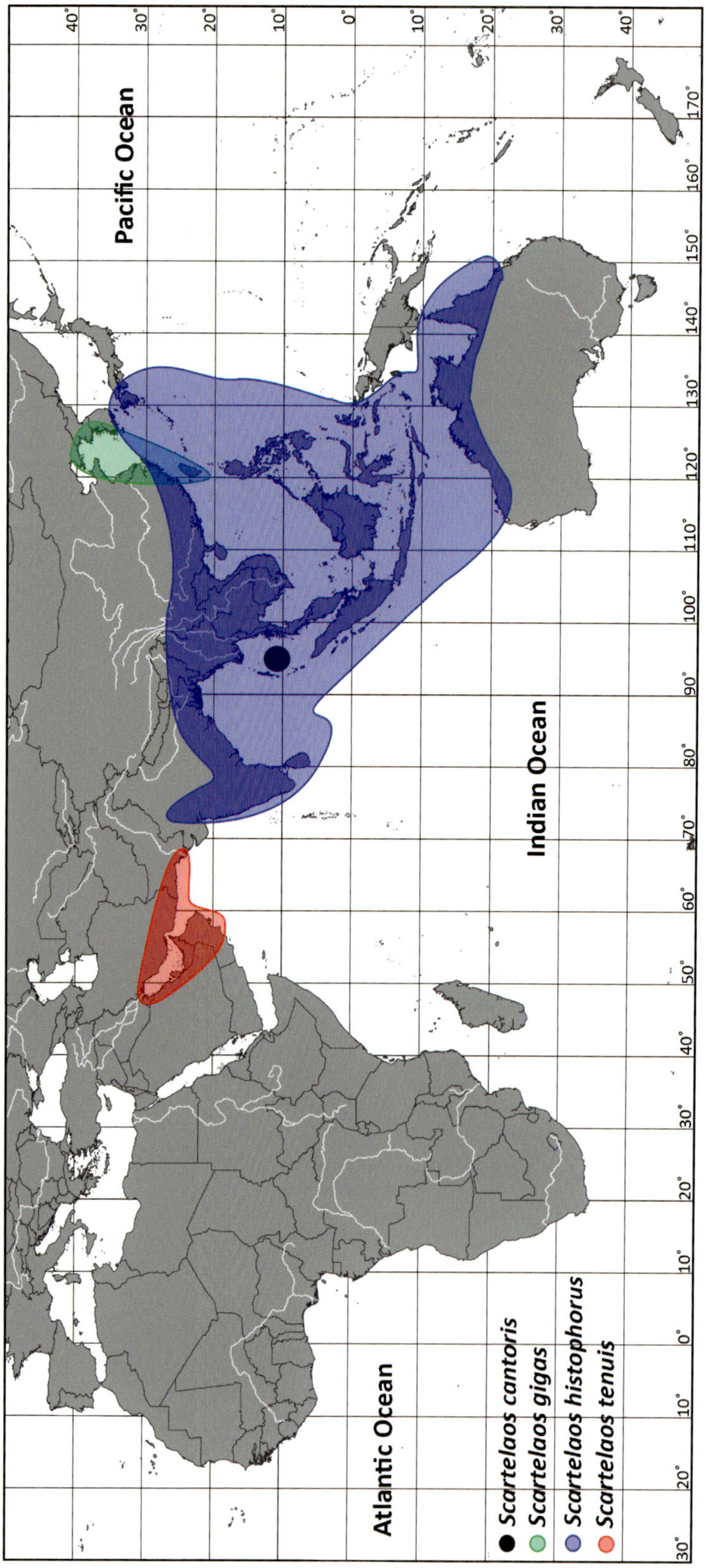

Figure 2.12 Distribution of species of the genus *Scartelaos*: *S. cantoris*, *S. gigas*, *S. histophorus*, and *S. tenuis*.

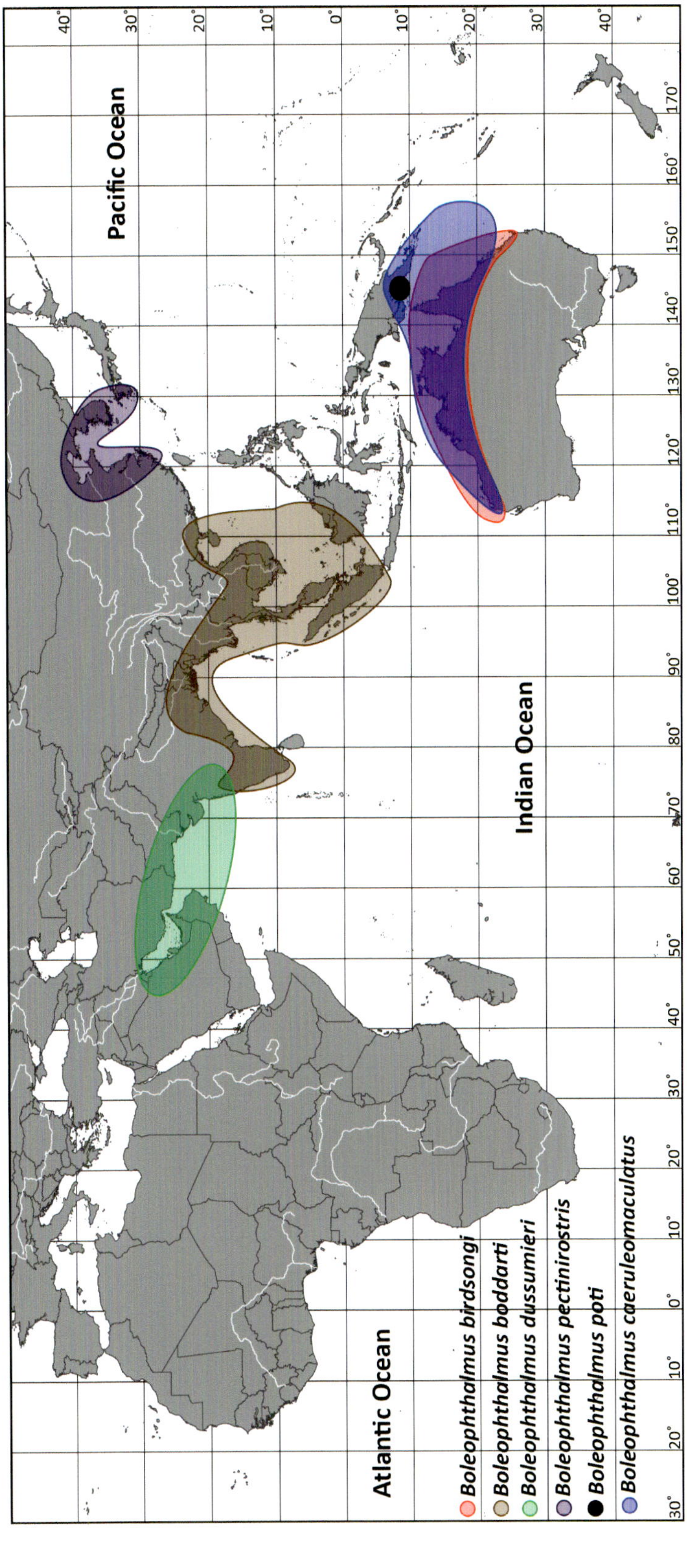

Figure 2.13 Distribution of species of the genus *Boleophthalmus*: *B. birdsongi*, *B. boddarti*, *B. caeruleomaculatus*, *B. dussumieri*, *B. pectinirostris*, and *B. poti*.

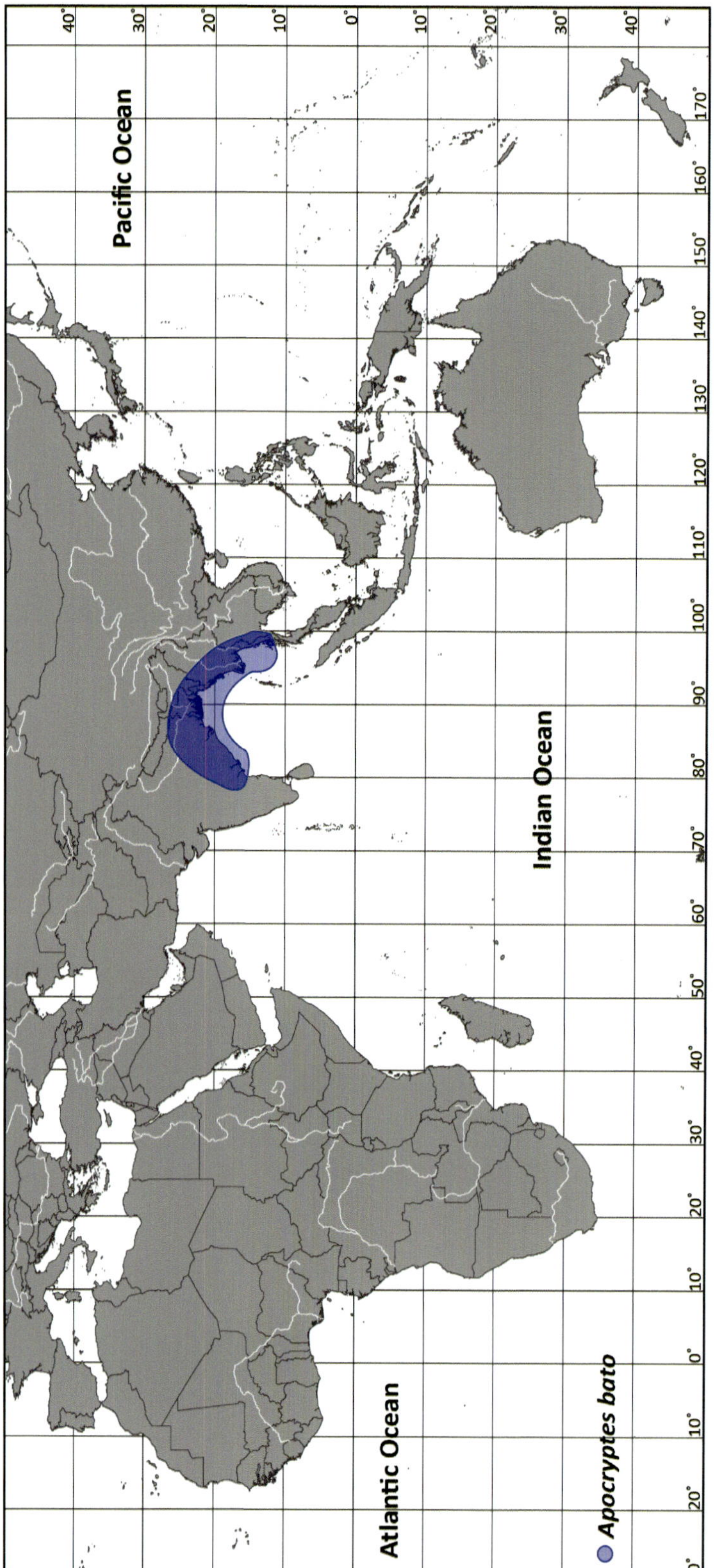

Figure 2.14 Distribution of the monotypic genus *Apocryptes*: *As. bato*.

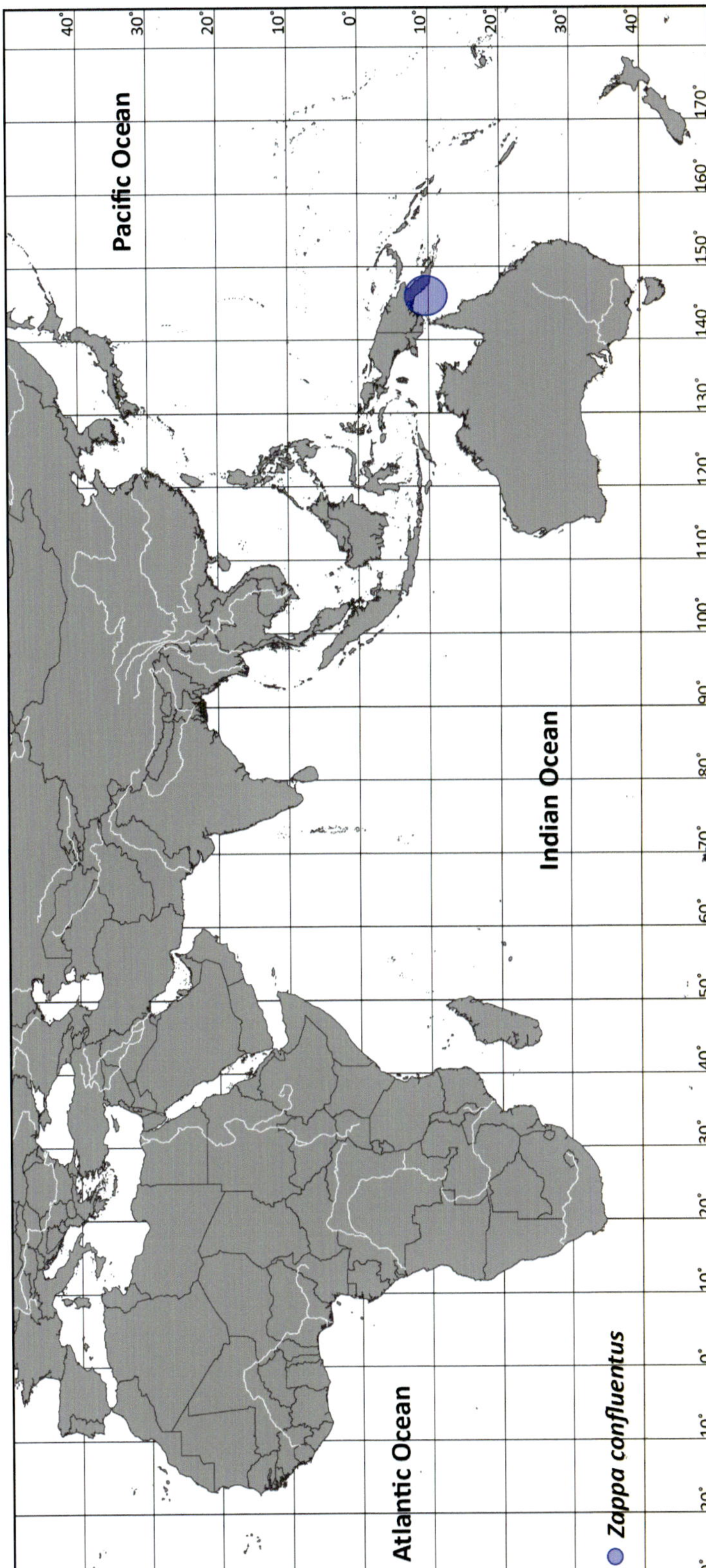

Figure 2.15 Distribution of the monotypic genus *Zappa*: *Z. confluentus*.

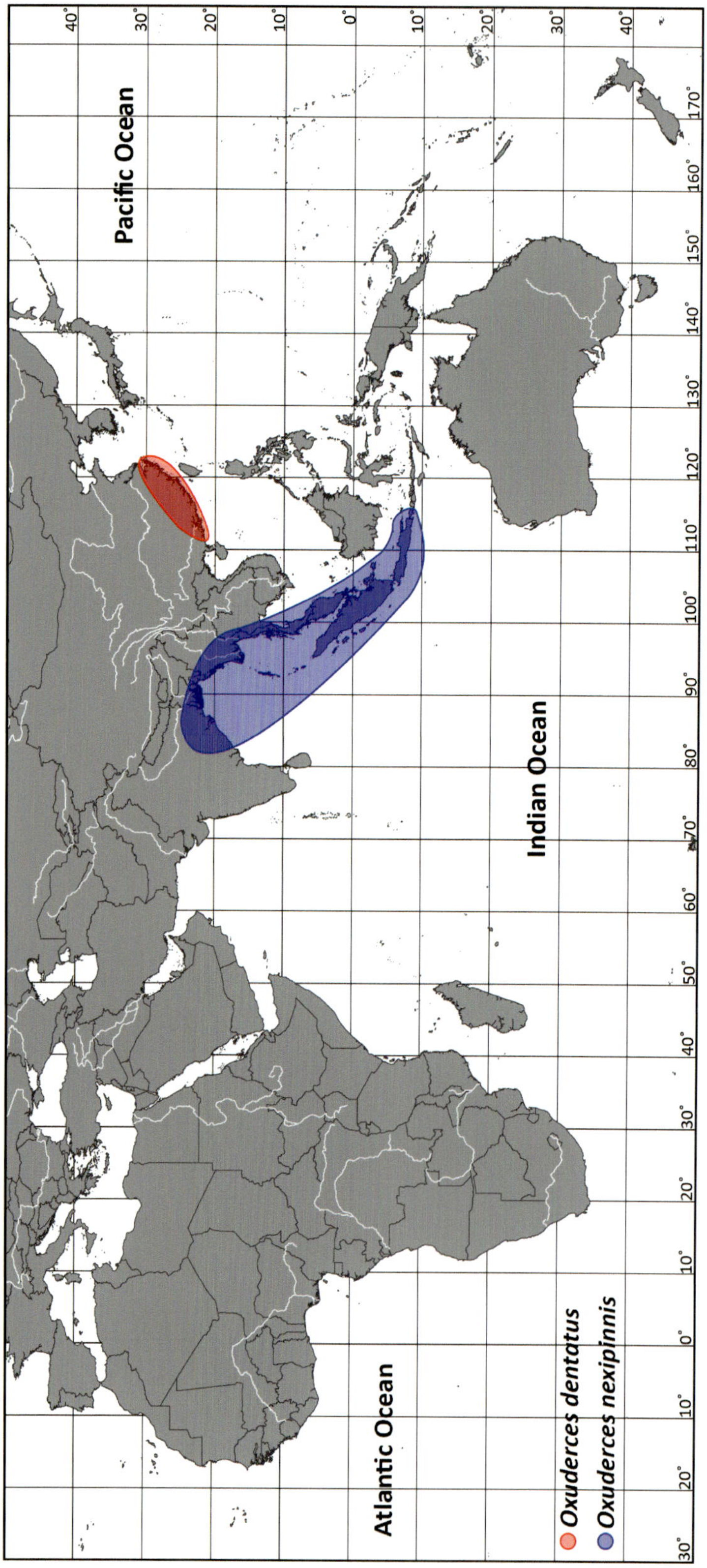

Figure 2.16 Distribution of species of the genus *Oxuderces*: *O. dentatus* and *O. nexipinnis*.

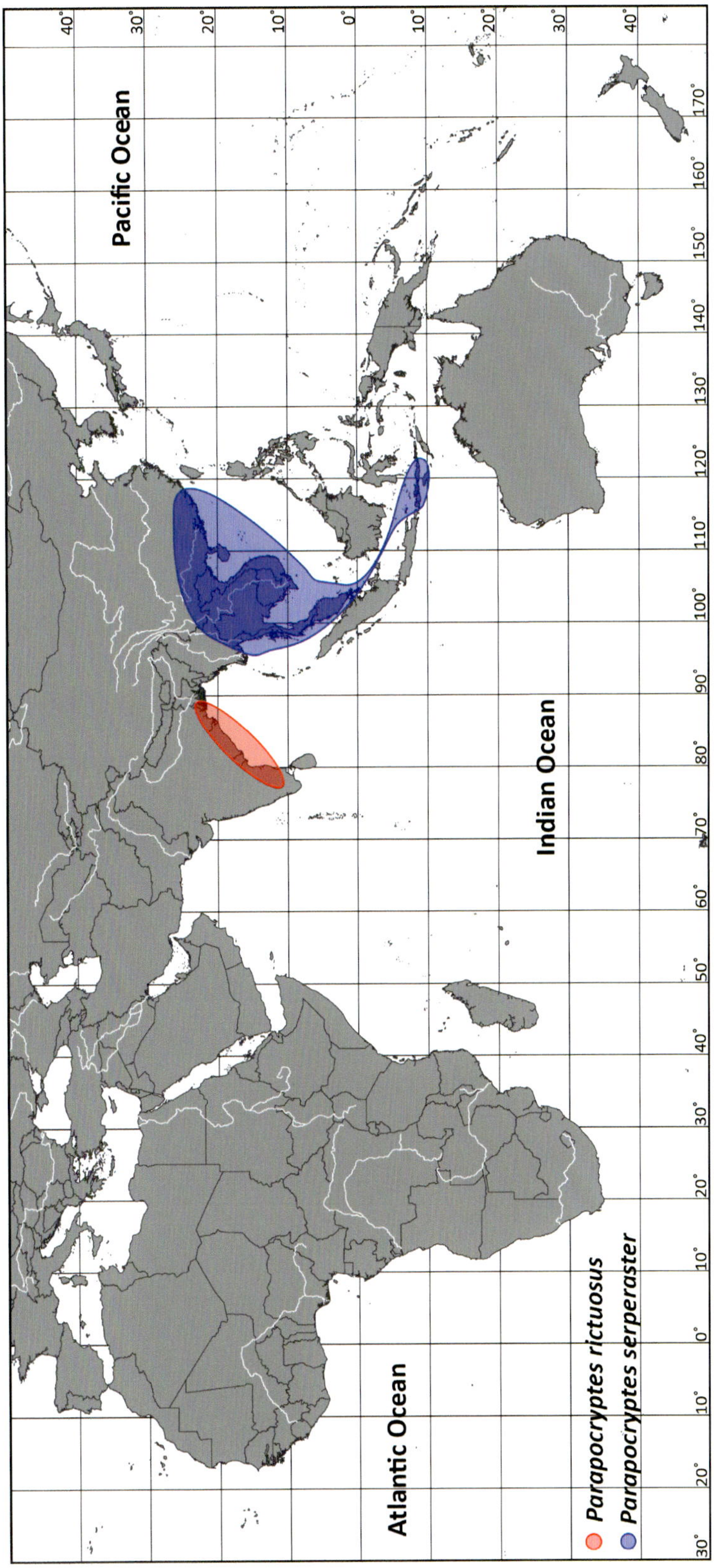

Figure 2.17 Distribution of species of the genus *Parapocryptes*: *Pa. rictuosus* and *Pa. serperaster*.

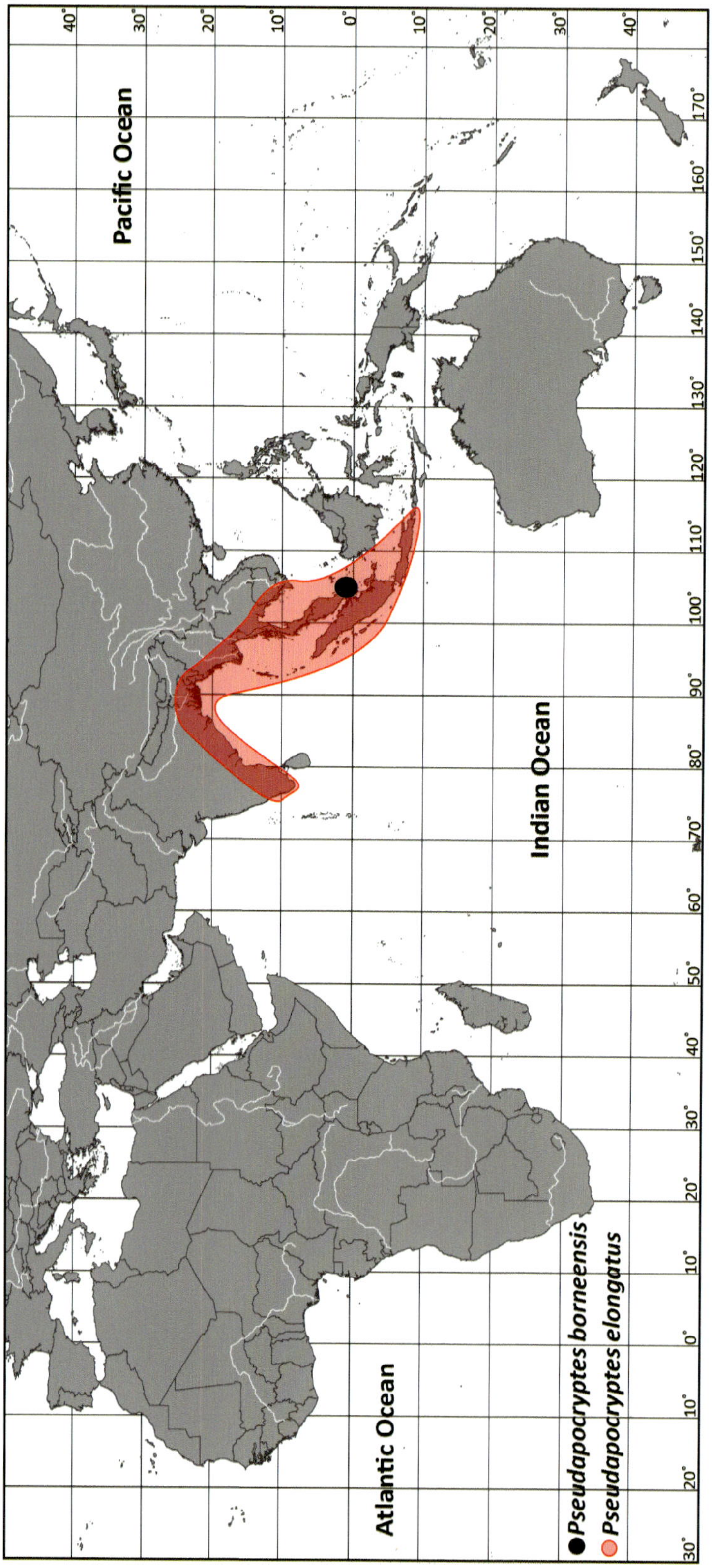

Figure 2.18 Distribution of species of the genus *Pseudapocryptes*: *Pd. borneensis* and *Pd. elongatus*.

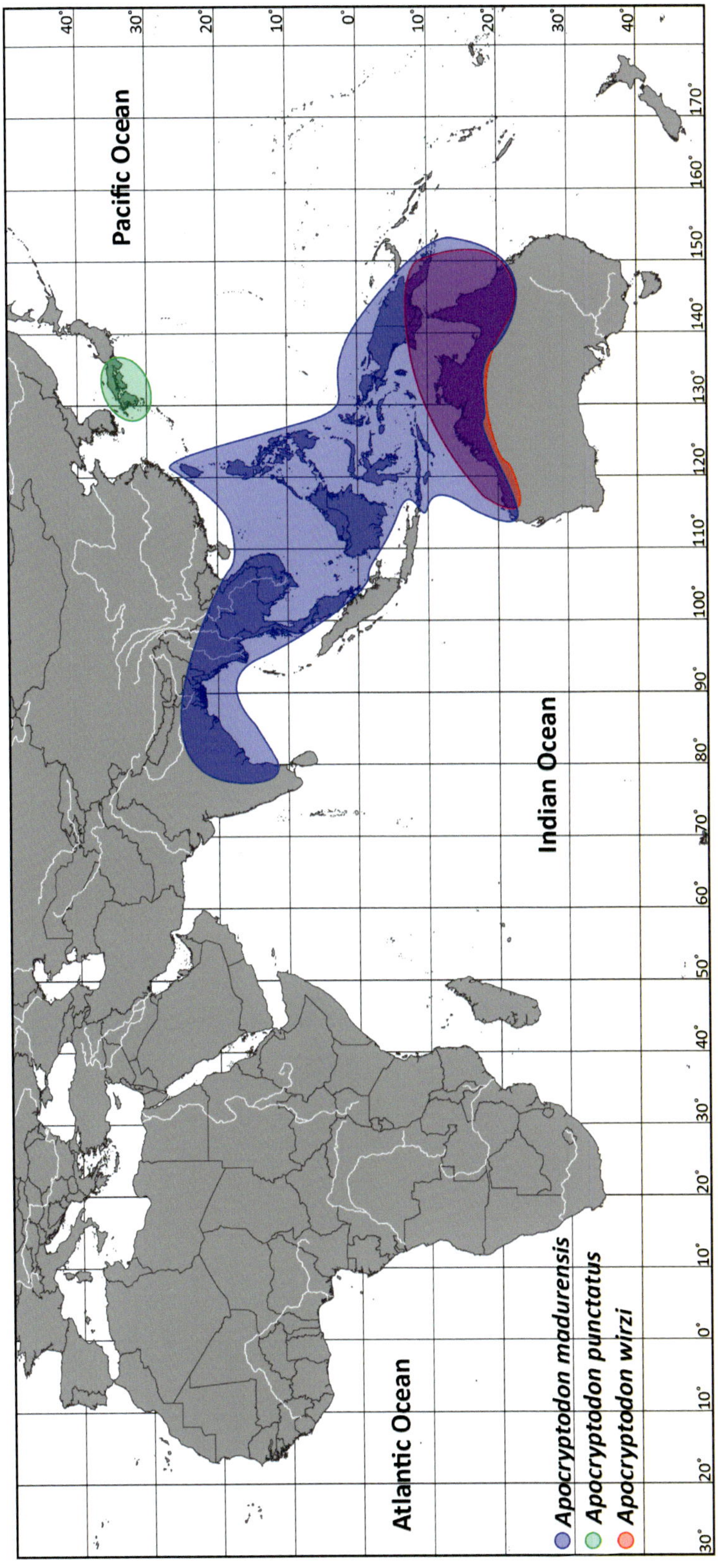

Figure 2.19 Distribution of species of the genus *Apocryptodon*: *An. madurensis*, *An. punctatus*, and *An. wirzi*.

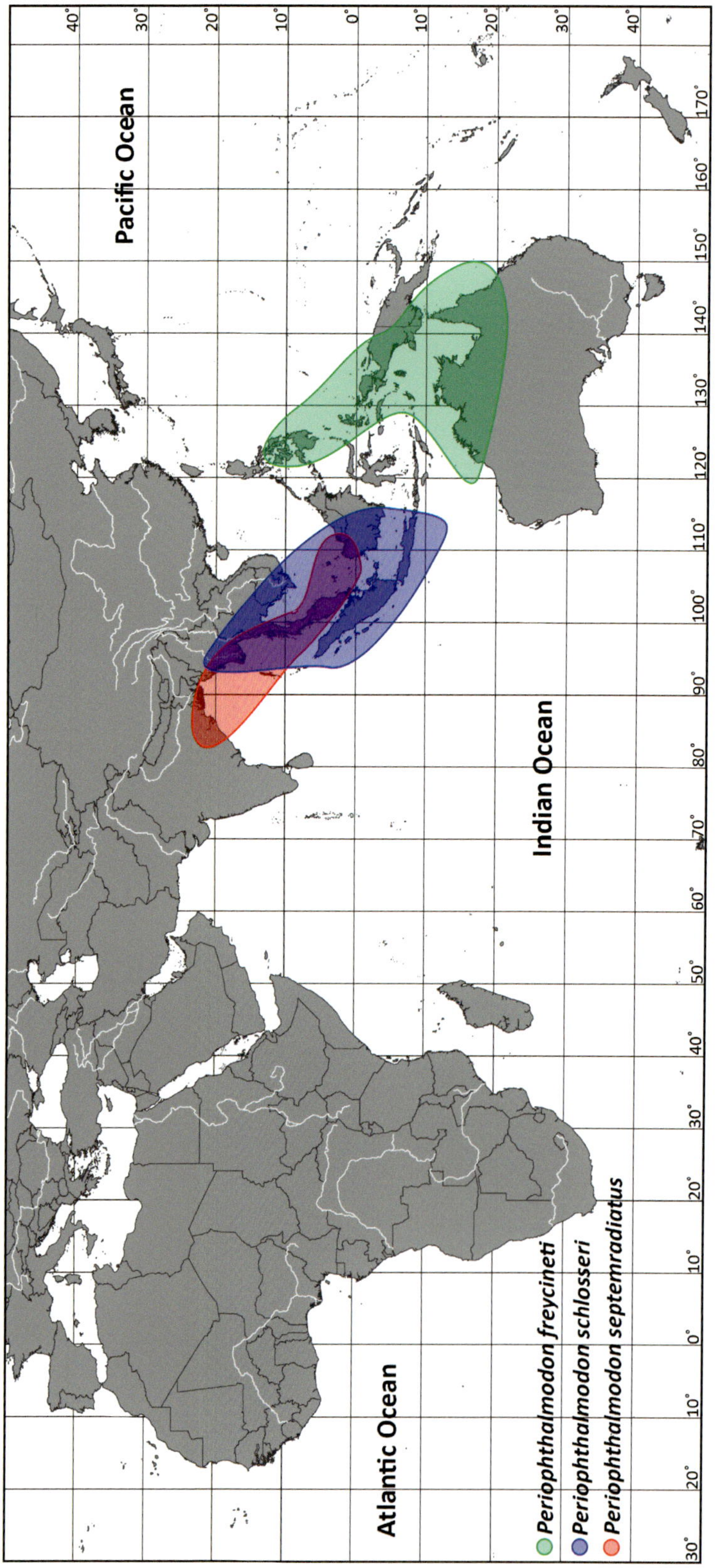

Figure 2.20 Distribution of species of the genus *Periophthalmodon*: *Pn. freycineti*, *Pn. schlosseri*, and *Pn. septemradiatus*.

Apocryptes (Figure 2.14) and *Zappa* (Figure 2.15), are each monotypic. Three genera—*Oxuderces* (Figure 2.16), *Parapocryptes* (Figure 2.17), and *Pseudapocryptes* (Figure 2.18)—comprise two species each. *Apocryptodon* (Figure 2.19) and *Periophthalmodon* (Figure 2.20) comprise three species each, whereas *Scartelaos* comprises four species (Figure 2.12), and *Boleophthalmus* comprises six species (Figure 2.13).

2.3.1 Oxudercine Goby Areas of Endemism

We recognize eight oxudercine goby areas of endemism, as defined below (see Figure 2.4): (1) Tropical Eastern Atlantic Ocean; (2) Southwestern Indian Ocean; (3) Arabian Sea; (4) Bay of Bengal; (5) South China Sea; (6) East China Sea; (7) Banda Sea; and (8) Oceania. Thirty of 43 species of oxudercine gobies live in only one area (see Figures 2.16 through 2.19 and Table 2.1). We assign these area names of oceans and seas, rather than land masses, because the distributions span oceans, not continents (see Parenti 1991, on sicydiine goby distribution). Equally important, the larvae of mudskippers, and many other gobies, are passively carried out to sea where they live as plankton for some 50 days prior to their transformation and ultimately return to the mangrove/mudflat adult habitats (Maeda and Tachihara 2014; Kobayashi et al. 1972). Therefore, the natural range of each species comprises the distribution of larvae as well as adults although the larval distribution of most gobies is unknown. Names and ranks of areas will be reconciled with others broadly in use, following the protocols of the International Code of Area Nomenclature (ICAN) (Ebach et al. 2008) when we test the areas.

2.3.1.1 Tropical Eastern Atlantic Ocean

This area comprises the western African coast from Morocco to Angola and the volcanic islands of Cape Verdes, Sáo Tomé, Principe, and Ascension. Just one species of mudskipper, the endemic *Ps. barbarus*, lives in this area.

The islands of St. Helena and Ascension form their own province in the tropical Atlantic realm in the classification of marine ecoregions by Spalding et al. (2007). Seventy-one species of near-shore fishes were reported from Ascension Island, collected on SCUBA up to a depth of 60 m (Lubbock 1980) but *Ps. barbarus* is not on the list. We examined a single collection of *Ps. barbarus* from Ascension Island, 15 specimens in the Museum of Comparative Zoology (MCZ 13307), collected by Waldren in the mid-1880s. We question the locality information because mangrove areas, and other suitable habitats for *Ps. barbarus*, have never been identified as a component of the littoral ecosystem of this isolated volcanic island (Price and John 1980).

The Tropical Eastern Atlantic Ocean is equivalent to the "West Africa" division of Murdy (1989: 74) and "Western Africa" of Duke et al. (1998). It comprises the West African Transition and Gulf of Guinea provinces in the classification of marine ecoregions by Spalding et al. (2007).

2.3.1.2 Southwestern Indian Ocean

This area comprises the east coast of the African continent south of the Red Sea, coastal habitats at the mouth of the Gulf of Aden, the Mozambique Channel, and the coasts of Madagascar and nearby islands.

Two widespread species of mudskippers, *Ps. argentilineatus* and *Ps. kalolo,* live in this area. There is molecular evidence, as described above, that the populations of *Ps. argentilineatus* in this area represent a distinct lineage.

This area falls within the larger "Eastern Africa and Madagascar" region of Duke et al. (1998) and is similar to the "East Africa" division of Murdy (1989: 74).

2.3.1.3 Arabian Sea

This area comprises the Arabian Sea, including the Gulf of Khambhat, the Gulf of Kutch, and coastal areas at the mouth of the Gulf of Oman. Six species of oxudercines in three genera live in this area: *B. boddarti*, *B. dussumieri*, *Ps. argentilineatus*, *Ps. waltoni*, *S. histophorus*, and *S. tenuis*. Three of the six species are endemic: *Ps. waltoni*, *B. dussumieri*, and *S. tenuis*. The gulfs of Khambhat and Kutch, along the western coast of the Indian subcontinent, may be the western-most limit of the distributions of *B. boddarti* and *S. histophorus*, respectively. This area is included within the larger "Eastern Africa and Madagascar" region of Duke et al. (1998) and mirrors the "Arabian Gulf" division of Murdy (1989: 74) which he defined as ranging from Iraq to Bombay (present day Mumbai).

2.3.1.4 Bay of Bengal

This area comprises the Bay of Bengal, Laccadive Sea, Andaman Sea, and the islands within these water bodies. Fourteen species of oxudercine gobies in nine genera live in this area: *As. bato*, *An. madurensis*, *B. boddarti*, *O. nexipinnis*, *Pa. rictuosus*, *Pn. schlosseri*, *Pn. septemradiatus*, *Ps. argentilineatus*, *Ps. kalolo*, *Ps. novemradiatus*, *Ps. walailakae*, *Pd. elongatus*, *S. cantoris*, and *S. histophorus*. Four of these species are endemic to the Bay of Bengal: *As. bato*, *Pa. rictuosus*, *Ps. novemradiatus*, and *S. cantoris*.

The Bay of Bengal area falls within both the larger "Indo-Malesia and Asia" region of Duke et al. (1998) and the "Indo-Malaya" division of Murdy (1989: 74), the latter defined as areas on the "west coast of India to Borneo, including Java and Vietnam." Murdy (1989) considered that the areas on the western coast of India south of Bombay lack suitable oxudercine habitats. Eggert (1935) named a subspecies *Periophthalmus vulgaris ceylonensis* based on specimens collected from Galle, Sri Lanka. Murdy (1989: 36) reported specimens of *Ps. chrysospilos* from Madras, on the east coast of India, but these were reidentified as *Ps. walailakae* by G. Polgar (pers. comm.). Murdy (1989) also reported *O. dentatus* from India; Jaafar and Parenti (2016) reclassified these specimens as *O. nexipinnis*.

2.3.1.5 South China Sea

This area comprises the Straits of Malacca, western South China Sea (including the gulfs of Thailand and Tonkin), Java Sea, and includes coastal areas of the islands of Sumatra, Java, and Borneo. Eighteen species of oxudercines in eight genera live in this area: *An. madurensis*, *B. boddarti*, *B.* cf. *pectinirostris*, *O. nexipinnis*, *Pa. serperaster*, *Pn. schlosseri*, *Pn. septemradiatus*, *Ps. argentilineatus*, *Ps. chrysospilos*, *Ps. gracilis*, *Ps. kalolo*, *Ps. minutus*, *Ps. spilotus*, *Ps. variabilis*, *Ps. walailakae*, *Pd. borneensis*, *Pd. elongatus*, and *S. histophorus*. Five of these species are endemic: *B.* cf. *pectinirostris*, *Ps. chrysospilos*, *Ps. spilotus*, *Ps. variabilis*, and *Pd. borneensis*.

The South China Sea area falls within the larger "Indo-Malesia and Asia" region of Duke et al. (1998) and "Indo-Malaya" division of Murdy (1989: 74). The specimens identified as *B. pectinirostris* by Polgar and Bartolino (2010) and Polgar et al. (2013) likely represent an undescribed species, and are here included as *B.* cf. *pectinirostris* (see Zhang et al. 2015). Murdy (1989) reported *O. dentatus* from this area, which Jaafar and Parenti (2016) reclassified as *O. nexipinnis*.

2.3.1.6 East China Sea

This area spans the East China Sea north of the Gulf of Tonkin and includes the Yellow Sea, the Sea of Japan, and the Taiwan Strait. The East China Sea is also an area of relatively high endemism. Six of the nine species of oxudercine gobies that live in this area are endemic: *An. punctatus*, *B. pectinirostris*, *O. dentatus*, *Ps. magnuspinnatus*, *Ps. modestus*, and *S. gigas*. The East China Sea also represents the northern limit of three widely distributed species: *Pa. serperaster*, *Ps. argentilineatus*, and *S. histophorus*. This area falls within the larger "Indo-Malesia and Asia" region of Duke et al. (1998) and mirrors the "China Sea" division of Murdy (1989: 75).

2.3.1.7 Banda Sea

This area comprises the Philippine Sea, Luzon Sea (eastern South China Sea), Sibuyan Sea, Sulu Sea, Visayan Sea, Bohol Sea, Celebes Sea, Banda Sea, Flores Sea, Molucca Sea, western Ceram Sea, Halmahera Sea, and Savu Sea. Ten species of oxudercines in five genera live in the area: *An. madurensis*, *B. boddarti*, *Pn. freycineti*, *Ps. argentilineatus*, *Ps. gracilis*, *Ps. kalolo*, *Ps. malaccensis*, *Ps. minutus*, *Ps. pusing*, and *S. histophorus*. Two species, *Ps. malaccensis* and *Ps. pusing*, are endemic.

The Banda Sea area falls within the larger "Indo-Malesia and Asia" and "Australasia" regions of Duke et al. (1998); the island of Maluku, but not North Maluku, was included in the Australasia region. In contrast, Murdy (1989: 74) considered the Banda Sea as part of the "Australia/New Guinea" division. His decision to include the Philippines, Timor, and Moluccas in the "Australia/New Guinea" division stemmed from the distribution of *Ps. malaccensis* (Figure 2.11) and *Pn. freycineti* (Figure 2.20). Murdy (1989) further supported this hypothesis citing Vari (1978) who showed that the terapon perches, family Terapontidae, are distributed in the Philippines, Sulawesi, New Guinea, and Australia. In 1989, no scientific collections of mudskippers were known from Sulawesi; therefore, Murdy (1989: 74) did not assign that island complex to any biogeographic area. We examined the specimen that Murdy (1989) identified as *Ps. novemradiatus* from the Philippines and reidentified it as *Ps. malaccensis*.

2.3.1.8 Oceania

This broad area comprises the Arafura Sea (including the Gulf of Carpentaria and the Torres Straits), Timor Sea, Gulf of Papua, Cendrawasih Bay, Philippine Sea, Bismarck Sea, Solomon Sea, and the Coral Sea east to 175° W from the western margin of the Pacific Plate to Samoa (Springer 1982; Murdy 1989). Sixteen species of oxudercines in six genera live in this area: *An. madurensis*, *An. wirzi*, *B. birdsongi*, *B. caeruleomaculatus*, *B. poti*, *Pn. freycineti*, *Ps. argentilineatus*, *Ps. darwini*, *Ps. gracilis*, *Ps. kalolo*, *Ps. minutus*, *Ps. novaeguineaensis*, *Ps. takita*, *Ps. weberi*, *S. histophorus*, and *Z. confluentus*. Nine of these species are endemic to this area, the highest number of endemic species of oxudercine gobies in any area. The endemic species are *An. wirzi*, *B. birdsongi*, *B. caeruleomaculatus*, *B. poti*, *Ps. darwini*, *Ps. novaeguineaensis*, *Ps. takita*, *Ps. weberi*, and *Z. confluentus*.

For mangrove forests, Duke et al. (1998) defined the "Australasia" region as the area east of New Guinea and Australia that encompasses areas in the western Pacific Ocean, such as the island of Maluku. Murdy (1989) included the Philippine Islands, the Moluccas Islands, and the island of Timor in his "Australia/New Guinea" division. The species that Murdy (1989) reported as *O. wirzi* was reclassified in *Apocryptodon* by Jaafar and Parenti (2016). Murdy (1989) also recognized a separate Oceania, from the western margin of the Pacific Plate eastward to Samoa, but the two oxudercine species in this area, *Ps. argentilineatus* and *Ps. kalolo*, are widespread and not endemic. We have no basis for recognizing this as a separate area of endemism for oxudercines and thus preliminarily combine two areas recognized by Murdy (1989), his "Oceania" and "Australia/New Guinea", into one large area: Oceania.

2.3.2 Distributional Gradients

The broad-scale patterns of species distribution of oxudercine gobies mirror those of mangrove plant communities, with species diversity higher in the IWP than in the Atlantic Ocean (Figure 2.1). Collectively, mangrove plants display a unimodal distribution pattern as their diversity decreases to the east and to the west of Southeast Asia (Ellison et al. 1999). The number of species of mangrove plants is highest in Southeast Asia (Indo-Malesia and Asia region), in which there are 30 species and lowest in the Caribbean in which there are fewer than five species (Ellison et al. 1999). In contrast, oxudercine gobies display a bimodal species richness distribution pattern, with peaks in the South China Sea (18 species) and in Oceania (16 species). The Banda Sea area lies between these two species rich areas, and is home to 10 species of mudskippers. In contrast to the South China Sea and Oceania, the Banda Sea area is relatively poorly sampled and studied. Further exploration throughout the Philippines, East and West Nusa Tenggara, Sulawesi, Timor, and the Moluccas Islands, is expected to lead to discovery of more species in the Banda Sea area and, consequently, alter our understanding of the broad species distribution patterns of oxudercine gobies.

Longitudinal variation is observed in many tropical marine taxa; species richness is generally lower in the Atlantic and eastern Pacific oceans when compared to the IWP (Hoeksema 2007; Carpenter and Springer 2005; Paulay 1997). The disjunct distribution of *Periophthalmus*—with one or more species in the Tropical Eastern Atlantic and congeners in the Indo-Pacific—is also observed in other tropical marine teleost genera: *Drepane* (Springer 1982: 40), *Grammoplites* (Springer 1982: 71), *Lethrinus* (Springer 1982: 40), *Monodactylus* (Springer 1982: 61), and *Psettodes* (Springer 1982: 81). We will explore the generality of this disjunct distribution in our larger study of the historical biogeography of mudskippers and the associated mangrove biota.

2.4 SUMMARY

Oxudercine gobies are distributed throughout tropical and subtropical coastal areas of the IWP and the eastern Atlantic. We recognize eight distinct areas for oxudercine gobies, each with at least one endemic species: Tropical Eastern Atlantic Ocean (one species), southwestern Indian Ocean (two species), Arabian Sea (six species), Bay of Bengal (14 species), South China Sea (18 species), East China Sea (nine species), Banda Sea (10 species), and Oceania (16 species). Oxudercine gobies exhibit a bimodal distribution pattern with highest species diversity in the South China Sea and Oceania. Latitudinal variation is also observed; two subtropical communities, Arabian Sea and East China Sea, include fewer species than do adjacent tropical communities.

Thirty of 43 oxudercine species are endemic to a single area. The highest number of endemic species is in Oceania (nine species). *Periophthalmus argentilineatus* is the most widely distributed species; it is absent only from the Tropical Eastern Atlantic Ocean. Using molecular data, Polgar et al. (2014) recognized three distinct populations within the widespread *Ps. argentilineatus*, which may be recognized at the species level. Thus, these areas will likely be revised further as more data become available. Oxudercine gobies are under-collected, hence under-represented in natural history museum collections globally. Data are sparse from many areas. Future investigations into poorly sampled areas and peripheral zones of biogeographic areas, along with fine-scale differentiation of populations, will further our understanding of the distribution patterns of oxudercine gobies and the extent of their endemism.

ACKNOWLEDGMENTS

We thank Anthony Gill, Edward Murdy, Ken Maeda, and Richard Mleczko for their comments on earlier versions of this manuscript. We are grateful for logistical support from Jeffrey Clayton, Sandra Raredon, Kris Murphy, and Diane Pitassy, who helped to locate material used in previous studies for us to reexamine. Dan Cole provided base maps used to illustrate the distributions. We also thank Gianluca Polgar for providing Figure 2.5 and for confirming the identity of *Periophthalmus walailakae* in a museum collection that he examined.

REFERENCES

Carpenter K.E. and Springer V.G. 2005. The center of the center of marine shore fish biodiversity: The Philippine Islands. *Environmental Biology of Fishes* 72:467–480.

Clayton D.A. 1993. Mudskippers. *Oceanography and Marine Biology Annual Review* 31:507–577.

Darumas U. and Tantichodok P. 2002. A new species of mudskipper (Gobiidae: Oxudercinae) from Southern Thailand. *Phuket Marine Biological Centre Research Bulletin* 64:99–106.

Duke N.C. 1992. Mangrove floristics and biogeography. In *Tropical Mangrove Ecosystems* (eds. A.I. Robertson and D.M. Alongi) pp. 63–100. Washington, DC: American Geophysical Union.

Duke N.C., Ball M.C. and Ellison J.C. 1998. Factors influencing biodiversity and distributional gradients in mangroves. *Global Ecology and Biogeography Letters* 7:27–47.

Ebach M.C., Morrone J.J., Parenti L.R. and Viloria Á.L. 2008. International Code of Area Nomenclature. *Journal of Biogeography* 35:1153–1157.

Eggert B. 1935. Beitrag zur Systematik, Biologie und geographischen Verbreitung der Periophthalminae. Ergebnisse einer durch die Notgemeinschaft der Deutschen Wissenschaft ermöglichten Reise nach Niederländisch-Indien 1926–1927 und der Sundaexpedition der Notgemeinschaft der Deutschen Wissenschaft 1929–1930. *Zoologische Jahrbücher, Abtheilung für Systematik, Geographie und Biologie der Thiere, Jena* 67:29–116.

Ellison A.M., Farnsworth E.J. and Merkt R.E. 1999. Origins of mangrove ecosystems and the mangrove biodiversity anomaly. *Global Ecology and Biogeography* 8:95–115.

Gill A.C. and Kemp J.M. 2002. Widespread Indo-Pacific shore-fish species: A challenge for taxonomists, biogeographers, ecologists, and fishery and conservation managers. *Environmental Biology of Fishes* 65:165–174.

Giri C., Ochieng E., Tieszen L.L. et al. 2010. Status and distribution of mangrove forests of the world using earth observation satellite data. *Global Ecology and Biogrography* 20:154–159.

Harold A.S. and Mooi R.D. 1994. Areas of endemism: Definition and recognition criteria. *Systematic Biology* 43:261–266.

Hoeksema B.W. 2007. Delineation of the Indo-Malayan centre of maximum marine biodiversity: The coral triangle. In *Biogeography, Time, and Place: Distributions, Barriers, and Islands* (ed. W. Renema) pp. 117–178. The Netherlands: Springer.

Hogarth P.J. 2001. Mangrove ecosystems. In *Encyclopedia of Biodiversity 3* (ed-in-chief S. A. Levin) pp. 853–870. San Diego, CA: Academic Press.

Humphries C.J. and Parenti L.R. 1986. *Cladistic Biogeography*. Oxford: Oxford University Press, 98 pp.

Jaafar Z. and Larson H.K. 2008. A new species of mudskipper, *Periophthalmus takita* (Teleostei: Gobiidae: Oxudercinae) from Australia with a key to the genus. *Zoological Science* 25:946–952.

Jaafar Z. and Parenti L.R. 2016. Systematics of the mudskipper genus *Oxuderces* Eydoux & Souleyet 1948 (Teleostei: Gobiidae: Oxudercinae) with resurrection from synonymy of *O. nexipinnis* (Cantor 1849). *Zoological Journal of the Linnean Society*. doi:10.1111/zoj.12482.

Jaafar Z., Perrig M. and Chou L.M. 2009. *Periophthalmus variabilis* (Teleostei: Gobiidae: Oxudercinae), a valid species of mudskipper, and a re-diagnosis of *Periophthalmus novemradiatus*. *Zoological Science* 26:309–324.

Jaafar Z., Polgar G. and Zamroni Y. 2016. Description of a new species of *Periophthalmus* (Teleostei: Gobiidae) from the Lesser Sunda Islands. *Raffles Bulletin of Zoology* 64:278–283.

Kobayashi T., Dotsu Y. and Miura N. 1972. Egg development and rearing experiments of the larvae of the mud skipper, *Periophthalmus cantonensis* (in Japanese). *Bulletin of the Faculty of Fisheries of Nagasaki University* 33:49–62.

Larson H.K. and Takita T. 2004. Two new species of *Periophthalmus* (Teleostei: Gobiidae: Oxudercinae) from northern Australia, and a re-diagnosis of *Periophthalmus novaeguineaensis*. *The Beagle, Records of the Museums and Art Galleries of the Northern Territory* 20:175–185.

Lee Y.J., Choi Y. and Ryu B-S. 1995. A taxonomic revision of the genus *Periophthalmus* (Pisces: Gobiidae) from Korea with description of a new species. *Korean Journal of Ichthyology* 7:120–127.
Lubbock R. 1980. The shore fishes of Ascension Island. *Journal of Fish Biology* 17:283–303.
Maeda K. and Tachihara K. 2014. Larval fish fauna of a sandy beach and an estuary on Okinawa Island, focusing on larval habitat utilization by the suborder Gobioidei. *Fisheries Science* 80:1215–1229.
Murdy E.O. 1989. A taxonomic revision and cladistic analysis of the oxudercine gobies (Gobiidae: Oxudercinae). *Records of the Australian Museum Supplement* 11:1–93.
Murdy E.O. and Takita T. 1999. *Periophthalmus spilotus*, a new species of mudskipper from Sumatra (Gobiidae: Oxudercinae). *Ichthyological Research* 46:367–370.
Nelson G. and Platnick N. 1981. *Systematics and Biogeography: Cladistics and Vicariance*. New York, NY: Columbia University Press, 567 pp.
Ozawa T., Yin W., Fu C., Claremont M., Smith L. and Reid D.G. 2015. Allopatry and overlap in a clade of snails from mangroves and mud flats in the Indo-West Pacific and Mediterranean (Gastropoda: Potamididae: *Cerithideopsilla*). *Biological Journal of the Linnean Society* 114:212–228.
Parenti L.R. 1991. Ocean basins and the biogeography of freshwater fishes. *Australian Systematic Botany* 4:137–149.
Parenti L.R. and Ebach M.C. 2009. *Comparative Biogeography: Discovering and Classifying Biogeographical Patterns of a Dynamic Earth*. Berkeley, CA: University of California Press, 295 pp.
Parenti L.R., Hadiaty R.K. and Lumbantobing D.N. 2015. Report on a collection of freshwater and coastal fishes from Sulawesi Tenggara, Indonesia. *Jurnal Iktiologi Indonesia* 14:1–19.
Paulay G. 1997. Diversity and distribution of reef organisms. In *Life and Death of Coral Reefs* (ed. C.E. Birkeland) pp. 298–353. New York, NY: Chapman & Hall.
Polgar G. and Bartolino V. 2010. Size variation of six species of oxudercine gobies along the intertidal zone in a Malayan coastal swamp. *Marine Ecology Progress Series* 409:199–212.
Polgar G., Jaafar Z. and Konstantinidis P. 2013. A new species of mudskipper, *Boleophthalmus poti* (Teleostei: Gobiidae: Oxudercinae) from the Gulf of Papua, Papua New Guinea, and a key to the genus. *Raffles Bulletin of Zoology* 61:311–321.
Polgar G., Zane L., Babbucci M. et al. 2014. Phylogeography and demographic history of two widespread Indo-Pacific mudskippers (Gobiidae: *Periophthalmus*). *Molecular Phylogenetics and Evolution* 73:161–176.
Price J.H. and John D.H. 1980. Ascension Island, south Atlantic: A survey of inshore benthic macroorganisms, communities and interactions. *Aquatic Botany* 9:251–278.
Spalding M.D., Fox H.E., Allen G.R. et al. 2007. Marine ecoregions of the world: A bioregionalization of coastal and shelf areas. *BioScience* 57:573–583.
Springer V.G. 1982. Pacific plate biogeography, with special reference to shore fishes. *Smithsonian Contributions to Zoology* 367:1–182.
Vari R.P. 1978. The terapon perches (Percoidei, Teraponidae): A cladistic analysis and taxonomic revision. *Bulletin of the American Museum of Natural History* 159:175–340.
Zhang J., Takita T., Muchtar A., Chen C. and Ishimatsu A. 2015. Mitochondrial genome of *Boleophthalmus* sp. nov. (Osteichthyes: Gobiidae). *Mitochondrial DNA* 27:3381–3382. doi:10.3109/19401736.2015.1018230.

CHAPTER **3**

Early Development of Mudskippers

Ken Maeda, Yuko Tsuhako, and Katsunori Tachihara

CONTENTS

3.1 INTRODUCTION

Mudskippers are one of the most important fish faunas in intertidal areas of tropical and subtropical Indo-Pacific regions. Early life histories of mudskippers, however, are not well understood. There are 43 known species of oxudercine gobies in 10 genera. Descriptions of early stages are available only for eight species in five genera: *Periophthalmus argentilineatus* (Okiyama 2014; Brillet 1976), *Periophthalmus modestus* (Kobayashi et al. 1972; Kimura 1958), *Periophthalmus magnuspinnatus* (Zhou et al. 2012), *Periophthalmodon schlosseri* (Tsuhako et al. 2003), *Boleophthalmus pectinirostris* (Okiyama 2014; Yang and Zhang 1990; Zhang et al. 1989; Hong et al. 1988), *Scartelaos gigas* (Kim et al. 2011), *Scartelaos histophorus* (Okiyama 2014), and *Apocryptodon punctatus* (Dôtu 1961). In this chapter, we provide descriptions of the early development for two *Periophthalmus* species, *Ps. argentilineatus* and *Ps. modestus*, and review the state of knowledge pertaining to the early life history of oxudercine gobies.

Periophthalmus argentilineatus and *Ps. modestus* are both found in the Ryukyu Archipelago, Japan (Figure 3.1a,b). The former species is distributed from the southern Red Sea and east coast of Africa, eastward to Japan and Oceania, while the latter is distributed from Hong Kong to Japan and Korea (Lee et al. 1995; Murdy 1989). In the middle and southern parts of the Ryukyu Archipelago, *Ps. argentilineatus* is found on most of the major islands, while *Ps. modestus* occurs only on Okinawa Island, the largest island in the archipelago, where it inhabits certain habitats (Tsuhako 2005). Where the two species co-occur, *Ps. argentilineatus* is predominantly found in mangrove forests, while *Ps. modestus* is usually found on exposed mudflats (Oshiro et al. 2005). Both species

Figure 3.1 Adult mudskippers, whose early development stages are described in this chapter, collected on Okinawa Island, Japan. (a) *Periophthalmus argentilineatus*, male, 70.6 mm SL; (b) *Periophthalmus modestus*, female, 59.2 mm SL. (Courtesy of Ken Maeda.)

Figure 3.2 Turreted burrow entrances of *Periophthalmus argentilineatus* on a mudflat in Oh-ura, Okinawa Island, Japan, April 23, 2006. (Courtesy of Akiko Kameyama.)

construct burrows within the muddy substrata (Figure 3.2); eggs are deposited on the inner wall of an egg chamber in the burrow (Brillet 1976; Kobayashi et al. 1971).

3.2 EGG CHAMBER

Burrows of *Ps. argentilineatus* and *Ps. modestus* were examined on Okinawa Island in 1999. Both species constructed burrows in the soft substrata within the mid-intertidal zone. Burrow shapes were of the "YL type" (Kobayashi et al. 1971), in which two turreted openings on the substrata (Figure 3.2) join an inclined tunnel, and subsequently meet to form a vertical tunnel. At the end of the vertical tunnel, the burrow continues at a right angle for a short distance, before finally continuing upward to form the egg chamber (Figure 3.3). The vertical tunnel in burrows of *Ps. argentilineatus* extended 30–75 cm below the surface, while the egg chambers were 4–12 cm in diameter. Burrow dimensions of *Ps. modestus* were slightly smaller; the vertical tunnel extended 25–30 cm below the surface, and the egg chamber was 1–2 cm in diameter. Within a mud-filled glass tank in the laboratory, we could observe eggs of *Ps. argentilineatus* being deposited on the wall of the air-filled egg chamber (Figure 3.4). Although only the eggs of *Ps. modestus* and *Pn. schlosseri* were reported to develop in air within the egg chamber, this strategy is assumed to be prevalent in all mudskipper species (Ishimatsu and Gonzales 2011; Ishimatsu et al. 2009, 2007, 1998). Aerial incubation enables the eggs to develop under considerable hypoxic conditions of the burrow.

Burrow structures vary between mudskipper species. Burrows of *Ps. chrysospilos*, *Ps. magnuspinnatus*, and *Ps. minutus* are known to have a similar structure to those of *Ps. argentilineatus* and *Ps. modestus* (Takita and Ishimatsu 2015; Brillet 1976; Kobayashi et al. 1971). *Pn. schlosseri* constructs a U-shaped burrow usually with two openings, that is bigger and deeper than most

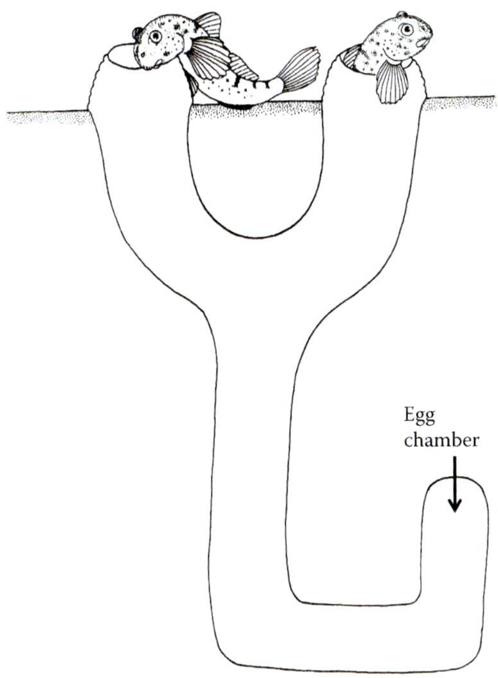

Figure 3.3 The "YL type" burrow structure of *Periophthalmus modestus*. (Courtesy of Katsunori Tachihara.)

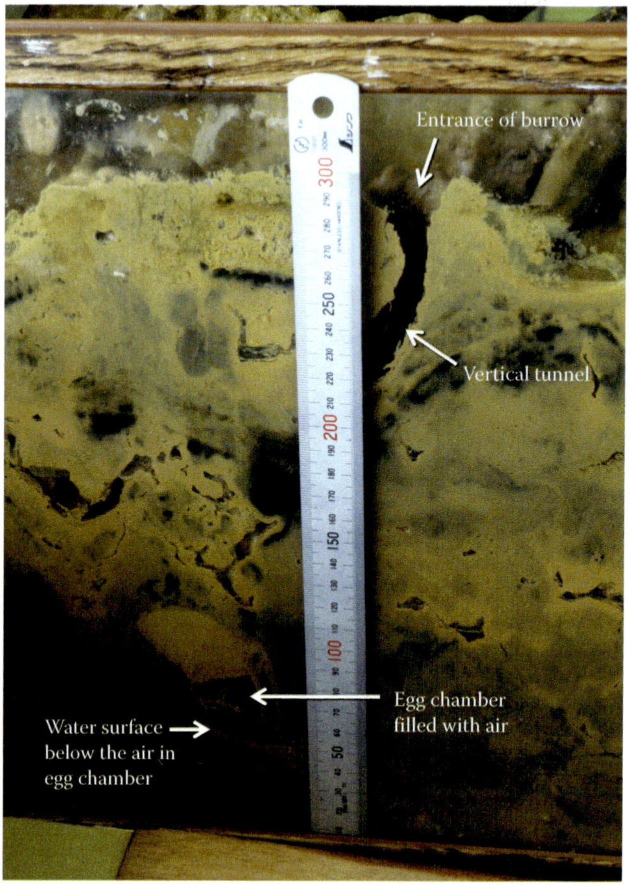

Figure 3.4 Burrow of *Periophthalmus argentilineatus* constructed in a glass tank. Vertical tunnel near the entrance and an egg chamber are visible through the side of the tank. (Courtesy of Katsunori Tachihara.)

of the burrows made by *Periophthalmus*. The deepest horizontal tunnel often extends more than 1 m below the mudflat surface. An air-filled hollow on the ceiling of this horizontal tunnel is the egg chamber (Takita and Ishimatsu 2015; Ishimatsu et al. 2009). The burrows of *B. pectinirostris* are complex in shape and branch out multiple times; the egg chamber is prepared as a horizontal branch (Toba and Ishimatsu 2014; Koga and Noda 1992). Burrows of *S. histophorus* and *S. gigas* are constructed as vertical shafts and horizontal tunnels; the egg chamber is within one of the horizontal tunnels (Kim et al. 2011; Okinawa Prefectural Government 1997).

3.3 EGG MASS

Egg masses were encountered in 12 burrows observed on Okinawa Island; six burrows each of *Ps. argentilineatus* and *Ps. modestus*, between May and August 1999. In nine of the 12 burrows, only one male was observed. In one instance, both male and female *Ps. argentilineatus* were observed in the burrow. In two instances, neither the male nor the female of *Ps. modestus* was observed in the burrow (Table 3.1). Males are generally responsible for the care of egg masses in gobioid fishes (Miller 1984). This is hypothesized to be the case for mudskippers because a solitary male is usually observed guarding burrows containing an egg mass. Furthermore, males of *Ps. modestus* have been reported to carry fresh air from the surface into the egg chamber to aerate the eggs (Ishimatsu et al. 2007).

Table 3.1 Number of Eggs and Adults in Each Burrow Found on Okinawa Island, Japan, in 1999

Species	Date	Location	Number of Eggs	Number of Adults	
				Male	Female
Periophthalmus argentilineatus	May 29	Manko	2,033	1	1
	July 12	Manko	3,018	1	0
			9,348	1	0
	Aug. 22	Gesashi	2,826	1	0
			2,530	1	0
	Aug. 24	Gesashi	1,692	1	0
Periophthalmus modestus	May 28	Manko	288	0	0
	June 12	Yagaji	602	1	0
			1,500	1	0
	July 12	Manko	944	0	0
			3,000	1	0
			1,140	1	0

Eggs of known oxudercine species, including *Ps. argentilineatus* and *Ps. modestus*, are demersal and deposited densely in a single layer on the ceiling and sidewalls of the egg chamber (Hosoya 2014; Toba and Ishimatsu 2014; Ishimatsu et al. 2009, 2007; Tsuhako et al. 2003; Okinawa Prefectural Government 1997). Fecundity (number of eggs in an ovary) of *Ps. magnuspinnatus* ranged from 2,249 to 3,472 (Zhou et al. 2012). Over 5 years, Noda et al. (1993) investigated 131 egg masses of *B. pectinirostris* from artificial spawning nests and counted a total of 767,530 eggs. The annual mean number of eggs in a spawning ranged from 4,610 to 7,002 (mean = 5,859). The maximum number of eggs found in one *B. pectinirostris* nest was 18,000. In *S. histophorus*, several thousand to nearly 30,000 eggs were found in an egg chamber (Hosoya 2014; Okinawa Prefectural Government 1997). We found 1,692–9,348 eggs in the burrows of *Ps. argentilineatus* and 288–3,000 eggs in the burrows of *Ps. modestus* (Table 3.1). Clutch sizes varied greatly within species, and the variation may be attributed to female body size.

3.4 EGG-CAPSULE MORPHOLOGY

Eggs of *Ps. argentilineatus* and *Ps. modestus* on Okinawa Island were oval (prolate spheroid) with a clump of adhesive filaments at the micropylar end. In eggs of *Ps. argentilineatus* (n = 10), the vertical diameter with the micropyle located at the top, ranged from 0.70 to 1.00 mm (mean ± SD, 0.87 ± 0.10 mm), and the horizontal diameter ranged from 0.60 to 0.73 mm (0.66 ± 0.04 mm). The eggs of *Ps. modestus* (n = 21) were 0.93–1.05 mm (1.01 ± 0.04 mm) and 0.60–0.70 mm (0.64 ± 0.02 mm).

Egg capsules of known oxudercine species are always oval (Zhou et al. 2012; Kim et al. 2011; Tsuhako et al. 2003; Okinawa Prefectural Government 1997; Zhang et al. 1989; Hong et al. 1988; Kobayashi et al. 1972; Dôtu 1961; Kimura 1958; Jones 1937). In *Periophthalmus* and *Periophthalmodon*, the ratio of horizontal-to-vertical diameter of eggs is 0.6–0.8 (Table 3.2). Egg sizes of *Periophthalmus* spp. are 0.7–1.1 and 0.6–0.8 mm, respectively, for vertical and horizontal diameters (Table 3.2). Although eggs of *Ps. magnuspinnatus* were reported to be more than two times larger in both diameters when compared to congeners (Zhou et al. 2012; Table 3.2), this needs to be verified. The eggs of *Pn. schlosseri* are larger than those of *Periophthalmus* spp. (Tsuhako et al. 2003; Table 3.2). Eggs of *Boleophthalmus*, *Scartelaos*, and *Apocryptodon* are more elongate (the ratios are 0.4–0.5) than *Periophthalmus* and *Periophthalmodon* (Table 3.2), and are generally larger (0.9–1.5 mm in the vertical diameter) than those of *Periophthalmus* (Table 3.2).

Table 3.2 Vertical and Horizontal Diameters of the Oxudercine Eggs Measured with the Micropyle Located at the Top

Species	Locality	Diameter (mm)		Ratio	Reference
		Vertical (*v*)	Horizontal (*h*)	(*h*/*v*)	
Periophthalmus argentilineatus	Okinawa Island, Japan	0.70–1.00 (mean 0.87)	0.60–0.73 (mean 0.66)	0.76	This study
Periophthalmus argentilineatus	Madagascar	1.00	–	–	Brillet 1976
Periophthalmus modestus	Okinawa Island, Japan	0.93–1.05 (mean 1.01)	0.60–0.70 (mean 0.64)	0.63	This study
Periophthalmus modestus	Ariake Sound, Japan	0.85–1.04 (mean 0.94)	0.60–0.66 (mean 0.64)	0.68	Kobayashi et al. 1972
Periophthalmus modestus	Miyazaki, Japan	0.76	0.61	0.80	Kimura 1958
Periophthalmus magnuspinnatus	Qiantan River estuary, China	2.42–3.43 (mean 2.95)	2.08–2.29 (mean 2.23)	0.76	Zhou et al. 2012
Periophthalmodon schlosseri	Penang, Malaysia	0.83–1.43 (mean 1.02)	0.56–1.13 (mean 0.73)	0.72	Tsuhako et al. 2003
Boleophthalmus pectinirostris	Zhanpu, Fujian, China	0.92–1.26	0.54–0.63	–	Hong et al. 1988; Zhang et al. 1989
Boleophthalmus pectinirostris	Ariake Sound, Japan	1.30–1.50 (mean 1.44)	0.70–0.75 (mean 0.74)	0.51	Koga et al. 1989
Scartelaos histophorus	Okinawa Island, Japan	1.3	0.5	0.38	Okinawa Prefectural Government 1997; Hosoya 2014
Scartelaos gigas	Jung-do Island, South Korea	1.23–1.48 (mean 1.37)	0.67–0.71 (mean 0.69)	0.50	Kim et al. 2011
Apocryptodon punctatus (unfertilized egg)	Ariake Sound, Japan	1.0	0.5	0.50	Dôtu 1961

3.5 EMBRYONIC DEVELOPMENT

Embryonic and larval developments of *Ps. argentilineatus* and *Ps. modestus* were serially observed for eggs collected on May 28, 1999 from Manko Wetlands and on June 12, 1999 from Haneji-naikai Inlet (Yagaji). Egg masses were brought to the laboratory with the substrata and placed into a plastic case for observation. They were maintained without water and exposed to the air for 68 hours and 50 minutes and 62 hours and 20 minutes after the collection, respectively, for *Ps. argentilineatus* and *Ps. modestus*. The eggs were then submerged in seawater (34–36 ppt, 24–25°C) with strong aeration (1.2 L/min). Results follow.

For *Ps. argentilineatus*, eggs were at early blastula stage 6 hours 20 minutes after collection (abbreviated to a. c.) (Figure 3.5a). Yolk appeared as translucent, yellowish sphere, 0.39–0.49 mm (0.45 ± 0.03 mm, n = 10) in diameter, and contained many small oil globules measuring 0.01–0.04 mm in diameter. The perivitelline cavity had been formed. At 11 hours 20 minutes a. c., the eggs had reached the late blastula stage, and the blastoderm covered three quarters of the yolk (Figure 3.5b). The oil globules fused, increasing in size, and decreasing in number. At 15 hours 10 minutes a. c. (Figure 3.5c), they reached the gastrula stage, and the germ ring began to invaginate. Embryos began to form eyes and Kupffer's vesicles appeared at 24 hours 50 minutes a. c. (Figure 3.5d). At 31 hours 10 minutes a. c. (Figure 3.5e), each embryo had 10 myomeres and the tail was detached from the yolk; the auditory vesicles, lenses of the eyes, and a pericardial cavity appeared. At 46 hours a. c. (Figure 3.5f), each embryo had 26 myomeres and an elongate tail; otoliths were formed in auditory vesicles and the heart began to beat. Xanthophores appeared over the yolk sac and the sides of the abdomen. Blanched melanophores appeared ventrally on the yolk, along the dorsal midline on the posterior part of the tail. During this time, the embryos began to turn over in the egg capsules. The alimentary canal and pectoral-fin buds appeared. The brain divided into three sections, and xanthophores appeared on the eyes at 55 hours 50 minutes a. c. At 58 hours 50 minutes a. c., the gas bladder formed, eyes were pigmented with melanophores and iridophores. At 61 hours 50 minutes a. c., the alimentary canal and pectoral fins developed (Figure 3.5g). At 68 hours 50 minutes a. c., branched melanophores distributed dorsally over the gas bladder and along the ventral and dorsal midlines of the body. After that, eggs were transferred into seawater. At 75 hours 30 minutes a. c., the mouth formed (Figure 3.5h). Hatching glands appeared on the snout and the top of the head at 84 hours 30 minutes a. c. At 95 hours 50 minutes a. c., eyes were fully pigmented with melanophores, nasal pits formed, and melanophores were distributed ventrally on the alimentary canal, the belly, and ventrally on the tail. Embryos often moved the tail, eyes, pectoral fins, and turned actively at this stage. At 104 hours a. c., the hatching glands increased in number and were distributed from the snout to the abdomen. By 122 hours a. c., oil globules had further decreased in number, and branched melanophores had developed along the dorsal midline on the tail and the belly (Figure 3.5i). Embryos began to hatch at 91 hours 40 minutes a. c., but the majority hatched at about 123 hours a. c. The hatching rate was 388 larvae/2,033 eggs, equivalent to 19.1%.

For *Ps. modestus*, at 4 hours 40 minutes a. c., each embryo had 10 myomeres (Figure 3.6a). Yolk was a translucent, yellowish sphere with many small oil globules (0.02–0.05 mm in diameter). The tail of each embryo was detached from the yolk; the auditory vesicles and lenses of the eyes were formed. Xanthophores were distributed over the yolk sac and the abdomen of the embryo. The heart was beating. By 14 hours 20 minutes a. c. (Figure 3.6b), 16 myomeres formed, oil globules fused, and otoliths appeared in the auditory vesicles. The alimentary canal appeared, and the brain divided into three sections. Xanthophores were distributed above and below the eyes, on the auditory vesicles, on the top of the head, and dorsally along the notochord in the tail. Melanophores

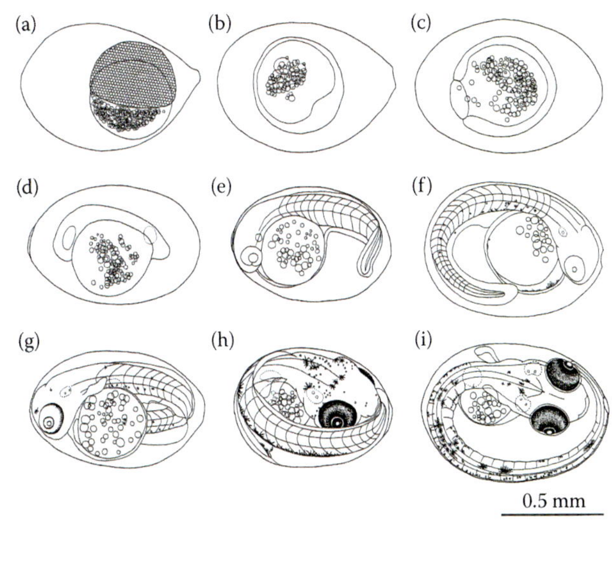

Figure 3.5 Embryonic development of *Periophthalmus argentilineatus* (a) 6 hours 20 minutes after collection (a. c.) of the egg mass; (b) 11 hours 20 minutes a. c.; (c) 15 hours 10 minutes a. c.; (d) 24 hours 50 minutes a. c.; (e) 31 hours 10 minutes a. c.; (f) 46 hours a. c.; (g) 61 hours 50 minutes a. c.; (h) 75 hours 30 minutes a. c.; (i) 122 hours a. c. (Courtesy of Yuko Tsuhako and Katsunori Tachihara.)

appeared on the eyes, on the belly, and along the ventral and dorsal midlines on the tail, together with the xanthophores. Melanophores also appeared on the yolk sac. Embryos actively moved their tails and turned in the egg capsules. At 25 hours 40 minutes a. c. (Figure 3.6c), the number of myomeres increased to 25 and the mouth opened. Hatching glands appeared on the head, and the pectoral fins and alimentary canal formed. The eyes were fully pigmented with melanophores and iridophores. Branched melanophores were distributed along the dorsal midline of the tail. Embryos moved their pectoral fins and eyes. At 62 hours 20 minutes a. c., eggs were transferred into seawater. At 65 hours 20 minutes a. c., hatching glands from the snout to the head increased in number. The nasal pits and a gas bladder formed. Melanophores were distributed dorsally over the gas bladder, and xanthophores appeared on the eyes (Figure 3.6d). At 118 hours 10 minutes a. c. (Figure 3.6e), the number of myomeres reached 26, oil globules decreased in number, and branched melanophores were distributed ventrally on the alimentary canal. Embryos began to hatch at 65 hours 20 minutes a. c., but most embryos hatched around 118 hours a. c. The hatching rate was 775 larvae/1,500 eggs, equivalent to 51.7%.

The embryonic development of *Ps. argentilineatus* and *Ps. modestus* described above is similar to those in previous reports for *Ps. modestus*, *Ps. magnuspinnatus*, and *B. pectinirostris*, which were observed with eggs incubated in water (Zhou et al. 2012; Zhang et al. 1989; Hong et al. 1988; Kobayashi et al. 1972; Kimura 1958). It was reported that the duration from fertilization to hatching of mudskippers differs depending on temperature and salinity of the water (Zhou et al. 2012; Koga et al. 1989). *Ps. modestus* hatched at 104 hours after fertilization with incubation in water at 25–26°C (Kimura 1958) and at 170–175 hours at 19–20°C (Kobayashi et al. 1972). *Ps. magnuspinnatus* hatched at 159 hours at 27°C (Zhou et al. 2012), and *B. pectinirostris* hatched at 88 hours at 27–29°C (Zhang et al. 1989; Hong et al. 1988) and at 5–6 days at 21–29°C (Koga et al. 1989).

3.6 HATCHING

The burrow-guarding male of *Ps. modestus* removes air from the egg chamber with its mouth preceding hatching events. The embryos hatch once the air is removed and the egg mass is immersed in water (Ishimatsu et al. 2007).

To investigate the time of hatching, larvae of *Ps. argentilineatus* and *Ps. modestus* were trapped as they were exiting the burrows on Okinawa Island from May 20 to July 8 in 2000. The trap was made from a plastic cylinder (9 cm in diameter, 25 cm in length) and a fine mesh was attached at one end. Traps were set at the entrances of burrows before they submerged during the rising tide and collected during the falling tide.

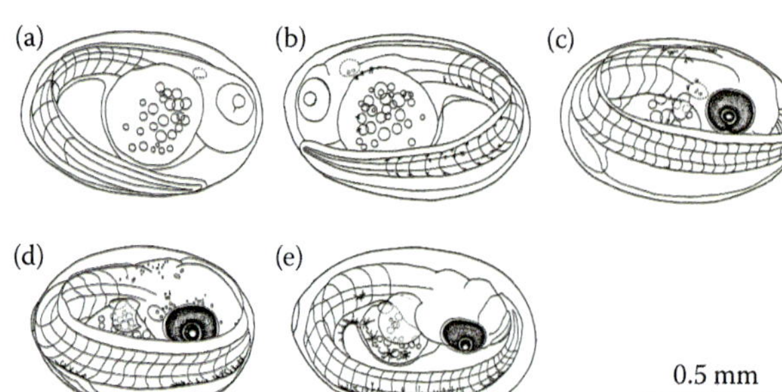

Figure 3.6 Embryonic development of *Periophthalmus modestus* (a) 4 hours 40 minutes after collection (a. c.) of the egg mass; (b) 14 hours 20 minutes a. c.; (c) 25 hours 40 minutes a. c.; (d) 65 hours 20 minutes a. c.; (e) 118 hours 10 minutes a. c. (Courtesy of Yuko Tsuhako and Katsunori Tachihara.)

Table 3.3 **Number of Larvae Hatched Out from Spawning Burrows of *Periophthalmus argentilineatus* and *Periophthalmus modestus***

				Larvae[d]		
Tide[a]	Date	Time[b]	Burrow[c]	Range	Total	Adult[e]
			Periophthalmus argentilineatus			
Spring	June 1	14:30–18:30	1	2,467	2,467	0
Spring	June 2	15:40–19:30	1	900	900	1
Spring	June 4	17:30–22:00	2	127–334	461	2
Spring	June 16	16:00–0:00	4	822–7,017	10,406	1
Spring	June 17	16:00–20:00	1	6,084	6,084	0
Spring	June 19	16:30–23:30	1	867	867	1
Spring	July 1	15:30–19:30	3	900–8,033	14,767	1
Spring	July 2	16:10–20:20	3	434–4,933	6,351	3
Spring	July 3	16:30–22:00	2	8,317–8,433	16,750	2
			Periophthalmus modestus			
Mid	May 31	15:00–18:30	1	100	100	1
Spring	June 1	15:45–20:00	1	167	167	1
Spring	June 2	17:15–21:00	3	767–2,967	4,701	2
Spring	June 4	16:20–23:00	2	1,484–3,100	4,584	2
Mid	June 7–8	19:20–1:00	2	33–1,033	1,066	1
Spring	June 17	17:30–21:10	1	1,033	1,033	0
Mid	June 30	15:40–20:00	1	4,367	4,367	0
Spring	July 2	15:00–22:10	1	5,267	5,267	1

Larvae were collected with a trap set at the entrances of the burrows during a rising tide and collected during an ebbing tide on Okinawa Island, Japan, from May 20 to July 8 in 2000.

[a] Tide of the day; midtide means a tide between spring and neap tides.
[b] Times of the day when the traps are set at the burrow entrances.
[c] Number of burrows from which larvae were released.
[d] Number of larvae captured by each trap (range) and total number of larvae collected (total).
[e] Number of traps in which an adult male was captured with the larvae.

Newly hatched larvae were collected from 23 burrows of *Ps. argentilineatus* and 37 burrows of *Ps. modestus* at about the time of spring tides (days with the greatest difference between high and low tides) (Table 3.3); larvae always exited the burrow at night. An evening spring high tide is a likely cue for the male to remove air from the egg chamber, immersing the egg mass in water to stimulate hatching. The number of larvae per trap ranged from 127 to 8,433 (total 59,053) and from 33 to 5,267 (total 21,285) for *Ps. argentilineatus* and *Ps. modestus*, respectively (Table 3.3). In 11 instances for *Ps. argentilineatus* and in eight instances for *Ps. modestus*, an adult male was also captured together with the larvae (Table 3.3). Newly hatched larvae of *Ps. argentilineatus* (n = 130) collected were 1.6–2.5 mm (mean ± SD = 2.1 ± 0.2 mm) in notochord length (NL) and 1.6–2.6 mm (2.2 ± 0.2 mm) in total length (TL), and those of *Ps. modestus* (n = 60) were 1.8–2.4 mm NL (2.1 ± 0.2 mm) and 1.8–2.5 mm TL (2.3 ± 0.2 mm).

3.7 LARVAL DEVELOPMENT

The hatched embryos above were maintained in a 30-L polycarbonate tank filled with seawater and fed rotifers, *Brachionus* spp., and *Artemia salina* nauplii. The behavior of the larvae of *Ps. argentilineatus* and *Ps. modestus* was observed while specimens were alive and in the tank. Changes in morphology were observed using specimens fixed in 5% formalin.

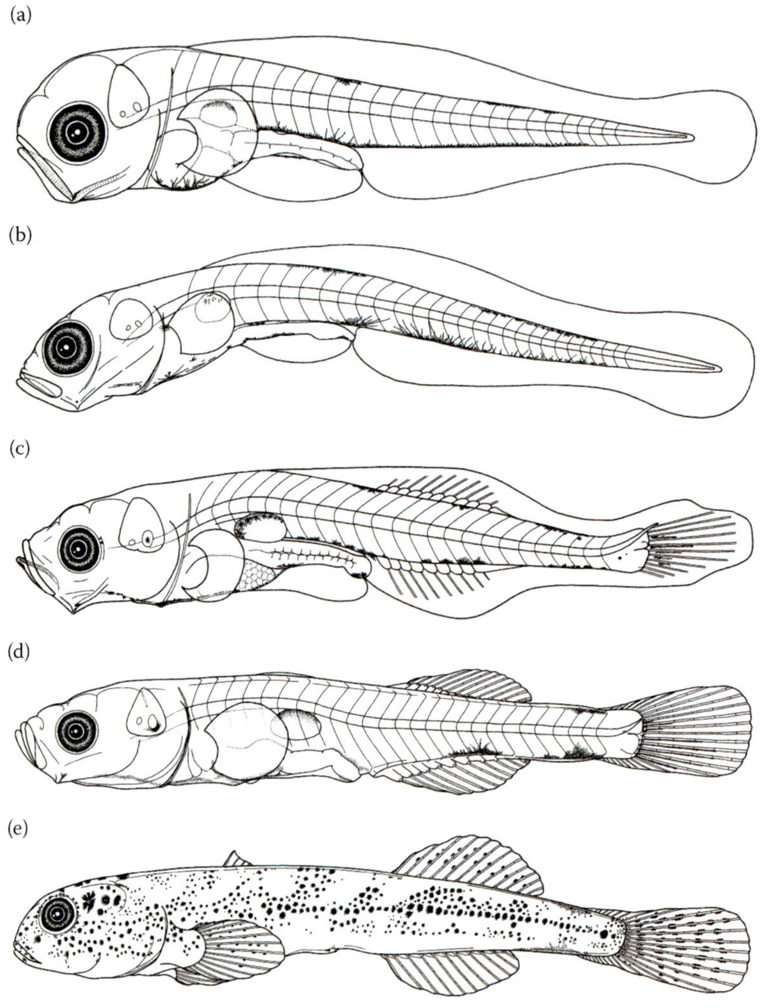

Figure 3.7 Reared larvae of *Periophthalmus argentilineatus* (a) immediately after hatching (2.1 mm NL); (b) 4 days after hatching (2.7 mm NL); (c) 10 days after hatching (4.1 mm NL); (d) 22 days after hatching (6.6 mm SL); (e) 30 days after hatching (8.3 mm SL). (Courtesy of Yuko Tsuhako and Katsunori Tachihara.)

The description of larval morphology for *Ps. argentilineatus* is based on 31 specimens of newly hatched larva and one individual each at 4, 10, 22, and 30 days after hatching. Newly hatched larvae were 1.6–2.2 mm NL and 1.7–2.3 mm TL. Each larva had a mouth, pigmented eyes, rudimentary pectoral fins, and a yolk sac (Figure 3.7a). The alimentary canal was straight and the anus opened below the 10th myomere. The gas bladder with melanophores was located above the yolk sac. Melanophores were distributed ventrally on the yolk sac, ventrally and dorsally on the abdominal cavity, and ventrally on the tail. Melanophores also occurred along the dorsal midline at the posterior part of the trunk and midtail. Larvae showed positive phototaxis.

At 4 days post-hatching, the larva measures 2.7 mm NL and 2.8 mm TL. The yolk was completely ingested and single melanophores appeared on angles of the lower jaw (Figure 3.7b).

The larva entered the flexion stage (4.1 mm NL, 4.8 mm TL), in which the posterior end of the notochord began to flex upward for hypural formation, 10 days after hatching (Figure 3.7c). Rudimentary rays appeared on the caudal, second dorsal, and anal fins.

By 22 days after hatching, the larva completed notochord flexion (i.e., post-flexion: 6.6 mm in standard length [SL], 7.7 mm TL; Figure 3.7d). Almost all fin rays were present on the caudal, second dorsal, and anal fins. Small finfolds remained at the position where the first dorsal fin forms in adults and at the belly. Larva had conspicuous melanophores ventrally and dorsally on the caudal peduncle. The ventral melanophore was located slightly more anterior than the dorsal one. Another conspicuous melanophore was on the posterior anal-fin base. Melanophores were also distributed at the angle of the lower jaw, below the sagittal otoliths, dorsally on the gas bladder, ventrally from the isthmus to the rudimentary pelvic-fin base, around the anus, and along the caudal-fin rays.

Metamorphosis commenced 30 days after hatching (8.3 mm SL, 10.2 mm TL; Figure 3.7e). The eyes moved to a dorsal position and the mouth moved to a ventral position. Pelvic fins developed. Pectoral- and pelvic-fin rays appeared. The first dorsal fin appeared, but it was small and had only four spines. The body surface was heavily pigmented. Larvae primarily attached to the tank wall.

Description of larval morphology for *Ps. modestus* is based on 22 specimens of newly hatched larva and one individual each at 5, 10, 20, and 29 days after hatching. Newly hatched larvae were 2.0–2.8 mm NL and 2.1–2.9 mm TL. Each larva had a mouth, pigmented eyes, rudimentary pectoral fins, and a yolk sac (Figure 3.8a). A gas bladder with melanophores was located above the yolk sac. Melanophores were distributed ventrally on the yolk sac, ventrally and dorsally on the abdominal cavity, and ventrally on the tail. Melanophores also occurred along the dorsal midline at the posterior part of the trunk and midtail. Larvae showed positive phototaxis.

At 5 days post-hatching, the larva, at 3.2 mm NL and 3.3 mm TL, had exhausted the yolk (Figure 3.8b). A rudiment of the hypural occurred near the tip of the notochord, which was accompanied by a melanophore. Single melanophores appeared on the angle of the lower jaw.

At 10 days after hatching, the larva was at the post-flexion stage (4.3 mm SL, 5.0 mm TL), at which time notochord flexion had been completed and the caudal-fin rays were formed (Figure 3.8c). Rudimentary rays appeared on the second dorsal and anal fins. A single melanophore appeared on the chin.

The larva observed at 20 days after hatching (6.1 mm SL, 7.7 mm TL) appeared somewhat emaciated (Figure 3.8d). Almost all fin rays were present on the second dorsal, anal, and caudal fins. Finfolds disappeared. The larva had conspicuous melanophores ventrally and dorsally on the caudal peduncle. The ventral one was located slightly anterior to the dorsal one. Another conspicuous melanophore occurred at the middle of the anal-fin base. Melanophores were also distributed at the angle of the lower jaw, dorsally on the gas bladder, ventrally from the isthmus to the anterior part of the belly, around the anus, and on the lower part of the caudal fin.

By 29 days after hatching (10.9 mm SL, 13.2 mm TL), metamorphosis commenced (Figure 3.8e). The eyes moved to a dorsal position and the mouth moved to a ventral position. Pectoral-fin rays appeared. The larva had pelvic fins with complete fin rays. The first dorsal fin was formed, but it was small and only had three spines. The body surface was heavily pigmented. Larvae primarily attached to the tank wall.

Larvae of *Ps. argentilineatus* and *Ps. modestus* are similar and it is difficult to distinguish them by their morphology. Newly hatched larvae of *Ps. modestus* vary in length and range from 2.3 mm TL (Kimura 1958) and 2.1–2.9 mm TL (this study; mean = 2.5 mm TL) to 2.8–2.9 mm TL (Kobayashi et al. 1972). Such length variation is thought to be environmentally induced. Embryonic development and the timing of hatching vary with handling or conditions during incubation (Zhou et al. 2012; Koga et al. 1989). Lengths of larvae trapped from burrows may be a better indication of natural sizes. The TL of larvae released from the burrows of *Ps. argentilineatus* and *Ps. modestus* was almost identical (mean = 2.2 and 2.3 mm, respectively; see the preceding section for details).

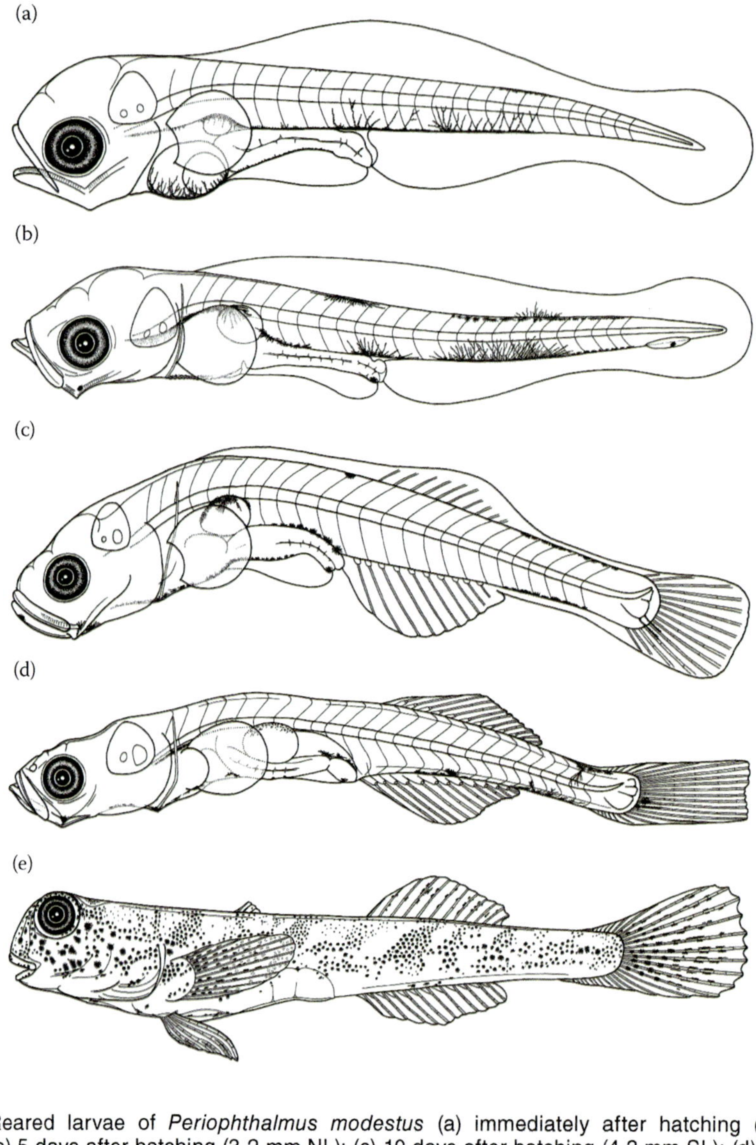

Figure 3.8 Reared larvae of *Periophthalmus modestus* (a) immediately after hatching (2.1 mm NL); (b) 5 days after hatching (3.2 mm NL); (c) 10 days after hatching (4.3 mm SL); (d) 20 days after hatching (6.1 mm SL); (e) 29 days after hatching (10.9 mm SL). (Courtesy of Yuko Tsuhako and Katsunori Tachihara.)

In conclusion, sizes at hatching do not differ between these two species and the TL varies from 1.7 to 2.9 mm.

Newly hatched larvae of the following species have been reported: *Pn. schlosseri* at 2.1–2.6 mm NL (Tsuhako et al. 2003); *B. pectinirostris* at 2.4–2.6 mm TL (Zhang et al. 1989; Hong et al. 1988) and 3.1–3.4 mm TL (Koga et al. 1989); *S. histophorus* at 2.2–2.3 mm TL (Okiyama 2014); and *S. gigas* at 2.6–3.2 mm TL (Kim et al. 2011).

Morphologies of larval stages have been described for *B. pectinirostris* (Okiyama 2014; Koga et al. 1989; Zhang et al. 1989; Hong et al. 1988), *S. histophorus* (Okiyama 2014), *An. punctatus* (Dôtu 1961), *Ps. argentilineatus* (this study), and *Ps. modestus* (Kobayashi et al. 1972, this study). Larval development of these species is similar, but in *B. pectinirostris*, the posterior part of the gut begins to coil at the flexion stage and the number of gut folds increases with maturation (Zhang et al. 1989; Hong et al. 1988).

Figure 3.9 Beaches on Okinawa Island where the larvae of *Periophthalmus argentilineatus* were collected for the age estimation study (a) sandy beach on Sedake; (b) stony beach on Aritsu. (Courtesy of Ken Maeda.)

3.8 MORPHOLOGY AND AGE AT SETTLEMENT

Larvae of *Ps. argentilineatus* occur at beaches on Okinawa Island (Figure 3.9a,b) from May to October (Maeda and Tachihara 2014). Those collected along the shoreline with a small seine net (Figure 3.10) are always post-flexion, nearing the end of the pelagic phase, and are 9–12 mm SL (Figure 3.11). The larval body is compressed and transparent, with a conspicuous gas bladder (Figure 3.12a). Fins possess all spines and segmented rays, except for the first dorsal fin, in which most spines have not formed. Left and right pelvic fins unite into a disc with a frenum (Figure 3.12b). Eyes are located laterally on the head and the mouth position is terminal. The posterior margin of the caudal fin is slightly concave (Figure 3.12a). Melanophores are located ventrally and dorsally on the caudal peduncle, on anterior (paired) and posterior (single) parts of the anal-fin base, along the ventral midline from the isthmus to the pelvic-fin base, on the belly posterior to the pelvic-fin base (V-shape), at the symphysis and angle of the lower jaw, dorsally on the gas bladder, dorsally on the intestine just above the anus, and lower part of the caudal fin (Okiyama 2014). Many

Figure 3.10 Small seine net used to collect mudskipper larvae. (Courtesy of Ken Maeda.)

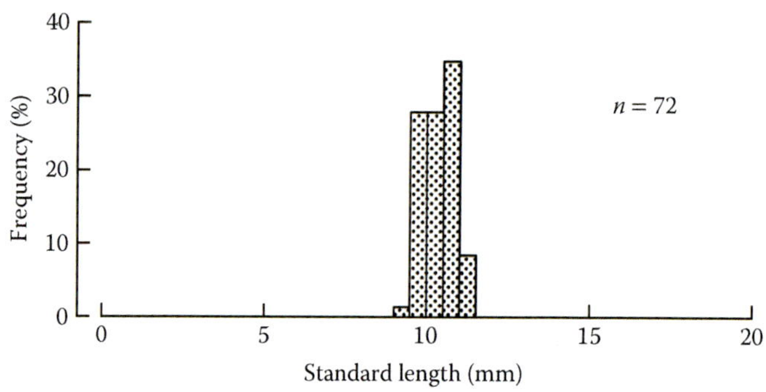

Figure 3.11 Size frequency distribution of *Periophthalmus argentilineatus* collected from Sedake Beach and an estuary of Teima Stream with a small seine net, from May to October 1999. (Modified from Maeda K. and Tachihara K., *Fisheries Science*, 80, 1215–1229, 2014.)

larval characteristics are shared with larvae of other gobioid species co-occurring at these localities (Maeda and Tachihara 2014).

Estuaries and beaches are not the primary habitats of pelagic larvae; larvae of *Ps. argentilineatus* born in the estuary migrate to a marine larval habitat and grow there (Maeda and Tachihara 2014). At the end of the pelagic phase, the larvae approach the beaches around the estuarine mouth before returning to estuaries. On mudflats within the estuaries, we found many juveniles of similar body sizes as larvae on the beaches (Figure 3.13). The larvae are thought to metamorphose to juveniles and settle immediately after the recruitment into the estuary.

Pelagic larval duration can be determined from the age of larvae collected on the beach because this phase immediately precedes settlement. Larval ages are estimated by otolith examination. Because growth of fish otoliths usually displays daily periodicity, ages are generally estimated from the number of daily increments visible on otoliths (Campana 2005). Daily deposition of the otolith

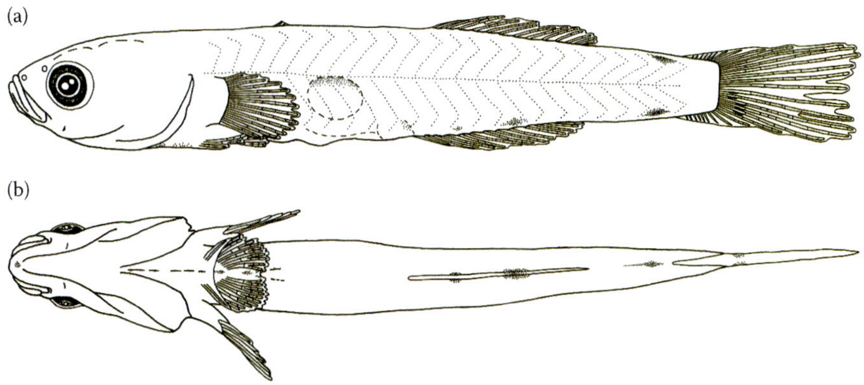

Figure 3.12 Post-flexion larva of *Periophthalmus argentilineatus* collected on Sedake Beach on September 19, 1999 (10.0 mm SL). (a) Lateral view; (b) ventral view. (Courtesy of Ken Maeda.)

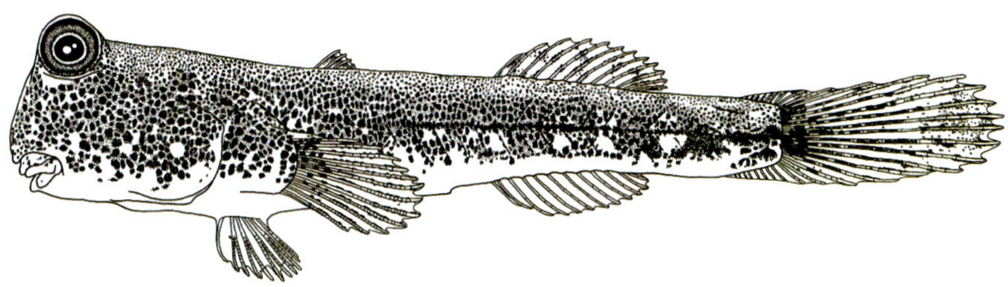

Figure 3.13 Settled juvenile of *Periophthalmus argentilineatus* collected at Manko, Okinawa Island on October 14, 2015 (12.1 mm SL). (Courtesy of Ken Maeda.)

increments has been validated for various gobioid species (Iida et al. 2010; Hoareau et al. 2007; Maeda et al. 2007; Hernaman et al. 2000; Shafer 2000; Iglesias et al. 1997; Radtke et al. 1988), although it has not been validated for mudskippers. We thus counted the number of increments from the core to the margin of sagittal otoliths (Figure 3.14) of *Ps. argentilineatus*; each increment was assumed to correspond to a day and provide an estimated pelagic larval duration. We examined 15 individuals of *Ps. argentilineatus* larvae (9.5–10.8 mm SL) collected from Sedake and Aritsu Beaches (Figure 3.9) on Okinawa Island, in October 2003, from June to October 2004, June 2006 and June 2007, and counted from 23 to 37 increments (mean ± SD = 26.5 ± 3.8) in their otoliths. Therefore, the pelagic larval duration of *Ps. argentilineatus* is approximately 1 month, and corroborates data obtained from reared larvae.

Pelagic larval durations of reared mudskippers, *Ps. modestus*, *B. pectinirostris*, and *S. histophorus*, are also approximately 1 month (Hosoya 2014; Koga et al. 1989; Kobayashi et al. 1972). The pelagic larval durations of gobioid fishes vary by species, but a duration of 1 month has been commonly reported in members of the suborder Gobioidei (Maeda et al. 2015).

3.9 EARLY LIFE HISTORY OF MUDSKIPPERS

Adult mudskippers possess unique characters to cope with being amphibious. Larval morphologies of mudskippers, however, are typical of gobioids; their morphologies change considerably after settlement, and they exhibit unique metamorphosis.

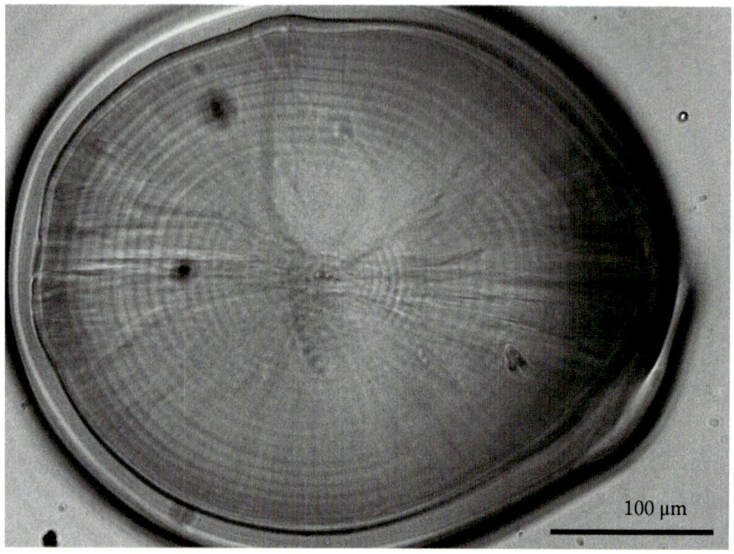

Figure 3.14 Otolith (sagitta) of a *Periophthalmus argentilineatus* post-flexion larva (10.1 mm SL) collected at Sedake Beach on October 3, 2004. (Courtesy of Ken Maeda.)

First dorsal fin: Most gobioid taxa possess two dorsal fins. All spines and segmented rays in these fins are usually formed before settlement, although the last spine in the first dorsal fin is sometimes absent or very small (Maeda et al. 2011; Maeda and Tachihara 2005). First dorsal-fin spines number six to eight (usually six) in adults of most gobioid taxa, whereas the first dorsal fin of adult *Periophthalmus* usually has more than eight spines (Murdy 1989). In *Ps. argentilineatus* and *Ps. modestus*, spines in adults reach 16 or 17 (Murdy 1989), but the first dorsal fin is small in juveniles, with only three to five spines when they settle on the mudflat (Figures 3.12 and 3.13). The number of spines typically exceeds 10 after 50 mm SL (Kobayashi et al. 1972). Thus, these mudskippers are unique in that the fin elements continue to increase after settlement, though larval development is otherwise quite normal.

Pelvic fin: The pelvic fins of many gobiid species unite into a disc with a frenum, and such fin morphology is completed prior to settlement (Figure 3.15a). The pelvic fins of adult *Ps. modestus* have a frenum, and the left and right fins join with a membrane between the innermost rays, while those of *Ps. argentilineatus* adult lack the frenum and the left and right fins are almost separate (Murdy 1989). This feature is used as one of the keys to distinguish these two species (Nakabo 2002). However, a frenum is briefly present in *Ps. argentilineatus* before settlement, at which time, a membrane connects the left and right pelvic fins (Figure 3.12b).

Pelvic fin morphologies of gobioid fish are diverse. For example, the left and right pelvic fins in *Eleotris* species are separated throughout the stages before and after settlement (Maeda and Tachihara 2005; Figure 3.15b). Another example, the united pelvic fins of sicydiine gobies, function as a strong sucker to climb up streams and waterfalls; the fleshy sucking-disc structure forms before settlement (Figure 3.15c), although it becomes more functional as a climbing aid only when the pelvic-fin rays have more branches, and they are directed more radially (Keith et al. 2015, 2008). The pelvic fins of adult *Eviota* and *Valenciennea* lack a frenum and are separate, but a short basal membrane often connects the innermost rays of the left and right fins. The basal membrane, considered vestigial in these fishes, persists in larvae and adults; the membrane does not diminish once developed (Okiyama 2014; Figure 3.15d,e). Thus, basic structures of various pelvic-fin types

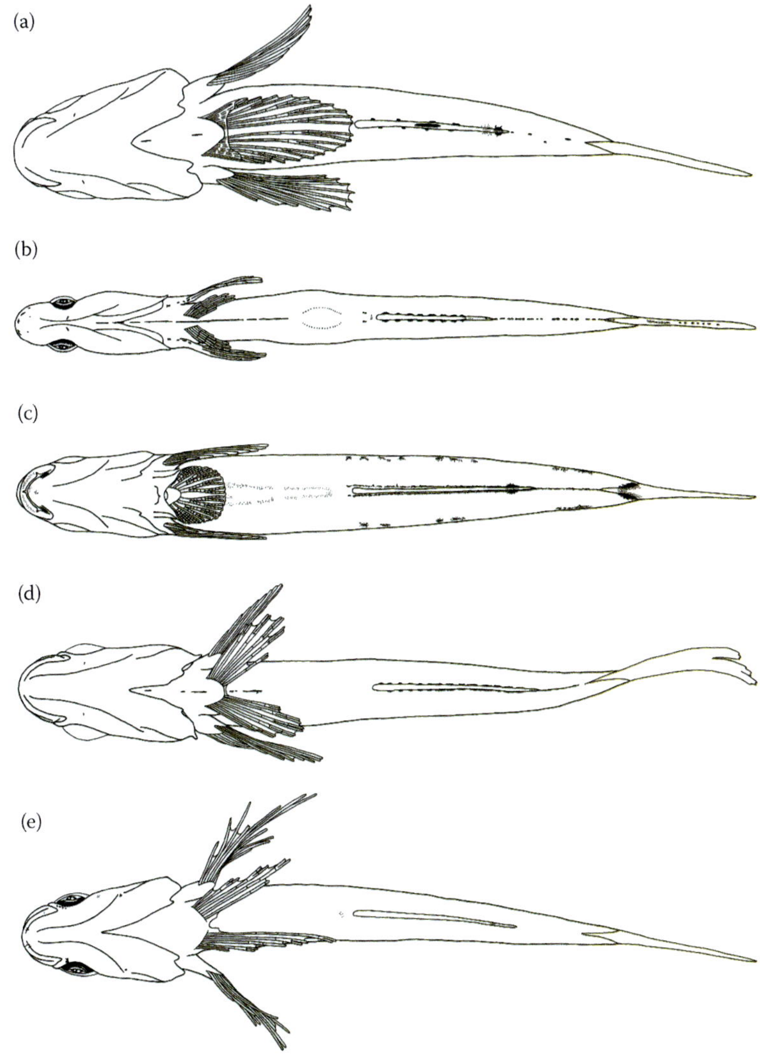

Figure 3.15 Post-flexion larvae of gobioid fishes collected on beaches or at estuarine mouths, showing pelvic-fin morphology in ventral view (a) *Bathygobius* sp. (7.0 mm SL); (b) *Eleotris acanthopoma* (12.1 mm SL); (c) *Sicyopterus lagocephalus* (25.7 mm SL); (d) *Valenciennea* sp. (16.9 mm SL); (e) *Eviota* sp. (8.3 mm SL). (Courtesy of Ken Maeda.)

are usually defined during the pelagic larval phase. The pelvic-fin formation process in *Ps. argentilineatus* is unique, as it develops the disc shape and then loses the frenum and the membrane between the innermost rays. Though morphological change of the pelvic fins in *Ps. modestus* is not as significant as that in *Ps. argentilineatus*, the round disc becomes heart shaped as the membrane between the innermost rays is notched after settlement (Kobayashi et al. 1972).

Morphological development of the mudskippers mentioned above shows that their larvae maintain a generalized morphology throughout the pelagic phase, and unique characteristics of the amphibious adults appear after settlement, suggesting that they share larval habitats and ecology with other gobioids. Unfortunately, the larval habitat and ecology of mudskippers have been poorly studied, in contrast to the more extensive research effort devoted to the adults.

Occurrences of larvae and juveniles of *B. pectinirostris* were investigated along river courses, and the sea around some river mouths in Ariake Sound, western Kyushu, Japan (Yuzuriha and Koga 1990; Igita 1986, 1985; Onohara 1980). At these localities, newly hatched larvae (2.5–3.5 mm TL; Yuzuriha and Koga 1990) and abundant transition-stage individuals (more than 14 mm TL) just before and after metamorphosis (Igita 1986, 1985) were found. However, only a few pelagic larvae of 4–14 mm TL were encountered (Yuzuriha and Koga 1990; Igita 1986, 1985; Onohara 1980). Consequently, the main habitat of pelagic larvae of *B. pectinirostris* has not yet been elucidated.

Larvae of *S. histophorus* grow in the water column (where the water depth is 20–30 m) in the middle and southern areas of Nakagusuku Bay, Okinawa Island, a bay approximately 20-km long from north to south (Tachihara 2015; Hosoya 2014). Among islands in the Ryukyu Archipelago, this species is found only on Okinawa Island, and Nakagusuku Bay is its only substantial habitat within the island. The mudflats around this bay provide ideal adult and juvenile habitat, while the bay provides the larval habitat. The distribution of *S. histophorus* is restricted not only by availability of adult habitat, but also by larval habitat as well.

The early development of *Ps. argentilineatus* and *Ps. modestus* is very similar, thus it is assumed that the habitat of their larvae is also similar. However, their distribution patterns are quite different. *Periophthalmus argentilineatus* is widespread in the Indo-Pacific and is found in all major islands in the Ryukyu Archipelago, whereas *Ps. modestus* is restricted to East Asia and occurs at few localities in the archipelago (see Section 3.1 for details). What caused this difference in distribution? There is no clear answer to this question at present. Distance and frequency of the pelagic larval dispersal may differ due to unknown causes. Or, greater availability of suitable adult habitat for *Ps. argentilineatus* may provide more stepping stones for this species. We hope that future studies of larval ecology and population genetics may answer this question.

REFERENCES

Brillet C. 1976. Structure du terrier, reproduction et comportement des jeunes chez le poisson amphibie *Periophthalmus sobrinus* Eggert. *Revue d'Ecology: La Terre et la Vie* 30:465–483 (in French with English summary).

Campana S.E. 2005. Otolith science entering the 21st century. *Marine and Freshwater Research* 56:485–495.

Dôtu Y. 1961. The bionomics and life history of the gobioid fish, *Apocryptodon bleekeri* (Day). *Bulletin of the Faculty of Fisheries, Nagasaki University* 10:133–139, plus 1 plate (in Japanese with English abstract).

Hernaman V., Munday P.L. and Schläppy M.L. 2000. Validation of otolith growth-increment periodicity in tropical gobies. *Marine Biology* 137:715–726.

Hoareau T.B., Lecomte-Finiger R., Grondin H., Conand C. and Berrebi P. 2007. Oceanic larval life of La Réunion 'bichiques', amphidromous gobiid post-larvae. *Marine Ecology Progress Series* 333:303–308.

Hong W., Dai Q., Zhang Q., Zhang J. and Cai Y. 1988. Observations on the early development of the mud skipper, *Boleophthalmus pectinirostris* (Linnaeus). *Tropic Oceanology* 2:1–8 (in Chinese with English abstract).

Hosoya S. 2014. Tokagehaze (*Scartelaos histophorus*): The conservation of intertidal mudflats ecosystems in Nakagusuku Bay, Okinawa Island. *Japanese Journal of Ichthyology* 61:109–114 (in Japanese).

Igita K. 1985. Ecological study of *Boleophthalmus pectinirostris*—II: Distribution of the juveniles and young stage in the river. *Bulletin of Saga Prefectural Ariake Fisheries Experimental Station* 9:55–59 (in Japanese).

Igita K. 1986. Distribution of larvae and juveniles in tidal compartment of Rokkaku River. *Bulletin of Saga Prefectural Ariake Fisheries Experimental Station* 10:35–45 (in Japanese).

Iglesias M., Brothers E.B. and Morales-Nin B. 1997. Validation of daily increment deposition in otoliths. Age and growth determination of *Aphia minuta* (Pisces: Gobiidae) from the northwest Mediterranean. *Marine Biology* 129:279–287.

Iida M., Watanabe S. and Tsukamoto K. 2010. Validation of otolith daily increments in the amphidromous goby *Sicyopterus japonicus*. *Coastal Marine Science* 34:39–41.

Ishimatsu A. and Gonzales T. 2011. Mudskippers: Front runners in the modern invasion of land. In *The Biology of Gobies* (eds. R.A. Patzner, J.L. Van Tassel, M. Kovačić and B.G. Kapoor) pp. 609–638. Boca Raton, FL: CRC Press, Taylor and Francis.

Ishimatsu A., Hishida Y., Takita T. et al. 1998. Mudskippers store air in their burrows. *Nature* 391:237–238.

Ishimatsu A., Takeda T., Tsuhako Y., Gonzales T.T. and Khoo K.H. 2009. Direct evidence for aerial egg deposition in the burrows of the Malaysian mudskipper, *Periophthalmodon schlosseri*. *Ichthyological Research* 56:417–420.

Ishimatsu A., Yoshida Y., Itoki N., Takeda T., Lee H.J. and Graham J.B. 2007. Mudskippers brood their eggs in air but submerge them for hatching. *Journal of Experimental Biology* 210:3946–3954.

Jones B.A. 1937. Observations on the breeding habits and development of certain brackish water fishes of Adyar, Madras. *Proceedings of the Indian Academy of Sciences—Section B* 5:261–289.

Keith P., Hoareau T.B., Lord C. et al. 2008. Characterisation of post-larval to juvenile stages, metamorphosis and recruitment of an amphidromous goby, *Sicyopterus lagocephalus* (Pallas) (Teleostei: Gobiidae: Sicydiinae). *Marine and Freshwater Research* 59:876–889.

Keith P., Lord C. and Maeda K. 2015. *Indo-Pacific Sicydiine Gobies: Biodiversity, Life Traits and Conservation*. Paris: Société Française d'Ichtyologie, 256 pp.

Kim J.-K., Baek H.-J., Kim J.-W., Chang D.-S. and Kim J.-I. 2011. Sexual maturity and early life history of the mudskipper *Scartelaos gigas* (Pisces, Gobiidae): Implications for Conservation. *Fisheries and Aquatic Sciences* 14:403–410.

Kimura M. 1958. Early development of the mud-skipper, *Periophthalmus cantonensis* (Osbeck). *Bulletin of the Japanese Society of Scientific Fisheries* 23:754–757.

Kobayashi T., Dotsu Y. and Miura N. 1972. Egg development and rearing experiments of the larvae of the mud skipper, *Periophthalmus cantonensis*. *Bulletin of the Faculty of Fisheries, Nagasaki University* 33:49–62 (in Japanese with English abstract).

Kobayashi T., Dotsu Y. and Takita T. 1971. Nest and nesting behavior of the mud skipper, *Periophthalmus cantonensis* in Ariake Sound. *Bulletin of the Faculty of Fisheries, Nagasaki University* 32:27–40.

Koga H. and Noda S. 1992. Seasonal changes of the burrows' form of mudskipper (short paper). *Bulletin of Saga Prefectural Ariake Fisheries Experimental Station* 14:81–84 (in Japanese).

Koga H., Noda S., Noguchi T. and Aoto I. 1989. Studies on artificial propagation of mud skipper *Boleophthalmus pectinirostris* (Linnaeus)—III: Hatching and larval rearing. *Bulletin of Saga Prefectural Ariake Fisheries Experimental Station* 11:17–28 (in Japanese with English abstract).

Lee Y.-J., Choi Y. and Ryu B.-S. 1995. A taxonomic revision of the genus *Periophthalmus* (Pisces: Gobiidae) from Korea with description of a new species. *Korean Journal of Ichthyology* 7:120–127.

Maeda K. and Tachihara K. 2005. Recruitment of amphidromous sleepers *Eleotris acanthopoma, Eleotris melanosoma*, and *Eleotris fusca* into the Teima River, Okinawa Island. *Ichthyological Research* 52:325–335.

Maeda K. and Tachihara K. 2014. Larval fish fauna of a sandy beach and an estuary on Okinawa Island, focusing on larval habitat utilization by the suborder Gobioidei. *Fisheries Science* 80:1215–1229.

Maeda K., Iida M. and Kondo M. 2015. Diversity of larval dispersal strategies in freshwater and estuarine gobies. In *Early Life History of Fishes* (eds. N. Mochioka, I. Kinoshita and T. Minami) pp. 89–101. Tokyo: Kouseisha-Kouseikaku (in Japanese).

Maeda K., Yamasaki N. and Tachihara K. 2007. Size and age at recruitment and spawning season of sleeper, genus *Eleotris* (Teleostei: Eleotridae) on Okinawa Island, southern Japan. *Raffles Bulletin of Zoology* (Suppl. 14):199–207.

Maeda K., Yamasaki N., Mukai T. and Tachihara K. 2011. Morphology and habitats of two gobiid species, *Pandaka trimaculata* and *P. lidwilli*, on Okinawa Island. *Japanese Journal of Ichthyology* 58:127–140 (in Japanese with English abstract).

Miller P.J. 1984. The tokology of gobioid fishes. In *Fish Reproduction: Strategies and Tactics* (eds. G.W. Potts and R.J. Wooton) pp. 119–152. London: Academic Press.

Murdy E.O. 1989. A taxonomic revision and cladistic analysis of the oxudercine gobies (Gobiidae: Oxudercinae). *Records of the Australian Museum* (Suppl. 11):1–93.

Nakabo T. 2002. *Fishes of Japan with Pictorial Keys to the Species, English Edition*. Tokyo: Tokai University Press, 1749 pp.

Noda S., Koga H. and Ohkuma H. 1993. Studies on artificial propagation of mudskipper *Boleophthalmus pectinirostris* (Linnaeus)—VI: Spawning condition and relationship between spawning and environmental factors. *Bulletin of Saga Prefectural Ariake Fisheries Experimental Station* 15:53–60 (in Japanese with English abstract).

Okinawa Prefectural Government. 1997. *Tokage-haze no hanashi*. Naha: Okinawa Prefectural Government, 11 pp (in Japanese).

Okiyama M. 2014. *An Atlas of Early Stage Fishes in Japan Second Edition*. Hadano: Tokai University Press, 1639 pp (in Japanese).

Onohara T. 1980. Ecological study of *Boleophthalmus pectinirostris*—I: Fisheries, distribution and growth. *Bulletin of Saga Prefectural Ariake Fisheries Experimental Station* 7:123–150 (in Japanese).

Oshiro M., Tsuhako Y. and Tachihara K. 2005. Habitat selection for burrow making by two species of mudskipper, *Periophthalmus argentilineatus* and *P. modestus* on the Okinawa Island, Japan. *Journal of Okinawa Association of Biology Education* 37:9–18 (in Japanese with English abstract).

Radtke R.L., Kinzie R.A. III and Folsom S.D. 1988. Age at recruitment of Hawaiian freshwater gobies. *Environmental Biology of Fishes* 23:205–213.

Shafer D.J. 2000. Evaluation of periodic and aperiodic otolith structure and somatic-otolith scaling for use in retrospective life history analysis of a tropical marine goby, *Bathygobius coalitus*. *Marine Ecology Progress Series* 199:217–229.

Tachihara K. 2015. *Scartelaos histophorus*. In *Red Data Book 2014—Threatened Wildlife of Japan—Volume 4, pisces—Brackish and Fresh Water Fishes* (ed. Ministry of the Environment) pp. 104–105. Tokyo: Gyosei Corporation (in Japanese with English abstract).

Takita T. and Ishimatsu A. 2015. *Fish Emerging from Water—The Mudskipper's Challenge*. Tokyo: Kaiyusha Publishers, 166 pp (in Japanese).

Toba A. and Ishimatsu A. 2014. Roles of air stored in burrows of the mudskipper *Boleophthalmus pectinirostris* for adult respiration and embryonic development. *Journal of Fish Biology* 84:774–793.

Tsuhako Y. 2005. *Periophthalmus modestus*. In *Threatened Wildlife in Okinawa, Second Edition (Animals): Red Data Okinawa* (ed. Department of Cultural and Environmental Affairs, Okinawa Prefectural Government) pp. 168–169. Naha: Department of Cultural and Environmental Affairs, Okinawa Prefectural Government (in Japanese).

Tsuhako Y., Ishimatsu A., Takeda T., Huat K.K. and Tachihara K. 2003. The eggs and larvae of the giant mudskipper, *Periophthalmodon schlosseri*, collected from a mudflat in Penang, Malaysia. *Ichthyological Research* 50:178–181.

Yang K. and Zhang Q. 1990. Developmental stages of larval, juvenile and young mudskipper (*Boleophthalmus pectinirostris*). *Journal of Xiamen University (Natural Science)* 29:76–80 (in Chinese with English abstract).

Yuzuriha M. and Koga H. 1990. Ecological study of mud skipper *Boleophthalmus pectinirostris* (Linnaeus)—III: Appearance and distribution of larvae and juveniles. *Bulletin of Saga Prefectural Ariake Fisheries Experimental Station* 12:15–20 (in Japanese with English abstract).

Zhang Q.-Y., Hong W.-S., Dai Q.-N., Zhang J., Cai Y.-Y. and Huang J.-L. 1989. Studies on induced ovulation, embryonic development and larval rearing of the mudskipper (*Boleophthalmus pectinirostris*). *Aquaculture* 83:375–385.

Zhou T.-S., Wang L., Tang W.-Q., Song X.-J. and Sun S.-S. 2012. Embryonic development and salinity tolerance of mudskipper (*Periophthalmus magnuspinnatus*). *Acta Hydrobiologica Sinica* 36:913–921 (in Chinese with English abstract).

CHAPTER 4

Age and Growth

Atsushi Nanami

CONTENTS

4.1 INTRODUCTION

Estimation of age and growth of organisms is essential in understanding the variations in growth rate and life span and also provides information on other related biological parameters such as reproductive characteristics. In fishes, data on age and growth can be applied for effective fisheries management and conservation. Some species of oxudercine gobies, also known as mudskippers, are used as local fisheries resources (Akinrotimi et al. 2013; Minh et al. 2010; Etim et al. 1996). Mudflats and mangrove forests are areas highly susceptible to human activities. As environmental changes both acutely and chronically affect the life history of fishes, examining the age and growth of mudskippers, and their spatial and temporal variation, is essential for understanding their biology, and their role in ecosystems prone to anthropogenic perturbation. Although numerous studies have reported the age and growth for marine fishes (e.g., Choat and Robertson 2002; Manooch 1987), there are relatively few studies available on the age and growth of mudskippers. In this chapter, the details of methods for age determination of mudskippers and the related literature are reviewed.

4.2 AGE DETERMINATION

Age and growth are usually estimated by three methods: (1) counting the number of increments on calcareous structures (age characters); (2) analyzing length–frequency data; and (3) growth estimation by mark-and-recapture (Manooch 1987). Among these basic methods, the first and second methods have been previously applied for age determination in some species of mudskippers.

In order to determine age, mudskippers can be collected either by hand (Sarafraz et al. 2011), hand net (Kruitwagen et al. 2006), hook-and-line (Washio et al. 1991), trap (Nanami and Takegaki 2005; Etim et al. 2002), or combinations of these sampling methods (Mazlan and Rohaya 2008; Park et al. 2008). Mudskippers should be sampled throughout the year so that the seasonal periodicity of the increment pattern of age characters (otolith or pectoral-radial bone) can be evaluated (see Section 4.3.4). As some mudskipper species become dormant during winter (e.g., Park et al. 2008; Jeong et al. 2004), year-round sampling is unlikely for such cases. Regular sampling intervals (e.g., monthly sampling) are also recommended to evaluate the seasonal periodicity of increment formation pattern. Sampling of individuals with various size ranges should be considered.

4.3 CALCAREOUS STRUCTURES AS AGE CHARACTERS

Age characters are the calcareous structures used for age determination (schlerochronology). Calcareous structures such as otoliths, vertebrae, scales, spines, and bones are useful as age characters in fishes. In this section, age characters that have been used for age determination of mudskippers are detailed.

4.3.1 Sagittal Otolith

The sagittal otolith is widely used for age determination in fishes (Figure 4.1a,b). The sagittal otolith has been used for age determination in *Boleophthalmus dussumieri* (reported as *Boleophthalmus dentatus*; Soni and George 1986), *Periophthalmus argentilineatus* (Kruitwagen et al. 2006), and *Periophthalmodon schlosseri* (Mazlan and Rohaya 2008). Soni and George (1986) observed increments on the sagittal otolith without any preparation. In contrast, Kruitwagen et al. (2006) and Mazlan and Rohaya (2008) mounted, sectioned, and polished the sagittal otolith in order to detect increments and the core.

These latter studies counted the number of microincrements (daily rings) and used the microincrements for age determination. Kruitwagen et al. (2006) reported that the maximum age was 6 years for *Ps. argentilineatus*. Mazlan and Rohaya (2008) showed that the standard length (SL) of *Pn. schlosseri* reached over 20 cm at 365 days.

4.3.2 Second Pectoral-Radial Bone

The method using the second pectoral-radial bone is specific for age determination of mudskippers (Figure 4.1c,d) and was first introduced by Tsukamoto et al. (1983). Six calcareous structures (pectoral-radial bones, coracoid, pelvic girdle, scale, sagittal otolith, and vertebra) were investigated in *B. pectinirostris* to select the most suitable character for age determination (Tsukamoto et al. 1983). Of the six structures, pectoral-radial bones were chosen for ease of increment reading and core detection. The pectoral fin of mudskippers has four pectoral-radial bones (Figure 4.1c). Tsukamoto et al. (1983) studied these four pectoral-radial bones in detail and found the second pectoral-radial bone to be the most reliable for increment reading and core detection (Figure 4.1d). They also confirmed a significant positive relationship between the radius of the second pectoral-radial bone and the total length (TL) of specimens. As a result, the second pectoral-radial bone was considered as the most appropriate age character for age estimation for *B. pectinirostris* (Tsukamoto et al. 1983).

Age estimation using the second pectoral-radial bone has been widely used (Sarafraz et al. 2011; Park et al. 2008; Nanami and Takegaki 2005; Jeong et al. 2004; Washio et al. 1991) due to relative ease of preparation. Procedures for preparation are as follows: (1) cut away the pectoral fin from the body and boil for 10–20 minutes; (2) extract the second pectoral-radial bone and dip it into 5% KOH or 3% NaOH aqueous solution to remove the muscle, rays, and other connective tissues; (3) clean the remaining second pectoral-radial bone in distilled water. After these three procedures, the increments and core can be detected without sectioning or polishing. Usually, the increments and core can be easily detected under a profile projector with a magnification of 10–20× (Nanami and Takegaki 2005; Washio et al. 1991; Zhang et al. 1990).

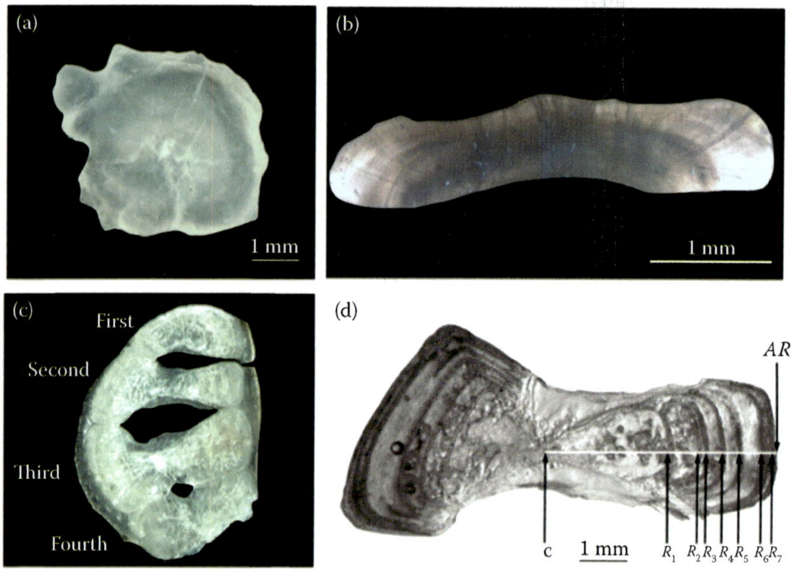

Figure 4.1 Age characters that have been used for age determination of mudskippers. The age characters in the photographs were obtained from *Boleophthalmus pectinirostris*. (a) Whole sagittal otolith; (b) Sectioned and polished sagittal otolith; (c) Four pectoral-radial bones; (d) Second pectoral-radial bone showing core (C), seven increment radii (R_1–R_7) and age character radius (AR); image. (Modified from *Fisheries Research*, 74, Nanami A. and Takegaki T, Age and growth of the mudskipper *Boleophthalmus pectinirostris* in Ariake Bay, Kyushu, Japan, 24–34, Copyright 2005, with permission from Elsevier.)

4.3.3 Increment Counting

For age determination, the number of increments on the age characters should be assessed. After sample preparation, opaque and translucent zones are alternately determined for the age character under transmitted light (e.g., Kruitwagen et al. 2006; Nanami and Takegaki 2005); opaque zones are usually considered as increments. Some samples have obscure increments that are difficult to read; such samples should be excluded from the analyses. Another problem for increment counting is precision and accuracy. Some samples showing many increments are difficult to count, especially for increments close to the edge, where they become narrower. To ensure precision and accuracy, multiple independent counts should be conducted. For example, Nanami and Takegaki (2005) counted the number of increments on the second pectoral-radial bones twice for 1356 samples and only the results that were in agreement were used for the analysis (999 samples: 74%). Kruitwagen et al. (2006) counted the number of increments in the otolith three times and the results were used if all three counts were the same. In the case of only two counts being the same, a fourth count was conducted. In this case, the results were included if three of the counts were the same (Kruitwagen et al. 2006).

4.3.4 Increment Periodicity

Increment periodicity can be assessed observing marginal increment patterns of age characters. This can be done with marginal increment ratio (MIR) analysis or direct observations of the increment at the edge of the age character. The MIR is calculated as follows (e.g., Park et al. 2008; Jeong et al. 2004): MIR = (AR – R_n)/(R_n – R_{n-1}), where AR is the age character radius (AR) (distance between core and edge) and R_n is the nth increment radius (distance between the core and nth increment).

The frequent occurrence of increments next to the edge in a specific time of the year indicates that this is when the increments are formed or the "increment formation period." Therefore, MIR ~ 0 during the increment formation period (i.e., AR ~ R_n).

4.3.4.1 Inter- and Intraspecific Difference in Increment Formations: Annual and Biannual

Nanami and Takegaki (2005) studied the marginal increment patterns of the second pectoral-radial bones of *B. pectinirostris* collected from the Tojin River (32°51' N, 130°34' E) in Japan between 2001 and 2002 and found that the increments were formed once a year (between September and November). The first increment was formed ca. 4 months after spawning (between May and August). The authors thus assumed that the first increment was formed at age 0.33 years (4/12) and the subsequent increments were formed at age 1.33 years (second increment), age 2.33 years (third increment), and so forth until age 7.33 years (eighth increment) (Figure 4.2a).

In contrast, Washio et al. (1991) found a biannual increment formation of the second pectoral-radial bones of *B. pectinirostris* collected from the Midori River (32°42' N, 130°36' E) in Japan between 1990 and 1991. They checked the marginal increment patterns of the second pectoral-radial bones and confirmed that the first, third, and fifth increments were formed during the autumn season (between October and November), whereas the second, fourth, and sixth increments were formed during the spawning season (between May and July). Individuals with second, fourth, and sixth increments were thus regarded as age 1, age 2, and age 3 individuals, respectively (Figure 4.2b).

Park et al. (2008) found a biannual increment formation of the second pectoral-radial bones in *Scartelaos gigas*. The first increment was formed in the winter season during dormancy, and the second increment was formed during the spawning season (around June). They assumed that the

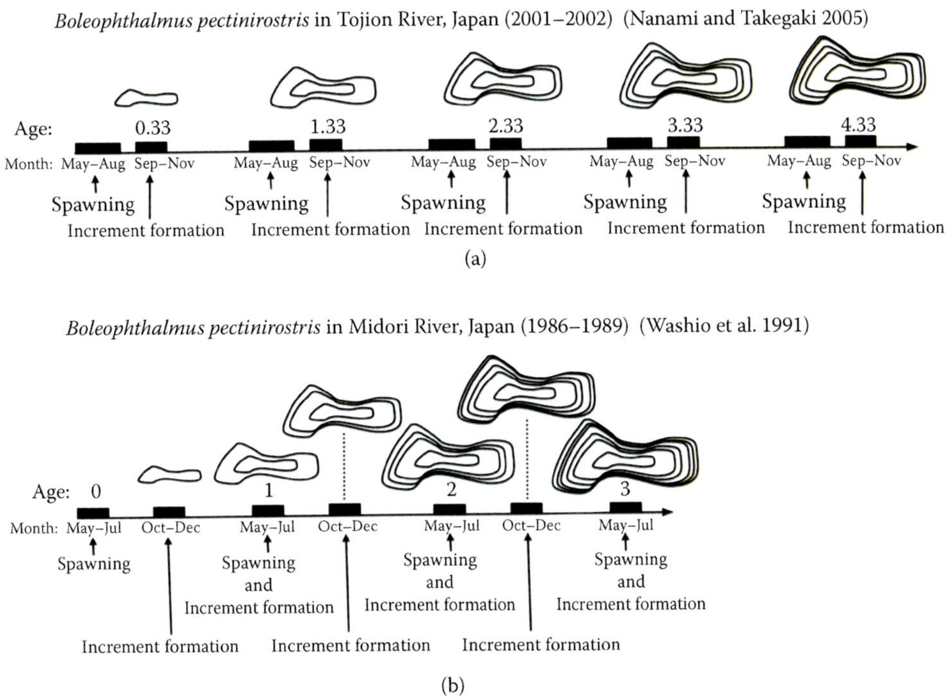

Figure 4.2 Examples of different increment formation patterns for *Boleophthalmus pectinirostris* at different sampling locations and different sampling periods. (Courtesy of Atsushi Nanami.)

first increment was formed at age 1 and subsequent increments were formed at age 1.5 (second increment), age 2 (third increment), age 2.5 (fourth increment), and so forth.

These studies demonstrate that the number of increments cannot be regarded as the age in years of the species.

4.3.4.2 Method Using Daily Microincrements

Counting daily microincrements between increments (opaque zones) is another method to examine whether the increments can be considered as annual rings. Kruitwagen et al. (2006) were unable to collect samples on a monthly basis and the MIR method could not be applied, so daily microincrements on the sagittal otolith for first, second, and third opaque zones were counted. As a result, the number of microincrements was 291, 388, and 840 for the first, second, and third opaque zones. Based on the results, Kruitwagen et al. (2006) assumed that the first opaque zone was the first annulus and the subsequent opaque zones were as annuli.

4.3.5 Back-Calculation

Back-calculation is a method to estimate the length of the fish at specific ages using the radii of the increments on the age character and the length of fish at time of collection. As mentioned above, Francis (1990) detailed the method for back-calculation in fishes; an example of back-calculation for mudskippers is shown in this section (Figure 4.3). First, each increment radius and the AR are measured. Then, the relationship between fish length and AR is estimated in order to estimate a linear regression. Finally, the back-calculated fish length at the specific age can be calculated using both the regression and the average increment radius of each increment.

Several studies have shown Lee's phenomenon, that is, the *n*th increment radius decreases with age (Nanami and Takegaki 2005; Jeong et al. 2004). Due to Lee's phenomenon, the average increment radius is underestimated if all samples are used for analysis. In order to minimize the effect of Lee's phenomenon, several studies have used the average radius of the most recent increment at each age for back-calculation (e.g., Nanami and Takegaki 2005; Uehara and Shimizu 1996; Masaki et al. 1986).

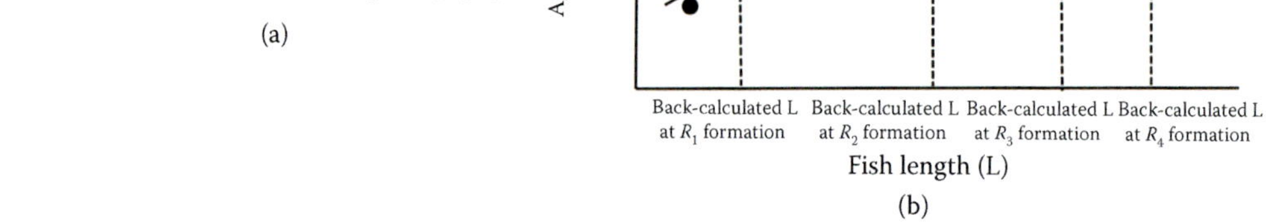

Figure 4.3 Schematic diagrams of the back-calculation method. The second pectoral-radial bone is used as the age character in this example. (a) Second pectoral-radial bone as age character with four increments showing the location of core (*C*), increment radii (R_1, R_2, R_3, and R_4) and the AR; (b) Fish length (*L*) versus AR, and linear regression. The back-calculated fish length at specific increment formation can be estimated from the regression, plotting average R_1, R_2, R_3, and R_4 radii on the *y*-axis. (Courtesy of Atsushi Nanami.)

4.4 LENGTH–FREQUENCY ANALYSIS

Age and growth can be estimated using size distributional data and the age-specific mode of size class. This method assumes that individuals found in the same size mode are considered to be in the same age class. Length–frequency analysis can be applied to species with a relatively short spawning period, short life spans, and fast growth (Manooch 1987).

Several studies have examined the age and growth of mudskippers by using length–frequency data, for example, for *Ps. barbarus* in Nigeria (Etim et al. 2002) and for *Pn. schlosseri* in Malaysia (Mazlan and Rohaya 2008). Parameters for age and growth were estimated by using FISAT software (Gayanilo and Pauly 2004).

4.5 GROWTH CURVE FUNCTION

In order to clarify the relationship between age and length for mudskippers, several growth curve functions have been applied to mudskippers.

4.5.1 von Bertalanffy Growth Function

$$L_t = L_\infty\{1 - \exp[-K(t + t_0)]\}$$

where L_t is the mean length at age t, L_∞ is the asymptotic mean length, K is the growth coefficient, and t_0 is the age at which the mean length is zero (von Bertalanffy 1938).

Von Bertalanffy growth function (VBGF) parameters can be estimated by applying the least-squares method, using the Newton algorithm. This calculation can be conducted using statistical software. L_∞, K, and t_0 are also estimated by the Walford method (Walford 1946):

$$L_{t+1} = L_\infty[1 - \exp(-K)] + \exp(-K)L_t$$

$$\exp[-K(t+t_0)] = 1 - L_t / L_\infty$$

L_t and L_{t+1} can be obtained from back-calculation for age class t and t+1 (see Section 4.3.5) and L_∞ [1 – exp (–K)] and exp (–K) are the regression intercept and regression slope, respectively; t_0 can be calculated from L_∞ and K.

4.5.2 Modified VBGF

The modification to the VBGF function is applied under the consideration of the effect of seasonal change on growth (Pauly et al. 1992):

$$L_t = L_\infty\{1 - \exp[-K(t - t_0) + (CK/2\pi)\times\sin 2\pi(t - t_s) - (CK/2\pi)\times\sin 2\pi(t_0 - t_s)]\}$$

where L_t is the mean length at age t, L_∞ is the asymptotic mean length, C is the amplitude of growth oscillations, K is the growth coefficient, t_0 is the age at which the mean length is zero, and t_s is the time between birth and onset of the first growth oscillations (Etim et al. 2002, 1996). The above-mentioned VBGF (see Section 4.5.1) is obtained if $C = 0$, that is, when there is no effect of seasonal change on growth.

4.5.3 Schnute Growth Function

This is an alternative growth function proposed by Schnute (1981). The growth function can be applied for non-asymptotic growth models such as linear, quadratic, exponential, and asymptotic models (Schnute 1981):

$$L_t = L_1^{\,b} + (L_2^{\,b} - L_1^{\,b}) \times [1 - \mathrm{e}^{-a(t-t_1)} / 1 - \mathrm{e}^{-a(t-a(t_2-t_1))}]^{1/b}$$

where $L_2 > L_1 > 0$ and $t_2 > t_1$. L_t is the fish length at age t, t_1 and t_2 are the ages of the youngest and the oldest fish in the sample, L_1 and L_2 are the mean fish length at age t_1 and age t_2, and a and b are relative growth rates (Kruitwagen et al. 2006; Schnute 1981). The method describing how to obtain the parameters appeared in Kruitwagen et al. (2006).

4.6 AGE AND GROWTH IN OXUDERCINE GOBIES

4.6.1 Age and Growth Estimated by VBGF

Boleophthalmus pectinirostris

Age and growth of *B. pectinirostris* were studied in China, Korea, and Japan. Zhang et al. (1990) estimated age and growth using the second pectoral-radial bone in specimens collected from southeastern China in March–May 1989 ($n = 449$). Annual increment formation was found in these specimens. Growth parameters were estimated as $L_\infty = 183.4$ mm SL, $K = 0.26$, and $t_0 = 1.25$ years for both sexes combined (SL). The maximum ages were estimated as 4 years (Figure 4.4).

Yamazaki (2001) also studied the age and growth of *B. pectinirostris* using pectoral-radial bones. Yamazaki (2001) sampled at two study sites in western Japan, in September and October 2000 (protected area: Rokkaku River [$n = 56$] and unprotected area: Tojin River [$n = 91$]; both sexes

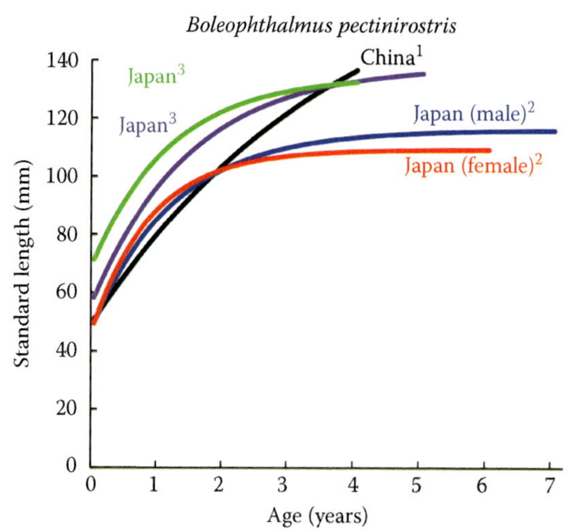

Figure 4.4 Age and growth estimated by the von Bertalanffy Growth Function (VBGF) for *Boleophthalmus pectinirostris*. (1) Zhang et al. (1990); (2) Nanami and Takegaki (2005); and (3) Yamazaki (2001). Purple and green plots for Yamazaki (2001) represent the data obtained at Rokkaku River (protected area) and Tojin River (unprotected area). (Courtesy of Atsushi Nanami.)

combined). This study found a biannual increment periodicity, as in Washio (1992). The first, second, fourth, and sixth increments were considered annual increments. Growth parameters were L_∞ = 138.4 mm SL, K = 0.67, and t_0 = 0.82 years for the protected area and L_∞ = 134.7 mm SL, K = 0.84, and t_0 = 0.90 years for the unprotected area. The maximum age was 5 and 4 years for the protected and unprotected areas, respectively (Figure 4.4).

Similarly, Nanami and Takegaki (2005) studied age and growth using pectoral-radial bones from *B. pectinirostris* collected in western Japan from April 2001 to March 2002 (n = 910). Growth parameters from back-calculated data were L_∞ = 116.3 mm SL, K = 0.80, and t_0 = 0.70 years for males and L_∞ = 109.5 mm SL, K = 1.10, and t_0 = 0.55 years for females. The maximum ages were 7 years for males and 6 years for females (Figure 4.4), which differ significantly from the results cited above. Although the reason for the significant difference in growth characteristics among the studies is still unknown, size-selective fishing pressure, climate change, and density-dependent effects might account for the observed differences among the locations (Nanami and Takegaki, 2005).

Based on *B. pectinirostris* collected in southwestern Korea in April–October 2001 (n = 193), Jeong et al. (2004) estimated age and growth using pectoral-radial bones. Increment periodicity was examined by the MIR method (Section 4.3.4). Growth parameters were estimated as L_∞ = 155.3 mm TL, K = 1.39, and t_0 = 0.35 years for males and L_∞ = 165.2 mm TL, K = 1.07, and t_0 = 0.23 years for females (TL). The maximum age was 2 years for males and 3 years for females (Figure 4.5a).

Washio et al. (1991) found a significant negative relationship between the first-year growth rate (distance between the first increment and core of the radial bone) and the second-year growth rate (distance between the first and the second increments), indicating that the fish having less growth in the first year grew faster in the second year (Washio et al. 1991).

Boleophthalmus boddarti

Ravi and Rajagopal (2007) studied the age and growth of *B. boddarti* using length–frequency data from fish collected in southern India. The growth parameters were estimated using the Walford method as L_∞ = 213.6 mm TL, K = 0.48, and t_0 = 0.82 for both sexes combined. The maximum age was estimated as 2 years (Figure 4.5a).

Periophthalmodon schlosseri

Mazlan and Rohaya (2008) estimated the age and growth for *Pn. schlosseri* collected in March 2003 and February 2004 in Malaysia, using length–frequency analysis (n = 467). Growth parameters were estimated as L_∞ = 29 cm TL, K = 1.4 (t_0 was not indicated) for both sexes combined (Figure 4.5b). The maximum age was not determined.

Periophthalmus argentilineatus

Kruitwagen et al. (2006) studied the age and growth of *Ps. argentilineatus* in Tanzania collected in November 2001 and June 2003 using otoliths (n = 96). Increment periodicity was examined by counting the daily microincrements, and the annual increment formation was found. Growth parameters were estimated for three sampling sites as L_∞ = 10.1–13.6 cm TL, K = 0.39–1.83, and t_0 = −1.26–0.87, for both sexes combined (Figure 4.5c). The maximum age was 6 years.

Periophthalmus waltoni

Sarafraz et al. (2011) studied the age and growth of *Ps. waltoni* collected in Iran in February 2007 (n = 192), using pectoral-radial bones. Increment periodicity was not determined and it was assumed that the number of increments equaled the number of years (age). Growth parameters were estimated as L_∞ = 208 mm TL, K = 0.19, and t_0 = 1.64 for males and L_∞ = 147 mm TL, K = 0.41, and t_0 = 0.89 for females. The maximum age was 4 years for both males and females (Figure 4.5c).

Scartelaos gigas

Park et al. (2008) studied the age and growth of *S. gigas* collected in Korea in March and September 2003 using second pectoral-radial bones (n = 560). Increment periodicity was examined by the MIR method and a biannual increment formation was found. Growth parameters were estimated as L_∞ = 179.4 mm TL, K = 0.78, and t_0 = 0.78 for combined sexes. The maximum age was 4 years for both sexes combined (Figure 4.5d).

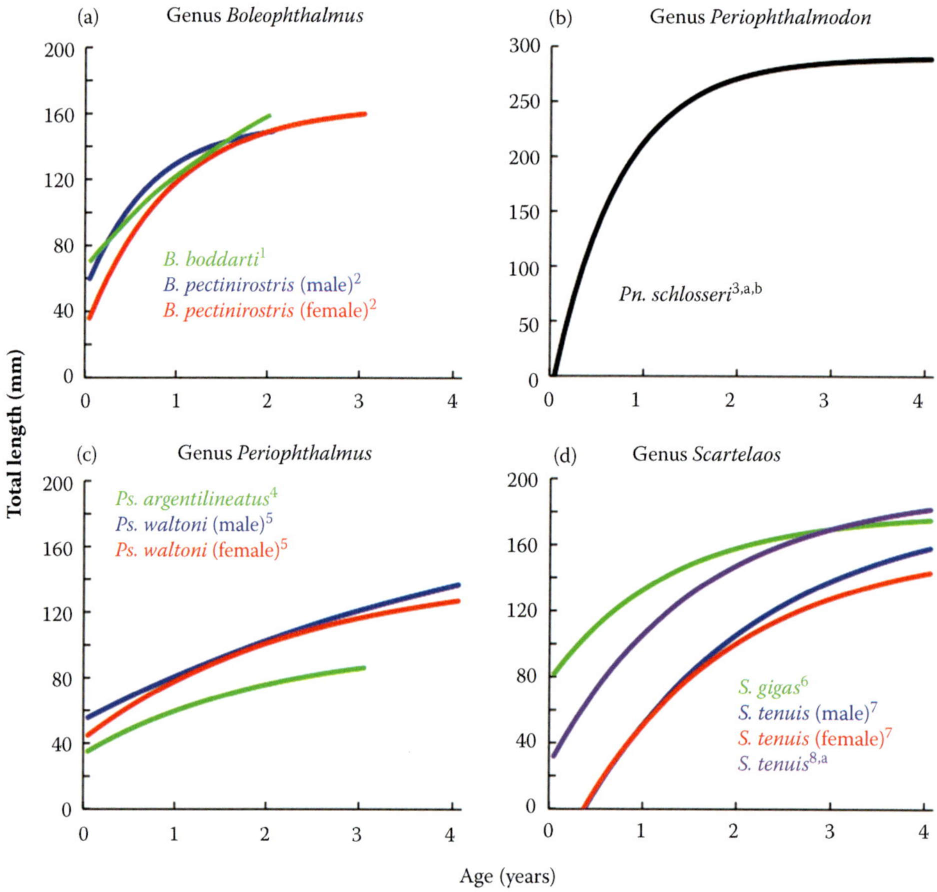

Figure 4.5 Age and growth estimated by the VBGF for seven species in four genera (a) *Boleophthalmus*; (b) *Periophthalmodon*; (c) *Periophthalmus*; and (d) *Scartelaos*. Numbered superscript corresponds to the following references: (1) Ravi and Rajagopal (2007); (2) Jeong et al. (2004); (3) Mazlan and Rohaya (2008); (4) Kruitwagen et al. (2006); (5) Sarafaz et al. (2011); (6) Park et al. (2008); (7) Abdoli et al. (2010); and (8) Salarpouri et al. (2013). [a]Since maximum ages were not indicated (Mazlan and Rohaya 2008; Salarpouri et al. 2013), the regressions are plotted up to 4 years; [b]Since t_0 was not indicated (Mazlan and Rohaya 2008), t_0 was regarded as 0 in this figure. (Courtesy of Atsushi Nanami.)

Scartelaos tenuis

Abdoli et al. (2010) studied the age and growth of *S. tenuis* in Iran in February 2006 and April 2008 using second pectoral-radial bones (n = 169). Also in this case, increment periodicity was not determined and it was assumed that the number of increments equaled number of years (age). Growth parameters were estimated as L_∞ = 185.7 mm TL, K = 0.52, and t_0 = −0.33 for male and L_∞ = 161.1 mm TL, K = 0.60, and t_0 = −0.32 for female. The maximum age was 4 years for both males and females (Figure 4.5d).

Salarpouri et al. (2013) also studied age and growth in *S. tenuis* in Iran; they used length–frequency analysis on fish collected in October 2008 and September 2009 (n = 2047). Growth parameters were estimated as L_∞ = 19.43 cm TL, K = 0.64, and t_0 = 0.28 for both sexes combined (Figure 4.5d). The maximum age was 4 years.

4.6.2 Age and Growth Estimated by Modified VBGF

Periophthalmus barbarus

Etim et al. (1996) estimated the age and growth of *Ps. barbarus* (reported as *Ps. papilio*) collected from the Cross River, Nigeria in February and June 1994 using length–frequency analysis (n is not indicated). Age and growth were estimated as L_∞ = 19.39 cm TL, K = 0.51, C = 0.3, and t_s = −0.1 (t_0 was not indicated).

Etim et al. (2002) also estimated age and growth for this species on fishes collected from the Imo River, Nigeria in April 1993 and March 1994 using length–frequency analysis (n = 1013). Age and growth were estimated as L_∞ = 21.6 cm TL, K = 0.55, C = 0.9, and t_s = −0.3 (t_0 was not indicated).

4.6.3 Age and Growth Estimated by Schnute Growth Function

Periophthalmus argentilineatus

Although Kruitwagen et al. (2006) studied age and growth for this species using VBGF for 96 individuals from three sampling sites (see Section 4.6.1 *Periophthalmus argentilineatus*), they also estimated the age and growth of this species using the Schnute growth function for 86 individuals collected at two other sampling sites. Growth parameters were estimated for two sampling sites as a = 0.62–1.08, b = −3.90–2.01, L_1 = 6.55–6.57 cm TL, L_2 = 14.29–15.23 cm TL, t_1 = 1.00, and t_2 = 5.00 for both sexes combined.

4.6.4 Length–Weight Relationship Studies

Boleophthalmus boddarti

Length–weight relationship was obtained in Malaysia for this species [BW: body weight (g), SL: (mm)] (Khaironizam and Norma-Rashid 2002): combined sex BW = $1.56 \times 10^{-5} \times SL^{3.00}$ (Figure 4.6a).

Boleophthalmus dussumieri

Length–weight relationships for this species were obtained in Iran by Abdoli et al. (2009): [BW (g), TL (mm)]: male BW = $6.7 \times 10^{-4} \times TL^{2.10}$, female BW = $5.2 \times 10^{-4} \times TL^{2.15}$, and combined sex BW = $2.0 \times 10^{-4} \times TL^{2.34}$ (Figure 4.7a) and by Ghanbarifardi et al. (2014): [BW (g), TL (cm)] combined sex BW = $3.73 \times 10^{-2} \times TL^{2.33}$ (Figure 4.7a).

Length–weight relationships were also obtained in the northern Arabian Sea [BW (g), SL (mm)] by Hoda (1987): Log_{10} (male BW) = −4.03 + 2.53 Log_{10} (SL) and Log_{10} (female BW) = −3.14 + 2.07 Log_{10} (SL).

The following data were also reported by Hoda (1987) on *B. dentatus* [BW (g), SL (mm)]: Log_{10} (male BW) = −3.86 + 2.45 Log_{10} (SL); Log_{10} (female BW) = −4.21 + 2.63 Log_{10} (SL). Murdy (1989) synonymized these two species as *B. dussumieri*. Mirza et al. (1996) again separated *B. dussumieri* from *B. dentatus*, whose taxonomic status is therefore still debated.

Boleophthalmus pectinirostris

Length–weight relationships were obtained in China as follows [BW (g), SL (mm)] (Zhang et al. 1990): combined sex BW = $1.47 \times 10^{-5} \times SL^{3.01}$ (Figure 4.6a). These relationships were also obtained in Korea [BW (g), TL (mm)] (Jeong et al. 2004): male BW = $4.76 \times 10^{-4} \times TL^{2.18}$ (Figure 4.7a); female BW = $5 \times 10^{-6} \times TL^{3.12}$ (Figure 4.7a).

Periophthalmodon schlosseri

The length–weight relationship was obtained in Malaysia [BW (g), SL (mm)] (Khaironizam and Norma-Rashid 2002): combined sex BW = $1.77 \times 10^{-5} \times SL^{3.06}$ (Figure 4.6b).

Periophthalmodon septemradiatus

The length–weight relationship was obtained in Malaysia [BW (g), SL (mm)] (Khaironizam and Norman-Rashid 2002): combined sex BW = $2.71 \times 10^{-5} \times SL^{2.90}$ (Figure. 4.6b).

Periophthalmus argentilineatus

The length–weight relationship was obtained in Malaysia [BW (g), SL (mm)] (Khaironizam and Norman-Rashid 2002): combined sex BW = $4.38 \times 10^{-6} \times SL^{3.34}$ (Figure 4.6c). The relationship was also obtained in Tanzania [BW (g), TL (cm)] (Kruitwagen et al. 2006): combined sex BW = $1.71 \times 10^{-2} \times TL^{2.83}$ (Figure 4.7b).

Periophthalmus barbarus

Length–weight relationships were obtained in Nigeria [BW (g), TL (cm)] (reported as *Ps. papilio*, King and Udo 1998): male BW = $9.3 \times 10^{-3} \times TL^{3.08}$ (Figure 4.7b); female BW =$1.11 \times 10^{-2} \times TL^{3.01}$ (Figure 4.7b). The relationships were also obtained in Nigeria [BW (g), TL (mm)] (Lawson 2011): Log_{10} (male BW) = −4.68 + 2.86 Log_{10} (TL); Log_{10} (female BW) = −4.79 + 2.92 Log_{10} (TL).

Periophthalmus chrysospilos

The length–weight relationship was obtained in Malaysia [BW (g), SL (mm)] (Khaironizam and Norman-Rashid 2002): combined sex BW = $6.55 \times 10^{-5} \times SL^{2.56}$ (Figure 4.6c).

Periophthalmus gracilis

The length–weight relationship was obtained in Malaysia [BW (g), SL (mm)] (Khaironizam and Norman-Rashid 2002): combined sex BW = $3.23 \times 10^{-5} \times SL^{2.80}$ (Figure 4.6c).

Periophthalmus magnuspinnatus

The length–weight relationship was obtained in Korea [BW (g), TL (cm)] (Baeck and Park 2015): male BW = $8.5 \times 10^{-3} \times TL^{3.01}$ (Figure 4.7b), female BW = $8.1 \times 10^{-3} \times TL^{3.06}$ (Figure 4.7b).

Periophthalmus modestus

The length–weight relationship was obtained in Korea [BW (g), TL (cm)] (Baeck and Park 2015): male BW = $6.4 \times 10^{-3} \times TL^{3.03}$ (Figure 4.7b); female BW = $8.8 \times 10^{-3} \times TL^{3.06}$ (Figure 4.7b).

Periophthalmus spilotus

The length–weight relationship was obtained in Malaysia [BW (g), SL (mm)] (Khaironizam and Norman-Rashid 2002): combined sex BW = $2.66 \times 10^{-6} \times SL^{3.50}$ (Figure 4.6c).

Periophthalmus variabilis

The length–weight relationship was obtained in Malaysia [BW (g), SL (mm)] (Khaironizam and Norman-Rashid 2002): combined sex BW = $5.12 \times 10^{-5} \times SL^{2.67}$ (Figure 4.6c).

Periophthalmus waltoni

Length–weight relationships were obtained at three sampling locations in Iran [BW (g), TL (mm)] (Sarafraz et al. 2011): combined sex BW = $2.28 \times 10^{-5} \times TL^{2.77}$ (one example is shown in Figure 4.7b). These relationships were also obtained by Ghanbarifardi et al. (2014) at seven sampling locations in Iran [BW (g), TL (cm)]: in one of these locations, the combined sex BW = $1.35 \times 10^{-2} \times TL^{2.84}$.

Pseudapocryptes elongatus

The length–weight relationship was obtained in Malaysia [BW (g), SL (mm)] (Khaironizam and Norman-Rashid 2002): combined sex BW = $2.54 \times 10^{-5} \times SL^{2.81}$ (Figure 4.6d). It was also obtained in Bangladesh [BW (g), TL (cm)] (Siddique et al. 2015): combined sex BW = $1.0 \times 10^{-2} \times TL^{2.66}$ (Figure 4.7c).

Scartelaos gigas

The length–weight relationship was obtained in Korea [BW (g), TL (mm)] (Park et al., 2008): combined sexes BW = $7 \times 10^{-6} \times TL^{2.98}$ (Figure 4.7d).

Scartelaos histophorus

The length–weight relationship was obtained in Malaysia [BW (g), SL (mm)] (Khaironizam and Norman-Rashid 2002): combined sex BW= $4.33 \times 10^{-5} \times SL^{2.62}$ (Figure 4.6e).

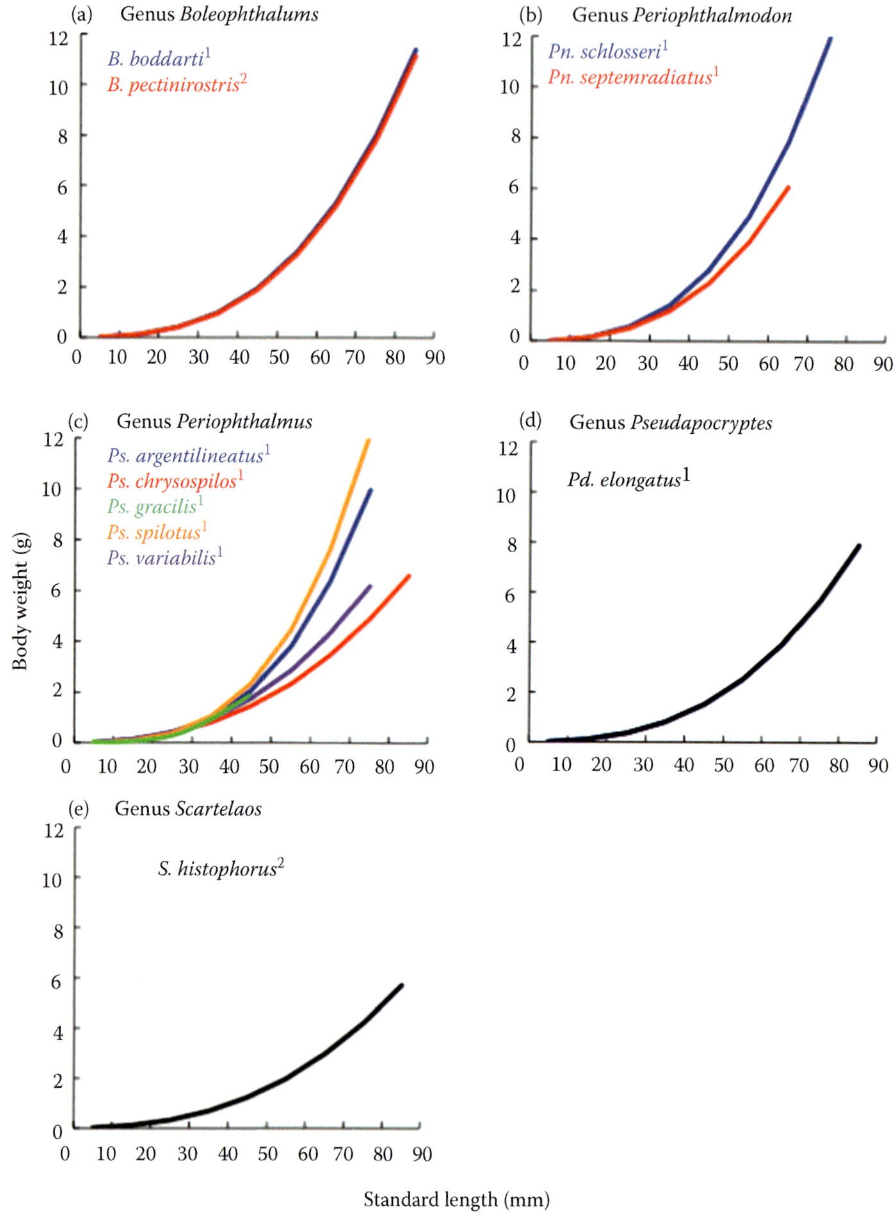

Figure 4.6 Standard length versus body weight for eleven oxudercine goby species in five genera (a) *Boleophthalmus*; (b) *Periophthalmodon*; (c) *Periophthalmus*; (d) *Pseudapocryptes*; and (e) *Scartelaos*. Numbered superscript corresponds to the following references: (1) Khaironizam and Norman-Rashid (2002); and (2) Zhang et al. (1990). (Courtesy of Atsushi Nanami.)

Scartelaos tenuis

Length–weight relationships were obtained in Iran by three studies [BW (g), TL (mm) for Abdoli et al. (2009) and TL (cm) for Ghanbarifardi et al. (2014) and Salarpouri et al. (2013)]: combined sexes BW = $6 \times 10^{-5} \times TL^{2.50}$ (Abdoli et al. 2009) (Figure 4.7d); combined sexes

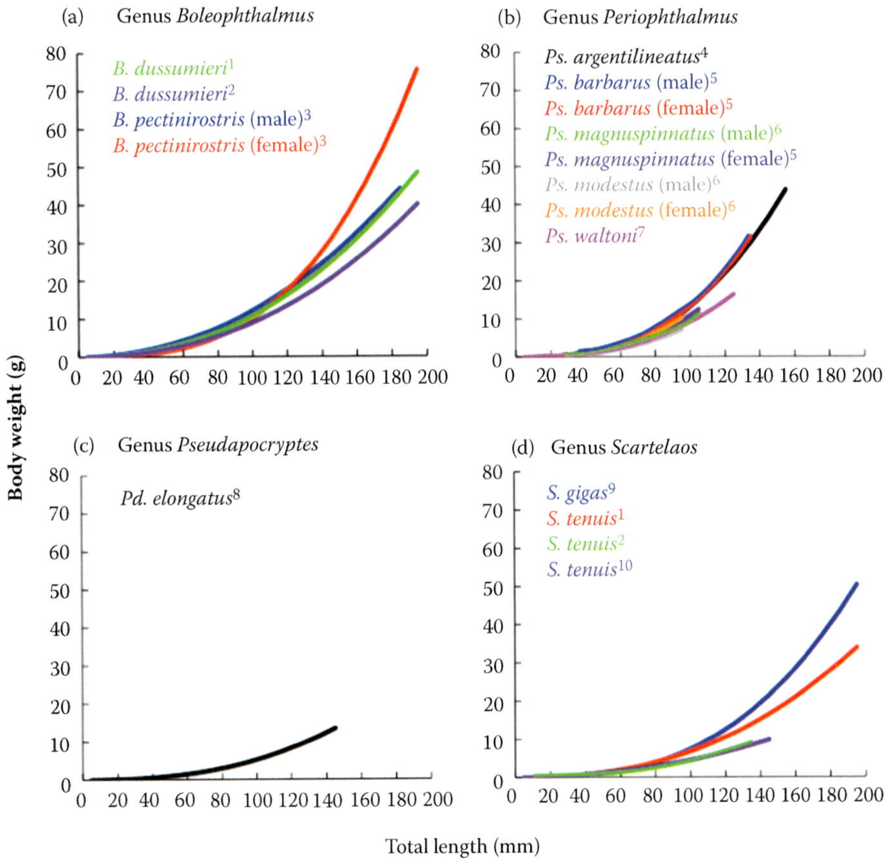

Figure 4.7 Total length versus body weight relationships for ten oxudercine goby species in four genera (a) *Boleophthalmus*; (b) *Periophthalmus*; (c) *Pseudapocryptes*; and (d) *Scartelaos*. Numbered superscript corresponds to the following references: (1) Abdoli et al. (2009); (2) Ghanbarifardi et al. (2014); (3) Jeong et al. (2004); (4) Kruitwagen et al. (2006); (5) King and Udo (1998); (6) Baeck and Park (2015); (7) Sarafraz et al. (2011); (8) Siddique et al. (2015); (9) Park et al. (2008); and (10) Salarpouri et al. (2013). (Courtesy of Atsushi Nanami.)

$BW = 1.29 \times 10^{-2} \times TL^{2.47}$ (Ghanbarifardi et al. 2014) (Figure 4.7d); combined sexes $BW = 2.74 \times 10^{-2} \times TL^{2.17}$ (Salarpouri et al. 2013) (Figure 4.7d).

4.6.5 Summary of Growth Characteristics

The growth characteristics of mudskippers estimated by VBGF are summarized in Table 4.1. Length–weight relationships are summarized in Tables 4.2 and 4.3. The maximum age of mudskippers is considered to be less than 7 years, which is almost equal to, or slightly longer than other non-oxudercine gobiid species (e.g., Pompei et al. 2015; Thompson and Simon 2015; Huo et al. 2014; Grul'a et al. 2012; Sokolowska and Fey 2011; Gümüs and Kurt 2009; Kovacic 2006).

Differences between sexes in L_∞ and maximum age have been studied for three oxudercine species (*B. pectinirostris*, *Ps. waltoni*, and *S. tenuis*). Consistent with most other gobioid species, males are found to be larger than females for all three species, except for the *B. pectinirostris* population in Korea (Jeong et al. 2004). Size differences may be attributed to differences in energy allocation between males and females. Females are thought to allocate greater energy for egg production

Table 4.1 Summary of Estimated Growth Parameters and Maximum Age of Mudskippers Obtained by VBGF

Species	Location	Sex	L_∞	Unit of Length	K	t_0	Max. Age	n	Method	Reference
Boleophthalmus pectinirostris	Korea	M	155.3	TL (mm)	1.39	0.35	2	106	Pectoral radial bone	Jeong et al. (2004)
		F	165.2	TL (mm)	1.07	0.23	3	87	Pectoral radial bone	Jeong et al. (2004)
	China	C	183.4	SL (mm)	0.26	1.25	4	449	Pectoral radial bone	Zhang et al. (1990)
	Japan	M	116.3	SL (mm)	0.80	0.70	7	427	Pectoral radial bone	Nanami and Takegaki (2005)
		F	109.5	SL (mm)	1.10	0.55	6	483	Pectoral radial bone	Nanami and Takegaki (2005)
		C	138.4	SL (mm)	0.67	0.82	5	56	Pectoral radial bone	Yamazaki (2001)
		C	134.7	SL (mm)	0.84	0.90	4	91	Pectoral radial bone	Yamazaki (2001)
Boleophthalmus boddarti	India	C	213.6	TL (mm)	0.48	0.82	2	(N.A.)	Length–frequency	Ravi and Rajagopal (2007)
Periophthalmodon schlosseri	Malaysia	C	29	TL (cm)	1.4	(N.A.)	(N.A.)	467	Length–frequency	Mazlan and Rohaya (2008)
Periophthalmus argentilineatus	Tanzania	C	13.6	TL (cm)	0.39	0.76	6	22	Otolith	Kruitwagen et al. (2006)
		C	12.9	TL (cm)	1.83	−1.26	5	14	Otolith	Kruitwagen et al. (2006)
		C	10.1	TL (cm)	0.49	0.87	3	60	Otolith	Kruitwagen et al. (2006)
Periophthalmus waltoni	Iran	M	208	TL (mm)	0.19	1.64	4	(N.A.)[a]	Pectoral radial bone	Sarafraz et al. (2011)
		F	147	TL (mm)	0.41	0.89	4	(N.A.)[a]	Pectoral radial bone	Sarafraz et al. (2011)
Scartelaos gigas	Korea	C	179.4	TL (mm)	0.78	0.78	4	560	Pectoral radial bone	Park et al. (2008)
Scartelaos tenuis	Iran	M	185.7	TL (mm)	0.52	−0.33	4	89	Pectoral radial bone	Abdoli et al. (2010)
		F	161.1	TL (mm)	0.60	−0.32	4	80	Pectoral radial bone	Abdoli et al. (2010)
		C	19.4	TL (cm)	0.64	0.28	4	2047	Length–frequency	Salarpouri et al. (2013)

M, male; F, female ; C, both sexes combined; SL, standard length; TL, total length; N.A., not available.
[a] Total number of samples (male + female) = 192.

during spawning, and thus have a decreased energy supply for growth than males (Hernaman and Munday 2005). Differences in maximum age between gender were reported for populations of *B. pectinirostris* (male:female = 2:3 years in Korea; 7:6 years in Japan) but not *Ps. waltoni* and *S. tenuis* (4 years for both sexes for both species).

There was a significant negative relationship between K and L_∞ (Figure 4.8) for *B. pectinirostris*, indicating that populations with lower growth rates eventually achieved a greater size. This result seems to be consistent with life-history traits of animals in which populations of smaller sized individuals have a more rapid growth rate compared with populations of larger sized individuals.

4.7 FUTURE PERSPECTIVES

Although oxudercine gobies comprise more than 40 species worldwide (Frose and Pauly 2015) and inhabit shallow coastal waters on three continents (Clayton 1993), age and growth for most of the species have not been studied. This might be attributed to the difficulty in capturing them, since mudskippers inhabit muddy environments difficult to access (Etim et al. 1996). Small size and cryptic coloration of mudskippers might be another reason for the difficulty of capturing them.

Two studies did not examine the increment periodicity, and instead, directly assumed that the number of increments equals the fish's age (Sarafraz et al. 2011; Abdoli et al. 2010). Increment periodicity is not only species-specific, but it is also highly variable at intraspecific level, even within the same geographic region (see Section 4.3.4.1). Detailed estimations of increment periodicity are therefore required. Such estimations would improve the accuracy of our understanding of growth parameters in these species.

Table 4.2 Summary of Length–Weight Relationships of Mudskippers, Which Relationships Were Obtained as Exponential Function

Species	Sex	n	Unit of Length	Size Range	BW = aL a	b	Study Location	Reference
Boleophthalmus boddarti	M + F	128	SL (mm)	30.0–128.4	1.56×10^{-5}	3.00	Malaysia	Khaironizam and Norman-Rashid (2002)
B. dussumieri	M	26	TL (mm)	79–283	6.7×10^{-4}	2.10	Iran	Abdoli et al. (2009)
	F	33	TL (mm)	86–289	5.2×10^{-4}	2.15	Iran	Abdoli et al. (2009)
	M + F	59	TL (mm)	79–289	2.0×10^{-4}	2.34	Iran	Abdoli et al. (2009)
	M + F	86	TL (cm)	12.3–22.5	3.73×10^{-2}	2.33	Iran	Ghanbarifardi et al. (2014)
B. pectinirostris	M + F	449	SL (mm)	52–140	1.47×10^{-5}	3.01	China	Zhang et al. (1990)
	M	106	TL (mm)	104.4–185.3	4.76×10^{-4}	2.18	Korea	Jeong et al. (2004)
	F	87	TL (mm)	108.3–195.7	5×10^{-6}	3.12	Korea	Jeong et al. (2004)
Periophthalmodon schlosseri	M + F	20	SL (mm)	79.5–207.0	1.77×10^{-5}	3.06	Malaysia	Khaironizam and Norman-Rashid (2002)
Pn. septemradiatus	M + F	22	SL (mm)	47.1–68.5	2.71×10^{-5}	2.90	Malaysia	Khaironizam and Norman-Rashid (2002)
Periophthalmus argentilineatus	M + F	15	SL (mm)	43.0–71.6	4.38×10^{-6}	3.34	Malaysia	Khaironizam and Norman-Rashid (2002)
	M + F	1052	TL (cm)	(N.A.)	1.71×10^{-2}	2.83	Tanzania	Kruitwagen et al. (2006)
Ps. barbarus	M	455	TL (cm)	3.3–14.3	9.3×10^{-3}	3.08	Nigeria	King and Udo (1998)
	F	602	TL (cm)	1.3–13.6	1.11×10^{-2}	3.01	Nigeria	King and Udo (1998)
Ps. chrysospilos	M + F	315	SL (mm)	18.2–83.2	6.55×10^{-5}	2.56	Malaysia	Khaironizam and Norman-Rashid (2002)
Ps. gracilis	M + F	110	SL (mm)	20.5–42.7	3.23×10^{-5}	2.80	Malaysia	Khaironizam and Norman-Rashid (2002)
Ps. magnuspinnatus	M	121	TL (cm)	4.9–10.3	8.5×10^{-3}	3.01	Korea	Baeck and Park (2015)
	F	99	TL (cm)	5.0–10.6	8.1×10^{-3}	3.06	Korea	Baeck and Park (2015)
Ps. modestus	M	170	TL (cm)	3.9–9.4	6.4×10^{-3}	3.03	Korea	Baeck and Park (2015)
	F	205	TL (cm)	3.1–9.9	8.8×10^{-3}	3.06	Korea	Baeck and Park (2015)
Ps. spilotus	M + F	29	SL (mm)	40.3–71.4	2.66×10^{-6}	3.50	Malaysia	Khaironizam and Norman-Rashid (2002)
Ps. variabilis	M + F	299	SL (mm)	18.3–70.2	5.12×10^{-5}	2.67	Malaysia	Khaironizam and Norman-Rashid (2002)
Ps. waltoni	M	229	TL (mm)	55–176	2×10^{-5}	2.86	Iran	Abdoli et al. (2009)
	F	250	TL (mm)	55–173	2×10^{-5}	2.86	Iran	Abdoli et al. (2009)
	M + F	479	TL (mm)	55–176	2×10^{-5}	2.86	Iran	Abdoli et al. (2009)
	M + F	77	TL (mm)	(N.A.)	2.28×10^{-5}	2.77	Iran	Sarafraz et al. (2011)
	M + F	47	TL (cm)	7.8–12.2	1.35×10^{-2}	2.84	Iran	Ghanbarifardi et al. (2014)
Pseudapocryptes elongatus	M + F	84	SL (mm)	20.1–133.6	2.54×10^{-5}	2.81	Malaysia	Khaironizam and Norman-Rashid (2002)
	M + F	161	TL (cm)	10.1–19.7	1.0×10^{-2}	2.66	Bangladesh	Siddique et al. (2015)
Scartelaos gigas	M + F	560	TL (mm)	111.6–201.2	7×10^{-6}	2.98	Korea	Park et al. (2008)
S. histophorus	M + F	97	SL (mm)	32.8–81.2	4.33×10^{-5}	2.62	Malaysia	Khaironizam and Norman-Rashid (2002)
S. tenuis	M	46	TL (mm)	86–186	4×10^{-5}	2.56	Iran	Abdoli et al. (2009)
	F	52	TL (mm)	73–191	4×10^{-5}	2.56	Iran	Abdoli et al. (2009)
	M + F	98	TL (mm)	73–191	6×10^{-5}	2.50	Iran	Abdoli et al. (2009)
	M + F	45	TL (cm)	6.7–13.4	1.29×10^{-2}	2.47	Iran	Ghanbarifardi et al. (2014)
	M + F	630	TL (cm)	(N.A.)	2.74×10^{-2}	2.17	Iran	Salarpouri et al. (2013)

M, male; F, female; C, both sex combined; SL, standard length; TL, total length; N.A., not available.

As some species of mudskippers have a wide geographic range (Frose and Pauly 2015), detailed geographic differences in growth parameters would be useful for understanding their life histories. Temporal differences in growth parameters in relation to changes in environmental characteristics (e.g., changes in climate, habitat quality, and population density) are also an interesting subject to explore and clarify. Kruitwagen et al. (2006) reported that the maximum age of *Ps. argentilineatus* was lower in a polluted area (about 2 years) than that in several unpolluted areas (3–6 years). Nanami and Takegaki (2005) found that the increment periodicity, maximum age, and growth rate of *B. pectinirostris* differed between 2001–2002 and 1990–1991 (Washio 1992). Changes in increment periodicity and maximum age might be due to climate change, specifically, the rise of annual average air temperature (Nanami and Takegaki 2005). Nanami and Takegaki (2005) also suggested

Table 4.3 Summary of Length–Weight Relationships in Oxudercine Gobies Obtained as Logarithmic Function

Species	Sex	*n*	Unit of Length	Size Range (mm)	BW = −*a* + *b* log₁₀ (*L*) *a*	*b*	Study Location	Reference
Boleophthalmus dussumieri	M	372	SL (mm)	58–99	4.03	2.53	Pakistan	Hoda (1987)
	F	517	SL (mm)	51–128	3.14	2.07	Pakistan	Hoda (1987)
(*Boleophthalmus dentatus*)[a]	M	324	SL (mm)	52–120	3.86	2.45	Pakistan	Hoda (1987)
	F	332	SL (mm)	51–128	4.21	2.63	Pakistan	Hoda (1987)
Periophthalmus barbarus	M	532	TL (mm)	69–180	4.68	2.86	Nigeria	Lawson (2011)
	F	746	TL (mm)	60–190	4.79	2.92	Nigeria	Lawson (2011)
					BW = −*a* + *b* log$_e$ (*L*) *a*	*b*		
Scartelaos tenius	M	89	TL (mm)	~156.5[b]	9.79	2.50	Iran	Abdoli et al. (2010)
	F	80	TL (mm)	~159.1[b]	9.58	2.47	Iran	Abdoli et al. (2010)

M, male; F, female; C, both sexes combined; SL, standard length; TL, total length.

[a] *B. dussumieri* was reported as *B. dentatus*.

[b] Only maximum TL was indicated.

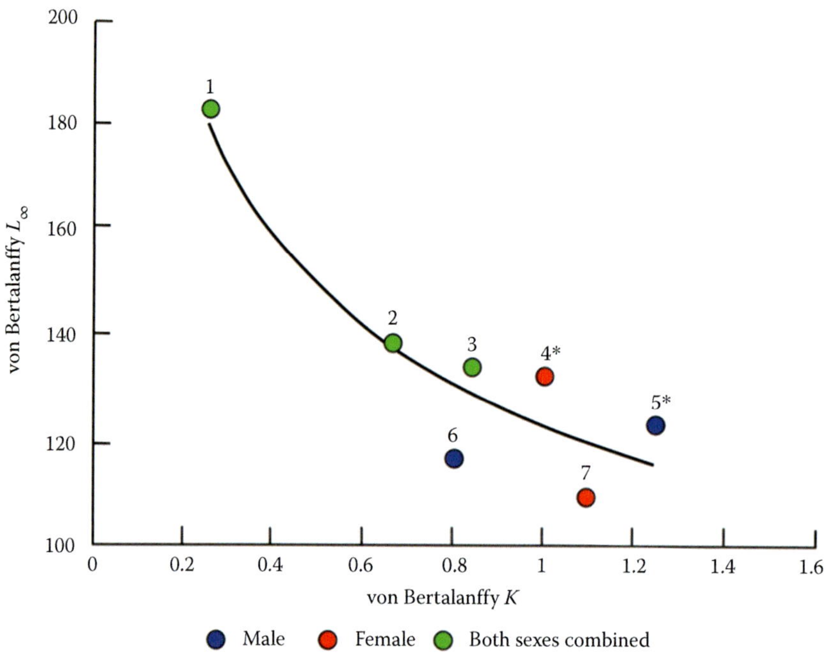

Figure 4.8 Negative correlation between K and L_∞ parameters for the VBGF in *Boleophthalmus pectinirostris* $L_\infty = 122.77 \times K^{-0.286}$ ($R^2 = 0.799$, $p < .01$). (1) Zhang et al. (1990); (2, 3) Yamazaki (2001); (4, 5) Jeong et al. (2004); (6, 7) Nanami and Takegaki (2005). (*) As Jeong et al. (2004) showed only total length (TL), standard length (SL) was estimated following Nanami and Takegaki (2005). (Courtesy of Atsushi Nanami.)

that increasing population density has led to a decrease in the growth rate and subsequent increase in maximum age. This might be due to the decrease of food availability with increasing density, as this species has feeding territories. These results suggest that monitoring the characteristics of age and growth would provide useful ecological information in coastal areas where mudskippers occur. The age and growth of mudskippers are some of the key bioindicators of environmental changes of coastal habitats, including estuaries, mudflats, and mangrove forests.

ACKNOWLEDGMENTS

I would like to express my sincere gratitude to Dr. T. Takegaki for inviting me to participate in the study of mudskippers and continuous support for writing this chapter. Constructive comments by Dr. Z. Jaafar, Dr. E. Murdy, Dr. G. Polgar, and Dr. C.P. Norman are much appreciated.

REFERENCES

Abdoli L., Kamrani E., Abdoli A. and Kiabi B. 2009. Length–Weight relationships for three species of mudskippers (Gobiidae: Oxudercinae) in the coastal area of the Persian Gulf, Iran. *Journal of Applied Ichthyology* 25:236–237.

Abdoli L., Kamrani E., Abdoli A. and Kiabi B. 2010. Age and growth of the mudskipper, *Scartelaos tenuis* (Day, 1876) (Gobiidae: Oxudercinae) in the coastal areas of the Persian Gulf at Bushehr province, Iran. *Zoology in the Middle East* 51:113–115.

Akinrotimi O.A., Edun O.M., Uka A. and Owhonda K.N. 2013. Public perception of mudskipper consumption in some fishing communities of Rivers State, Nigeria. *Journal of Fisheries and Aquatic Science* 8:208–212.

Baeck G.W. and Park J.M. 2015. Length–Weight and length-length relationships and seasonal condition factors for two mudskippers, *Periophtalmus modestus* (Cantor, 1842) and *P. magnuspinnatus* (Lee, Choi & Ryu, 1995) (Gobiidae), on tidal flats of Korea. *Journal of Applied Ichthyology* 31:261–264.

Choat J.H. and Robertson D.R. 2002. Age-based studies. In *Coral Reef Fishes; Dynamics and Diversity in a Complex Ecosystem* (ed. P.F. Sale) pp. 57–80. San Diego, CA: Academic Press, 549 pp.

Clayton D.A. 1993. Mudskippers. *Oceanography and Marine Biology Annual Review* 31:507–577.

Etim L., Brey T. and Arntz W. 1996. A seminal study of the dynamics of a mudskipper (*Periophthalmus papilio*) population in the Cross River, Nigeria. *Netherlands Journal of Aquatic Ecology* 30:41–48.

Etim L., King R.P. and Udo M.T. 2002. Breeding, growth, mortality and yield of the mudskipper *Periophthalmus barbarus* (Linneaus 1766) (Teleostei: Gobiidae) in the Imo River estuary, Nigeria. *Fisheries Research* 56:227–238.

Francis R.I.C.C. 1990. Back-calculation of fish length: A critical review. *Journal of Fish Biology* 36:883–902.

Frose R. and Pauly D. 2015. FishBase. http://www.fishbase.org (accessed April 2015).

Gayanilo F.C. and Pauly D. 2004. *FAO-ICLARM Stock Assessment Tools: FAO Computerised Information Series (Fisheries)* No. 8. Rome: FAO, 262 pp.

Ghanbarifardi M., Ghasemian S., Alibadian M. and Pehpuri A. 2014. Length–Weight relationships for three species of mudskippers (Gobiiformes: Gobionellidae) in the coastal areas of the Persian Gulf of Oman, Iran. *Iranian Journal of Ichthyology* 1:29–31.

Grul'a D., Balážová M., Copp G.H. and Kováč V. 2012. Age and growth of invasive round goby *Neogobius melanostomus* from middle Danube. *Central European Journal of Biology* 7:448–459.

Gümüs A. and Kurt A. 2009. Age structure and growth by otolith interpretation of *Neogobius melanostomus* (Gobiidae) from southern Black Sea. *Cybium* 33:29–37.

Hernaman V. and Munday P.L. 2005. Life-history characteristics of coral reef gobies. I. Growth and life-span. *Marine Ecology Progress Series* 290:207–221.

Hoda S.S.M. 1987. Relative growth of body parts and length–Weight relationships in *Boleophthalmus dussumieri* and *B. dentatus* of Karachi coast. *Indian Journal of Fisheries* 34:120–127.

Huo B., Madenjian C.P., Xie C.X., Zhao Y.M., O'Brien T.P. and Czesny S.J. 2014. Age and growth of round gobies in Lake Michigan, with preliminary mortality estimation. *Journal of Great Lakes Research* 40:712–720.

Jeong S.J., Han K.H., Kim J.K. and Sim D.S. 2004. Age and growth of the blue spot mudskipper (*Boleophthalmus pectinirostris*) in the mudflat of southwestern Korea. *Journal of the Korean Fisheries Society* 37:44–50 (in Korean, with English abstract).

Khaironizam M.Z. and Norma-Rashid Y. 2002. Length–weight relationship of mudskippers (Gobiidae: Oxudercinae) in the coastal areas of Selangor, Malaysia. *Naga, WorldFish Center Quarterly* 25:20–22.

King R.P. and Udo M.T. 1998. Dynamics in the length–weight parameters of the mudskipper *Periophthalmus barbarus* (Gobiidae) in Imo River estuary, Nigeria. *Helgoländer Meeresunters* 52:179–186.

Kovacic M. 2006. Age structure, growth and mortality of the striped goby, *Gobius vittatus* (Gobiidae) in the northern Adriatic Sea. *Scientia Marina* 70:635–641.

Kruitwagen G., Hecht T., Pratap H.B. and Wendelaar Bonga S.E. 2006. Changes in morphology and growth of the mudskipper (*Periophthalmus argentilineatus*) associated with coastal pollution. *Marine Biology* 149:201–211.

Lawson E.O. 2011. Length–weight relationships and fecundity estimates in mudskipper, *Periophthalmus papilio* (Bloch and Schneider 1801) caught from the mangrove swamps of Lagos Lagoon, Nigeria. *Journal of Fisheries and Aquatic Science* 6:264–271.

Manooch C.S. 1987. Age and growth of snappers and groupers. In *Tropical Snapper and Groupers: Biology and Fisheries Management* (eds. J.J. Polovina and S. Ralston) pp. 329–373. Boulder, CO: Westview.

Masaki Y., Ito H., Tokai T. and Yamaguchi Y. 1986. Age and growth of some flounder in Suo-Nada of Seto Inland Sea. *Nippon Suisan Gakkaishi* 52:435–445 (in Japanese with English abstract).

Mazlan A.G. and Rohaya M. 2008. Size, growth and reproductive biology of the giant mudskipper, *Periophthalmodon schlosseri* (Pallas, 1770), in Malaysian waters. *Journal of Applied Ichthyology* 24:290–296.

Minh T.H., Gallardo W.G. and Phuong N.T. 2010. Fishery and aquaculture of juvenile mudskipper *Pseudapocryptes elongatus* (Cuvier, 1816) in the coastal zone of Mekong Delta, Vietnam. *Asian Fisheries Science* 23:224–239.

Mirza M.R., Hameed S. and Naeem Javed M. 1996. Fishes of Shadi Khoar (Baluchistan). Second Symposium on Fish and Fisheries of Pakistan, November, 25–26, 1996, Department of Zoology, Government College, Lahore, 8–9 pp (abstract).

Murdy E.O. 1989. A taxonomic revision and cladistic analysis of the Oxudercine gobies (Gobiidae: Oxudercinae). *Records of the Australian Museum, Supplement* 11:93.

Nanami A. and Takegaki T. 2005. Age and growth of the mudskipper *Boleophthalmus pectinirostris* in Ariake Bay, Kyushu, Japan. *Fisheries Research* 74:24–34.

Park K.D., Kim J.K., Chang D.S., Kim J.I. and Oh C.W. 2008. Age and growth of the mudskipper, *Scartelaos gigas* (Perciformes, Gobiidae) from Korea. *Animal Cells and Systems* 12:305–311.

Pauly D., Soriano-Bartz M., Moreau J. and Jarre-Teichmann A. 1992. A new model accounting for seasonal cessation of growth in fishes. *Australian Journal of Marine and Freshwater Research* 43:1151–1156.

Pompei L., Giannetto D. and Lorenzoni M. 2015. Age and growth of Arno goby, *Padogobius nigricans* (Canestrini, 1867), in the Aggia River (Umbria, Central Italy). *Journal of Applied Ichthyology* 31:494–500.

Ravi V. and Rajagopal S. 2007. Age and growth of the mudskipper, *Boleophthalmus boddarti* (Pallas, 1770). *Journal of Aquatic Biology* 22:123–128.

Salarpouri A., Behzadi S., Mortazawi M., Darvishi M., Taherizadeh M. and Armond K. 2013. Population dynamics and reproduction biology of the slender mudskipper (*Scartelaos tenuis*) from Hormuzgan province creeks, Persian Gulf. *Iranian Scientific Fisheries Journal* 22:37–48.

Sarafraz J., Abdoli A., Kiabi B.H., Karrani E. and Akbarian M.A. 2011. Determination of age and growth of the mudskipper *Periophthalmus waltoni* Koumans, 1955 (Actinopterygii: Perciformes) on the mudflats of Qeshm Island and Bandar-Abbas, Iran. *Progress in Biological Sciences* 1:25–30.

Schnute J. 1981. A versatile growth model with statistically stable parameters. *Canadian Journal of Fisheries and Aquatic Science* 38:1128–1140.

Siddique M.A.M., Khan M.S.K. and Aktar M. 2015. Length–weight relationship for two gobioid fishes *Odontamblyopus rubicundus* (Hamilton, 1822) and *Pseudapocryptes elongatus* (Cuvier, 1816) from the Bakkhali River estuary, Bay of Bengal. *Zoology and Ecology* 25:106–109.

Sokolowska E. and Fey D.P. 2011. Age and growth of the round goby *Neogobius melanostomus* in the Gulf of Gdansk several years after invasion. Is the Baltic Sea a new promised land? *Journal of Fish Biology* 78:1993–2009.

Soni V.C. and George B. 1986. Age determination and length–weight relationship in the mudskipper *Boleophthalmus dentatus*. *Indian Journal of Fisheries* 33:231–234.

Thompson H.A. and Simon T.P. 2015. Age and growth of round goby *Neogobius melanostomus* associated with depth and habitat in the western basin of Lake Erie. *Journal of Fish Biology* 86:558–574.

Tsukamoto K., Onohara T., Ushirogawa K. and Matsuura S. 1983. Study on age characters of the mudskipper *Boleophthalmus chinensis*. *Bulletin of the Japanese Society of Scientific Fisheries* 49:69–74.

Uehara S. and Shimizu M. 1996. Age and growth of stone flounder *Kareius bicoloratus* in Tokyo Bay, Japan. *Fisheries Science* 62:897–901.

Von Bertalanffy L. 1938. A quantitative theory of organic growth. *Human Biology* 10:181–213.

Walford L.A. 1946. A new method of describing the growth of animal. *Biological Bulletin* 90:141–147.

Washio M. 1992. Biological studies of mudskipper, *Boleophthalmus pectinirostris*, in Ariake Sound, western Kyushu, Japan. Doctoral thesis, Nagasaki University (in Japanese with English abstract).

Washio M., Tsutsui M. and Takita T. 1991. Age and growth of the mudskipper *Boleophthalmus pectinirostris* distributed in the mudflat of the Midori River Kumamoto Prefecture. *Nippon Suisan Gakkaishi* 57:637–644 (in Japanese, with English abstract).

Yamazaki K. 2001. Hogoku to hi-hogoku ni okeru mutsugorou no nenreisosei to seiiku zyoutai. B.Sc. thesis, Nagasaki University (in Japanese).

Zhang Q., Sheng S. and Zhang W. 1990. Age, growth and fecundity of the mudskipper *Boleophthalmus pectinirostris*. *Journal of Fisheries of China* 14:179–188 (in Chinese, with English abstract).

CHAPTER 5

Respiratory and Circulatory Adaptations

Atsushi Ishimatsu

CONTENTS

5.1 INTRODUCTION

Among present day fishes, mudskippers are the most active on land. The two most terrestrial genera, *Periophthalmodon* and *Periophthalmus*, spend the majority of time out of water. Some species of these genera remain out of water even during high tide and show stress responses, such as a lowered heart rate, and a substantial drop in blood oxygen levels, when forcibly submerged in water. Remarkably, these mudskippers retain the typical piscine respiratory and circulatory anatomy despite their ability to endure periods out of water. The amphibious abilities of mudskippers have prompted their comparison to early stem tetrapods. Also known as tetrapodomorphs, these are the first vertebrates to transition from water to land in the Late Devonian, approximately 365–385 million years ago (Schoch 2014; Clack 2012). Even though both early stem tetrapods and mudskippers may represent early stages of transition from aquatic to terrestrial mode of life, these are animals of very different morphology; first, the body sizes and structures of the extinct tetrapodomorphs and of extant mudskippers are vastly different. The body lengths of tetrapodomorphs were 1 m or longer, while that of the largest mudskipper, *Periophthalmodon schlosseri*, is approximately 30 cm and most other mudskippers are much smaller. Second, the skeletons of tetrapodomorphs such as *Pandarichthys* or *Tiktaalik* show substantial deviations from the typical fish design, whereas the mudskipper skeleton retains the basic fish architecture (Lee 1990). Third, ancient vertebrates are thought to have had the lungs for aerial respiration (Schoch 2014; Clack 2012); ventilation of the lungs is hypothesized to occur through the spiracles, as recently demonstrated for *Polypterus* (Graham et al. 2014). Mudskippers lack lungs and their aerial gas exchange takes place across the gills, the bucco-opercular epithelia and the skin. Mudskippers gulp air through the mouth to fill their enlarged bucco-opercular cavity (here defined as the space in the buccal, pharyngeal, and opercular cavities combined) but the aerial ventilatory mechanisms in this group are not well studied (Aguilar et al. 2000).

Still, mudskippers are suitable extant vertebrates for better understanding the process of land invasion of vertebrates (Kutschera and Elliot 2013). Despite the limits of their body design and structure, mudskippers are able to exploit resources beyond the water's edge. It is tempting to imagine that there may have been multiple attempts at land invasion by ancient fishes before the appearance of stem tetrapods. Should this be true, the anatomy of these ancient fishes, larger or smaller, would have deviated little from a purely aquatic fish design. Even if the fossils of extinct mudskippers were found, their amphibious or terrestrial capabilities would be difficult to determine. This chapter reviews current knowledge on the anatomy and physiology of cardiorespiratory systems of mudskippers, and suggests future research directions.

5.2 ANATOMY

5.2.1 Respiratory System

The respiratory surfaces of mudskippers consist of the gills, the bucco-opercular cavity, and the skin. Most air-breathing fishes that live in hypoxic freshwater bodies throughout their life will not emerge on land, and will have specialized accessory respiratory organs; they are known as "aquatic air-breathing fishes" (Graham 1997). Mudskippers, on the other hand, neither develop any specialized aerial respiratory organ in the buccal or pharyngeal cavity, nor do they show any respiratory modifications of the gas bladder or in the digestive tract. In fact, mudskippers in the genera *Periophthalmodon* and *Periophthalmus* lack a gas bladder (Polgar et al. 2011; Ishimatsu, personal observation). The bucco-opercular cavity and the skin of mudskippers are richly vascularized, but lack the structural complexities present in the air-breathing organs of many aquatic air-breathing fishes. Of the three respiratory surfaces in mudskippers, the gills and the bucco-opercular cavity are ventilated actively by the actions of the muscles and skeletal machinery surrounding them. The skin is passively ventilated by body movements or a current of ambient air or water (Feder and Burggren 1985).

5.2.1.1 Gills

All mudskippers studied with respect to respiratory anatomy have four pairs of gill arches in the pharynx (Figure 5.1a; Graham 1997; Milward 1974). Each gill arch bears two rows of gill filaments

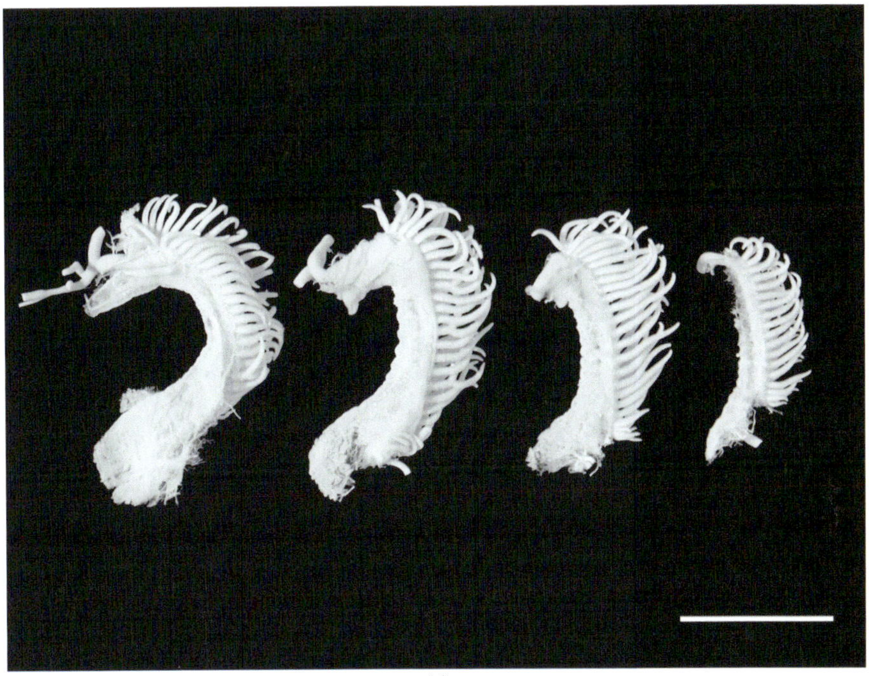

(a)

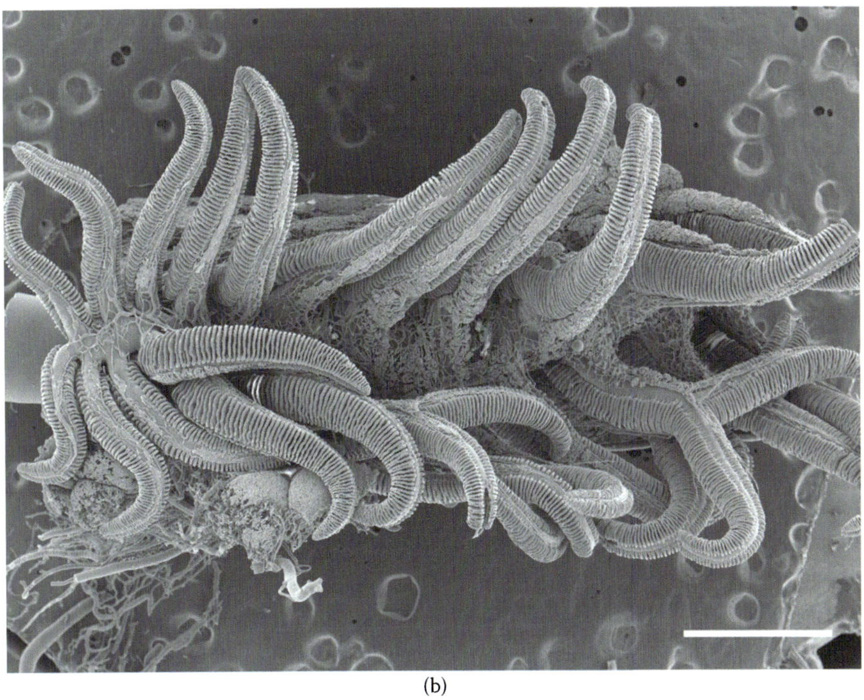

(b)

Figure 5.1 (a) Macro-photograph of a vascular cast of the left gill arches (I–IV) of *Periophthalmodon schlosseri* as viewed laterally (bar = 1 cm). (b) Low-magnification scanning electron micrograph of the first gill arch viewed ventrally, note branching of the filament (bar = 1 mm). (Reproduced from Gonzales T.T. et al., *Journal of Morphology*, 272, 629–640, 2011. With permission.)

on its anterior and posterior sides (holobanch), as in most other teleosts. Individual gill filaments bear thin secondary folds, termed "secondary lamellae," on the dorsal and ventral surfaces of the filaments. Epithelia covering the two sides of the secondary lamellae are anchored by column-shaped cells (pillar cells), and the oxygen-poor venous blood ejected from the heart flows through the space between the pillar cells, during which gas exchange takes place (Ishimatsu 2012; Al-Kadhomiy and Hughes 1988).

Although mudskippers retain the general organization of teleost gills, the dimensions of the gills vary between genera; the gills of more terrestrial mudskippers are usually smaller in size (Graham et al. 2007; Graham 1997; Low et al. 1990; Schöttle 1931). This is also supported by the meta-analysis of the published data of gill surface areas in mudskippers by Clayton (1993). Low et al. (1990) found that *Boleophthalmus boddarti* has more filaments than *Pn. schlosseri* and *Periophthalmus chrysospilos* of comparable body size (note: the *F* value given for *Pn. schlosseri* in tab. 2 of Low et al. (1990) is incorrect. Based on the regression equation for the species, it should be 312, not 547 as given in the table). Gill filaments in *B. boddarti* are more densely spaced along the gill arch, and the length of each filament is longer than those in *Pn. schlosseri* and *Ps. chrysospilos* (Low et al. 1990). Schöttle (1931) classified the lamellae of *Periophthalmus* mudskippers into three types. Type I lamellae are similar to those of typical water-breathing fishes with the lamellar surface consisting of thin pavement cells and a small number of granule-containing cells (probably mucus cells). Mitochondria-rich (MR) cells occur only at the base of the lamellae. This type was found for *Periophthalmus argentilineatus* (reported as *Periophthalmus dipus* and subspecies of *Periophthalmus vulgaris*) and *Periophthalmus kalolo* (reported as *Periophthalmus koelreuteri fuscatus*). Type II lamellae differ from the type I in that the lamellar epithelium is thicker and inundated with a large number of MR cells. Thus, in the cross section of the secondary lamellae of *Ps. argentilineatus* (Schöttle 1931: fig. 6), many apical crypts open at the lamellar surface. This type was found in *Periophthalmus modestus* (reported as *Periophthalmus cantonensis*), *Ps. chrysospilos* and two subspecies of *Ps. argentilineatus* (reported as *Ps. vulgaris*). Type III is the most specialized, seen only in *Pn. schlosseri* (reported as *Periophthalmus schlosseri schlosseri*). The neighboring lamellae are largely but not completely fused together, leaving only limited water channels between the lamellae. The lamellar epithelium is even thicker than in the type II and composed of a large number of MR cells.

Mazlan et al. (2006) showed scanning electron microscope (SEM) photomicrographs of the secondary lamellae of *Ps. chrysospilos*, which have a number of openings over the lamellar surface. Graham et al. (2007) also found openings over the lamellar surface of *Ps. chrysospilos* and *Ps. modestus*. These SEM observations agree with the histological findings of Schöttle (1931), and indicate these openings to be apical crypts of MR cells. *Periophthalmus* species also show extra-branchial distribution of MR cells; *Ps. modestus* show a high MR cell density in the skin covering the body surface that adjoins the base of the pectoral fin when the fin is adducted (Sakamoto et al. 2000; Yokota et al. 1997). In contrast, *Periophthalmus minutus* show a very high MR cell density in the inner operculum (Itoki et al. 2012).

Schöttle (1931) categorized *B. boddarti, Scartelaos histophorus* (reported as *Boleophthalmus viridis*) and *Pseudapocryptes elongatus* (reported as *Apocryptes lanceolatus*) into the least specialized lamellar type (type I), with a thin epithelium containing no mucus or MR cells, which was confirmed by later studies (Graham et al. 2007; Al-Kadhomiy and Hughes 1988).

In addition to the reduced size and lamellar fusion, the gill filaments of most terrestrial mudskippers may bifurcate, as observed in *Pn. schlosseri* (Figure 5.1b; Gonzales et al. 2011; Low et al. 1988) and *Periophthalmodon freycineti* (reported as *Pn. schlosseri*, Milward 1974). To my knowledge, the gill morphology of *Periophthalmodon septemradiatus* has not been reported. The gills of *Periophthalmus* species do not show filament bifurcation but lamellar fusion was observed for *Ps. modestus* (Gonzales et al. 2008a) and *Ps. minutus* at filament tips (Graham et al. 2007).

Graham (1997) characterized the gills of mudskippers as having (1) significantly reduced fourth gill arches; (2) short gill filaments (the term primary lamellae is used in Graham 1997); (3) fewer and smaller secondary lamellae (Graham 1997: tab. 3.6); and as a result (4) smaller gill surface area (Graham 1997:

tab. 3.4). Among these, (2), (3), and (4) seem to have been generally supported, but (1) needs some scrutiny. Graham (1997) cited Tamura and Moriyama (1976) and Niva et al. (1981, cited incorrectly as Biswas et al. 1981) as sources of the information, but a closer look at Tamura and Moriyama (1976) defies the contention; in table 4, in which the authors compared the ratios of several morphometric parameters of the fourth to first gill arches, the ratios of the number of filaments, and the length and the density of secondary lamellae were comparable between mudskippers (*Boleophthalmus pectinirostris* reported as *B. chinensis* and *Ps. modestus* reported as *Ps. cantonensis*) and water-breathing fishes (*Anguilla japonica*, *Carassius auratus*, and *Cyprinus carpio*). It is only filament length that showed a smaller value in these mudskippers (62%) than in water-breathing fishes (79–108%). The allometric equations given by Niva et al. (1981) for *B. boddarti* show that the gill surface area in the fourth arch is 40% of the first arch (calculation based on body weight of 10 g, nearly the upper end of the body weight range they used), which is mainly due to the smaller bilateral surface area of each secondary lamella (45%) and smaller total filament length (70%) in the fourth arch as compared with the first arch. However, the number of filaments and the density of secondary lamellae are slightly larger in the fourth arch and the total number of secondary lamellae is comparable between them. The source of information that Graham (1997: tab. 3.6) used for the table is not clear, but it is probably premature to make conclusions on the size of the fourth gill arch from this study. Rather, I am of the opinion that the fourth gill arches of mudskippers are not particularly reduced compared with other water-breathing and air-breathing fishes. The reduction of the fourth gill arches is more pronounced in aquatic air-breathing fishes like *Anabas testudineus* (Olson et al. 1986) and *Channa* species (Olson et al. 1994; Ishimatsu and Itazawa 1983; Ishimatsu et al. 1979).

Clayton (1993) argued that a negative slope for the regression equation for the frequency (number/unit filament length) of secondary lamellae against body weight supports the idea of Graham (1976) that greater spacing between secondary lamellae is an adaptation to aerial exposure because it reduces the collapsing of the gills in air. However, negative slopes of this relationship were consistently found for water-breathing fishes (Wegner et al 2010; Chapman and Hulen 2001; Kunzmann 1990; Hughes et al. 1986; Roy and Munshi 1986; Hughes 1972), and therefore they cannot be taken as supporting data for the adaptation to an amphibious mode of life.

Mudskippers from the temperate zones overwinter in the mud for several months (Takegaki et al. 2006), during which they were reported to remain in water-filled burrows (Kobayashi et al. 1971). During the overwintering period, the fishes have no atmospheric air supply and presumably take up whatever oxygen they need from burrow or interstitial water through the gills and the skin or possibly from an air pocket that might have been created before overwintering by gulping air and storing it in a burrow. The fishes probably experience aquatic hypoxia, and it is unlikely that they can store a large enough volume of air as an oxygen supply to last for several months. It would be interesting to examine if the mudskipper gills alter their morphology during overwintering to improve oxygen uptake efficacy in a cold hypoxic subterranean environment.

Aquatic hypoxia is known to induce alterations in gill morphology, which is termed "gill remodeling" (see Nilsson et al. 2012 for review), first reported by Sollid et al. (2003) in *Carassius carassius*. The secondary lamellae of *C. carassius* were embedded in a mass of tissue filling the interlamellar space and had limited free gas exchange surface when the fish was kept in normoxic water. However, when the fish was acclimated to hypoxic water, the interlamellar tissue mass drastically shrank, through an increased occurrence of apoptosis and a reduction in cell division, which resulted in a 7.5-fold increase in lamellar surface area. The change was completely reversible (Sollid et al. 2003).

Gill remodeling has also been shown during ontogenetic growth of an air-breathing fish, *Arapaima gigas*, a gigantic aquatic air-breathing fish from the Amazon. Brauner et al. (2004) demonstrated for *A. gigas* that the secondary lamellae were gradually embedded in filamental tissue with growth. The gills of 100 g fish still had protruding lamellae, which were covered by a thick epithelium containing numerous MR cells, whereas the lamellae of a 1 kg fish were no longer discernible from the filamental tissues and the MR cells were aligned to the filamental surface. In 1 kg fish, the filaments appeared like rods with a smooth surface and lamellae were no longer visible by SEM observation. Ramos et al. (2013) confirmed and extended the findings of Brauner et al. (2004), showing that both the thickness of the filament and lamellae increased with growth, gradually obstructing the interlamellar space from 25 to 30 μm in 100 g fish to only 8 μm in 5000 g fish. MR cells constituted a relatively small portion of both filamental and lamellar epithelia when fish were less than 100 g, but the portion increased markedly as the gills underwent the morphological alteration with growth. The morphological alteration of gills of *A. gigas* with growth resembles the growth-related branchial changes reported for *Pn. schlosseri* by Schöttle (1931). She found that the gills of small *Pn. schlosseri* (body length 1.5 cm) resembled those of *Ps. chrysospilos* and *Ps. argentilineatus* (reported as *Periophthalmus vulgaris*), both of which had protruding secondary lamellae (i.e., type II), whereas larger specimens had extensively fused lamellae with the interlamellar space almost completely occluded by numerous MR cells (type III).

Gill remodeling is also observed in another amphibious air-breathing fish, *Kryptolebias marmoratus*, which voluntarily emerges from water to capture prey, and to avoid unfavorable water conditions or intraspecfic aggression (see Turko et al. 2011). Tissue proliferation occurred in the interlamellar tissues after air exposure (Ong et al. 2007), but in contrast to MR cell positions seen in large *A. gigas* and *Pn. schlosseri*, MR cells in this species occurred deep inside the cell mass and were not exposed to ambient water (LeBlanc et al. 2010).

Several species of air-breathing fishes including mudskippers were alleged to use the gills for aerial gas exchange. However, no experimental evidence exists to verify or refute this assumption. The well-developed gills of *Boleophthalmus* and *Scartelaos* are likely efficient in aquatic gas exchange, but their efficiency is questionable in air. Graham (1997) pointed out that the role of the gills in aerial respiration has been inferred from the absence of respiratory vasculature in tissues adjacent to the gills when oxygen uptake in the head region was measured in a partitioned respirometer. Carter and Beadle (1931) presumed the role of the gills in aerial respiration from the ratio of the height to the thickness of the secondary lamellae in *Brachyhypopomus* (formerly *Hypopomus*), which was larger than water-breathing fishes. *Synbranchus* is often cited as a typical example of the fish using gills for aerial respiration, but this is based on the absence of specialized accessory organs and morphological alteration of gill filaments, which may help maintain their configuration in air (i.e., the asymmetrical linear arrangement of the holobranchs and the hypertrophy of the gill rays, Liem 1987) or on the large size of the secondary lamellae (Johansen 1966). To verify the role of the gills in aerial respiration separately from that through the adjacent epithelia, it is necessary to determine the partial pressure of oxygen (P_{O_2}) and oxygen content of the pre- and postbranchial blood and cardiac output, which is challenging for most amphibious air-breathing fishes including mudskippers due to their relatively small sizes.

5.2.1.2 Bucco-Opercular Cavity

The bucco-opercular cavity of mudskippers is especially enlarged and can accommodate a large volume of the ambient medium (water or air) within the cavity. Often on the mudflats, mudskippers bulge their "cheeks" either with water before foraging microalgae on the mudflat surface (e.g., *Boleophthalmus* species) or with air for gas exchange (all genera, Aguilar et al. 2000 and personal observation). *Boleophthalmus* (Toba and Ishimatsu 2014), *Periophthalmodon* (Ishimatsu et al. 2009, 1998), and *Periophthalmus* (Ishimatsu et al. 2007) also gulp a mouthful of air to store inside their burrows. The unusual size of a mudskipper's bucco-opercular cavity is best exemplified by comparing their volumes to other gobiid fishes (Ishimatsu and Gonzales 2011; Graham et al. 2007). This volume (as measured by a gas volume held in the cavity) made up 14% of the total body volume in *S. histophorus* (Graham et al. 2007), 16% in *B. pectinirostris* (Graham et al. 2007) and *Pn. schlosseri* (Aguilar et al. 2000), and 18% in *Ps. modestus* (Ishimatsu et al. 2007). These values are considerably higher than the 2–4% reported for non-oxudercine gobies, *Arenigobius* (2 spp.), *Cryptocentroides* (1 sp.), *Favonigobius* (1 sp.), *Chlamydogobius* (1 sp.), *Mugilogobius* (3 spp.), and *Pseudogobius* (1 sp.) (Gee and Gee 1991). These nine non-oxudercine gobies are not amphibious, but use the buoyancy gained by holding an air bubble to aid in aquatic surface respiration. According to Gee and Gee (1991), the bubble did not fill the buccal chamber completely, and its position oscillated with gill ventilation. This observation also supports the idea that the air bubble was used to obtain positive buoyancy but not for aerial gas exchange. *Gillichthys mirabilis* is another non-oxudercine goby species that uses the buccal cavity for aerial respiration (Todd and Ebeling 1966). This fish is a facultative aquatic air breather but also has the ability to locomote on land (Todd 1968).

Apart from gobiid fishes, a few air-breathing fishes have evolved a gas exchange surface inside the mouth. But again the relative volume of the bucco-opercular cavity in these species is much smaller than in mudskippers. Farber and Rahn (1970) demonstrated that *Electrophorus electricus* holds a gas volume that is 1.3% of the body volume (assuming a body density of one), and Bicudo and Johansen (1979) reported 1.3–2% for *Synbranchus marmoratus*.

It is not immediately clear why the bucco-opercular cavity has evolved to such an unusual degree in mudskippers. A large volume held in an air-breathing organ would be beneficial for aquatic air-breathing fishes to reduce the frequency of surfacing. However, once a fish has emerged from water, it may not have a great respiratory significance because air is readily available. The very large bucco-opercular volume could have evolved before mudskippers became amphibious. Another possibility is that it is related to the ability for mudskippers to store air in burrows. The air stored in mudskipper burrows is used as an oxygen source for developing embryos incubated in an air-filled spawning chamber and/or for adult fish during high-tide burrow confinement (see Chapter 9). A large bucco-opercular volume reduces the energetic costs to replenish air in the burrow during low tide. It could also be an adaptation for excavating burrows. All the amphibious mudskippers (Dinh et al. 2014a; Takita et al. 2011), several other oxudercine (*Parapocryptes serperaster*, Dinh et al. 2014b; *Pd. elongatus*, Das 1934), and non-oxudercine (*Odontamblyopus lacepedii*, Gonzales et al. 2008b) gobies excavate burrows with their mouths. Thus, a larger mouth volume might have been selected for the reduced time needed for completing the task, rather than to improve gas exchange. Graham et al. (2007) reported that males of *Ps. modestus* had a larger bucco-opercular volume than females of the same species. This might be an adaptation for the male's role in embryonic care by repeatedly adding mouthfuls of air in the burrow during low tide. However, the difference is relatively small (18.3% for males vs. 15.9% for females), which makes the hypothesis less convincing.

The bucco-opercular cavities of mudskippers are lined with vascularized epithelia (Gonzales et al. 2011; Zhang 2000; Graham 1997; Milward 1974). The vascularization appears more intense in *Periophthalmus* than in *Boleophthalmus* (Schöttle 1931), yet the degree of vascularization varies among species. Schöttle (1931) observed no respiratory capillaries in the buccal cavity of *Ps. modestus* and in most subspecies of *Ps. argentilineatus* (reported as *Ps. vulgaris*) that she studied. This is in accord with the finding of Gonzales et al. (2011) that the density of respiratory capillaries (within 10 μm from the surface) was much higher in the inner operculum than in the buccal cavity in *Ps. modestus*. Among mudskippers, *Periophthalmodon* shows the highest capillary density inside the bucco-opercular cavity; in *Pn. schlosseri*, a dense capillary network covers the entire inner surface of the cavity (Figure 5.2a,b; Gonzales et al. 2011). The surfaces of the gill arches and the leading edge of the gill filaments are also covered with capillary networks in *Pn. schlosseri* (Gonzales et al. 2011), whereas vascularized epithelia are absent on the gill arches of *Ps. argentilineatus* and *Ps. modestus* (Schöttle 1931). Zhang (2000) reported rich vascularization in the inner opercular surface in *Oxuderces dentatus*, *Pd. elongatus*, and *S. histophorus*, in addition to *Boleophthalmus*, *Periophthalmodon*, and *Periophthalmus*. She also found sparse vascularization in the inner opercular surface in *Apocryptodon punctatus*.

5.2.1.3 Skin

There is relatively ample information on the morphology of mudskipper skin. Schöttle (1931) was the first to describe the skin histology of mudskippers in detail. More recently, skin structure has been reported for *B. pectinirostris* (Park et al. 2003; Zhang et al. 2000; Yokoya

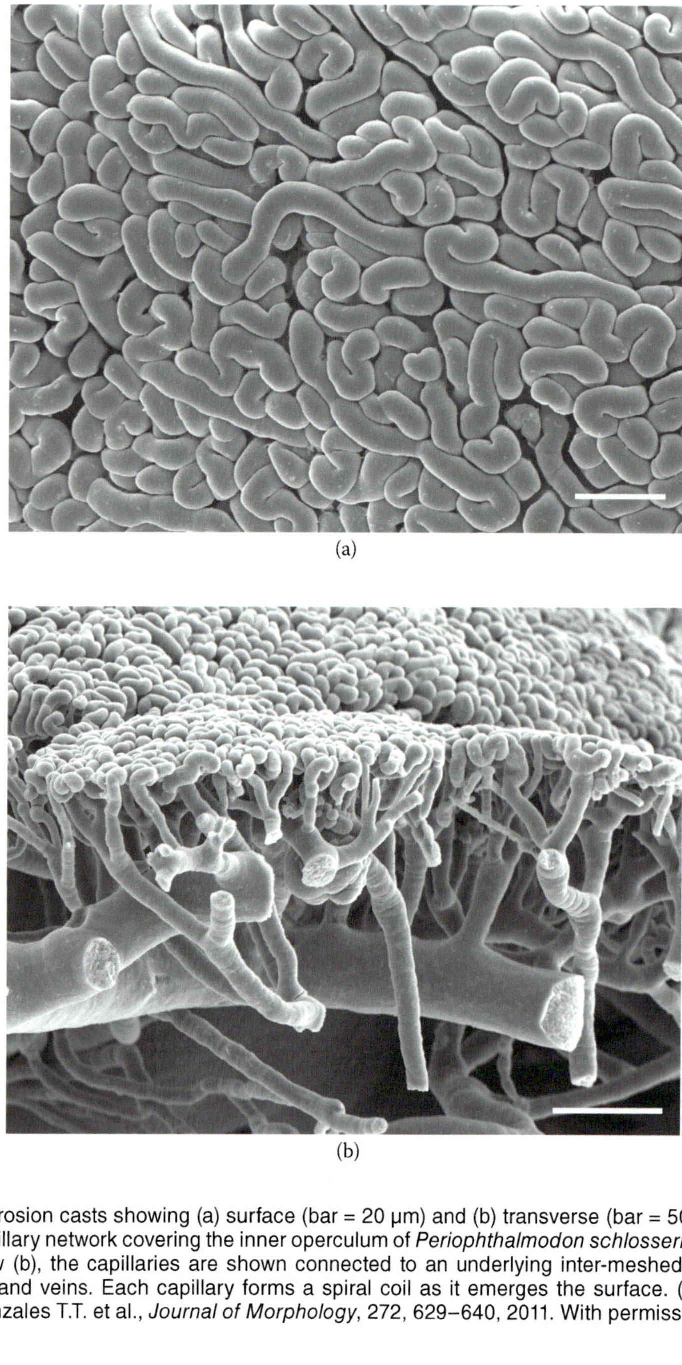

(a)

(b)

Figure 5.2. Corrosion casts showing (a) surface (bar = 20 μm) and (b) transverse (bar = 50 μm) views of the capillary network covering the inner operculum of *Periophthalmodon schlosseri*. In the transverse view (b), the capillaries are shown connected to an underlying inter-meshed network of arteries and veins. Each capillary forms a spiral coil as it emerges the surface. (Reproduced from Gonzales T.T. et al., *Journal of Morphology*, 272, 629–640, 2011. With permission.)

and Tamura 1992), *Pn. schlosseri* (Zhang et al. 2003), *Ps. modestus* (Yokoya and Tamura 1992), *Periophthalmus magnuspinnatus* (Park et al. 2006; Park 2002), *Scartelaos gigas* (Beon et al. 2013), and *S. histophorus* (Zhang et al. 2000). All mudskippers studied have capillaries in the skin with a diffusion distance of less than 10 μm (Beon et al. 2013). Species of *Boleophthalmus* have a characteristic structure in their skin for gas exchange, often termed dermal bulges (Figure 5.3a through c). The dermal bulges are papillary protrusions in the skin with a diameter of 100–500 μm (Park et al. 2003). Inside a dermal bulge, the caudal periphery of each scale pocket

(a)

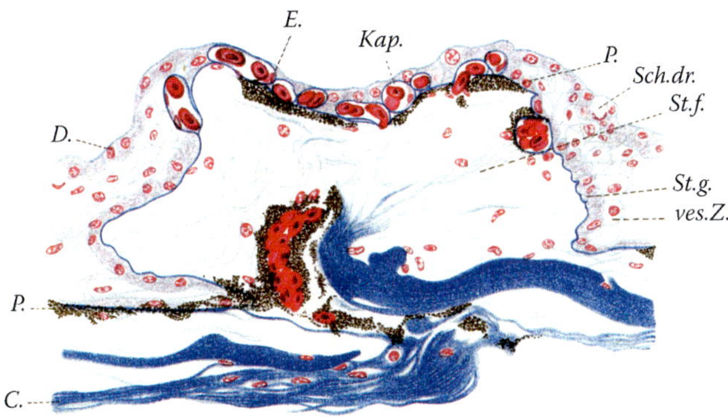

(b)

(c)

Figure 5.3 The dermal bulge of *Boleophthalmus*. (a) Surface view of the skin of live *Boleophthalmus pectinirostris* (bar = 1 mm). (b) The original drawing of the cross section of a dermal bulge of *Boleophthalmus boddarti* by Schöttle (1931); *C*, corium (= dermis); *D*, surface layer; *E*, endothelium; *Kap*, capillaries; *P*, pigment; *Sch.dr.*, mucus gland; *St.f.*, supporting fibers; *St.g.*, stratum germinativum; *ves.Z.*, vesicular cell. (c) Surface view of a vascular cast of the dermal bulge of *Boleophthalmus pectinirostris*; Kurita, Noma, and Ishimatsu (unpublished).

(Guerreiro et al. 2013) is expanded, protrudes toward the skin surface, and forms a round projection. At the apical surface of a dermal bulge, the epidermis is only a few cell-layers thick, considerably thinner than in the skin between the dermal bulges. An artery and a vein course together perpendicularly toward the skin surface at the center of the expanded scale pocket. The artery branches into numerous capillaries, which radiate from the center of the pocket over its apical surface (Schöttle 1931). The blood from these capillaries is drained by peripheral veins running around the pocket that return the blood to the more central veins. Transmission electron microscopy observations showed that the capillaries are located immediately below the basal lamina, which separates the dermis (where blood vessels occur) and the epidermis, in *B. pectinirostris* (Yokoya and Tamura 1992). *Scartelaos* also has dermal bulges (Beon et al. 2013; Zhang et al. 2000). Dermal bulges are more numerous on the dorsal and lateral parts of the body, and most are exposed to air when the fish emerges.

Species of *Periophthalmus* and *Periophthalmodon* have dense capillary networks in the epidermis, but they do not have dermal bulges (Park et al. 2006; Zhang et al. 2003; Park 2002; Suzuki 1992; Yokoya and Tamura 1992). As in *B. pectinirostris*, the capillaries found in the epidermis of *Periophthalmus* and *Periophthalmodon* are "not actually intraepithelial, but a kind of finger-shaped protrusion of the dermis" (Schöttle 1931: 42).

5.2.2 Circulatory System

The cardiovascular anatomy of mudskippers has drawn little attention, probably because Schöttle (1931) had already shown that the central circulatory system of mudskippers retains the typical teleost design and there is little deviation from it. This is not surprising because mudskippers have not developed any specialized respiratory organ but rely upon capillary beds developed in the body skin and mucous membranes lining the bucco-opercular cavity and the gill arches for aerial gas exchange. The same is true for other amphibious fishes inhabiting coastal areas, such as air-breathing non-oxudercine gobies (e.g., *Gillichthys*) and blennies (e.g., *Alticus*, see Graham 1997). In sharp contrast, a large number of aquatic air-breathing fishes have developed intricate accessory respiratory organs in various sites along the alimentary tract, which in several species accompanies alterations and reorganizations of vascular pathways, with a few showing the capacity of partial separation of oxygen-rich and oxygen-poor blood in the central circulatory system (Ishimatsu 2012; Graham 1997). With increasing terrestriality, mudskippers show an increasing degree of vascularization over the body surface, and an accompanying reduction of gill structure, but no trend toward separation of oxygen-rich and oxygen-poor blood. The fact that mudskippers are the most terrestrial among fishes testifies that morphological alterations in the central cardiovascular system are not a required trait at least for the early stages of land invasion by fishes.

5.2.2.1 Heart

Schöttle (1931) briefly described the anatomy of the mudskipper heart. It consists of four chambers as in all other teleosts; from posterior to anterior, the sinus venosus, the atrium, the ventricle and the bulbus arteriosus. Between all these chambers, valves prevent backflow of blood. She stated that there was a trend in the structural complexity of the luminal septa protruding from the bulbus wall with the development of respiratory function in the mouth. Icardo (2006) studied the morphology of the outflow tract in 28 teleost species, and Icardo and Colvee (2011) studied the morphology of the atrioventricular region in 29 teleost species, both including *Pn. schlosseri*.

5.2.2.2 Central Vascular Connection

The entire cardiac output is directed to the four pairs of gill arches through the ventral aorta, which gives off the afferent branchial arteries to perfuse the gill arches. Running through each gill arch, the afferent branchial artery branches off the afferent filamental arteries, which in turn perfuse the secondary lamellae through the afferent lamellar arterioles. Efferent blood from the secondary lamellae is collected sequentially by the efferent lamellar arterioles, the efferent filamental arteries, and the efferent branchial arteries. The blood from the efferent branchial arteries is collected dorsally by the lateral dorsal aorta to perfuse most of the head and trunk regions. The efferent branchial arteries of the first and second gill arches give off additional branches on the ventral side to perfuse the inner epithelium of the opercular cavity and the lower jaw region, while the significance of this vascular supply to the corresponding regions varies between species (see Section 5.2.2.4.). Venous blood is returned to the heart by the anterior cardinal vein from the head region, the posterior cardinal vein from the trunk, and by the hepatic vein from the visceral organs. Thus, the heart pressurizes the blood only once during one cycle of circulation through the body (termed single circulation, Figure 5.4a,b), which is typical of fishes. In mudskippers, the oxygen-rich blood from the respiratory surfaces of the bucco-opercular cavity and the skin is drained into the systemic veins, which most likely results in mixing of oxygen-rich and oxygen-poor blood in these veins. There seems to have been no selection pressure toward separation of two types of blood streams among mudskippers as seen in lungfishes (Burggren and Johansen 1986) and *Channa argus*

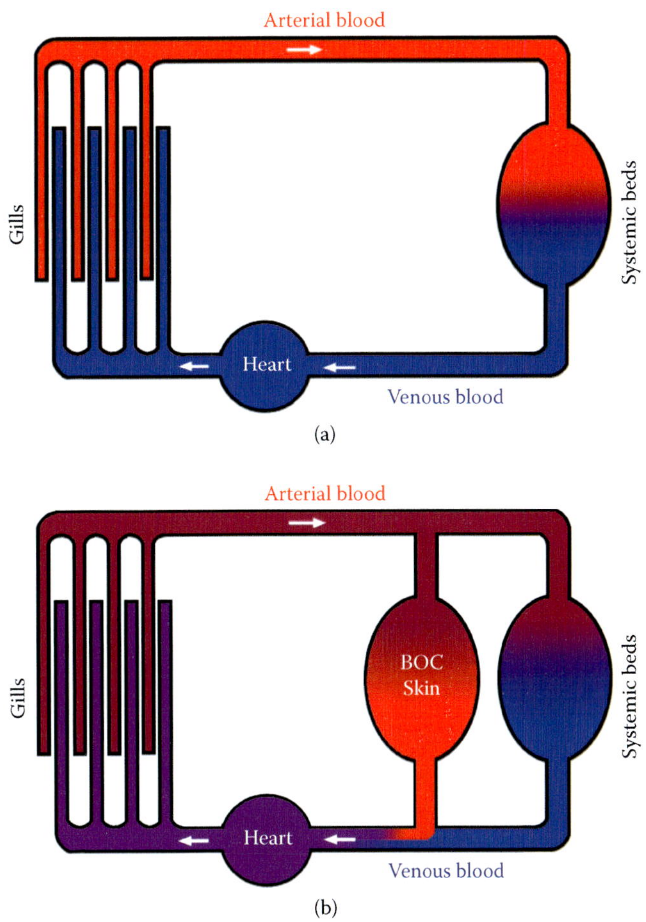

Figure 5.4 Diagrammatic representation of the circulatory system of (a) a generalized fish and (b) mudskippers. Red represents oxygen-rich blood, blue represents oxygen-poor blood, and purple represents a mixture of the two types of blood. Arrows represent the directions of blood flow. Note that the efferent (outflow) side of the bucco-opercular (BOC) cavity and the skin is brighter red than the afferent (inflow) side. (Figure prepared by Mizuri Murata.)

(Ishimatsu et al. 1983), which is considered as a precursor for the evolution toward double circulation. A direct vascular connection between the ventral and dorsal aortae, as seen in lungfishes or *Monopterus* (Ishimatsu 2012; Graham 1997), has not been reported for mudskippers.

Older literature reported anomalous vascular configurations in a few species of mudskippers. Schöttle (1931) provided a figure of *S. histophorus* (reported as *Boleophthalmus viridis*) in which the ventral aorta extended further anteriorly after giving off the first afferent branchial arteries, and bifurcated into another pair of arteries, which are connected with the first efferent branchial arteries posteriorly, send off the lingual artery anteriorly, and continue as the hypomandibular artery posterolaterally. The lingual artery perfused the respiratory vasculature on the ventral floor of the buccal cavity and the hyomandibular artery constituted the major blood supply to the buccal and opercular cavities. If this is true, respiratory capillaries of the bucco-opercular cavity, at least in part, are perfused with oxygen-poor venous blood, which among teleosts, had been indisputably demonstrated only for *Monopterus* (Munshi et al. 1990); a direct connection of prebranchial blood vessels to capillaries of the accessory respiratory organ in *Electrophorus electricus* figured in Carter (1935) is questionable and must be reexamined.

Anomalous vascular architecture has been also reported for *B. boddarti* and *Pd. elongatus*. Niva et al. (1981) depicted the bucco-opercular artery branching off from the first afferent branchial artery to perfuse the buccal and opercular cavities in *B. boddarti* (reported as *B. boddaerti*). We did not find such an anomaly in *B. boddarti* using the vascular-casting technique. Instead, we found the anterior extremity of the first efferent branchial artery bifurcated, and the lateral branch turned sharply to the posterior direction after leaving the first gill arch and seems to perfuse the operculum (Ishimatsu, unpublished observation). Vascular architecture similar to that in *B. boddarti* was also reported for *Pn. schlosseri* (Gonzales et al. 2011), in which the lateral branch was confirmed to be the hyomandibular artery. The medial branch of the first efferent branchial artery in *B. boddarti* was connected with the second efferent branchial artery. The bifurcation of the first efferent branchial artery lies very close to the dorsal surface of the first afferent branchial artery, and it was extremely difficult for us to discern through dissection of preserved specimens. Like Niva et al. (1981), Das (1934) also showed in the vascular diagram of *Pd. elongatus* (reported as *Pseudapocryptes lanceolatus*) that the first afferent branchial artery gives off the bucco-opercular artery to perfuse the inner surfaces of the buccal and the opercular cavities. In addition, he described a connecting vessel between the first and second afferent branchial arteries. We also examined the branchial vasculature of *Pd. elongatus* using the vascular-casting technique and found the same vascular connection as in *B. boddarti*, that is, bifurcation of the anterior end of the first efferent branchial artery with the lateral branch perhaps perfusing the operculum (Figure 5.5; Noma and

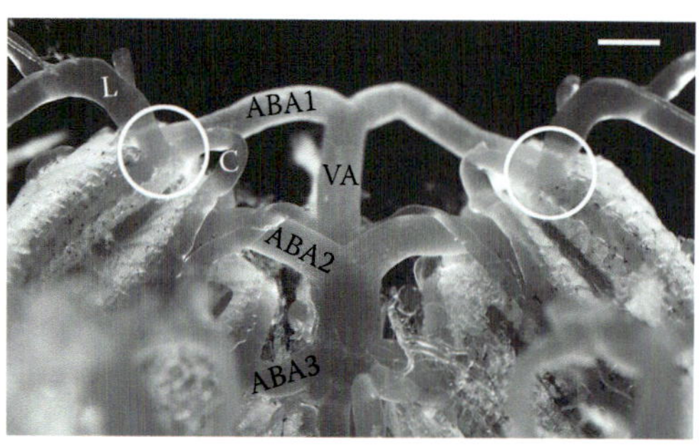

Figure 5.5 Vascular cast of *Pseudapocryptes elongatus* to show the bifurcation at the anterior extremity of the efferent branchial artery of the first gill arch (encircled). Dorsal view. ABA 1–3, afferent branchial arteries of the first to third gill arches; C, connecting artery between the efferent branchial arteries of the first and second gill arches; L, the lateral branch of the first efferent branchial artery; VA, ventral aorta (bar = 1 mm). (Noma and Ishimatsu, unpublished.)

Ishimatsu, unpublished). Our observations indicated that the reported vascular anomalies in the first gill arches of *B. boddarti* and *Pd. elongatus* were due to misidentification by Niva et al. (1981) and Das (1934), of the lateral branch of the first efferent branchial artery as originating from the first afferent branchial artery, which was closely juxtaposed to each other. The possible vascular anomaly of *S. histophorus* reported by Schöttle (1931) needs verification.

5.2.2.3 Gill Circulation

A vascular shunt, in which blood bypasses the secondary lamellae during transit through the gills, has not been reported in the gills of mudskippers. The lamellar vasculature of *Pn. schlosseri* shows a different type of specialization, which is most likely related to the diminishing role of the gills as a respiratory organ during transition from an aquatic to terrestrial way of life. The lamellar vasculature of this species consists of five to eight parallel blood channels, unlike the lacunar type blood space seen in most fishes, whose internal diameter is larger than the size of red blood cells (Gonzales et al. 2011). Channelization of lamellar vasculature has also been reported in freshwater air-breathing fishes such as *Amia calva* (Olson 1981), *Anabas testudineus* (Olson et al. 1986), and three species of *Channa* (Olson et al. 1994). Studying the branchial vasculature of other mudskippers will prove insightful especially for species such as *Pn. septemradiatus*, which was observed to stay completely out of water on a half-dried substratum for longer than 30 minutes (Zhang 2000), and one population of *Ps. minutus*, which inhabits a high intertidal mudflat that may remain exposed for more than 20 days during some neap tides (Itoki et al. 2012; Takeda et al. 2011). Our unpublished observation demonstrated that the lamellar vasculature in *B. pectinirostris* also showed channelization, but to a lesser degree than in *Pn. schlosseri* (Noma and Ishimatsu, unpublished). Al-Kadhomiy and Hughes (1988) reported the histology of the gills and skin of *B. dussumieri* (note that the authors identified the fish as *B. boddarti*, but the site of collection, Kuwait Bay, indicates that the fish was likely to be *B. dussumieri*) and found no specializations in the gills. The anatomy of lamellar vasculature has not been reported for other species of *Boleophthalmus*, or any species of *Periophthalmus* or *Scartelaos*.

5.2.2.4 Circulation through the Bucco-Opercular Cavity

The respiratory surface of the bucco-opercular cavity in mudskippers is perfused with systemic arterial blood (possibly also with prebranchial blood in *S. histophorus*, see Section 5.2.2.2). The hyomandibular artery arising from the anterior end of the first efferent branchial artery constitutes the largest conduit of blood supply to the inner epithelium of the opercular cavity, while the lingual and hypobranchial arteries both make relatively minor contributions, sending blood to the anterior ventral region and medial and ventral wall of the buccal cavity, respectively, in *B. boddarti* (Schöttle 1931; Figure 5.6a). In comparison, the subclavian, not hyomandibular, artery is the major blood supply conduit to the inner epithelium of the opercular cavity in *Pn. schlosseri* (Figure 5.6b). The hypobranchial, lingual, mandibular, and nasal arteries supply blood to the buccal cavity in *Pn. schlosseri*. Schöttle (1931) also stated that arteries supplying the inner epithelium of the buccal cavity were poorly developed in the subspecies of *Ps. argentilineatus* (reported as *Ps. vulgaris*) that did not develop vascular epithelium in the cavity. The blood from the bucco-opercular cavity is drained dorsally by the anterior cardinal vein and ventrally by the inferior jugular vein.

These early observations on the vascular supply to the epithelium of the opercular cavity in *Pn. schlosseri* by Schöttle (1931) were largely confirmed by the vascular casting study of the same species by Gonzales et al. (2011). These authors demonstrated a very high density of respiratory capillaries in the inner surface of the opercular cavity (ca. 70 capillaries per mm in cross section, Figure 5.2). The specimens studied by Gonzales et al. (2011) were prepared using fishes with deflated bucco-opercular cavities, although future studies should include observations on capillary

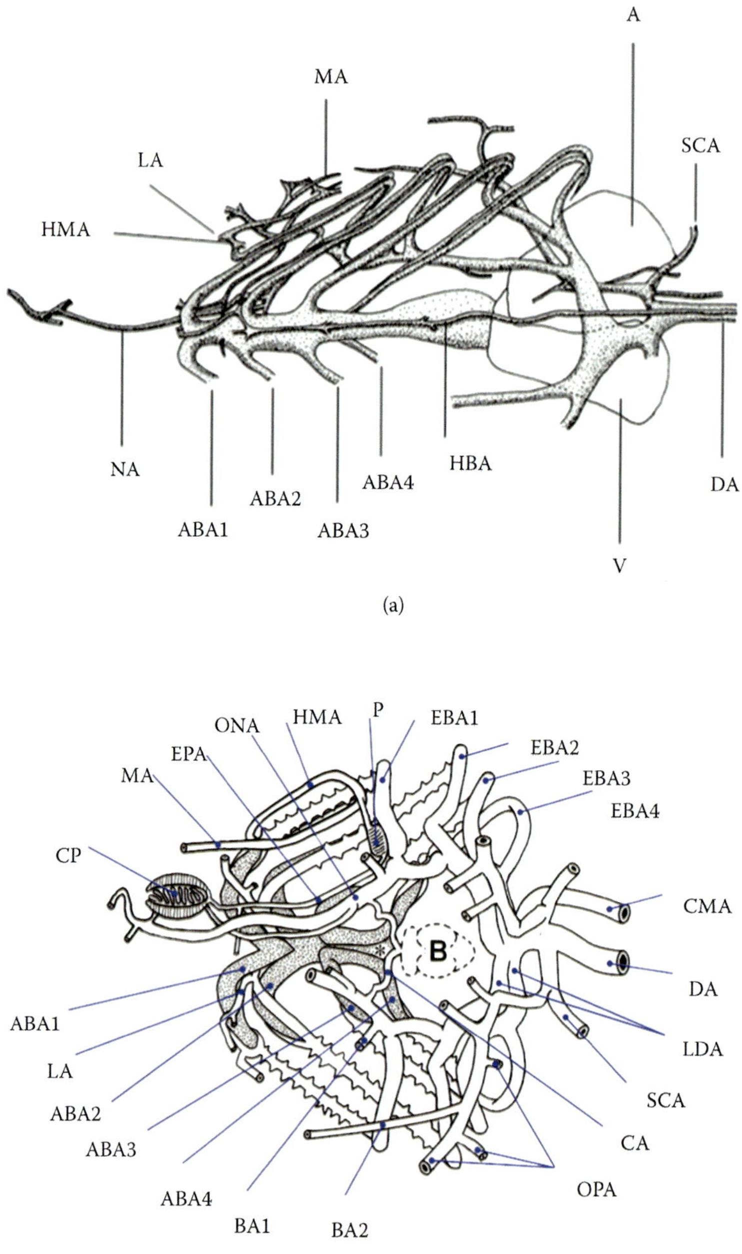

Figure 5.6 Central arterial system of (a) *Boleophthalmus boddarti* (ventral view: Modified from Schöttle E., *Zeitschrift für wissenschaftliche Zoologie*, 140, 1–114, 1931.) and (b) *Periophthalmodon schlosseri.* (dorsal view: Modified from Gonzales T.T. et al., *Journal of Morphology*, 272, 629–640, 2011.) A, atrium; ABA1–4, afferent branchial arteries of the first to fourth gill arches; B, brain; BA1, buccal artery from the orbitonasal artery; BA2, buccal artery from the dorsal aorta; CA, cerebral artery; CMA, coeliacomesenteric artery; CP, choroid plexus; DA, dorsal aorta; EBA 1–4, efferent branchial arteries of the first to fourth gill arches; EPA, efferent pseudobranchial artery; HBA, hypobranchial artery; HMA, hyomandibular artery; LA, lingual artery; LDA, lateral dorsal aorta; MA, mandibular artery; NA, nasal artery; ONA, orbitonasal artery; OPA, opercular artery; P, pseudobranch; SCA, subclavian artery; V, ventricle. (Figures prepared by Mizuri Murata.)

orientation with inflated cavities. Gonzales et al. (2011) found an unusually large left inferior jugular vein in *Pn. schlosseri*, which perhaps relates to a relatively large perfusion rate through the respiratory surfaces of the tongue and other parts of the lower jaw.

5.2.2.5 *Skin Circulation*

The respiratory surfaces of the skin are perfused with systemic arterial blood, as are the epithelia in the bucco-opercular cavity. Drawings of the microvasculature of the dermal bulge (reported as a papilla, see Figure 5.3) by Schöttle (1931) remain one of the best descriptions of the structure and its blood supply. In *B. boddarti*, the respiratory dermal bulges of the head are perfused with the blood from the nasal (anteriormost skin), hyomandibular (lateral skin), and subclavian (dorsal and lateral skin posterior to the eyes) arteries. Of these three arteries, the hyomandibular artery, the largest arterial blood supply to the inner epithelium of the opercular cavity (see Chapter 5.2.2.4), also forms the largest blood supply route to the head skin vasculature. Blood from the bulges is drained by the nasal, external jugular, and opercular veins. These veins return blood to the anterior cardinal and inferior jugular veins, which also receive blood from the respiratory epithelia of the bucco-opercular cavity and on the gill arches. We have found that the dermal bulges of the trunk region are supplied blood from the segmental arteries and drained by the segmental veins in *B. pectinirostris* (Kurita, Noma and Ishimatsu, unpublished). The vascular casts showed that in each dermal bulge, there are about 70 capillaries radiating from the central artery to the peripheral veins draining the blood oxygenated at the dermal bulge (Figure 5.3c). At a higher magnification, the capillaries are seen to form spiral coils. Similar spiral capillary morphology has been observed for the inner opercular lining of *Pn. schlosseri* (Gonzales et al. 2011) and the aerial respiratory surfaces (buccopharynx and air sac) of *Monopterus cuchia* (Munshi et al. 1990).

Dermal bulges do not occur in the skin of *Periophthalmodon* or *Periophthalmus*, but the capillaries are found in the outermost layer of the epidermis (Park et al. 2006; Zhang et al. 2003; Park 2002). The skin capillaries in the head are supplied blood from nasal (anteriormost skin), external carotid (lateral skin), and subclavian (lateral skin posterior to the eyes and back) arteries. The subclavian artery is the most important in blood perfusion of the skin respiratory surface in *Periophthalmodon* and *Periophthalmus*, whereas it is the hyomandibular artery in *Boleophthalmus* (Schöttle 1931).

5.3 PHYSIOLOGY

5.3.1 Respiratory Gas Exchange

Respiratory gas exchange in mudskippers in water and on land has been extensively reviewed. Considering that not much has been added to the existing data in recent years, readers are advised to refer to these detailed reviews (see Ishimatsu et al. 2011; Martin and Bridges 1999; Graham 1997, 1976; Clayton 1993). In experimental setups, mudskippers are generally able to maintain oxygen uptake at a comparable level both in water and on land. Carbon dioxide was released efficiently in air, resulting in the gas exchange ratio (carbon dioxide elimination rate/oxygen uptake rate) of 0.7–0.8, indicating that these fishes did not suffer accumulation of carbon dioxide in air, as it happens when aquatic air-breathing fishes are experimentally exposed to air (in many species, the gas exchange ratio is 0.1–0.5 in air, Graham 1997). The contribution of cutaneous respiration varied among species of mudskippers and the medium in which measurements were made, but the value was generally quite high (>40% of the total oxygen uptake rate). These data demonstrate that a fish can effectively exchange respiratory gases with air without developing a specialized organ for aerial respiration.

Jew et al. (2013) compared exercise endurance and recovery of *Ps. modestus* on a terrestrial treadmill under hyperoxic (35% O_2), normoxic (21% O_2), and hypoxic (7% O_2) conditions, and found that endurance and excess post-exercise oxygen consumption (the immediate repayment of the oxygen debt incurred during exercise) were both increased with oxygen availability. They hypothesized that this was due to increased venous oxygen content. Because oxygen-rich blood from the skin and the bucco-opercular mucosa will mix with oxygen-poor systemic venous blood while returning to the heart (see Figure 5.4), an elevated oxygen concentration in hyperoxic air would enhance oxygen supply to cardiac muscles. Most fishes lack the coronary circulation and receive the majority of their oxygen need from the blood contained in the systemic venous return. The cardiac anatomy of mudskipper hearts has been little studied, but our observations on preserved specimens could not find the coronary arteries in *B. boddarti* or *Ps. chrysospilos* (Ishimatsu, personal observation). The hypothesis is difficult to test in small-sized fishes such as *Ps. modestus* (body weight 3–5 g), in which chronic cannulation of the central veins can be challenging; but the technique may be administered to larger species such as *Pn. schlosseri* (body weight up to >200 g, see Section 5.3.2). Other candidate species for studies of this nature include *Boleophthalmus dussumieri* that reaches a maximum total length of 28 cm (Abdoli et al. 2009) and can weigh approximately 56 g (Ghanbarifardi et al. 2014).

5.3.2 Blood Gas Levels

Little has been published on the circulatory physiology of mudskippers largely because of their small size. As stated in Section 5.3.1, adults of *Pn. schlosseri* may weigh over 200 g, which allows for the chronic cannulation of an artery and/or a vein. Ishimatsu et al. (1999) cannulated a branch of the coeliaco-mesenteric artery of this species for measurements of blood gas levels, blood pressure and heart rate. The arterial P_{O_2} in resting conditions was 4.17 kPa (30 mmHg) with the hemoglobin oxygen saturation of 62%. Under resting conditions, where the fish was undisturbed in normoxic water with free access to air, it breathed air every few minutes, but did not show aquatic ventilation. During breath holding, the opercula were kept inflated, and both the mouth and the gill openings were closed, as we usually see in the species' natural habitats. The low arterial saturation probably reflects the supposedly low gas exchange capacity of the gills (see Section 5.2.1.1). The arterial P_{CO_2} was 1.47 kPa (11 mmHg), which is higher than in purely water-breathing fishes (<5 mmHg), but lower than in terrestrial air-breathing animals (>20–30 mmHg, Ultsch 1996; Rahn 1966). Emersion did not affect arterial blood P_{O_2} or P_{CO_2}, but there was a slight but significant rise of arterial blood pH. Forced submersion drastically reduced arterial P_{O_2} from 4.5 kPa to less than 2 kPa within 1.5 hours, which remained at this low level until submersion was terminated 12 hours afterward (Figure 5.7a through f). Arterial blood saturation decreased from about 70% to 30%. Arterial P_{CO_2} decreased significantly, but with no change in arterial blood pH. These results strongly indicate that the cardiorespiratory system of *Pn. schlosseri* is better adapted to air than water. This conclusion is further supported by the findings that *Pn. schlosseri* can repay oxygen debt by forced exercise only when the fish is in air but not in water (Takeda et al. 1999).

5.3.3 Blood Circulation

Garey (1962) was the first to test the effect of submersion on the heart rate of *Pn. freycineti* (reported as *Periophthalmus australis*), and found a development of bradycardia when the mudskipper was submerged or placed in a burrow (detailed methods not given). Gordon et al. (1969) showed only a small (<10 bpm) change in heart rate when *Ps. argentilineatus* (reported as *Periophthalmus sobrinus*) was submerged in normoxic water and then emerged. Similarly, Martin and Bridges (1999) showed that heart rate in *Ps. argentilineatus* was unaffected by aquatic hypoxia (water P_{O_2} < 10.7 kPa) or hyperoxia (>40 kPa) or emersion. Ishimatsu et al. (1999) reported effects of emersion, aquatic hypoxia (with access to air), and forced submersion on blood pressure and heart rate

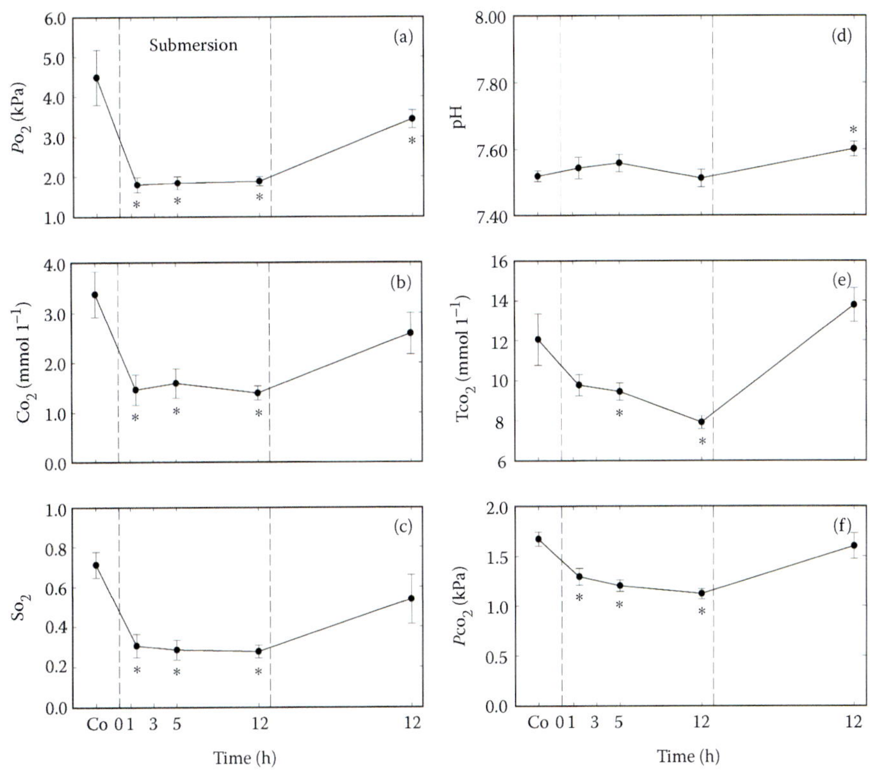

Figure 5.7 Changes in arterial (a) partial pressure of oxygen *(P_{O_2})*; (b) oxygen content; (c) hemoglobin oxygen-saturation; (d) pH; (e) total CO_2 concentration *(T_{CO_2})*; and (f) partial pressure of carbon dioxide (P_{CO_2}) of *Periophthalmodon schlosseri* under control (Co) conditions and during 12 hours of forced submersion in normoxic water without access to air and 12 hours of recovery. Before and after the submersion period, fishes were partially immersed in normoxic 50% seawater with free access to air. Values are means ± S.E.M. ($N = 6$). Asterisks indicate significant differences from the corresponding control value ($p < .05$). (Reproduced from Ishimatsu A. et al., *Journal of Experimental Biology*, 202, 1753–1762, 1999. With permission.)

of chronically cannulated *Pn. schlosseri*. Emersion did not have any effect on blood pressure and it resulted in only a small but transient rise in heart rate. Forced submersion in normoxic water rapidly and significantly depressed both blood pressure and heart rate. Upon re-emersion, the heart rate increased from about 40 to 100 bpm immediately at the first air breath, which indicates that the response was triggered by some neural mechanism (Figure 5.8a,b). Kok et al. (1998) also reported the development of bradycardia upon submersion in *Pn. schlosseri*.

5.3.4 Ecological Interpretation of Data

The observed experimental blood gas and circulatory responses to submersion in *Pn. schlosseri* agree well with the field observations that the fish avoid water during flood tide and remain emerged during high tide (Chua et al. 2011; Polgar and Crosa 2009). How this species cares for developing embryos within burrows during high tide remains a mystery. It has been demonstrated by endoscope observations that embryos of *Pn. schlosseri* develop in the air-filled spawning chamber of a burrow (Ishimatsu et al. 2009), as in *Ps. modestus* (Ishimatsu et al. 2007; also see Chapter 9). It has not been ascertained where the burrow-guarding males of *Ps. modestus* occur during tide high, but we speculated that the fish may spend high tide near the burrow openings where, while intercepting

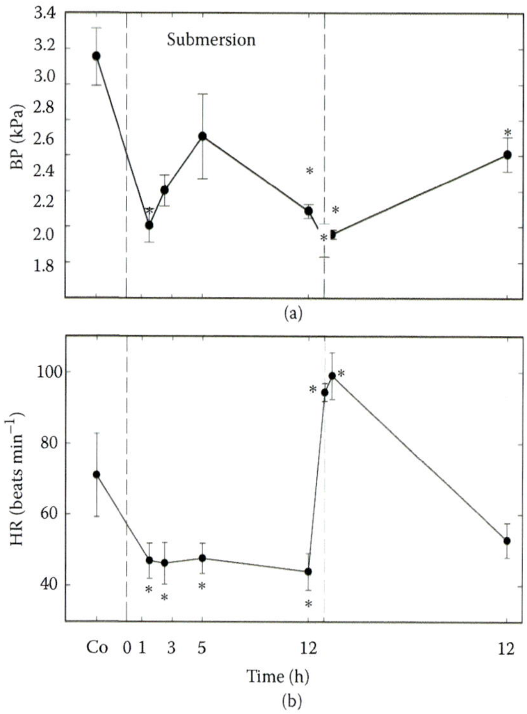

Figure 5.8 Changes in (a) arterial pressure, and (b) heart rate of *Periophthalmodon schlosseri* under control (Co) conditions and during 12 hours of forced submersion in normoxic water without access to air and 12 hours of recovery. For further details, see Figure 5.7. Values are means ± S.E.M. (*N* = 6). (Reproduced from Ishimatsu A. et al., *Journal of Experimental Biology*, 202, 1753–1762, 1999. With permission.)

potential nest predators, it could respire aquatically (Ishimatsu et al. 2007). However, the physiological responses to forced submersion seem to refute this mode of burrow-guarding behavior during high tide in *Pn. schlosseri*. Even though the fish were able to survive 12 hours of forced submersion in normoxic water, the fish were stressed; experimental fishes immediately started vigorous gill ventilation that persisted at rates of 40–50 per minute throughout the submersion period (Ishimatsu et al. 1999). If the burrow-guarding fish respires air stored in its burrow, then additional amounts of oxygen must be stored. Nevertheless, the fact that gas was collected from the burrows of *Pn. schlosseri* in Malaysia during November and December (Ishimatsu et al. 1998), which are the months outside of the species' breeding season in that country (Mazlan and Rohaya 2008), might indicate burrow confinement and aerial respiration therein of the species during high tide (see Table 9.1).

5.4 FUTURE STUDIES

5.4.1 Central Vascular Anatomy of *Scartelaos*

The vascular architecture depicted by Schöttle (1931) for *S. histophorus* (reported as *B. viridis*) is unusual for a teleost and needs to be reexamined by vascular casting (see Section 5.2.2.2). *Scartelaos histophorus* is a relatively small species with a body length of about 10 cm and, thus, may not be a suitable candidate for vascular casting. Assuming the same vascular architecture, *S. gigas* may be the species of choice as it is larger (maximum body length about 20 cm) and is commercially available along the southern coast of Korea.

5.4.2 Regulation of Cutaneous Blood Flow

The surprisingly high reliance of mudskippers (40–80% of the total oxygen uptake in resting in *Periophthalmus* and *Boleophthalmus* species, see Graham 1997: 171, but can be up to 84% during post-exercise recovery, Jew et al. 2013) would become a threat to survival during burrow confinement unless the fish has a regulatory capacity for skin blood flow. This is because mudskipper burrows are filled with severely hypoxic water (Ishimatsu et al. 2000; El-Ziady et al. 1979; Gordon et al. 1978), and blood oxygen may be lost through the skin capillaries, if blood flow persists during burrow confinement. Therefore, vasoconstriction upon submersion in burrow hypoxic water and vasodilation upon emersion of the skin capillaries should be expected. Effects of environmental conditions on skin blood flow and possible control mechanisms for cutaneous gas exchange have drawn little attention of comparative animal physiologists (Feder and Burggren 1985). Though data are limited, there is evidence that these factors influence cutaneous gas exchange and blood flow particularly in amphibians. The mechanisms controlling skin blood flow include direct local vasomotor actions of external O_2 and CO_2 (Malvin and Hlastala 1986), and autonomic nervous control (Jennings and Donald 2008; Malvin and Riedel 1990). Mudskippers resemble amphibians in their reliance on the skin as a gas-exchange organ and are expected to be capable of regulating cutaneous blood perfusion according to external conditions and internal demands.

5.4.3 Swimming Ecophysiology of Mudskippers

We tend to assume that after settlement on a mudflat, mudskippers lead sedentary lives within a relatively small home range or territory. This may not be the case for at least some species of mudskippers. A study demonstrated that a large number of tagged *B. pectinirostris* individuals were recaptured at a distance of 1–2 km from their release sites during the summer on the mudflats of the Ariake Sea, Japan (Onohara and Koga 1992). Onohara (1980) conjectured, on the basis of fisheries catch analyses and interviews with mudskipper fishermen, that *B. pectinirostris* migrate toward offshore areas of the mudflat in April and May, shortly before their breeding season, during ebb tide. These fishes purportedly migrate back toward nearshore areas in August when the breeding season is over. These observations were partly supported by the average daily catches of *B. pectinirostris* that peaked mid-May, using nets designed to catch fishes and shrimps migrating with flood and ebb tides. Another burrow-dwelling gobiid species, *O. lacepedii*, is also caught in large numbers by the same type of set net in the Ariake Sea (Gonzales et al. 2006; Figure 5.9).

Figure 5.9 *Odontamblyopus lacepedii* caught in bag nets designed to catch fishes and shrimps migrating with flood and ebb tides in the Ariake Sea, Japan. (Image courtesy of Kango Nakao.)

In aquatic-air breathing fishes, swimming augments aerial oxygen uptake (Lefevre et al. 2014). It is currently unknown if swimming stimulates aerial oxygen uptake of amphibious air-breathing fishes. Some mudskippers including *B. boddarti*, *B. pectinirostris*, *Oxuderces nexipinnis* and *S. histophorus* retain swim bladders (Ishimatsu, personal observation), contrary to prevailing reports of the absence of this organ in mudskippers. This may be adaptation to swimming mode of life, at least in some phase of their lives.

5.4.4 Plasticity of Cardiorespiratory System during Ontogenetic Shift to Amphibious Life

We have recently found that emersion from water was not indispensable for the survival of the juveniles of *Ps. modestus* under laboratory conditions (Inaba and Ishimatsu, unpublished). We reared two groups of *Ps. modestus* postlarvae in the last planktonic phase in accordance to Kobayashi et al. (1972), approximately 27–29 days after hatching. We repeated this experiment three times and found that the survival was nearly identical between the control group, which had access to land, and the experimental group, which was maintained in normoxic water without access to land or air until the termination of the experiment (70 days after hatching): the mean (±SE) survival was 82 ± 10% for control and 78 ± 9% for the experimental group ($p > .05$). We found the body length to be slightly longer in the experimental group than in the control group ($p < .05$), although there were no obvious differences in gross morphology.

All factors and conditions within the laboratory were kept constant, such as normoxic water supply, sufficient food, and a constant temperature of 26°C. These same conditions are not readily encountered in the natural habitat. However, the data still indicate that *Ps. modestus* has the physiological capacity to survive in water without emergence. Longer-term experiments are necessary to assess the effect of emersion/water confinement on survival and growth, and to compare the respiratory and circulatory anatomy and physiology of the mudskippers reared under these conditions.

5.4.5 Cardiorespiratory Anatomy and Physiology of Overwintering Mudskippers

Mudskippers living in the temperate zone (e.g., *B. pectinirostris*, *Ps. magnuspinnatus*, and *Ps. modestus*) overwinter for several months within the mud (Baeck et al. 2008; Takegaki et al. 2006). Almost nothing is known of the physiology of overwintering mudskippers. In the Ariake Sea, Japan, *B. pectinirostris* and *Ps. modestus* overwinter from early December to early March when the air temperature is below 13°C (Takegaki et al. 2006). Baeck et al. (2008) reported that *Ps. magnuspinnatus* overwinters from mid-October to April in the Suncheonman Bay, Korea. Both *B. pectinirostris* and *Ps. modestus* remain in burrows while overwintering (Kobayashi et al. 1971; Toba and Ishimatsu, unpublished observation), but it is not known how these fishes satisfy their oxygen requirements while in their burrows as there are no data on oxygen concentration of burrow water in overwintering periods. The small volumes of air (<150 mL) within the burrows of *B. pectinirostris* during the overwintering months seem not to be a sole oxygen source (Toba and Ishimatsu 2014). Filling in such information gaps would be much easier if the metabolic rates of mudskippers at temperatures they experience during overwintering were known. If these mudskippers do indeed rely on aquatic respiration while overwintering, they might undergo gill remodeling (see Section 5.2.1.1). Alternatively, it is possible that overwintering mudskippers suppress metabolic rates, which may be reduced to 5–40% of resting levels (Guppy and Withers 1999).

ACKNOWLEDGMENTS

The author thanks Ms. Mizuri Murata for her help in preparing the figures.

REFERENCES

Abdoli L., Kamrani E., Abdoli A. and Kiabi B. 2009. Length-weight relationships for three species of mudskippers (Gobiidae: Oxudercinae) in the coastal areas of the Persian Gulf, Iran. *Journal of Applied Ichthyology* 25:236–237.

Aguilar N.M., Ishimatsu A., Ogawa K. and Khoo K.H. 2000. Aerial ventilatory responses of the mudskipper, *Periophthalmodon schlosseri*, to altered aerial and aquatic respiratory gas concentrations. *Comparative Biochemistry and Physiology* 127A:285–292.

Al-Kadhomiy N.K. and Hughes G.M. 1988. Histological study of different regions of the skin and gills in the mudskipper, *Boleophthalmus boddarti* with respect to their respiratory function. *Journal of the Marine Biological Association of the United Kingdom* 68:413–422.

Baeck G.W., Takita T. and Yoon Y.H. 2008. Lifestyle of Korean mudskipper *Periophthalmus magnuspinnatus* with reference to a congeneric species *Periophthalmus modestus*. *Ichthyological Research* 55:43–52.

Beon M.S., Oh M.K., Lee Y.J., Kim C.H. and Park J.Y. 2013. A comparative study on vascularization and the structure of the epidermis of an amphibious mudskipper fish, *Scartelaos gigas* (Gobiidae, Teleostei), on different parts of the body and the appendages. *Journal of Applied Ichthyology* 29:410–415.

Bicudo J.E.P.W. and Johansen K. 1979. Respiratory gas exchange in the airbreathing fish, *Synbranchus marmoratus*. *Environmental Biology of Fishes* 4:55–64.

Brauner C.J., Matey V., Wilson M., Bernier N.J. and Val A.L. 2004. Transition in organ function during the evolution of air-breathing; insights from *Arapaima gigas*, an obligate air-breathing teleost from the Amazon. *Journal of Experimental Biology* 207:1433–1438.

Burggren W.W. and Johansen K. 1986. Circulation and respiration in lungfishes (Dipnoi). *Journal of Morphology* 190 (Suppl. 1):217–236.

Carter G. S. 1935. Respiratory adaptations of the fishes of the forest waters, with descriptions of the accessory respiratory organs of *Electrophorus electricus* (Linn.) (=*Gymnotus electricus* auctt.) and *Plecostomus plecostomus* (Linn.). *Journal of Linnean Society* 39:219–233.

Carter G.S. and Beadle L.C. 1931. The fauna of the Paraguayan Chaco. *Zoological Journal of the Linnean Society of London* 37:327–368.

Chapman L.J. and Hulen K.G. 2001. Implications of hypoxia for the brain size and gill morphometry of mormyrid fishes. *Journal of Zoology* 254:461–472.

Chua J.L.T., Jaafar Z. and Sivasothi N. 2011. Tidal migration of the giant mudskipper, *Periophthalmodon schlosseri* (Pallas, 1770) in a Singapore mangrove. *Undergraduate Research Opportunities U063111B*. National University of Singapore, Singapore.

Clack J.A. 2012. *Gaining Ground: The Origin and Evolution of Tetrapods*, 2nd Edition. Bloomington, IN: Indiana University Press, 523 pp.

Clayton D.A. 1993. Mudskippers. *Oceanography and Marine Biology: An Annual Review* 31:507–577.

Das B.K. 1934. The habits and structure of *Pseudapocryptes lanceolatus*, a fish in the first stage of structural adaptation to aerial respiration. *Proceedings of the Royal Society of London* 115:422–430.

Dinh Q.M., Giang T.T.N., Duy N.L.N., Dong H.D. and Hau T.L. 2014a. Burrow configuration and utilization of the blue-spotted mudskipper, *Boleophthalmus boddarti* caught in Soc Trang, Vietnam. *Kasetsart University Fishery Research Bulletin* 38:1–9.

Dinh Q.M., Qin J.G., Dittmann S. and Tran D.D. 2014b. Burrow morphology and utilization of the goby (*Parapocryptes serperaster*) in the Mekong Delta, Vietnam. *Ichthyological Research* 61:332–340.

El-Ziady S., Eissa S.M. and Al-Naqi Z. 1979. Ecological studies on the mud-skipper fish, *Periophthalmus chrysospilos* (Bleeker) in Kuwait. *Bulletin of Faculty of Science, Cairo University* 48:113–133.

Farber J. and Rahn H. 1970. Gas exchange between air and water and the ventilation pattern in the electric eel. *Respiration Physiology* 9:151–161.

Feder M.E. and Burggren W.W. 1985. Cutaneous gas exchange in vertebrates: Design, patterns, control and implications. *Biological Reviews* 60:1–45.

Garey W.F. 1962. Cardiac responses of fishes in asphyxic environments. *Biological Bulletin* 122:362–368.

Gee J.H. and Gee P.A. 1991. Reactions of gobioid fishes to hypoxia: Buoyancy control and aquatic surface respiration. *Copeia* 1991:17–28.

Ghanbarifardi M., Ghasemian S., Aliabadian M. and Pehpuri A. 2014. Length-weight relationships for three species of mudskippers (Gobiiformes: Gobionellidae) in the coastal areas of the Persian Gulf and Gulf of Oman, Iran. *Iranian Journal of Ichthyology* 1:29–31.

Gonzales T.T., Katoh M. and Ishimatsu A. 2006. Air breathing of aquatic burrow-dwelling eel goby, *Odontamblyopus lacepedii* (Gobiidae: Amblyopinae). *Journal of Experimental Biology* 209:1085–1092.

Gonzales T.T., Katoh M. and Ishimatsu A. 2008a. Respiratory vasculatures of the intertidal air-breathing eel goby, *Odontamblyopus lacepedii* (Gobiidae: Amblyopinae). *Environmental Biology of Fishes* 82:341–351.

Gonzales T.T., Katoh M. and Ishimatsu A. 2008b. Intertidal burrows of the air-breathing eel goby, *Odontamblyopus lacepedii* (Gobiidae: Amblyopinae). *Ichthyological Research* 55:303–306.

Gonzales T.T., Katoh M., Mazlan G. and Ishimatsu A. 2011. Gross and fine anatomy of the respiratory vasculature of the mudskipper, *Periophthalmodon schlosseri* (Gobiidae: Oxudercinae). *Journal of Morphology* 272:629–640.

Gordon M.S., Boëtius I., Evans D.H., McCarthy R. and Oglesby L.C. 1969. Aspects of the physiology of terrestrial life in amphibious fishes I. The mudskipper, *Periophthalmus sobrinus*. *Journal of Experimental Biology* 50:141–149.

Gordon M.S., Ng W.W. and Yip A.Y. 1978. Aspects of the physiology of terrestrial life in amphibious fishes. III. The Chinese mudskipper *Periophthalmus cantonensis*. *Journal of Experimental Biology* 72:57–75.

Graham J.B. 1976. Respiratory adaptations of marine air-breathing fishes. In *Respiration of Amphibious Vertebrates* (ed. G.H. Hughes) pp. 165–187. New York, NY: Academic Press.

Graham J.B. 1997. *Air-Breathing Fishes: Evolution, Diversity and Adaptation*. San Diego, CA: Academic Press, 299 pp.

Graham J.B., Lee H.J. and Wegner N.C. 2007. Transition from water to land in an extant group of fishes: Air breathing and the acquisition sequence of adaptations for amphibious life in oxudercine gobies. In *Fish Respiration and Environment* (eds. M.N. Fernandes, F.T. Rantin, M.L. Glass and B.G. Kapoor) pp. 255–288. Enfield: Science Publisher.

Graham J.B., Wegner N.C., Miller L.A. et al. 2014. Spiracular air breathing in polypterid fishes and its implications for aerial respiration in stem tetrapods. *Nature Communications* 5:3022. doi:10.1038/ncomms4022

Guerreiro P.M., Costa R. and Power D.M. 2013. Dynamics of scale regeneration in seawater-and brackish water-acclimated sea bass, *Dicentrarchus labrax*. *Fish Physiology and Biochemistry* 39:917–930.

Guppy M. and Withers P. 1999. Metabolic depression in animals: Physiological perspectives and biochemical generalizations. *Biological Reviews* 74:1–40.

Hughes G.M. 1972. Morphometrics of fish gills. *Respiration Physiology* 14:1–25.

Hughes G.M., Perry S.F. and Piiper J. 1986. Morphometry of the gills of the elasmobranch *Scyliorhinus stellaris* in relation to body size. *Journal of Experimental Biology* 121:27–42.

Icardo J.M. 2006. Conus arteriosus of the teleost heart: Dismissed, but not missed. *Anatomical Record* 288:900–908.

Icardo J.M. and Colvee E. 2011. The atrioventricular region of the teleost heart. A distinct heart segment. *Anatomical Record* 294:236–242.

Ishimatsu A. 2012. Evolution of the cardiorespiratory system in air-breathing fishes. *Aqua-BioScience Monographs* 5:1–28.

Ishimatsu A., Aguilar N.M., Ogawa K. et al. 1999. Arterial blood gas levels and cardiovascular function during varying environmental conditions in a mudskipper, *Periophthalmodon schlosseri*. *Journal of Experimental Biology* 202:1753–1762.

Ishimatsu A. and Gonzales T.T. 2011. Mudskippers: Front runners in the modern invasion of land. In *The Biology of Gobies* (eds. R. Patzner, J.L. Van Tassell, M. Kovačić and B.G. Kapoor) pp. 609–638. Enfield: Science Publishers.

Ishimatsu A., Hishida Y., Takita T. et al. 1998. Mudskippers store air in their burrows. *Nature* 391:237–238.

Ishimatsu A. and Itazawa Y. 1983. Difference in blood oxygen levels in the outflow vessels of the heart of an air-breathing fish, *Channa argus*: Do separate blood streams exist in a teleostean heart? *Journal of Comparative Physiology* 149B:435–440.

Ishimatsu A., Itazawa Y. and Takeda T. 1979. On the circulatory systems of the snakeheads *Channa maculata* and *C. argus* with reference to bimodal breathing. *Japanese Journal of Ichthyology* 26:167–180.

Ishimatsu A., Takeda T., Kanda T., Oikawa S. and Khoo K.H. 2000. Burrow environment of mudskippers in Malaysia. *Journal of Bioscience* 11:17–28.

Ishimatsu A., Takeda T., Tsuhako Y., Gonzales T.T. and Khoo K.H. 2009. Direct evidence for aerial egg deposition in the burrows of the Malaysian mudskipper, *Periophthalmodon schlosseri*. *Ichthyological Research* 56:417–420.

Ishimatsu A., Yoshida Y., Itoki N., Takeda T., Lee H.J. and Graham J.B. 2007. Mudskippers brood their eggs in air but submerge them for hatching. *Journal of Experimental Biology* 210:3946–3954.

Itoki N., Sakamoto T., Hayashi M., Takeda T. and Ishimatsu A. 2012. Morphological responses of mitochondria-rich cells to hypersaline environments in the Australian mudskipper, *Periophthalmus minutus*. *Zoological Science* 29:444–449.

Jennings B.L. and Donald J.A. 2008. Neurally-derived nitric oxide regulates vascular tone in pulmonary and cutaneous arteries of the toad, *Bufo marinus*. *American Journal of Physiology-Regulatory, Integrative and Comparative Physiology* 295:R1640–R1646.

Jew C.J., Wegner N.C., Yanagitsuru Y., Tresguerres M. and Graham J.B. 2013. Atmospheric oxygen levels affect mudskipper terrestrial performance: Implications for early tetrapods. *Integrative and Comparative Biology* 53:248–257.

Johansen K. 1966. Air breathing in the teleost *Symbranchus marmoratus*. *Comparative Biochemistry and Physiology* 18:383–395.

Kobayashi T., Dotsu Y. and Miura N. 1972. Egg development and rearing experiments of the larvae of the mud skipper, *Periophthalmus cantonensis*. *Bulletin of the Faculty of Fisheries, Nagasaki University* 33:49–62 (in Japanese with English abstract).

Kobayashi T., Dotsu Y. and Takita T. 1971. Nest and nesting behavior of the mud skipper, *Periophthalmus cantonensis* in Ariake Sound. *Bulletin of the Faculty of Fisheries, Nagasaki University* 32:27–40 (in Japanese with English abstract).

Kok W.K., Lim C.B., Lam T.J. and Ip Y.K. 1998. The mudskipper *Periophthalmodon schlosseri* respires more efficiently on land than in water and vice versa for *Boleophthalmus boddaerti*. *Journal of Experimental Zoology* 280:86–90.

Kunzmann A. 1990. Gill morphometrics of two Antarctic fish species *Pleuragramma antarcticum* and *Notothenia gibberifrons*. *Polar Biology* 11:9–18.

Kutschera U. and Elliott J. 2013. Do mudskippers and lungfishes elucidate the early evolution of four-limbed vertebrates? *Evolution: Education and Outreach* 6:8. doi:10.1186/1936-6434-6-8

LeBlanc D.M., Wood C.M., Fudge D.S. and Wright P.A. 2010. A fish out of water: Gill and skin remodeling promotes osmo- and ionoregulation in the mangrove killifish *Kryptolebias marmoratus*. *Physiological and Biochemical Zoology* 83:932–949.

Lee C.L. 1990. Osteological study of the mudskipper, *Periophthalmus cantonensis* (Perciformes, Gobiidae) from Korea. *Korean Journal of Zoology* 33:402–410 (in Korean with English abstract).

Lefevre S., Domenici P. and Mckenzie D. 2014. Swimming in air-breathing fishes. *Journal of Fish Biology* 84:661–681.

Liem K.F. 1987. Functional design of the air ventilation apparatus and overland excursion by teleosts. *Fieldiana Zoology, New Series* 37:1–29.

Low W.P., Ip Y.K. and Lane D.J.W. 1990. A comparative study of the gill morphometry in the mudskippers-*Periophthalmus chrysospilos, Boleophthalmus boddaerti* and *Periophthalmodon schlosseri*. *Zoological Science* 7:29–38.

Low W.P., Lane D.J.W. and Ip Y.K. 1988. A comparative study of terrestrial adaptations of the gills in three mudskippers—*Periophthalmus chrysospilos, Boleophthalmus boddaerti*, and *Periophthalmodon schlosseri*. *Biological Bulletin* 175:434–438.

Malvin G.M. and Hlastala M.P. 1986. Regulation of cutaneous gas exchange by environmental O_2 and CO_2 in the frog. *Respiration Physiology* 65:99–111.

Malvin G.M. and Riedel C. 1990. Autonomic regulation of cutaneous vascular resistance in the bullfrog *Rana catesbeiana*. *Journal of Experimental Biology* 152:425–439.

Martin K. and Bridges C. 1999. Respiration in water and air. In *Intertidal Fishes: Life in Two Worlds* (eds. M. Horn, K. Martin and M. Chotkowski) pp. 54–78. San Diego, CA: Academic Press.

Mazlan A.G., Masitah A. and Mahani M.-C. 2006. Fine structure of gills and skins of the amphibious mudskipper, *Periophthalmus chrysospilos* Bleeker, 1852, and a non-amphibious goby, *Favonigobius reichei* (Bleeker, 1853). *Acta Ichthyologica et Piscatoria* 36:127–133.

Mazlan A.G. and Rohaya M. 2008. Size, growth and reproductive biology of the giant mudskipper, *Periophthalmodon schlosseri* (Pallas, 1770), in Malaysian waters. *Journal of Applied Ichthyology* 24:290–296.

Milward N.E. 1974. Studies on the taxonomy, ecology, and physiology of Queensland mudskippers. PhD thesis, University of Queensland.

Munshi J.S.D., Olson K.R., Ghosh T.K. and Ojha J. 1990. Vasculature of the head and respiratory organs in an obligate air-breathing fish, the swamp eel *Monopterus* (=*Amphipnous*) *cuchia*. *Journal of Morphology* 203:181–201.

Nilsson G.E., Dymowska A. and Stecyk J.A. 2012. New insights into the plasticity of gill structure. *Respiration Physiology and Neurobiology* 184:214–222.

Niva B., Ojha J. and Munshi J.S.D. 1981. Morphometrics of the respiratory organs of an estuarine goby, *Boleophthalmus boddaerti*. *Japanese Journal of Ichthyology* 27:316–326.

Olson K.R. 1981. Morphology and vascular anatomy of the gills of a primitive air-breathing fish, the bowfin (*Amia calva*). *Cell and Tissue Research* 218:499–517.

Olson K.R., Munshi J.S.D., Ghosh T.K. and Ojha J. 1986. Gill microcirculation of the air-breathing climbing perch, *Anabas testudineus* (Bloch): Relationships with the accessory respiratory organs and systemic circulation. *American Journal of Anatomy* 176:305–320.

Olson K.R., Roy P.K., Ghosh T.K. and Munshi J.S.D. 1994. Microcircultion of gills and accessory respiratory organs from the air-breathing snakehead fish, *Channa punctata, C. gachua*, and *C. marulius*. *Anatomical Record* 238:92–107.

Ong K.J., Stevens E.D. and Wright P.A. 2007. Gill morphology of the mangrove killifish (*Kryptolebias marmoratus*) is plastic and changes in response to terrestrial air exposure. *Journal of Experimental Biology* 210:1109–1115.

Onohara T. 1980. Mutsugoro no seitai—I. Gyogyo seisan, bunpu oyobi seicho nitsuite (The ecology of *Boleophthalmus pectinirostris*—I. Fishery catch, distribution and growth). *Bulletin of Saga Prefectural Ariake Fisheries Research and Development Center* 7:123–150 (in Japanese).

Onohara T. and Koga H. 1992. Mutsugoro no seitai—V. hyoshiki horyu kara mita kotai seicho to idou (Ecological study of mud skipper—V. Growth and movement of adult by tagged release). *Bulletin of Saga Prefectural Ariake Fisheries Research and Development Center* 14:1–8 (in Japanese).

Park J.Y. 2002. Structure of the skin of an air-breathing mudskipper, *Periophthalmus magnuspinnatus*. *Journal of Fish Biology* 60:1543–1550.

Park J.Y., Kim I.S. and Lee Y.J. 2006. A study on the vascularization and structure of the epidermis of the air-breathing mudskipper, *Periophthalmus magnuspinnatus* (Gobiidae, Teleostei), along different parts of the body. *Journal of Applied Ichthyology* 22:62–67.

Park J.-Y., Lee Y.-J., Kim I.-S. and Kim S.-Y. 2003. A comparative study of the regional epidermis of an amphibious mudskipper fish, *Boleophthalmus pectinirostris* (Gobiidae, Pisces). *Folia Zoologica* 52:431–440.

Polgar G. and Crosa G. 2009. Multivariate characterisation of the habitats of seven species of Malayan mudskippers (Gobiidae: Oxudercinae). *Marine Biology* 156:1475–1486.

Polgar G., Malavasi S., Cipolato G., Georgalas V., Clack J.A. and Torricelli P. 2011. Acoustic communication at the water's edge: Evolutionary insights from a mudskipper. *PLoS One* 6:e21434. doi:10.1371/journal.pone.0021434.

Rahn H. 1966. Aquatic gas exchange: Theory. *Respiration Physiology* 1:1–12.

Ramos C.A., Fernandes M.N., da Costa O.T.F. and Duncan W.P. 2013. Implications for osmorespiratory compromise by anatomical remodeling in the gills of *Arapaima gigas*. *Anatomical Record* 296:1664–1675.

Roy P. and Munshi J.D. 1986. Morphometrics of the respiratory organs of a freshwater major carp, *Cirrhinus mrigala* in relation to body weight. *Japanese Journal of Ichthyology* 33:269–279.

Sakamoto T., Yokota S. and Ando M. 2000. Rapid morphological oscillation of mitochondrion-rich cell in estuarine mudskipper following salinity changes. *Journal of Experimental Zoology* 286:666–669.

Schoch R.R. 2014. *Amphibian Evolution: The Life of Early Land Vertebrates*. Chichester: John Wiley & Sons, 264 pp.

Schöttle E. 1931. Morphologie und Physiologie der Atmung bei Wasser-, Schlamm- und Landlebenden Gobiiformes. *Zeitschrift für wissenschaftliche Zoologie* 140:1–114.

Sollid J., De Angelis P., Gundersen K. and Nilsson G.E. 2003. Hypoxia induces adaptive and reversible gross morphological changes in crucian carp gills. *Journal of Experimental Biology* 206:3667–3673.

Suzuki N. 1992. Fine structure of the epidermis of the mudskipper, *Periophthalmus modestus* (Gobiidae). *Japanese Journal of Ichthyology* 38:379–397.

Takeda T., Hayashi M., Toba A., Soyano K. and Ishimatsu A. 2011. Ecology of the Australian mudskipper, *Periophthalmus minutus*, an amphibious fish inhabiting a mudflat in the highest intertidal zone. *Australian Journal of Zoology* 59:312–320.

Takeda T., Ishimatsu A., Oikawa S., Kanda T., Hishida Y. and Khoo K.H. 1999. Mudskipper *Periophthalmodon schlosseri* can repay oxygen debts in air but not in water. *Journal of Experimental Zoology* 284:265–270.

Takegaki T., Fujii T. and Ishimatsu A. 2006. Overwintering habitat and low-temperature tolerance of the young mudskipper *Boleophthalmus pectinirostris*. *Nippon Suisan Gakkaishi* 72:880–885 (in Japanese with English abstract).

Takita T., Larson H.K. and Ishimatsu A. 2011. The natural history of mudskippers in northern Australia, with field identification characters. *The Beagle, Record of the Museums and Art Galleries of the Northern Territory* 27:189–204.

Tamura O. and Moriyama T. 1976. On the morphological feature of the gill of amphibious and air breathing fishes. *Bulletin of the Faculty of Fisheries, Nagasaki University* 41:1–8.

Toba A. and Ishimatsu A. 2014. Roles of air stored in burrows of the mudskipper *Boleophthalmus pectinirostris* for adult respiration and embryonic development. *Journal of Fish Biology* 84:774–793.

Todd E.S. 1968. Terrestrial sojourn of the long-jaw mudsucker, *Gillichthys mirabilis*. *Copeia* 1968:192–194.

Todd E.S. and Ebeling A.W. 1966. Aerial respiration in the longjaw mudsucker *Gillichthys mirabilis* (Teleostei: Gobiidae). *Biological Bulletin* 130:265–288.

Turko A., Earley R. and Wright P. 2011. Behaviour drives morphology: Voluntary emersion patterns shape gill structure in genetically identical mangrove rivulus. *Animal Behaviour* 82:39–47.

Ultsch G.R. 1996. Gas exchange, hypercarbia and acid-base balance, paleoecology, and the evolutionary transition from water-breathing to air-breathing among vertebrates. *Palaeogeography, Palaeoclimatology, Palaeoecology* 123:1–27.

Wegner N.C., Sepulveda C.A., Bull K.B. and Graham J.B. 2010. Gill morphometrics in relation to gas transfer and ram ventilation in high-energy demand teleosts: Scombrids and billfishes. *Journal of Morphology* 271:36–49.

Yokota S., Iwata K., Fujii Y. and Ando M. 1997. Ion transport across the skin of the mudskipper *Periophthalmus modestus*. *Comparative Biochemistry and Physiology* 118A:903–910.

Yokoya S. and Tamura O.S. 1992. Fine structure of the skin of the amphibious fishes, *Boleophthalmus pectinirostris* and *Periophthalmus cantonensis*, with special reference to the location of blood vessels. *Journal of Morphology* 214:287–297.

Zhang J. 2000. A comparative study on the morphological adaptations in oxudercine gobies to their terrestrial life. PhD thesis, Nagasaki University.

Zhang J., Taniguchi T., Takita T. and Ali A.B. 2000. On the epidermal structure of *Boleophthalmus* and *Scartelaos* mudskippers with reference to their adaptation to terrestrial life. *Ichthyological Research* 47:359–366.

Zhang J., Taniguchi T., Takita T. and Ali A.B. 2003. A study on the epidermal structure of *Periophthalmodon* and *Periophthalmus* mudskippers with reference to their terrestrial adaptation. *Ichthyological Research* 50:310–317.

CHAPTER 6

Structure and Function of Sensory Organs

Michał Kuciel, Krystyna Żuwała, Eugenia R. Lauriano, Gianluca Polgar, Stefano Malavasi, and Giacomo Zaccone

CONTENTS

6.1 INTRODUCTION

Mudskippers have many morphological adaptations that enable them to cope with the different conditions encountered in and between aquatic and terrestrial habitats. This is challenging especially for fishes with a body plan suitable for the aquatic medium. In addition to essential processes such as osmoregulation, excretion, respiration, thermal balance, and locomotion, basic functioning senses such as photoreception, chemoreception, and mechanoreception must now accommodate both habitats (Sayer 2005).

At a position usually directly on or close to the body surface, receptor cells can readily receive impulses from the surrounding environment but at the same time, expose these receptors to external conditions. The adaptations to these demanding environments have made mudskippers a frequently studied fish group and make them possible extant models for the Paleozoic evolution of vertebrate transition from water to land (Graham and Lee 2004).

6.2 ANATOMY OF THE EYE

Vision is a basic and important sense for most fish species. The diversity and variations in the anatomy and function of the fish eye have been described in many studies (Lamb et al. 2007; Douglas and Djamgoz 1990). Fish eyes are similar to the eyes of other vertebrate animals in their primary functions: collecting light, focusing images on the retina, and transforming images to neural signals. Typically, eyes of bony fishes are large, hemispheric, and without eyelids. There are three main concentric layers: the external sclera transforming into a transparent cornea, the medial uveal layer, and the retina, which is the inner photosensitive layer of nerve tissues (Figure 6.1a,b).

The anatomy of amphibious fish eyes has been relatively well studied (Sayer 2005). The difference in refractive indices of air and water (1.00–1.33) causes light to behave differently in these two media. The refractive indices for water and the cornea are the same. While in air, however, the refractive properties of the cornea are added to those of the lens and cause myoptic "short-sighted" vision. Other concerns adapting to vision out of water include an increased amount of light reaching the eye, high probability of desiccation, and removing excess water from the cornea (Graham 1971).

Draining the cornea of excess water improves the quality of vision but at the same time increases the rate of water loss. A dermal cup, or skin pocket, is present at the ventral margin of the

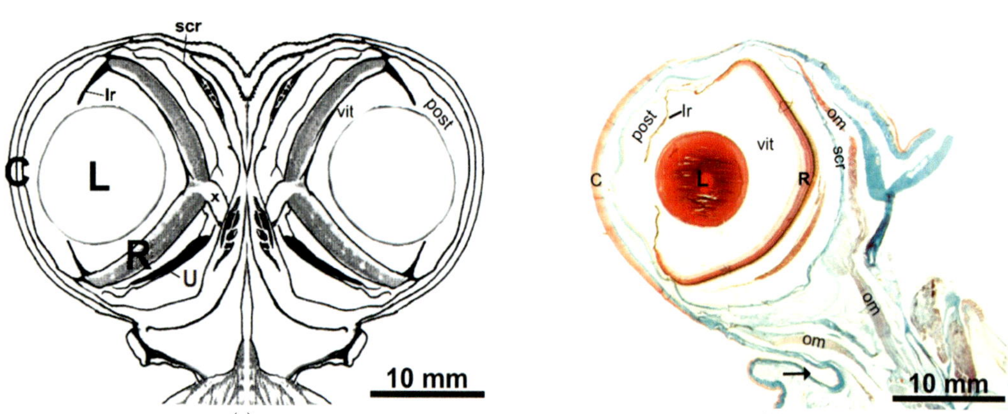

Figure 6.1 Longitudinal sections of the eyes of *Periophthalmus barbarus*. (a) Entire eye region. (b) Right eye, with Gomori's trichrome stain; arrow points to dermal cup. C, cornea; lr, iris; L, round lens; om, ocular muscle; post, posterior chamber; R, retina; scr, sclera; U, uveal; vit, vitreous body; x, olfactory nerve.

corneal surfaces just beneath the eye (Figures 6.1b and 6.2a,c,d), which retains some moisture in the amphibious blenny *Mnierpes macrocephalus* (Graham and Rosenblatt 1970) and most mudskippers (Graham 1971). Many mudskipper species are also able to retract their eyes into the cranial cavity, thereby simultaneously removing excess water and moisturizing the cornea with the residual water. In other amphibious fishes such as *Anabas testudineus*, the eyes are mounted within deep bony orbits (Davenport and Matin 1990). Additionally, in many amphibious fishes, corneal coverings or thickened conjunctivae protect the eyes from drying out (Jermann and Senn 1992; Zander 1974; Graham 1971). Morphological changes to eyes in amphibious fishes do not automatically translate to superior vision while on land. Observations on terrestrial foraging in *A. testudineus* and *Gobionellus sagittala*, for example, suggest that they have poor vision, and successful prey capture appears random rather than targeted (Todd 1976). In some amphibious fishes, superior visual acuity allows precision during prey capture (Sponder and Lauder 1981) and avoidance of predators (Munk 1970), while on land. Mudskippers, for example, prey on flying insects, an ability that not only requires good visual capabilities, but also precise coordination, motility, and reflex.

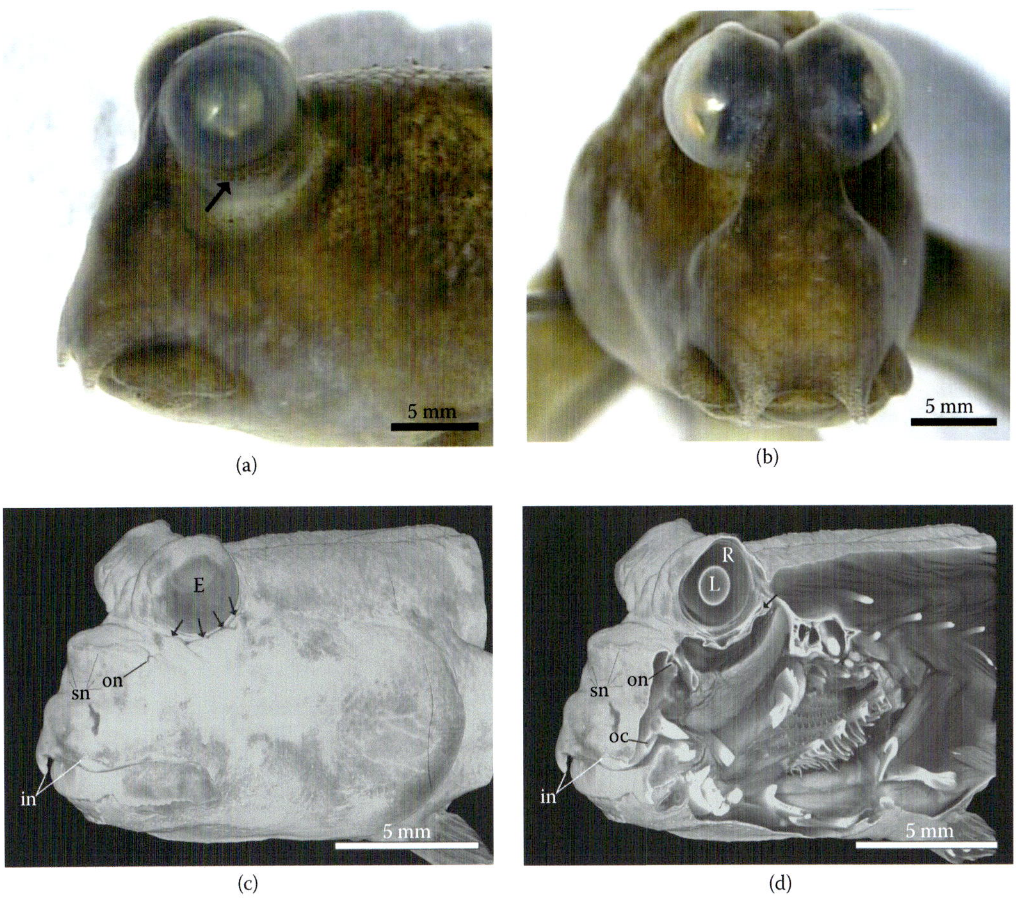

Figure 6.2 Heads of species of *Periophthalmus*. (a) Close-up external lateral view of head of *Periophthalmus minutus*. (b) Close-up external frontal view of head of *Periophthalmus minutus*. (c) Computed tomography (CT)-scan of head of *Periophthalmus barbarus*. (d) Longitudinal section (CT-scan) of head of *Periophthalmus barbarus*. Arrow points to dermal cup; E, eye; in, inlet naris; L, round lens; oc, olfactory canal (left olfactory canal mis-shapen at the anterior part); on, outlet naris; R, retina; sn, surface neuromasts.

The anatomy of mudskipper eyes has been described for *Periophthalmodon schlosseri* (Baumeister 1913; Volz 1905a, b, c), *Periophthalmus argentilineatus* (Munk 1970; Karsten 1923; Harms 1914), *Periophthalmus kalolo* (reported as *Periophthalmus koelreuteri*) (Munk 1970; Hess 1912), *Periophthalmus chrysospilos* (Yew and Wu 1979), *Periophthalmus modestus* (reported as *Periophthalmus cantonensis*) (Yamamoto 1931), *Boleophthalmus boddarti* (Munk 1970; Baumeister 1913; Volz 1905a), and *Boleophthalmus pectinirostris* (Franz 1910).

6.2.1 Gross Morphology

6.2.1.1 Cornea and Sclera

The external layer of the eye comprises the cornea and the sclera, which is internal to the cornea. The sclera is non-refractile and comprises collagen, fibroblasts, elastic fibers, and an external sheath of hyaline cartilage. It extends from the cornea to the optic nerve and protects the eye, maintains intraocular pressure, and provides attachment for surrounding muscles (Kerr 1999). The cornea is made up of non-pigmented squamous epithelia cells, membranous stroma, and a thin endothelium. The corneal epithelium is hydrated, multilayered, and contiguous with the integument of the head. The wall of the posterior portion of the eye is comprised of a thick fibroblastic sclera, which is contiguous with the corneal stroma (Genten et al. 2009).

The cornea in *Ps. barbarus*, like those of other vertebrates, is divided into distal and proximal lamellae (Figure 6.3a,b; Karsten 1923). The outer lamella (Figure 6.3b) is presumed to provide additional protection from desiccation and mechanical damage (Graham 1971).

6.2.1.2 Uveal Layer

In *Ps. barbarus*, the uveal structure is similar to those described in other vertebrate species (Figure 6.1a,b). It forms the middle coat of the eyeball and consists of three parts: the choroid, the ciliary process, and the iris. The choroid lies on the inner aspect of the sclera and ends anteriorly beyond the terminal margin of the retina at the ciliary process. The ciliary process supports the retina with essential nutrients. At the outer margin of the lens, the choroid is modified to contribute to the core of the ciliary processes, which are irregular finger-like processes covered by a double epithelial layer. The superficial ciliary cells are not pigmented; however, the cells of the inner layer next to the core of connective tissue are highly pigmented. The iris (Figures 6.1a and 6.3b) is a thin diaphragm of tissue arising from the ciliary body and rests on the lens (Figures 6.1a,b and 6.3a,c). The part of the iris in contact with the lens contains melanocytes. The central opening of the iris, the pupil, regulates the amount of light entering the eye (Kerr 1999).

6.2.1.3 Eye Lens

The eye lenses in teleost fishes are usually round and protrude into the aqueous chamber almost reaching the cornea, and together with the cornea, provide a wide field of vision by transmitting and focusing incoming light onto the retina. The lens is made up of an extracellular matrix (capsule) secreted by the cells contained within it. The lens cells are organized into two subpopulations: cells in the anterior chamber are single-layered epithelial cells, whereas cells in the posterior chamber, bathed with vitreous humor, are elongated fibers. Once light passes through the lens, it is transmitted through a transparent liquid medium and reaches the retina in which photoreceptors are contained (Helfman et al. 2009). Due to the different refraction indices of the cornea and liquid medium, the lens is responsible for the correction caused by refraction (Land and Nilsson 2012). In order to achieve this visual accommodation, the lens moves forward and backward via retractor muscles and suspensory ligaments (Genten et al. 2009; Helfman et al. 2009).

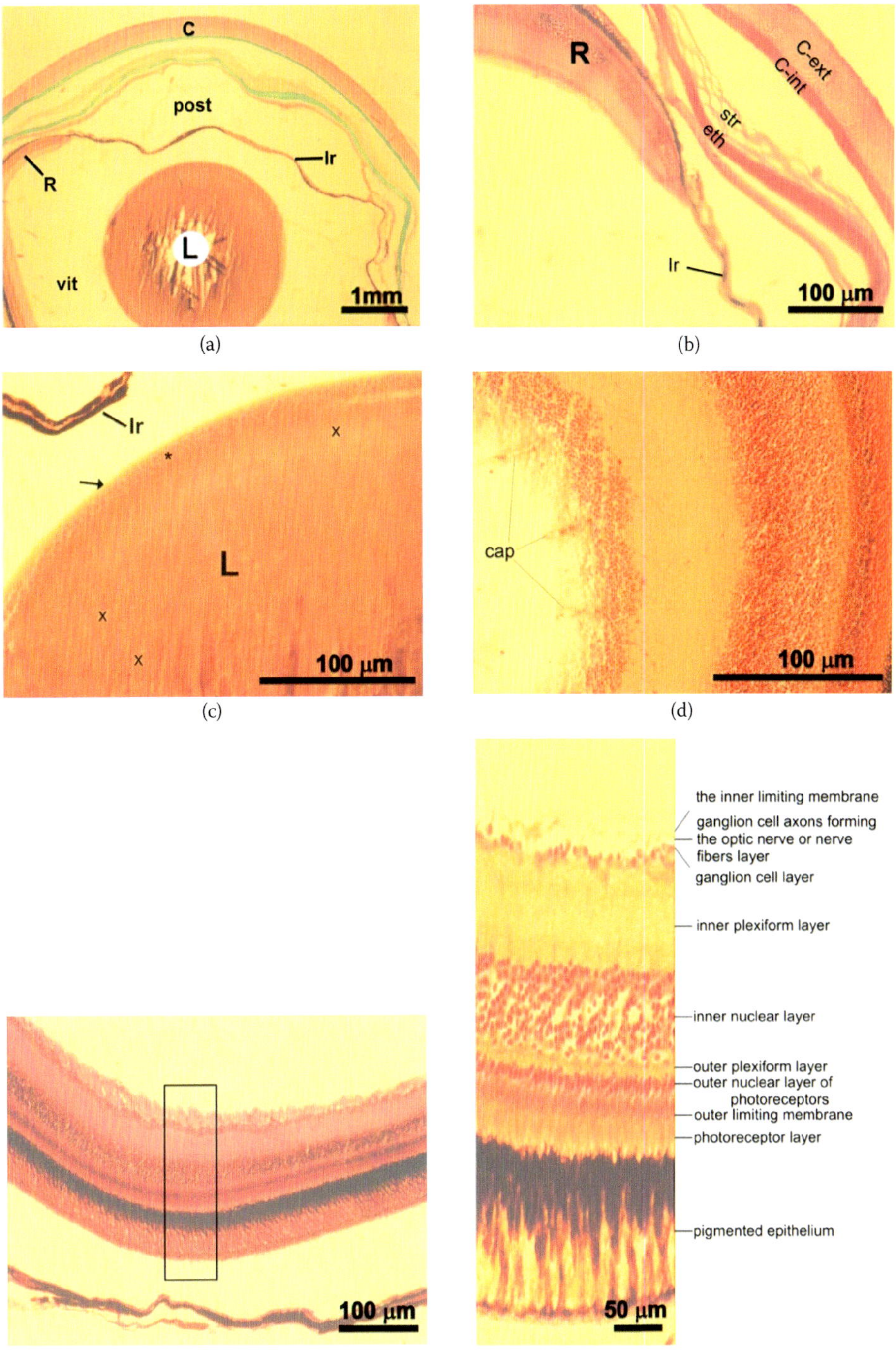

Figure 6.3 Longitudinal sections of the eye of *Periophthalmus barbarus* with Gomori's trichrome stain. (a) Close-up of the cornea and lens. (b) Close-up of the cornea and iris. (c) Close-up of the iris. (d) and (f) Close-up of the retina cross section (e). Arrow, lens capsule; cap, capillaries, C-ext, cornea external layer; C-int, cornea internal layer; eth, thin corneal endothelium; Ir, iris; L, round lens; post, posterior chamber; R, retina; *, monolayer of epithelial cells; str, squamous corneal epithelium; vit, vitreous body; x, elongated cells facing the chamber of the vitreous body.

Karsten (1923) first documented the muscular eye structure in mudskippers. He described *Ps. barbarus* (reported as *Ps. koelreuteri*) as possessing oval lenses supported by muscles similar in arrangement to other fishes, thus enabling accommodation for the movement of the lens. Yamamoto (1931) found the eye of *Ps. modestus* to be hypermetropic, slightly in air and strongly in water. The range for image accommodation is probably insufficient to restore emmetropic vision in water, and, thus, the *Periophthalmus* eye is likely adapted for aerial vision. The strongly curved cornea and, in comparison to other fishes, the flattened lens of *Periophthalmus* and *Boleophthalmus* also contribute to adaptation out of water (Graham 1971).

We determined that the lens of *Ps. barbarus* is composed of cells that are surrounded by a capsule with two different cell types: (1) a monolayer of external epithelial cells and (2) elongate cells without nuclei, facing the chamber of the vitreous body. The lens is circular in transverse (Figure 6.1b), longitudinal (Figure 6.1b), and sagittal (Figure 6.1a) sections, in contrast with previous studies reporting an oval lens to be characteristic for the genus *Periophthalmus* (Graham 1971).

6.2.1.4 Retina

The retina is a thin layer of neural tissue lining the inner eye. It is stratified with 10 distinct layers: (1) the pigment epithelium; (2) the elongate tips of cones and rods; (3) the external limiting membrane; (4) the outer nuclear zone containing the cell bodies and nuclei of the photoreceptors; (5) the outer plexiform layer; (6) the inner nuclear layer; (7) the inner plexiform layer; (8) the nuclei of ganglion cell layer; (9) nerve fiber layer; and (10) the inner limiting membrane (Genten et al. 2009). Most fishes have duplex retinas containing both rods and cones held by the external limiting membrane. Cone cells are conical and ellipsoid. Rod cells, responsible for detection of light during dim photic conditions, are usually longer and thinner. Subpopulations of rod cells detect red, green, and blue regions of the visible spectrum, making color vision possible. Some cartilaginous and deep-sea fishes have only rod cells.

Periophthalmus has a curved retina (Figure 6.1a,b) that is tilted away from the lens. It causes the retinal distance to increase continuously in a vertical direction (Munk 1970; Karsten 1923). Besides accommodation, this allows objects at different distances from the eye to be in focus. A curved retina allows objects on the ground to be in focus on the dorsal part of the retina at the same time that a distant object located higher in the visual field is in focus on the ventral part of the retina. The density of visual cells in the dorsal part of the retina is greater than that in the ventral part. Fine development of the retina in *Ps. barbarus* makes it easy to distinguish all 10 layers (Figure 6.3d,e) as described in other teleosts.

6.2.1.5 Conclusion

More studies are necessary to establish details concerning the ultrastructure of the eye in oxudercine gobies. Comparative studies between all 10 oxudercine genera would be useful to elucidate the evolution of vision from water to land.

6.3 ANATOMY OF CHEMOSENSORY ORGANS

6.3.1 Gustatory Organs

In the majority of vertebrates, gustatory organs are present in the form of taste buds. Attempts to understand gustatory organs began in the first half of the nineteenth century, with fishes being the first group studied (Leydig 1851; Weber 1827). Leydig (1851) was the first to describe the presence of taste buds in the skin of freshwater fishes, whereas Schulze (1863) distinguished sensory and supporting cells in taste buds present in the oral cavity lining. Afanasjev (1875) reported taste buds present not only in the skin and oral cavity mucosa, but also on the lips and barbels, the latter a site where taste buds are highest in density. The ultrastructure of taste buds in fishes was first described by Truillo-Cenoz (1961). In subsequent years, electron microscopy allowed other researchers to study the structure of taste buds in detail (e.g., Jakubowski 1983; Grover-Johnson and Farbman 1976; Reutter 1971; Hirata 1966).

There are numerous review papers on the distribution and structure of fish gustatory organs (primarily Teleostei) (Tagliafiero and Zaccone 2001; Jakubowski and Żuwała 2000; Jakubowski and Whitear 1990; Reutter 1986; Meyer-Rochow 1981; Kapoor et al. 1975; Hirata 1966) and the physiology of nerve endings connected to taste buds and signal transduction (Hara 1994; Kanwal and Finger 1992; Caprio 1982; Bardach and Atema 1971).

6.3.1.1 Gross Morphology

In vertebrates, taste buds occur in the lining of the oral cavity and pharynx. In fishes, taste buds also occur in the skin (Fishelson et al. 2004; Jakubowski and Żuwała 2000; Żuwała and Jakubowski 1993; Gomahr et al. 1992; Jakubowski and Whitear 1990; Jakubowski 1983; Reutter 1978; Whitear 1971; Leydig 1851). Additionally, fishes possess more taste buds than do most other vertebrate groups. In most fishes, taste buds are most numerous in the oral cavity lining, in the skin of the head especially anteriorly, and in structures such as barbels and lips (e.g., Ohkubo et al. 2005; Żuwała and Jakubowski 1993; Reutter et al. 1974). The trunk and fins usually have the fewest taste buds. The number of taste buds in fishes increases with size. A specimen of *Ictalurus melas* measuring 6 cm, for example, possessed 11,000 taste buds throughout the entire body. Another individual measuring 40 cm possessed about 680,000 taste buds (Finger et al. 1991). The concentration and density of taste buds are dependent on the diet and environment of a species; there are fewer taste buds in planktivorous, midwater fishes when compared to those that forage on the substrate (Gomahr et al. 1992). At the sub-microscopic level, however, taste buds reveal a common basic structure. Each taste bud is composed of sensory and non-sensory cells. Sensory cells, also known as receptor cells, include gustatory and basal cells, while non-sensory cells (supporting cells) are of only one type.

We describe, for the first time, the peripheral gustatory organs of several species of mudskippers—*Ps. argentilineatus*, *Ps. barbarus*, *Ps. gracilis*, *Ps. minutus*, *Ps. novemradiatus*, and *S. histophorus*. Taste buds were found in the skin of the head, mucosa of the oral cavity, and mucosa of the pharynx and branchial arches (Figures 6.4a through c and 6.5a through d). In the skin and on the tongue, taste buds most frequently occur singularly. In the mucosa of the anterior part of the oral cavity, near the teeth, and in the pharynx, these taste buds occur in various distinct groupings (Figure 6.5a through d). Groupings of taste buds have been also observed in other gobioid and teleost groups (Fishelson and Delarea 2004).

The relatively high number of taste buds in *Ps. barbarus* indicates that this is an important chemoreceptive organ. In the species we examined, the highest density of taste buds was observed in the lining of the pharynx, whereas the lowest density occurred near the anterior naris and in the skin at the base of the dorsal fin (Table 6.1). External gustatory organs make preliminary evaluation of food usability. The large number of taste buds in the pharynx of mudskippers indicates the location where food is subjected to final quality assessment (Fishelson et al. 2012; Linser et al. 1998).

In cross sections through fish skin and mucosa, the taste buds are mostly onion- or pear-shaped (Figures. 6.6a through c and 6.7a). Within a taste bud, elongated cells (gustatory receptor cells and supporting cells) that span from the apex to the base can be easily distinguished using a light microscope. The nuclei of these elongated cells are located in the lower, broader part of this organ. In the basal part of a taste bud, basal cells are perpendicular to those described above. The arrangement of these cells in mudskippers is similar to those present in other teleosts (Żuwała and Kuciel, unpublished data).

In *Ps. barbarus*, taste buds in the skin are located in the most external "compact" layer of epidermis (Figure 6.5b,c), while those in the mucosa of the oral cavity are distributed throughout the epidermis (Figure 6.5a).

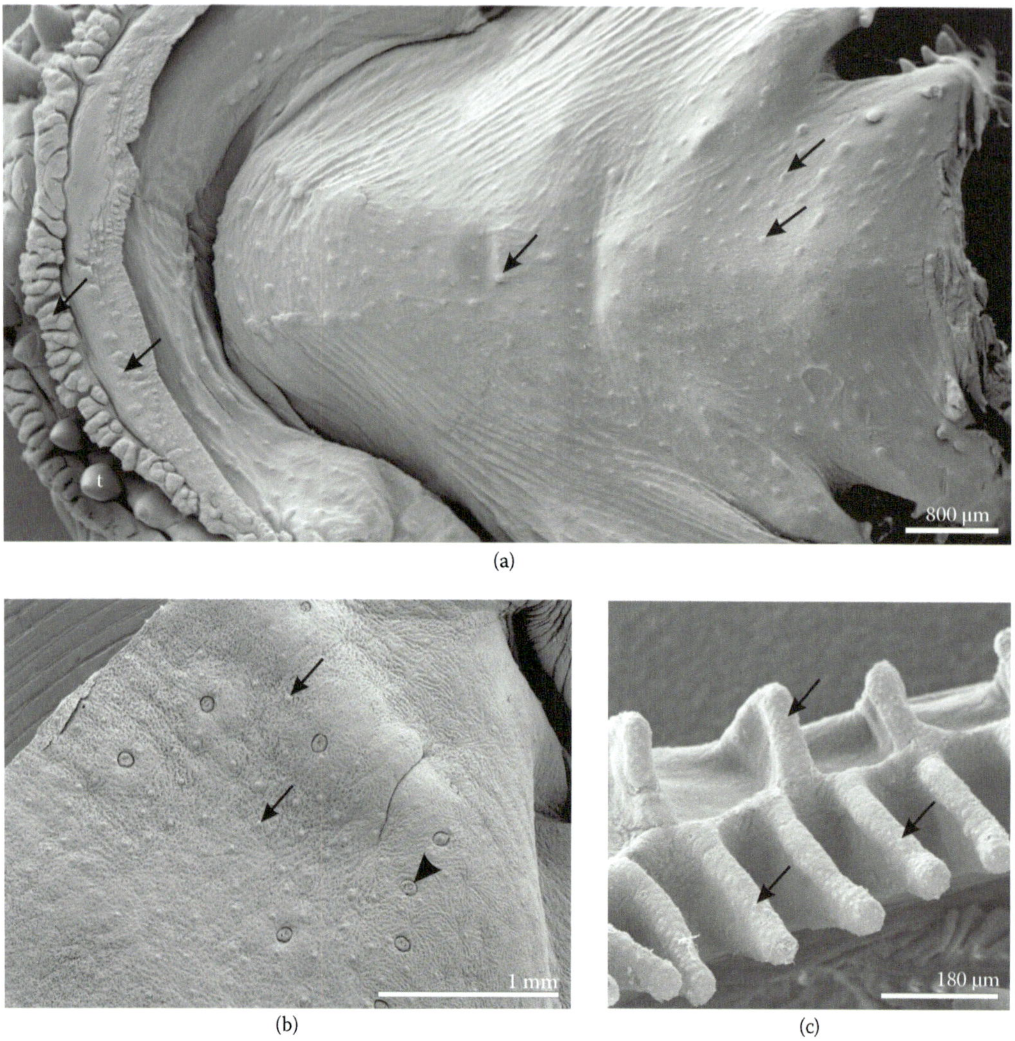

Figure 6.4 Distribution of taste buds, scanning electron microscopy (SEM). (a) Floor of oral cavity of *Periophthalmus barbarus.* (b) Skin on head of *Periophthalmus barbarus.* (c) Gill arches of *Periophthalmus argentilineatus.* Arrows, sensory zone of taste buds; arrowhead, surface neuromast; t, tooth.

In freshwater teleost fishes, blood vessels and nerve fibers in the dermal papillae innervate the taste buds; the blood vessel at the basal part of each taste bud forms a loop of capillary blood vessels (Jakubowski 1983, 1966, 1958). In *Ps. barbarus*, taste buds are located both on the dermal papillae (e.g., around teeth and on the jaw and mandible) and flatly distributed on the dermis (e.g., on the epithelium of the primary tongue).

Apical protrusions of gustatory and supporting cells usually form a small circular sensory zone (Figure 6.8a,b). Three types of taste buds have been distinguished in teleost fishes (Yashpal et al. 2006; Boudriot and Reutter 2001; Reutter 1991, 1978; Jakubowski and Whitear 1990, 1986; Ezeasor 1982; Grover-Johnson and Farbman 1976; Reutter et al. 1974; Whitear 1971): type I in which the taste buds are located above the epithelium, type II in which the taste buds are located at the level of the epithelium, and type III in which the taste buds are located below the epithelium, thus forming a "pore." All three taste bud types can be observed in two characid fish species, *Astyanax mexicanus* and *Astyanax jordani* (Boudriot and Reutter 2001), as well as in each of the mudskipper species examined.

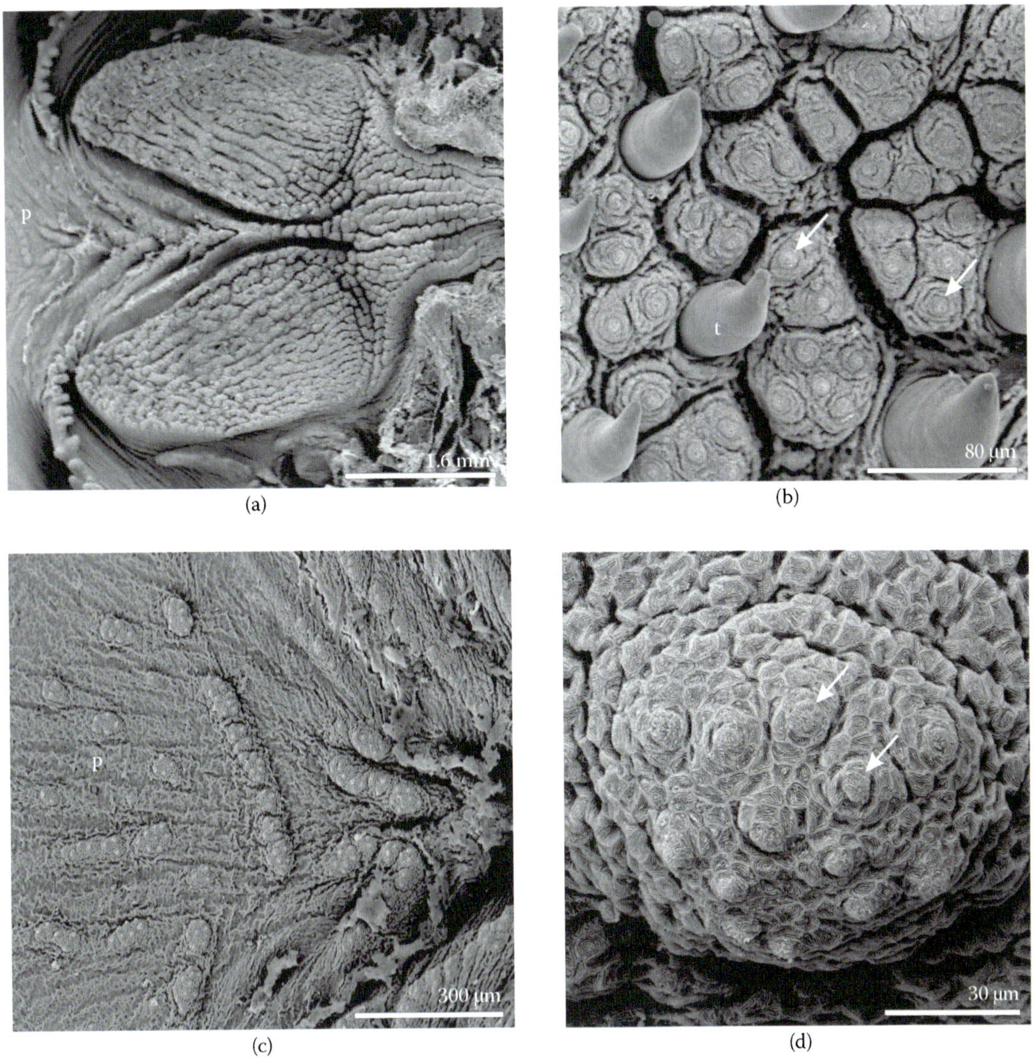

Figure 6.5 Grouping of taste buds, SEM. (a) Pharynx of *Periophthalmus minutus*, dorsal view. (b) Pharynx of *Periophthalmus minutus*, dorsal view. (c) Palate of *Periophthalmus minutus*, posteriorly. (d) Palatine of *Periophthalmus argentilineatus*. P, palatine; arrows, sensory zone of taste buds; t, tooth.

6.3.1.2 Ultrastructure of Gustatory Receptor Cells

In taste buds of *Ps. barbarus*, the apical part of gustatory cells forms thick finger-like protrusions with a core of tonofilaments, from the base to the apical cytoplasm. On transverse sections (close to the epithelial surface), gustatory cells are circular. The area between gustatory cells is filled with supporting cells of irregular shape. Close to the sensory area, gustatory cells form desmosomal connections with supporting cells surrounding them (as in other teleostean fishes) (e.g., Fishelson and Delarea 2004; Żuwała and Jakubowski 1993; Welsch and Storch 1969).

The cytoplasm of a gustatory cell contains numerous cisterns of smooth endoplasmic reticulum (Figure 6.7a), sparse microtubules, and numerous mitochondria. The nucleus of a gustatory cell is located in the basal part of the cytoplasm. In this part of the cell, cisterns of Golgi apparati can also

Table 6.1 The Number of Taste Buds per 1 mm² of Skin or Oral Cavity Mucosa in *Periophthalmus barbarus* (unpublished data)

Body Part	Number of Taste Buds per 1 mm²
Lips	220
Olfactory area	190
Crown forehead	255
Skin around anterior naris	120
Primary tongue (anterior)	195
Floor of pharynx	435
Skin at base of dorsal fin	90

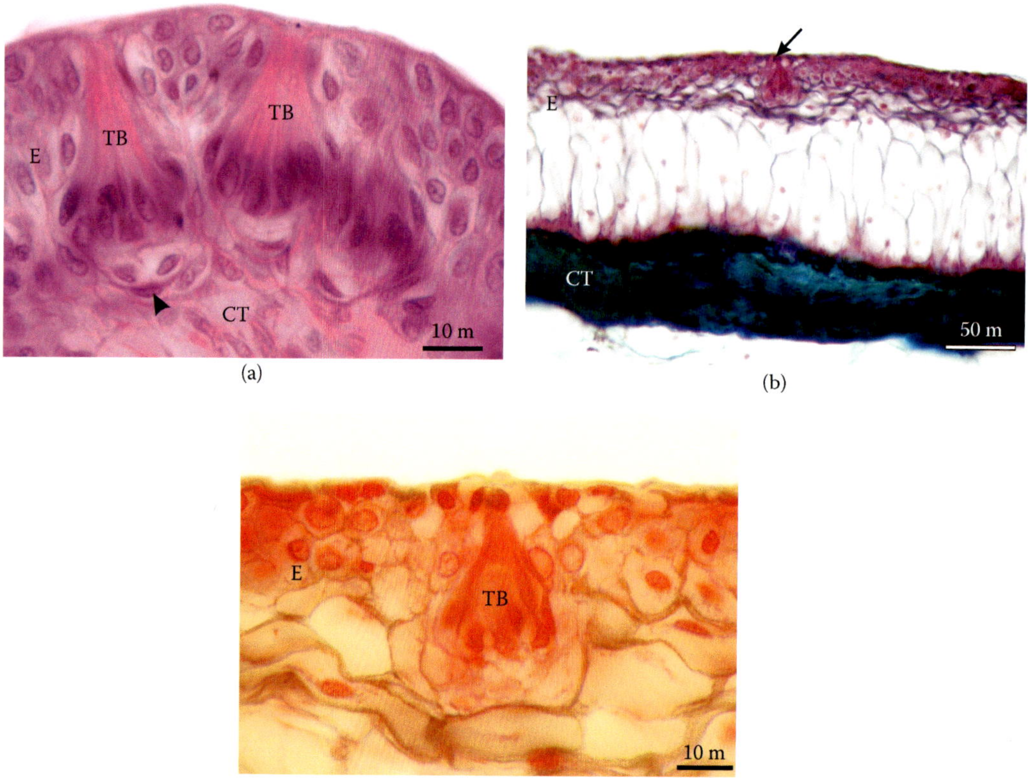

Figure 6.6 The cross section through a lining of oral cavity (a) and skin (b, c) of *Periophthalmus barbarus*. LM. TB, taste bud, arrow, sensory zone of taste buds; E, epithelium; CT, connective tissue.

be seen. In the cytoplasm of the basal part of the cell are sparse synaptic vesicles. Projections of gustatory cells do not form hemidesmosomes with the basal lamina.

6.3.1.3 *Ultrastructure of Basal Cells*

In *Ps. barbarus*, the discoid-shaped basal cell lies on the basal lamina (Figures 6.6a and 6.7b) and does not form hemidesmosomes. On the longitudinal sections through the taste buds, only one basal cell was observed. The central part of the cytoplasm of a basal cell is occupied by an indented large nucleus. In the cytoplasm, there are bundles of tonofilaments (Figure 6.7a), cisterns of Golgi

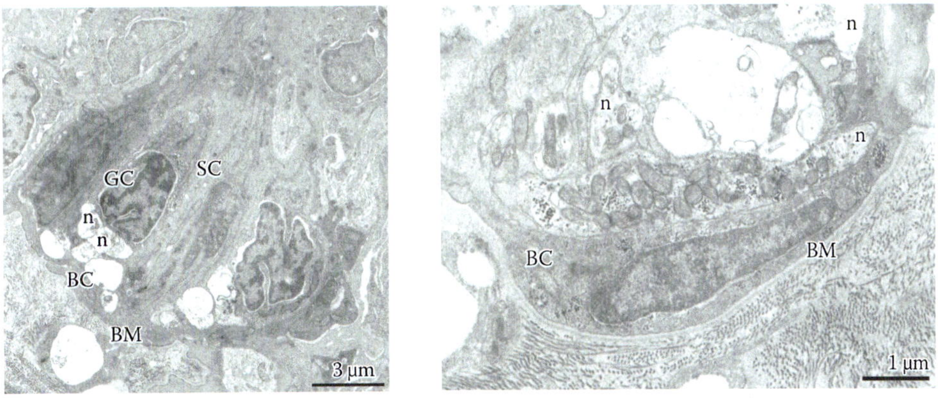

(a) (b)

Figure 6.7 The basal cell (BC) is covered by an expanded nerve ending (n) in *Periophthalmus barbarus*. (a) A taste bud from the oral cavity lining, vertical section. (b) Base of taste bud, vertical section. GC, gustatory receptorcell; SC, supporting cell; BM, basal lamina.

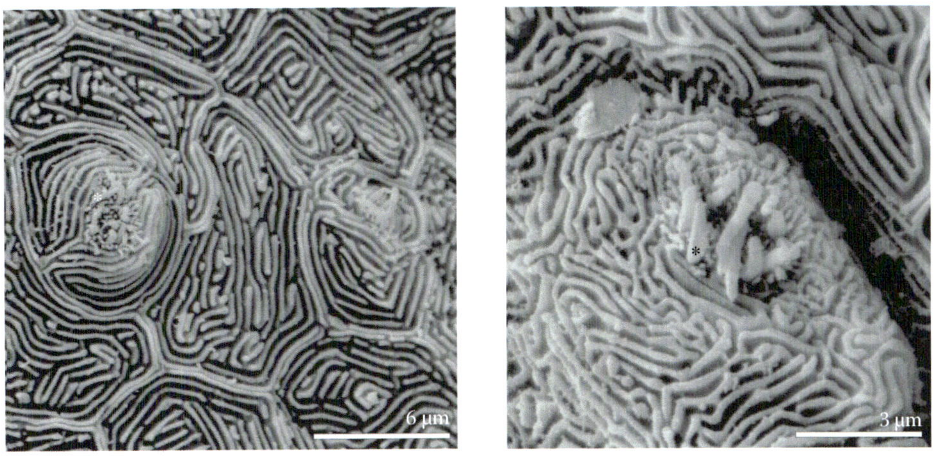

(a) (b)

Figure 6.8 Sensory zones of taste buds, SEM. (a) *Scartelaos histophorus*. (b) *Boleophthalmus boddarti*. *, finger-like terminals of gustatory receptor cells.

apparatus, smooth endoplasmic reticulum, and glycogen granules. Synaptic vesicles are also located there, most frequently seen in groups close to the synaptic connection of a basal cell with the nerve ending. The membrane of a basal cell from the side of a nerve plexus often forms a few microvillar processes with a tonofilament core (Jakubowski 1983, 1980).

The presence of a large nucleus with an irregular membrane in the cytoplasm of basal cells and finger-like protrusions with a tonofilament core resembles Merkel cells observed in the basal layer of the epidermis and in the mucosa lining the oral cavity of vertebrates including fishes (also in *Ps. barbarus*) and amphibians. Toyoshima (1994) suggested that in the development of amphibian gustatory organs, Merkel cells fulfill the functions of basal cells. Merkel cells, originally considered to function as mechanoreceptors, may have broader functions (e.g., Zachar and Jonz 2012; Toyoshima 1994; Fox and Whitear 1978).

Based on the ultrastructure of taste buds, Cordier (1964) suggested that basal cells are constant components of the taste bud. Published data indicate this to be true for taste buds present in the lining of the oropharyngeal cavity. However, in the taste buds in the dermal layer of the barbels in *Ciliata mustela* (Crisp et al. 1975), in some species of gadiform fishes, as well as in *Anguilla*, *Poecilia*, and *Colisa* (Jakubowski and Whitear 1990), basal cells have not been found. In mudskippers (e.g., *Ps. barbarus*), all sections through the taste buds show only one basal cell (Figures 6.6a and 6.7b). Nevertheless, further studies on taste buds present in the oral cavity and skin of other mudskipper species are necessary for better understanding and comparison.

6.3.1.4 Ultrastructure of Supporting Cells

Supporting cells surround gustatory cells (Figure 6.7a). The apical part of the supporting cell shows a few microvilli and secretory vesicles of various electron densities and sizes depending on the species. Their cytoplasm is rich in tonofilaments, cisterns of rough endoplasmic reticulum, and polyribosomes. In the teleost genera *Cobitis*, *Misgurnus*, and *Agonus*, the cytoplasm is also rich in microtubules (Jakubowski and Whitear 1990; Jakubowski 1983). Proximal protrusions of supporting cells may lean on a basal cell or on the basement membrane with which they form hemidesmosomes.

6.3.1.5 Ultrastructure of Nerve Plexus

Several nerve fibers (nerve fiber plexus) enter the taste bud through the basement membrane. At the base of a taste bud, individual nerve fibers are adjacent to proximal protrusions of supporting and gustatory receptor cells. In this region of a taste bud, it is sometimes difficult to attribute cell protrusion to a given cell. In *Ps. barbarus*, it was very difficult to find synaptic connections between basal cells and the nerve endings and between gustatory receptor cells and the nerve endings. Only "darkenings" of cell membranes or synaptic-like vesicles near the cell membrane in the cytoplasm of gustatory receptor cells and basal cells are present. The afferent and efferent synaptic connections between a nerve ending and gustatory cells and basal cells are easily observed in other fishes such as *Ameiurus* (Royer and Kinamon 1996; Desgranges 1966), *Phoxinus* (Jakubowski and Whitear 1990), *Lepisosteus* (Reutter et al. 2000), and *Astyanax* (Boudriot and Reutter 2001). In *Ps. barbarus*, nerve fibers entering the taste bud adhere to the basal cell covering it dorsally (Figure 6.7a), a configuration also observed in other teleost fish taxa such as *Agonus*, *Anguilla*, *Blennius*, *Gasterosteus*, and *Triglopterus* (Jakubowski and Whitear 1990).

6.3.1.6 Conclusions

Gustatory organs of mudskippers are similar to other fish groups. External and internal taste buds in mudskippers are characterized by high numbers and relatively small sizes (Table 6.2). Taste organs are well developed and can be assumed to be the dominant chemosensory organ in the aquatic life phase and while out of water.

Table 6.2 Size of Taste Buds in Species of *Periophthalmus*, *Scartelaos*, and Other Fishes

Genera	Species	Sensory Area Diameter (μm)	Taste Bud Diameter (μm)	Taste Bud Height (μm)	References
Periophthalmus	*argentilineatus*	2.5–4	–	–	This study
	barbarus	4–6	16–22	24–40	Unpublished data
	gracilis	3	–	–	This study
	minutus	3	–	–	This study
	novemradiatus	3–4	–	–	This study

(*Continued*)

Table 6.2 (*Continued*) **Size of Taste Buds in Species of *Periophthalmus*, *Scartelaos*, and Other Fishes**

Genera	Species	Sensory Area Diameter (µm)	Taste Bud Diameter (µm)	Taste Bud Height (µm)	References
Scartelaos	*histophorus*	2–4	–	–	This study
Apteronotus (Apteronotidae, Gymnotiformes)	*albifrons*	10–12	33–39	32–48	This study
Gyrinocheilus (Gyrinocheilidae, Cypriniformes)	*aymonieri*	15–19	25–29	28–41	This study
Amia (Amiidae, Amiiformes)	*calva*		45	65	Reutter et al. (2000)
Tinca (Cyprinidae, Cypriniformes)	*tinca*		25–65	60–95	Żuwała and Jakubowski (1993)
Lepisosteus (Lepisosteidae, Lepisosteiformes)	*oculatus*		45	65	Reutter et al. (2000)
Scyliorhinus (Scyliorhinidae, Carcharhiniformes)	*cancicula*		30	50	Reutter (1994); Whitear and Moate (1994)

6.3.2 Olfactory Organs

Olfactory organs of fishes are adapted to the niche they occupy (Meyer-Rochow and Coddington 2003). Published studies on olfactory organs over the past 100 years provide interesting, if fragmentary, insights into the diversity of olfactory organ structures (see Hansen and Zielinski 2005; Zeiske et al. 1992; Graham et al. 1985; Døving et al. 1977; Hara 1975; Kleerkoper 1969; Burne 1909). The olfactory system of fishes, as in most vertebrates, consists of the peripheral (with olfactory nerve formed by axons of the olfactory sensory neurons) and the central (olfactory bulbs and the brain) parts of the olfactory organ. The former is responsible for the detection of odor stimuli (e.g., food, environmental, reproduction), whereas the latter is involved in processing olfactory information (Hansen and Reutter 2004).

Hagfishes and lampreys possess a single olfactory organ comprising one chamber on the dorsal preorbital part of the head and an olfactory sensory epithelium that is located ventrally in the chamber. Elasmobranch fishes have paired olfactory organs located symmetrically on the ventral side of the rostrum except for members of Chlamydoselachidae in which these organs are on the dorsal side of rostrum (Jasiński 1976). In teleost fishes, the olfactory organ is also paired and is always located on the dorsal, preorbital side of head, in the vicinity of the inlet and outlet nares; the olfactory chamber and the olfactory rosette are usually present. The olfactory organ structures in a few species of *Boleophthalmus*, *Periophthalmus*, and *Scartelaos* were described (Kuciel et al. 2014, 2013, 2011; Kuciel 2012) and are herein figured (Figures 6.9c through g and 6.10a through c).

6.3.2.1 Gross Morphology

The gross morphology of olfactory organs in teleost fishes is conservative. However, there is considerable variation in the shape and structure of the peripheral olfactory organs depending on systematic groups and ecological habitats (see Hansen and Zielinski 2005; Hansen and Finger 2000; Zeiske et al. 1992; Hara 1992; Yamamoto and Ueda 1978; Kleerkoper 1969). The most frequent form is the olfactory rosette with lamellae radiating from a middle raphe. On the surface of the olfactory rosette is

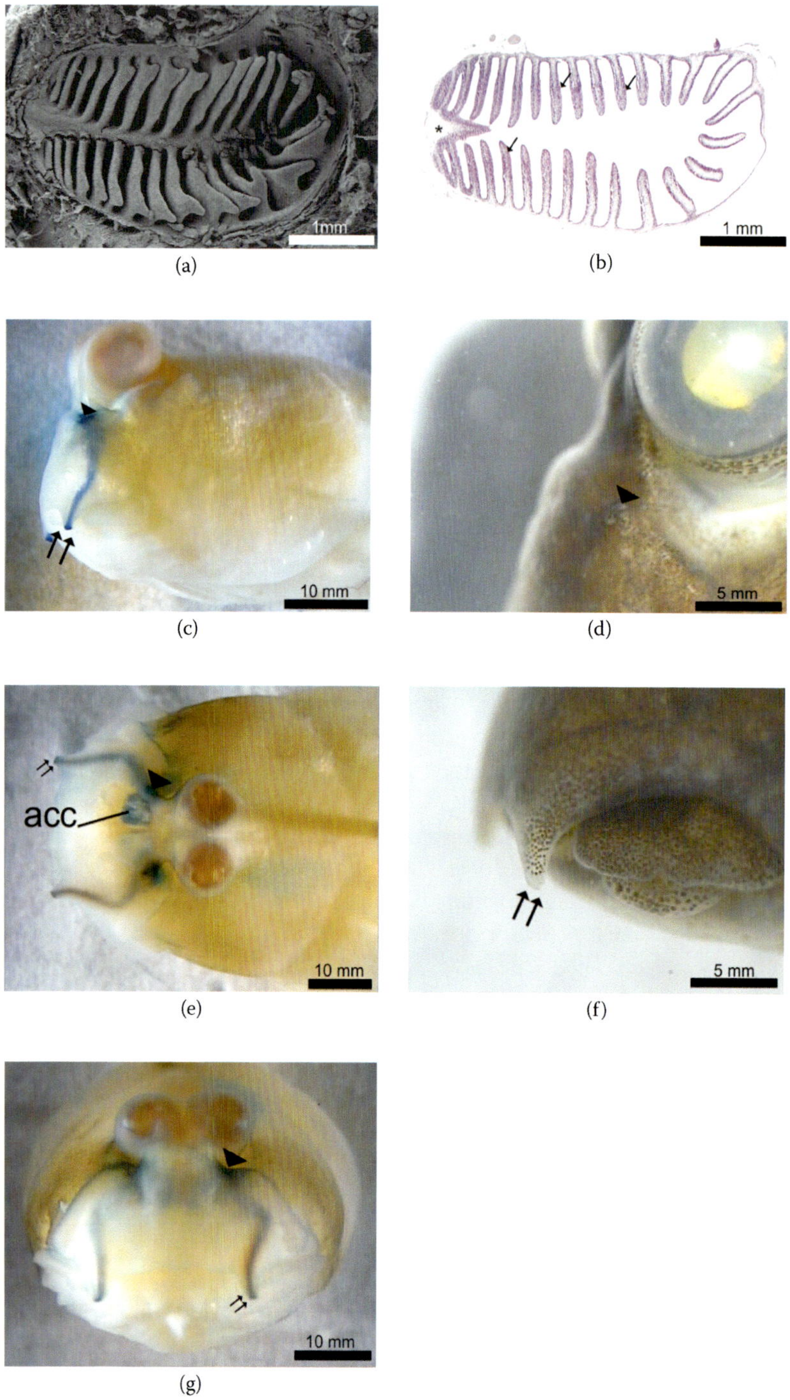

Figure 6.9 Gross morphology of the olfactory organ. (a) Olfactory rosette of *Amia calva*, as a representative of the most common olfactory organ structure in fishes, SEM, surrounding tissues dissected. (b) Olfactory rosette of *Amia calva*, sagittal section, stained with hematoxylin and eosin. (c–g) Canal-like olfactory organ in mudskippers. (c, e, g): *Periophthalmus chrysospilos*, (d, f): Periophthalmus minutus. Skin pigmentation in (c), (e), and (g) removed by perhydrol. Arrow, olfactory lamella; *, central strip; arrowhead, outlet naris; double arrow, inlet naris; acc, accessory nasal sac.

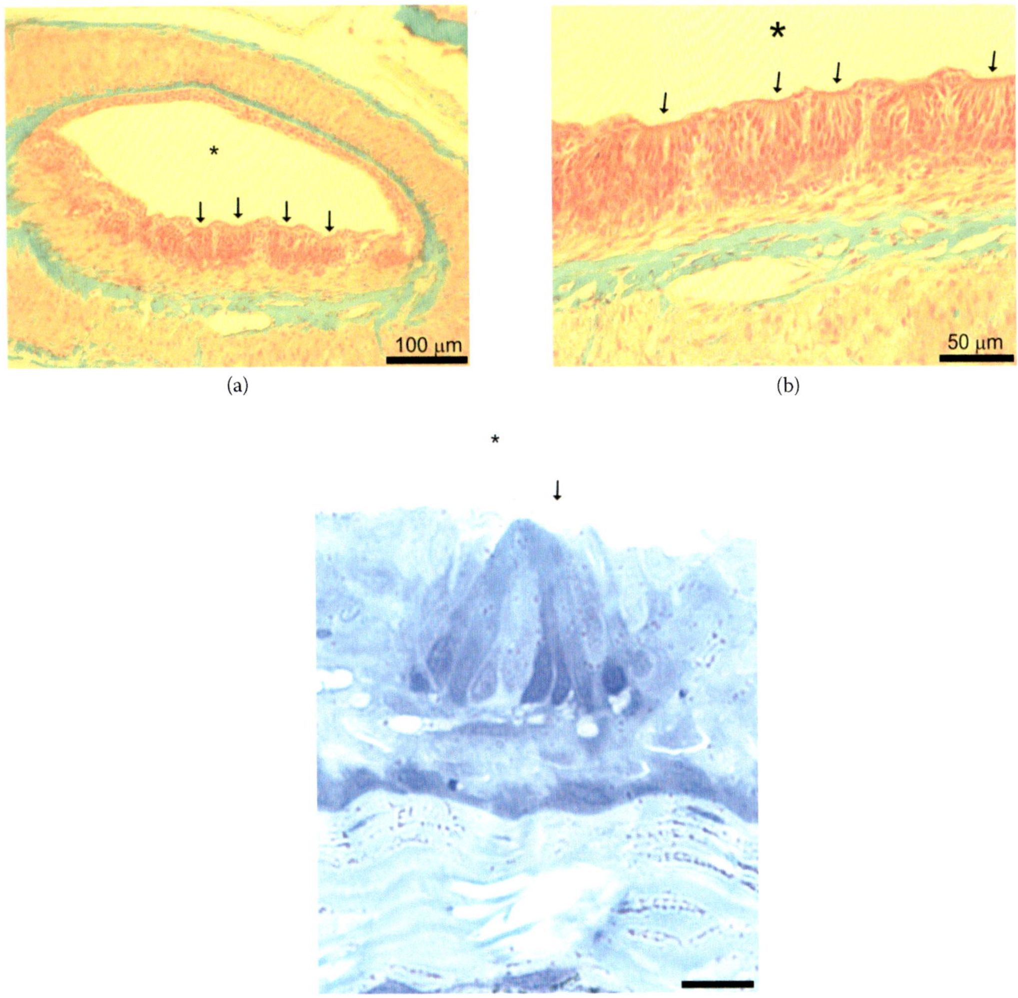

Figure 6.10 Olfactory epithelia in *Periophthalmus barbarus* located in the olfactory canal, cross section. (a) Olfactory canal cross section, Gomori's trichrome stain. (b) Olfactory islet/s, Gomori's trichrome stain. (c) Olfactory islet/s, methylene blue and azur III stains. *, olfactory canal; arrow, olfactory sensory epithelium islets.

the olfactory sensory epithelium (Figure 6.9a,b). Olfactory rosettes among species differ in the length, shape, and number of olfactory lamellas. Differences are also observed in the detailed structure of the olfactosensory epithelia of the olfactory lamellae, the olfactory chambers, and the olfactory canals.

A single olfactory lamella from the olfactory rosette can be described as follows: continuous form where the olfactory lamella is continuous except at the borders, for example, in *Oncorhynchus masou* (Pfeiffer 1963); zonal form, for example, in *Silurus glanis* (Jakubowski 1981); irregular form, for example, in gasterosteiform fishes (Yamamoto and Ueda 1978); and islet form, for example, in tetraodontiform fishes (Yamamoto and Ueda 1979).

In mudskippers, the olfactory organ structure is unique and probably resulted from an amphibious lifestyle. Unlike most fish species, the olfactory organ structure does not have an olfactory rosette. Kuciel et al. (2011) described the olfactory organ structure in *Ps. barbarus* as a paired canal-like organ with a round anterior naris in the depression of the upper lip and a slit-like posterior naris under the eye. A single canal (sensory part) widens in the ethmoid region of the head into an accessory nasal sac, which has a mechanical function of improving the flushing of the canal with water. In the centripetal part of the canal, the olfactory sensory epithelium is located and depending on the species, it may be formed in a few different ways (islets, stripped, zonal), which are one of many described in fishes.

In gobioid species lacking the olfactory rosette, several types of olfactosensory epithelia were observed in the olfactory canal or chamber: epithelia smooth at the base and on the lateral walls and ceiling of the olfactory chamber, for example, *Neogobius melanostomus* (Belanger et al. 2003) and *Neogobius gymnotrachelus* (Żuwała et al., unpublished data); epithelia smooth at the base and on the lateral walls of the olfactory chamber, for example, *Neogobius fluviatilis* (Żuwała et al., unpublished data); epithelia with islets, for example, *Ps. barbarus* (Kuciel et al. 2011), *Ps. chrysospilos*, and *Ps. variabilis* (Figure 6.11a; Kuciel et al. 2013); epithelia fascicular with finger-like ramifications for example, *Ps. argentilineatus*, and *Ps. minutus* (Figure 6.11b; Kuciel et al. 2013); and epithelia smooth at the base and on the lateral walls of the olfactory canal, for example, *B. boddarti*, *S. histophorus* (Kuciel et al. 2013), and *Parapocryptes rictuosus* and *Proterorhinus marmoratus* (Figure 6.11c; Żuwała et al., unpublished data).

In some teleost fish groups, the olfactory organ is equipped with accessory nasal sacs involved in the mechanics of improving the flushing of olfactory sensory epithelia either by pressure changes in the oral cavity or by muscular movements. One or two accessory nasal sacs are commonly observed in teleost fish species; these fishes are called "cyclosmates" (Døving et al. 1977) and are "equipped" with

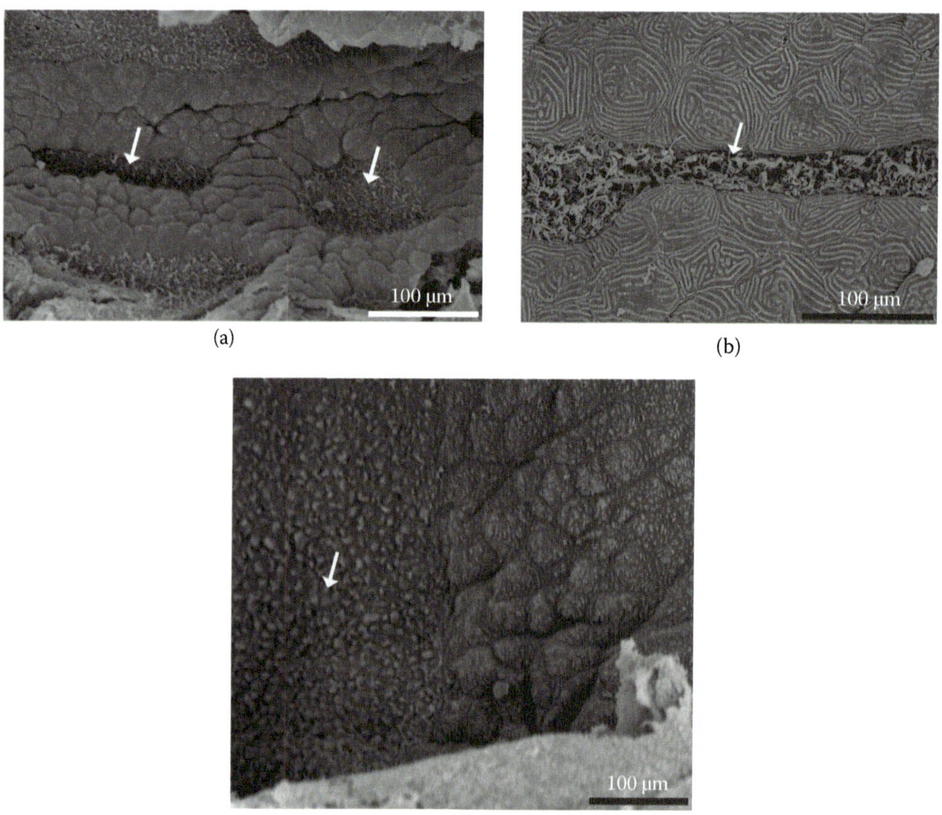

Figure 6.11 Variation in location of olfactory epithelia within the olfactory canal, SEM. (a) Islets form in *Periophthalmus chrysospilos*. (b) Fascicularin *Periophthalmus argentilineatus*. (c) Zonalin *Scartelaos histophorus*. Arrow, olfactory epithelia.

an active mechanism for olfactory organ ventilation. Species where the olfactory organ is ventilated during swimming and/or by water current and/or ciliated epithelia are called "isomates" (Døving et al. 1977). Three main types of sensitivity to olfaction have been distinguished (Burne 1909): (1) good or macrosmatic; (2) average or mesosmatic; and (3) weak or microsmatic. As described by Kuciel (2012), the mechanism of olfactory organ ventilation in *Ps. barbarus* actively involves the *levator arcus palatine* (*lap*) and *adductor mandibularis* (*a0*) muscles. The *a0* muscle on one side is attached to the skull and on the other to the maxilla. The *lap* muscle is attached on one side to the palatine and on the other to the ventral side of the vomer. Contraction of the *a0* muscle moves both the embedded maxilla and palatine in a dorsomedial direction. A ligament (palatinum–palatinum) transfers anteriorly an *a0* muscle contraction, causing the displacement of the ascending process of the premaxilla with the rostral cartilage in a caudal direction. Displacement of the dorsal part of the palatine in the direction of centripetal and rostral cartilages (embedded in the ascending process of the premaxilla) toward the caudal end causes compression of the space occupied by the accessory nasal sacs and the rise of pressure, which pushes the water outside. A new volume of water enters the olfactory organ by means of a negative pressure created in the organ after the disappearance of an *a0* muscle contraction and the return of displaced elements to their initial positions. In addition, *lap* muscle contraction moves the ventral part of the palatine toward the vomer and the dorsal part of the palatine (adjacent to the chamber-like sac) in the centrifugal direction increasing the space around the chamber-like sac. Negative pressure results due to the presence of a narrow, longitudinal slit-like posterior naris, which is permanently closed and serves as a kind of one-way valve (Table 6.3, Figure 6.9f).

6.3.2.2 Ultrastructure of Olfactory Epithelia

Three groups of cells have been identified in the olfactory epithelium of actinopterygian fishes: (1) olfactory sensory neurons—ciliated, microvillar, and crypt; (2) non-sensory cells—ciliated cells, supporting cells; and (3) basal cells (Figure 6.12a). In laboratory studies of mudskippers, all three types of olfactory sensory neurons were found and, typically, the olfactory epithelium ultrastructure of mudskippers resembles that of other species. Slight differences were observed mostly in the number of olfactory sensory neurons or non-sensory cells.

Olfactory sensory neurons are specialized nerve cells that have peripheral dendritic protrusions and a centripetal axon, classified as type I. The central part is a spindle-shaped cell body. In the dendritic protrusions are receptors, which are proteins responsible for specific reactions with ligands. On the opposite side, the axon together with axons of other olfactory sensory neurons of the olfactory nerve I, form the olfactory bulb. In *Lota lota*, for example, there are about 10^7 axons in the olfactory nerve.

In many teleost fish groups (e.g., *Anguilla*, *Esox*, *Salmo*), as well as in *Boleophthalmus*, *Periophthalmus*, and *Scartelaos*, the olfactory bulb is located adjacent to the telencephalon (Kuciel et al. 2013, 2011; Hara 1975). In other taxa such as *Coryphaena hippurus*, *Gymnothorax kidako*, and *Raniceps raninus*, the olfactory bulb is positioned between the brain and olfactory rosette. However,

Table 6.3 Number of Accessory Nasal Sacs in *Periophthalmus, Boleophthalmus, Scartelaos*, and *Parapocryptes*

Genus	Species	Nasal Sacs	Reference
	barbarus	1	Kuciel et al. (2011)
	argentilineatus	1	
Periophthalmus	*chrysospilos*	1	Kuciel et al. (2013)
	minutus	1	
	variabilis	1	
Boleophthalmus	*boddarti*	2	Kuciel et al. (2013)
Scartelaos	*histophorus*	2	
Parapocryptes	*rictuosus*	2	N/A

in other taxa (i.e., elasmobranch fishes, *Carassius* spp., *Ictalurus* spp., cobitid fishes, silurid fishes), the olfactory bulb is located next to the olfactory rosette (Uchihashi 1953).

In the olfactory bulb of all fishes, there are synaptic connections with the dendritic appendages of the mitral cells, constituting the second tract for olfactory impulses. Four distinct layers of cells process the incoming information and convey it to the higher brain centers: (1) the olfactory nerve layer; (2) the glomerular layer; (3) the mitral cell layer; and (4) the internal cell layer (Hansen and Reutter 2004).

6.3.2.3 Ultrastructure of Ciliated Olfactory Sensory Neurons

In mudskippers, the body of the ciliated olfactory sensory neuron is elongate with a spindle-like end on the basal side. There are dendritic protrusions, which are cilia localized radially around the olfactory knob (Figure 6.12b) that penetrate the space dorsal to the olfactory epithelium at the apical part of the cell. The olfactory knob rises above the surface of olfactory epithelium. The number and length of the cilia vary with size, age, and species.

The number of cilia typically ranges in fish species from 3 to 10. In *Danio rerio*, cilia length is 2–3 μm (Hansen and Zeiske 1993), whereas in *Acipenser baeri*, the length reaches 10 μm (Hansen and Zielinski 2005). At the base of each cilium, there are basal bodies, which are two centrioles perpendicular to each other. In ciliated olfactory sensory neurons, basal bodies are located just under the membrane surrounding the olfactory knob and have never been observed below the "neck" of the olfactory knob. The nucleus is typically located in the basal part of the cell. Above the nucleus, numerous mitochondria were observed. In some fish species, for example, *A. baeri*, ciliated olfactory sensory neurons have additional short microvillar appendages (Hansen and Zielinski 2005).

A specific group of ciliated olfactory sensory neurons are rod-like cells, which are equipped with a single thick protrusion. These cells are rarely seen in mudskippers and other fishes, and their function is not fully understood. Their construction is analogous to ciliated olfactory receptor cells, but the differences are apparent in the apical part (Figure 6.12d), where all the cilia are surrounded by an outer membrane, thereby forming a thick vertically oriented protrusion, which extends substantially above the surface of the olfactory epithelium (Hansen and Zielinski 2005).

6.3.2.4 Ultrastructure of Microvillar Olfactory Sensory Neurons

The body of microvillar olfactory sensory neurons in mudskippers is elongated and spindle-shaped at its base. The perikaryon reaches to about two-thirds of the olfactory epithelium thickness, which is typical for this type of olfactory sensory neuron. The olfactory knob is expanded and extends slightly above the olfactory epithelial surface (Figure 6.12c). Olfactory knobs in the microvillar neurons are less accentuated than those in the ciliated olfactory sensory neurons. On the surface of the olfactory knobs are some small projections, that is, 10–30 in *Dicentrarchus labrax* (Diaz et al. 2002), 30–60 in *Anguilla anguilla* (Schulte 1972), and 40–70 in *Carassius auratus* (Hansen et al. 1999). Centrioles are always located below the "neck" of the olfactory knob and may be present in great numbers (Hansen and Zielinski 2005). The nucleus and other organelles are located in the cytoplasm as well as in ciliated olfactory sensory neurons.

6.3.2.5 Ultrastructure of Crypt Olfactory Sensory Neurons

The most striking feature of the crypt receptor neuron type is the "crypt," which in mudskippers as well as in other fish species is located at the apical part of the cell that resembles a pocket and reaches one-third of the cell depth. Few cilia are present; they do not exceed the surface of sensory epithelium. The number of microvilli (usually seven or fewer) depends on the species and corresponds approximately to the number of cilia observed in the ciliated olfactory sensory neurons. Supportive cells with numerous microvilli surround the "pocket" (Hansen and Finger 2000). The shape of the crypt receptor

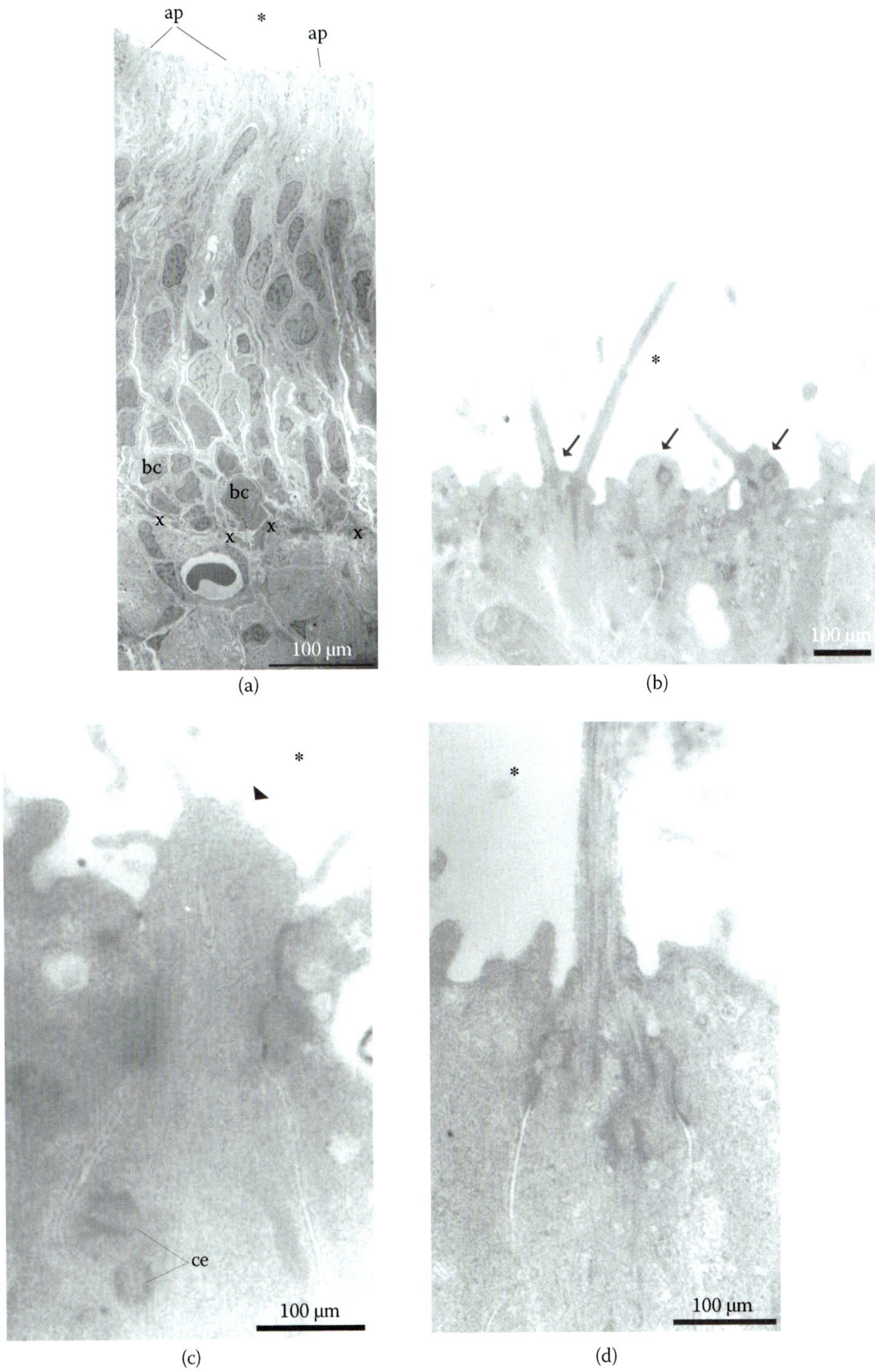

Figure 6.12 Ultrastructure of olfactory epithelia in mudskippers, transmission electron microscopy (TEM). (a) *Boleophthalmus boddarti*, longitudinal section (b) Apical portion of the ciliated olfactory sensory neurons (arrow), longitudinal section. (c) Microvillar olfactory sensory neuron (arrowhead). (d) "Giant cell," peculiar type of olfactory sensory neuron. *, olfactory canal; bc, basal cells; x, basal lamina; ap, apical parts of olfactory sensory neurons; ce, centriole.

cells is different from the ciliated and the microvillar olfactory sensory neurons; they are ovoid and extend to a depth of about one-third of the olfactory sensory epithelia. As in the previous two cell types, the nucleus of each cell is situated in the basal part of the cell body. Above the nucleus, there are organelles such as the Golgi apparatus, mitochondria, lysosomes, and rough endoplasmic reticulum.

Olfactory receptor cells are equipped with specific protein receptors on the membrane (Hansen et al. 2003; Nagai et al. 1996) reacting with a specific ligand causing a conformational change of the G-protein alpha (α) subunit. The result is a signal propagated along the olfactory receptor cell membrane to the telencephalon. Different types of olfactory sensory neurons are specific to particular ligands due to the different types of G-protein α subunits present in the membranes. Studies by Hara (1992), Laberge and Hara (2001), and Hansen et al. (2003) determined which G-protein α subunits are related to particular olfactory sensory neurons. Ciliated olfactory sensory neurons are equipped with $G\alpha_{olf/s}$ protein subunits and bind with amino acids and bile acids. Microvillar olfactory sensory neurons have $G\alpha_{q/11}$ protein subunits and react with amino acids and nucleotides. The crypt olfactory sensory neurons have $G\alpha_{o}$, protein subunits but it remains unclear to which substances they adhere. It is believed that the crypt olfactory sensory neurons are responsible for hormone identification and they may occur in the olfactory epithelium periodically, for example, during the reproductive season, in the post-reproduction period they are diminished.

Immunohistochemistry of the olfactory system of *Ps. barbarus* revealed a $G\alpha_{olf/s}$ protein subunit in the ciliated olfactory receptor neurons and a $G\alpha_{i3}$ protein subunit in the microvillar olfactory receptor neurons and crypt cells (Kuciel et al. 2014). The presence of $G\alpha_{i3}$ in these neurons is interesting as the closely associated $G\alpha_{i2}$ gene is included in the same family of putative pheromone receptor genes (Berghard et al. 1996).

6.3.2.6 Ultrastructure of Ciliated Non-Sensory Cells

Ciliated non-sensory cells are cylindrical and extend from the basal lamina to the surface of olfactory epithelia. They are present in the olfactory epithelium of all teleost fishes including mudskippers and are responsible for the movement of mucus and water within the epithelial surfaces where olfactory sensory neurons protrusions occur (Zeiske et al. 1992). In these cells, numerous mitochondria were observed. The underside of the relatively small nucleus is filled with dense endoplasmic reticulum; their position location depends on the species but can be categorized into broad areas: peripherally on the lamella of the olfactory rosette, evenly distributed in the whole olfactory area and within the islets (Hansen et al. 1999; Jakubowski 1981). Nuclei may have different shapes and typically do not contain condensed chromatin. Numerous cilia are present on the wide, flat top of the cell. The number and size of cilia are dependent on the species, for example, 40–60 in *Ps. barbarus* and *Ps. minutus*, 40 in *D. rerio* (Hansen and Zeiske 1998), and 140 in *A. anguilla* (Schulte 1972).

6.3.2.7 Ultrastructure of Supporting Cells

Supporting cells in mudskippers inhabit the spaces between receptor cells and act as a scaffold to hold the receptor cells in position. Supporting cells are cylindrical in shape, often with the base adhering to the basal lamina. A selvedge membrane of these cells forms microridges on their surface, creating a kind of "maze." In the apical part of the olfactory epithelia, neighboring supporting cells are connected together by desmosomes. In the cytoplasm, long tufts of tonofilaments perpendicular to the long axis of the cell are present.

6.3.2.8 Ultrastructure of Basal Cells

In mudskippers, each basal cell is relatively small with a large nucleus usually filling the entire cell space and is similar to other fish species studied to date. They are situated in the deepest part of the olfactory epithelia just dorsal to the basal lamina. They divide into new non-sensory and olfactory sensory neurons (Hansen et al. 1999; Zippel et al. 1997; Evans et al. 1982; Breipohl et al. 1976; Thornhill 1967). For the moment, it is unclear whether they can also give rise to supporting cells that are able to divide by mitosis (Hansen and Zielinski 2005).

6.3.2.9 *Conclusion*

New olfactory organ configurations are still being described despite more than a century of research into these organs in actinopterygian fishes. The olfactory organ structure in mudskippers is most peculiar even though its ultrastructure resembles that studied in other fish species. The unique structure of the olfactory organ in mudskippers is probably the result of adaptation for periodic transitions from water to land. In three mudskipper genera (*Boleophthalmus*, *Periophthalmus*, and *Scartelaos*), slight differences were observed in congeners, indicating a high level of genetic plasticity (e.g., number of accessory nasal sacs, location of olfactory epithelium, size, and length of olfactory canal). Olfactory abilities of mudskippers are thought to be weak when compared to other teleost fishes. But, taking into account structural complexity of the olfactory organ and the presence of an active ventilation mechanism (Kuciel 2012), it would be worthwhile to determine the role of an active ventilation mechanism with respect to an amphibious lifestyle. The inclusion of other oxudercine taxa, together with more study of the physiology of olfaction, will also enable a better understanding of olfactory organ evolution in this fish group.

6.3.3 Solitary Chemosensory Cells

In addition to the gustatory and olfactory systems, there is another chemosensory structure that is rarely observed or reported: the solitary chemosensory cells, described for the first time by Whitear (1992). The variability in their morphology and distribution renders them difficult for use in physiological experiments. Consequently, their function is poorly understood, but they probably serve to locate food and/or a nearby predator or conspecific (Hansen and Reutter 2004).

Solitary chemosensory cells are embedded in the epidermis and dispersed as isolated cells across the body surface. They also occur in the epithelia of the oropharyngeal cavity, the gills, and olfactory epithelium of *C. auratus* (Hansen et al. 1999; Hansen and Zippel 1995), with uneven distribution over the fish trunk. The most numerous solitary chemosensory cells per unit area were found on the

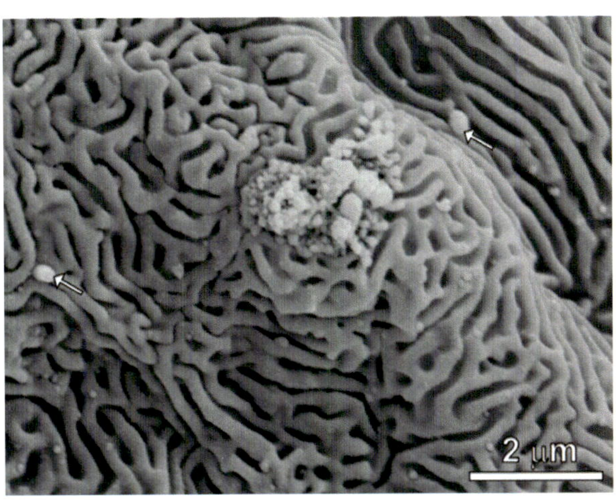

Figure 6.13 Solitary chemosensory cells in *Periophthalmus barbarus* with short protrusions (arrow), SEM.

anterior dorsal-fin rays of *Gaidropsarus mediterraneus* (Hansen and Reutter 2004). The number of solitary chemosensory cells varies, from 3 to 6 million in a 20 cm *G. mediterraneus* (Kotrschal et al. 1984) to poorly developed or absent in *Spinachia spinachia*, and *Ps. barbarus* (Figure 6.13; Kuciel et al. 2014; reported as *Ps. koelreuteri* in Whitear 1992). In *Ps. barbarus*, single solitary chemosensory cells were found in the dorsal, terminal part of the olfactory canal. There were a few processes 1–2 μm long projecting directly between the epithelial cells. Other structures beside epithelial cells with microridges were mucous glands that have been observed adjacent to solitary chemosensory cells.

These cells populate non-olfactory epithelia and may function as a monitor for possible irritants or toxins in the incoming water or air stream (Hansen and Finger 2009). As solitary chemosensory cells were observed in mudskippers, but only inside the olfactory canals, these also may be another amphibious adaptation to protect fragile sensory cells from harmful land conditions during fish emersion. If these cells were located externally, they would be subject to desiccation during emersion.

6.4 ACOUSTIC COMMUNICATION

The soniferous activities of gobioid fishes are well known and well documented (Horvatić et al. 2016; Malavasi et al. 2008, 2003; Amorim and Neves 2007; Myrberg and Lugli 2006; Bass and Mckibben 2003; Lugli and Torricelli 1999; Lugli et al. 1997, 1995; Torricelli et al. 1990; Takemura 1984). Twenty-two species of gobies produce both tonal and pulsatile communicative sounds in their social interactions (Zeyl et al. 2016). One of these soniferous species is the mudskipper *Periophthalmodon septemradiatus*, whose agonistic acoustic communication was described by Polgar et al. (2011).

Acoustic emissions of *Pn. septemradiatus* were transmitted through wet muddy substrata. While males were aggressively interacting out of water, a hydrophone was inserted into wet mud to capture the acoustic emissions. The acoustic bouts were found to be complex, being combinations of pulses and tonal components. Pulses were repeated at a low rate, with peak frequency below 100 Hz, and tonal components were amplitude- and frequency-modulated segments with a fundamental frequency of about 170 Hz. The relatively low dominant frequencies of these emissions, below 200 Hz, are typical of vocal communications in gobioid fishes. *Periophthalmodon septemradiatus* appears to share with other soniferous gobies the dominance of lower frequencies for sound production, and of higher frequencies for hearing (Zeyl et al. 2016; Lindemann and Mann 2004; Diesselhorst 1938). However, some of the sound traits produced by *Pn. septemradiatus* appear derived relative to basal gobioid fishes (Agorreta et al. 2013). For example, *Odontobutis obscura* produces pulsatile acoustic bouts (Takemura 1984) that are less complex than in *Pn. septemradiatus.*

Mudskippers associate vocal communications with aggressive visual displays, such as mouth gaping and erection of the dorsal fins (Brillet 1970). Whereas transmission through the emerged and wet muddy substratum is a specific novelty of the sound production in *Pn. septemradiatus*, other features appear similar to the basic patterns of sound production in water, in other gobioid fishes. Nevertheless, acoustic communication through the substratum has never been tested in benthonic aquatic gobies. The presence of substratum-transmitted components of gobioid sounds in aquatic gobies would support an exaptation hypothesis for acoustic communication in *Pn. septemradiatus* as suggested by Polgar et al. (2011). Accordingly, the acoustic mechanism evolved underwater in aquatic gobioids and acted as a preadaptation to acoustic communication in wet subaerial environments in *Pn. septemradiatus*. Future studies may test whether the acoustic features observed in *Pn. septemradiatus* are also found in related taxa of the "*Periophthalmus* lineage" (Agorreta et al. 2013). If these hypotheses are corroborated, this lineage would have conserved an aquatic acoustic gobioid mechanism and soniferous patterns, and used them for semi-terrestrial communications.

Short, rapid, and downward-directed head movements are associated with sound production in *Pn. septemradiatus* (Polgar et al. 2011) suggesting involvement of the head and the pectoral girdle in sound emission. The observed constraints and trade-offs between rate or frequency and duration, between frequency and rate of different types of units, and between frequency and time intervals suggest that the sound mechanism involves muscular contraction (Polgar et al. 2011). This hypothesis is consistent with the only existing experimental data on sound production mechanism in gobies, which indicate a cranio-pectoral mechanism of sound production in *Gobius paganellus* involving the *levator pectoralis* muscle (Parmentier et al. 2013). During sound emissions, *G. paganellus* rests on its pelvic fins and abducts its pectoral fins detaching them from the substrate. This suggests that the vibrations are transmitted to the cleithrum by the sonic muscle and are then transmitted to the enlarged radials of the pectoral fins; the latter acting like loudspeakers (Parmentier et al. 2013). *Periophthalmodon septemradiatus* contacts the substrate with its pectoral fins while emitting sounds (tripod locomotion; Polgar et al. 2011); therefore, if the mechanism is similar, vibrations would be transmitted directly into the substrate, acting as sound transducers.

The gas bladder is an important acoustic organ in several fish species and can be involved in both sound emission and acoustic reception (Ladich and Fine 2006). In gobioids, the gas bladder is not necessary to generate sounds (Lugli and Fine 2003). Whether the gas bladder, where present in soniferous species, can influence the sound structure remains to be investigated. Some goby species are without a gas bladder (Parmentier et al. 2013) and this condition is also found in *Pn. septemradiatus* (Polgar et al. 2011) and other mudskipper species whose soniferous traits were not investigated, such as *B. boddarti*, *Pa. rictuosus*, and *S. histophorus* (M. Kuciel, personal observation). This condition is also found in other soniferous gobies such as *Padogobius nigricans*, *P. marmoratus* (Zeyl et al. 2016), and *N. fluviatilis* (Horvatić et al. 2016). In contrast, many other goby species, including most of the soniferous species, have retained the gas bladder. This poses an interesting question about the evolutionary pattern of gas bladder loss or retention in this group and the relationships between sound structure and the presence/absence of the gas bladder.

A comparative analysis of the sound emission mechanism in gobies and mudskippers is urgently needed to understand the relationship between sound structure and the emission mechanism, and the evolutionary pattern of acoustic communication within this group. This would allow discriminating the basal acoustic gobioid patterns from the specific novelties evolved in the context of the "*Periophthalmus* lineage."

6.5 SUMMARY

The sensory organs of fishes are exposed to remarkably different types of stimuli. Fishes inhabiting specific ecological niches evolve complex sensory organs with specific adaptive structures. This resulted in specialization where one sense is clearly dominant: fishes with high visual acuity such as *Thunnus thunnina* (Sihna and Sihna 1990); "good smellers" such as *Holopagrus guentheri* (Pfeifer 1964), *Anguilla anguilla* (Schulte 1972), and *Silurus glanis* (Jakubowski 1981); and those with well-developed lateral line systems such as *Dasyatis sabina* (Maruska and Tricas 1998) and *Astyanax mexicanus* (Yoshizawa et al. 2014). In cases in which there is a lack of exposure to some stimuli types, complete or partial regression to a corresponding organ may occur.

Mudskippers are the most amphibious fishes known (Graham et al. 2007). Eye anatomy and behavioral ecology suggest that vision is the most important and dominant sensory cue while mudskippers are out of water. This claim is corroborated by the presence of protruding eyes that can be independently retracted into dermal cups, the unique shape of the retina with a flattened distal region, thickened cornea, round lenses, effective terrestrial predatory activity, subaerial courtship behavior, terrestrial predator avoidance, and terrestrial territorial behavior. Long terrestrial exposure may cause extended cornea desiccation, decreasing its transparency and impairing vision. While in the aquatic medium, chemoreception and mechanoreception may play more important roles. The sense of taste seems to be highly developed in mudskippers as high numbers of external and internal taste buds are present in the epithelia. It is reasonable to postulate that mudskippers are able to sense odors while on land and in water. Olfactory organs in mudskippers are sophisticated in structure and have a unique functioning mechanism. However, these seem to be not particularly important while on land as the olfactory epithelium inside the olfactory canals has a relatively small surface area.

The acoustic world of mudskippers is virtually unexplored. Observations have been made on only a single species, *Pn. septemradiatus* (Polgar et al. 2011). However, the lack of species-specific acoustic pattern and anatomical structures strongly suggests that acoustic production and perception are not novel adaptations in this group. Many gobies are soniferous and have highly developed hearing abilities; mudskippers may have inherited these traits from their aquatic ancestors. The semi-aquatic or semi-terrestrial habitats in which mudskippers live are characterized by water-saturated substrates and an abundance of capillary water upon emersion. Although mudskippers do come out of water with most of their body when performing subaerial activities, their "continuously wet" environments (Inger 1957) might offer physical conditions that are drastically different than dry land. Many sensory adaptations that evolved underwater, including soniferous activities, might work as well on wet substrates. This exaptation hypothesis (Polgar et al. 2011) is worth experimental testing as it might provide clear clues on the possible mechanisms that facilitated water-to-land transitions in vertebrates.

ACKNOWLEDGMENTS

Authors would like to thank Professor Dominique Adriaens (Centre for X-Ray Tomography, University of Ghent, Belgium) for providing the CT-scanning data and Dr. Dagmara Podkowa (Jagiellonian University, Crakow, Poland) for technical support.

REFERENCES

Afanasjev M. 1875. On the nerve endings in the tactile organs in fishes. *Zapiski Akademii Nauk* 25:1–25 (in Russian).

Agorreta A., San Mauro D., Schliewen U. et al. 2013. Molecular phylogenetics of Gobioidei and phylogenetic placement of European gobies. *Molecular Phylogenetics and Evolution* 69:619–633.

Amorim M.C.P. and Neves A.S.M. 2007. Acoustic signalling during courtship in the painted goby, *Pomatoschistus pictus*. *Journal of the Marine Biological Association of the United Kingdom* 87:1017–1023.

Bardach J.E. and Atema J. 1971. The sense of taste in fishes. In *Handbook of Sensory Physiology* Vol. IV/2 (ed. L.M. Beidler) pp. 293–336. Berlin: Springer-Verlag.

Bass A.H. and Mckibben J.R. 2003. Neural mechanisms and behaviors for acoustic communication in teleost fish. *Progress in Neurobiology* 605:1–26.

Baumeister L. 1913. Uber die Augen der Schlammspringer (*Periophthalmus* und *Boleophthalmus*). *Zoologische Jahrbucher, Abteilungfur Anatomieund Ontogenie der Tiere* 35:341–354.

Belanger R.M., Smith C.M., Corkum L.D. and Zielinski B.S. 2003. Morphology and histochemistry of the peripheral olfactory organ in the round goby, *Neogobius melanostomus* (Teleostei: Gobiidae). *Journal of Morphology* 257(1):62–71.

Berghard A., Buck L.B. and Liman E.R. 1996. Evidence for distinct signaling mechanisms in two mammalian olfactory sense organs. *Proceedings of the National Academy of Science USA* 93:2365–2369.

Boudriot F. and Reutter K. 2001. Ultrastructure of the taste buds in the blind cavefish *Astyanax jordani* ("*Anoptichthys*") and the sighted river fish *Astyanax mexicanus* (Teleostei: Characidae). *Journal of Comparative Neurology* 434:428–444.

Breipohl W., Zippel H.P., Ruckert K. and Oggolter H. 1976. Morphologische und elektrophysiologische Studien zur Struktur und Funktion des olfaktorischen Systems beim Goldfisch unter normalen und experimentellen Bedingungen. *Beitr Elektronenmikroskopische Direktabbildung Oberflächen* 9:561–584.

Brillet C. 1970. Relations entre comportement sexuel, territoire et aggressivitè chez le periophtalmes. *Comptes Rendus Hebdomadaires des Séances de l'Academie des sciences, Série D* 270:1507–1510.
Burne R.H. 1909. The anatomy of the olfactory organ of teleostean fishes. *Proceedings of the Zoological Society of London* 2:610–663.
Caprio J. 1982. High sensitivity and specificity of olfactory and gustatory receptors of catfish to amino acids. In *Chemoreception in Fishes* (ed. T.J. Hara) pp. 109–134. New York, NY: Elsevier Science.
Cordier R. 1964. Sensory cell. In *The cell: Biochemistry, Physiology, Morphology* (eds. J. Barchet and A.E. Mirsky) pp. 313–386. New York, NY: Academic Press.
Crisp M., Lowe G.A. and Laverack M.S. 1975. On the ultrastructure and permeability of taste buds of the marine teleost *Ciliata mustela*. *Tissue and Cell* 7:191–202.
Davenport J. and Matin A.K.M. 1990. Terrestrial locomotion in the climbing perch, *Anabas testudineus* (Bloch) (Anabantidae, Pisces). *Journal of Fish Biology* 37:175–184.
Desgranges J.C. 1966. Sur la double innervations des cellules sensorielles des bourgeons du gout des barbillons du Poisson chat. *Comptes Rendus de l'Académie des Sciences Paris* 263:1103–1106.
Diaz J.P., Prie-grane M., Noell T. and Connes R. 2002. Ultrastructural study of the olfactory organ in adult and developing European sea bass, *Dicentrarchus labrax*. *Canadian Journal of Zoology* 80:1610–1622.
Diesselhorst G. 1938. Horversuche an Fischen ohne Weberschen Apparat. *Zeitschrift für Vergleichende Physiologie* 25:748–783.
Douglas R.H. and Djamgoz M.B.A. 1990. *The Visual System of Fish*. London: Chapman and Hall, 506 pp.
Døving K.B., Dubois-Dauphin M., Holley A. and Jourdan F. 1977. Functional anatomy of the olfactory organ of fish and the ciliary mechanism of water transport. *Acta Zoologica* 58:245–255.
Evans R.E., Zielinski B. and Hara T.J. 1982. Development and regeneration of the olfactory organ in rainbow trout. In *Chemoreception in Fishes* (ed. T.J. Hara) pp. 15–37. New York, NY: Elsevier Science.
Ezeasor D.N. 1982. Distribution and ultrastructure of taste buds in the oropharyngeal cavity of the rainbow trout, *Salmo gairdneri* Richardson. *Journal of Fish Biology* 20:53–68.
Finger T.E., Drake S.K., Kotrschal K., Womble M. and Dockstader K.C. 1991. Postlarval growth of the peripheral gustatory system in the channel catfish, *Ictalurus punctatus*. *Journal of Comparative Neurology* 314:55–66.
Fishelson L., Baldwin C.C. and Hastings P.A. 2012. Comparison of the oropharyngeal cavity in the Starksiini (Teleostei: Blenniiformes: Labrisomidae): Taste buds and teeth, including a comparison with closely-related genera. *Journal of Morphology* 273(6):618–628.
Fishelson L. and Delarea Y. 2004. Taste buds on the lips and mouth of some blenniid and gobiid fishes: Comparative distribution and morphology. *Journal of Fish Biology* 65:651–665.
Fishelson L., Delarea Y. and Zverdling A. 2004. Taste bud form and distribution on lips and in the oropharyngeal cavity of cardinal fish species (Apogonidae, Teleostei), with remarks on their dentition. *Journal of Morphology* 259:316–327.
Fox H. and Whitear M. 1978. Observations on Merkel cells in amphibians. *Biologie Cellulaire* 32:223.
Franz V. 1910. Die japanischen Knockenfische der Sammlungen Harberer und Doflein. *Abhandlungen Bayerische Akademie der Wissenschaften, Mathematisch-Naturwissenschaftliche Klasse* (Suppl. 4):1–32.
Genten F., Terwinghe E. and Danguy A. 2009. *Atlas of Fish Histology*. Enfield, NH: Science Publishers, 223 pp.
Gomahr A., Palzenberg M. and Kotrschal K. 1992. Density and distribution of external taste buds in cyprinids. *Environmental Biology of Fishes* 33:125–134.
Graham J.B. 1971. Aerial vision in amphibious fishes. *Fauna Rancho Mirage, California, USA* 3:14–23.
Graham J.B., Jones C.B. and Rubinoff I. 1985. Behavioural, physiological and ecological aspects of the amphibious life of the pearl blenny, *Entomacrodus nigracans* Gill. *Journal of Experimental Marine Biology and Ecology* 89:255–268.
Graham J.B. and Lee H.J. 2004. Breathing air in air: In what ways might extant amphibious fish biology relate to prevailing concept about early tetrapods, the evolution of vertebrate air breathing, and the vertebrate land transition? *Physiological and Biochemical Zoology* 77:720–731.
Graham J.B., Lee H.J. and Wegner N.C. 2007. Transition from water to land in extant group of fishes: Air breathing and the acquisition sequence of adaptations for amphibious life in oxudercinae gobies. In: *Fish Respiration and Environment* (eds. M.N. Fernandes, F.T. Rantin, M.L. Glass and B.G. Kapoor) pp. 255–288. Enfield, NH: Science Publishers.

Graham J.B. and Rosenblatt R.H. 1970. Aerial vision: Unique adaptation in an intertidal fish. *Science* 168:586–588.

Grover-Johnson N. and Farbman A.J. 1976. Fine structure of taste buds in the barbel of the catfish, *Ictarulus punctatus. Cell and Tissue Research* 169:395–403.

Hansen A. and Finger T.E. 2000. Phyletic distribution of crypt-type olfactory receptor neurons in fishes. *Brain Behaviour Evolution* 55:100–110.

Hansen A. and Finger T.E. 2009. Solitary chemosensory cells: Phylogeny and ontogeny. In *Airway Chemoreceptors in the Vertebrates—Structure, Evolution and Function* (eds. G. Zaccone, E. Cutz, D. Adriaensen, C.A. Nurse and A. Mauceri) pp. 377–387. Enfield, NH: Science Publishers.

Hansen A. and Reutter K. 2004. Chemosensory systems in fish: Structural, functional and ecological aspects. In *The Senses of Fish. Adaptations for the Reception of Natural Stimuli* (eds. G. Von Der Emde, J. Mogdans and B.G. Kapoor) pp. 55–63. New Delhi: Narosa Publishing House.

Hansen A., Rolen S.H., Anderson K., Morita Y., Caprio J. and Finger T.E. 2003. Correlation between olfactory receptor cell type and function in the channel catfish. *Journal of Neuroscience* 23(28):9328–9339.

Hansen A. and Zeiske E. 1993. Development of the olfactory organ in the zebrafish, *Brachydanio rerio. Journal of Comparative Neurology* 333, 289–300.

Hansen A. and Zeiske E. 1998. The peripheral olfactory organ of the zebrafish, *Danio rerio*: An ultrastructural study. *Chemosensory Senses* 23:39–48.

Hansen A. and Zielinski B.S. 2005. Diversity in the olfactory epithelium of bony fishes: Development, lamellar arrangement, sensory neuron cell types and transduction components. *Journal of Neurocytology* 34:183–208.

Hansen A. and Zippel H.P. 1995. Solitary chemosensory cells in the olfactory organs in fish. *Chemosensory Senses* 20:105.

Hansen A., Zippel H.P., Sorensen P.W. and Caprio J. 1999. Ultrastructure of the olfactory epithelium in intact, axotomized, and bulbectomized goldfish, *Carassius auratus. Microscopy Research Technics* 45:325–338.

Hara T.J. 1975. Olfaction in fish. *Progress in Neurobiology* 5:271–335.

Hara T.J. 1992. Mechanisms of olfaction. In *Fish Chemoreception* (ed. T.J. Hara) pp. 151–170. London: Chapman and Hall.

Hara T.J. 1994. Olfaction and gustation in fish: An overview. *Acta Physiologica Scandinavica* 152:207–217.

Harms W.1914. Uber die Augen der am Grunde der Gewasser lebenden Fische. *Zoologische Anzeiger* 44: 35–41.

Helfman G., Collette B.B., Facey D.E. and Bowen B.W. 2009. *The Diversity of Fishes: Biology, Evolution and Ecology*, 2nd Edition. Chichester: Wiley-Blackwell, 720 pp.

Hess C. 1912. Untersuchungen zur vergleichenden Physiologie und Morphologie des Ciliarringes. *Zoologische Jahrbiicher* (Suppl. 15):155–175.

Hirata Y. 1966. Fine structure of the terminal buds in the barbels of some fishes. *Archiwum Histologicum Japonicum* 26:507–523.

Horvatić S., Cavraro F., Zanella D. and Malavasi S. 2016. Sound production in the Ponto-Caspian goby *Neogobius fluviatilis* and acoustic affinities within the *Gobius* lineage: Implications for phylogeny. *Biological Journal of the Linnean Society* 117:564–573.

Inger R.F. 1957. Ecological aspects of the origin of the tetrapods. *Evolution* 11:373–376.

Jakubowski M. 1958. The structure and vascularization of the skin in the pond-loach (*Misgurnus fossilis* L.) *Acta Biologica Cracoviensia Series Zoologica* 1:113–127.

Jakubowski M. 1966. The lateral-line organs in some Cobitidae. *Acta Biologica Cracoviensia Series Zoologica* 9:71–80.

Jakubowski M. 1980. Distribution and fine structure (SEM, TEM) of the taste buds in *Cobitis taenia* (Cobitidae, Pisces). Paper presented at the Second Congress of European Society for Comparative Skin Biology, Copenhagen, 25–28 pp.

Jakubowski M. 1981. Ultrastructure (SEM, TEM) of the olfactory epithelium in the Wels, *Silurus glanis* L. (Siluridae, Pisces). *Zeitschrift für Mikroskopisch-anatomische Forschung* 93:728–735.

Jakubowski M. 1983. New details of the ultrastructure (TEM, SEM) of taste buds in fishes. *Zeitschrift für Mikroskopisch-anatomische Forschung* 97:849–862

Jakubowski M. and Whitear M. 1986. Ultrastructure of taste buds in fishes. *Folia Histochemica Cytobiologica* 24:310–311.

Jakubowski M. and Whitear M. 1990. Comparative morphology and cytology of taste buds in teleosts. *Zeitschrift för Mikroskopisch-anatomische Forschung* 104:529–560.

Jakubowski M. and Żuwała K. 2000. Taste organs in lower vertebrates: Morphology of the gustatory organs in fishes. In *Vertebrate Functional Morphology Horizon of Research in the 21st Century* (eds. H.A. Dutta and J.S.D. Munshi) pp. 159–172. Enfield, NH: Science Publishers.

Jasiński A. 1976. Narządy zmysłów. In *Anatomia Porównawcza Kręgowców* (ed. H. Szarski) p. 432. Warszawa: PWN.

Jermann T. and Senn D.G. 1992. Amphibious vision in *Coryphoblennius galerita* L. (Perciformes). *Experientia* 48:217–218.

Kanwal J.S. and Finger T.E. 1992. Central representation and projections of gustatory systems. In *Fish Chemoreception* (ed. T.J. Hara) pp. 79–102. London: Chapman and Hall.

Kapoor B.C., Evans H.E. and Pevzner R.A. 1975. The gustatory system in fish. *Advances in Marine Biology* 13:53–108.

Karsten H. 1923. Das auge von *Periophthalmus koelreuteri. Zeitschrift fur Naturwissenschaft* 59:115–154.

Kerr J.B. 1999. *Atlas of Functional Histology*. London: Mosby International Lynton House, 374 pp.

Kleerkoper H. 1969. *Olfaction in Fishes*. Bloomington, IN: Indiana University Press, 222 pp.

Kotrschal K., Whitear M. and Adam H. 1984. Morphology and histology of the anterior dorsal fin of *Gaidropsarus mediterraneus* (Pisces Teleostei), a specialized sensory organ. *Zoomorphology* 104:365–372.

Kuciel M. 2012. The mechanism of olfactory organ ventilation in *Periophthalmus barbarus* (Gobiidae, Oxudercinae). *Zoomorphology* 132:81–85.

Kuciel M., Lauriano E.R., Silvestri G., Zuwała K., Pergolizzi S. and Zaccone D. 2014. The structural organization and immunohistochemistry of G-protein alpha subunits in the olfactory system of the air-breathing mudskipper, *Periophthalmus barbarus* (Linnaeus, 1766) (Gobiidae, Oxudercinae). *Acta Histochemica* 116:70–78.

Kuciel M., Żuwała K. and Jakubowski M. 2011. A new type of fish olfactory organ structure in *Periophthalmus barbarus* (Gobiidae, Oxudercinae). *Acta Zoologica Stockholm* 92(3):276–280.

Kuciel M., Żuwała K. and Satapoomin U. 2013. Comparative morphology (SEM) of the peripheral olfactory organ in the Oxudercinae subfamily (Gobiidae, Perciformes). *Zoologischer Anzeiger* 252:424–430.

Laberge F. and Hara T.J. 2001. Neurobiology of fish olfaction: A review. *Brain Research Review* 36:46–59.

Ladich F. and Fine M. L. 2006. Sound-generating mechanisms in fishes: A unique diversity in vertebrates. In *Communication in Fishes* Vol. 1 (eds. F. Ladich, S.P. Collin, P. Moller and B.G. Kapoor) pp. 1–42. Enfield, NH: Science Publishers.

Lamb T.D., Collin S.P. and Pugh E.N. 2007. Evolution of the vertebrate eye: Opsins, photoreceptors, retina and eye cup. *Nature Reviews Neuroscience* 8(12):960–76.

Land M.F. and Nilsson D-E. 2012. *Animal Eyes*. City: Oxford University Press, 271 pp.

Leydig F. 1851. Uber die Haut einiger SuBwasserfische. *Zeitschrift Für Wissenschaftliche Zoologie* 3:1–12.

Lindemann M.A. and Mann D.A. 2004. Hearing in the amphibious mudskipper, *Periophthalmus barbarus*. Paper presented at the twenty-seventh annual midwinter research meeting of the Association for Research in Otolaryngology, Daytona Beach, 21–26 pp.

Linser P.J., Carr W.E., Cate H.S., Derby C.D. and Netherton J.C. 1998. Functional significance of the co-localization of taste buds and teeth in the pharyngeal jaws of the largemouth bass, *Micropterus salmoides*. *The Biological Bulletin* 195(3):273–81.

Lugli M. and Fine M.L. 2003. Acoustic communication in two freshwater gobies: Ambient noise and short-range propagation in shallow streams. *Journal of Acoustic Society of America* 114:512–521.

Lugli M., Pavan G., Torricelli P. and Bobbio L. 1995. Spawning vocalizations in male freshwater gobiids (Pisces, Gobiidae). *Environmental Biology of Fishes* 43:219–231.

Lugli M. and Torricelli P. 1999. Prespawning sound production in Mediterranean sand-gobies. *Journal of Fish Biology* 54:691–694.

Lugli M., Torricelli P., Pavan G. and Mainardi D. 1997. Sound production during courtship and spawning among freshwater gobiids (Pisces, Gobiidae). *Marine and Freshwater Behaviour and Physiology* 29:109–126.

Malavasi S., Collatuzzo S. and Torricelli P. 2008. Interspecific variation of acoustic signals in Mediterranean gobies (Perciformes, Gobiidae): Comparative analysis and evolutionary outlook. *Biological Journal of the Linnean Society* 93:763–778.

Malavasi S., Torricelli P., Lugli M., Pravoni F. and Mainardi D. 2003. Male courtship sounds in a teleost with alternative reproductive tactics, the grass goby, *Zosterisessor ophiocephalus*. *Environmental Biology of Fishes* 66:231–236.

Maruska K.P. and Tricas T.C. 1998. Morphology of the mechanosensory lateral line system in the Atlantic stingray, *Dasyatis sabina:* The mechanotactile hypothesis. *Journal of Morphology* 238(1):1–22.

Meyer–Rochow V.B. 1981. Fish tongues—surface fine structure and ecological considerations. *Zoological Journal of the Linnean Society* 71:413–426.

Meyer-Rochow V.B. and Coddington P.E. 2003. Eyes and vision of the New Zealand torrentfish *Cheimarrichthys fosteri* Von Haast (1874): Histology, photochemistry and electrophysiology. In *Fish Adaptations* (eds. A.L. Val and B.G. Kapoor) pp. 339–383. Enfield, NH: Science Publishers.

Munk O. 1970. On the occurrence and significance of horizontal band-shaped retinal area in teleosts. *Videnskabelige Meddelelserfra Dansk Naturhistorisk Forening i Khobenhavn* 133:85–120.

Myrberg A.A. and Lugli M. 2006. Reproductive behavior and acoustical interactions. In *Communication in Fishes* Vol. I (eds. F. Ladich, S. Collin, P. Moller and B.G. Kapoor) pp. 149–176. Enfield, NH: Science Publishers.

Nagai T., Kim D.J., Delay R.J. and Roper S.D. 1996. Neuromodulation of transduction and signal processing in the end organs of taste. *Chemosensory Senses* 21:353–365.

Ohkubo Y., Masubuchi M., Fujioka K. et al. 2005. Distribution and morphological features of taste buds in the zebrafish, *Danio rerio*. *Journal of Oral Biosciences* 47:77–82.

Parmentier E., Kéver L., Boyle K., Corbisier Y., Sawelew L. and Malavasi S. 2013. Sound production mechanism in *Gobius paganellus* (Gobiidae). *Journal of Experimental Biology* 216:3189–3199.

Pfeiffer W. 1963. The morphology of the olfactory organ of the Pacific salmon, (*Oncorhynchus masou*). *Canadian Journal of Zoology* 41:1233–1336.

Pfeifer W. 1964. The morphology of the olfactory organ of *Holopagrus guentheri* Gill 1862. *Canadian Journal of Zoology* 42:235–237.

Polgar G., Malavasi S., Cipolato G., Georgalas V., Clack J.A. and Torricelli P. 2011. Acoustic communication at the water's edge: Evolutionary insights from a mudskipper. *PLoS One* 6:e21434.

Reutter K. 1971. Die Gesmacksknospen des Zwergwelses, *Ameiurus nebulosus* (Lesueur). *Morphologische und histochimische Untersuchungen* 120:280–308.

Reutter K. 1978. Taste organ in the bullhead (Teleostei). *Advances in Anatomy Embryology and Cell Biology* 55:1–98.

Reutter K. 1986. Chemoreceptors. In *Biology of the Integument 2-Vertebrates* (eds. J. Bereiter-Hann, A.G. Matoltsy and K.S. Richards) pp. 586–604. Berlin: Springer-Verlag.

Reutter K. 1991. Ultrastructure of taste buds in the Australian lungfish *Neoceratodus forsteri* (Dipnoi). *Chemical Senses* 16:404.

Reutter K. 1994. Ultrastructure of taste buds in the spotted dogfish *Scyliorhinus canicula* (Selachii). In *Olfaction and Taste XI* (eds. K. Kurihara, N. Suzuki and H. Ogawa) pp. 754. Tokyo: Springer.

Reutter K., Baudriot F. and Witt M. 2000. Heterogeneity of fish taste bud ultrastructure as demonstrated in the holosteans *Amia calva* and *Lepisosteus oculatus*. *Philosophical Transactions the Royal Society of London B* 355:1225–1228.

Reutter K., Breipohl W. and Bijvank G.J. 1974. Taste bud types of fishes. II. Scanning electron microscopical investigations of *Xiphophorus helleri* Heckel (Poecilidae, Cyprinodontiformes, Teleostei). *Cell and Tissue Research* 153:151–165.

Royer S. and Kinamon J.C. 1996. Comparison of high-pressure freezing/freeze substitution and chemical fixation of catfish barbel taste buds. *Microscopy Research and Technique* 35:385–412.

Sayer M.D.J. 2005. Adaptations of amphibious fish for surviving life out of water. *Fish and Fisheries* 6:186–211.

Schulte E. 1972. Untersuchungen an der Regio olfactoria des Aals, *Anguilla anguilla* L. *Zeitschrift für Zellforschung Mikroskopische Anatomie* 125:210–228.

Schulze F.E. 1863. Uber die becherf Ormigen Orange der Fische. *Zeitschrift für wissenschaftliche Zoologie* 12:218–222.

Sihna S.K. and Sihna R.K. 1990. Morphology and the anatomy of the olfactory organs of the marine fish *Thynnus thunnina* (Cuv et Val.). *Folia Morphologica* 2:169–173.

Sponder D.L. and Lauder G.V. 1981. Terrestrial feeding in the mudskipper *Periophthalmus* (Pisces: Teleostei): A cineradiographic analysis. *Journal of Zoology London* 193:517–530.

Tagliafiero G. and Zaccone G. 2001. Morphology and immunohistochemistry of taste buds in bony fishes. In *Sensory Biology of Jawed Fishes.* (eds. B.G. Kapoor and T.J. Hara) pp. 335–345. Enfield, NH: Science Publishers.

Takemura A. 1984. Acoustical behavior of the freshwater goby *Odontobutis obscura. Bulletin of the Japanese Society for the Science of Fish* 50:561–564.

Thornhill R.A. 1967. The ultrastructure of the olfactory epithelium of the lamprey *Lampetra fluviatilis. Journal of Cell Science* 2:591–602.

Todd E.S. 1976. Terrestrial grazing by the eastern tropical Pacific goby, *Gobionellus sagittula. Copeia* 1976:374–377.

Torricelli P., Lugli M. and Pavan G. 1990. Analysis of sound produced by male *Padogobius martensi* (Pisces: Gobiidae) and factors affecting their structural properties. *Bioacoustics* 2:261–275.

Toyoshima K. 1994. Role of Merkel cells in the taste organ morphogenesis of frog. In *Olfaction and Taste XI* (eds. K. Kurihara, N. Suzuki and H. Ogawa H) pp. 13–15. Tokyo: Springer-Verlag.

Truillo-Cenoz O. 1961. Electron microscope observations on chemo-and mechano-receptor cells of fishes. *Zeitschrift fur Zellforschung und Mikroskopische Anatomie* 54:654–676.

Uchihashi K. 1953. Ecological study of the Japanese teleosts in relation to the brain morphology. *Bulletin of the Japan Sea Region Fisheries Research Laboratory* 2:1–166.

Volz W. 1905a. Uber das Auge von *Periophthalmus* und *Boleophthalmus. Mitteilungen der Naturforschenden Gesellschaft in Bern* 108–111.

Volz W. 1905b. Zur Kenntnis des Auges von *Periophthalmus* und *Boleophthalmus. Zoologische Jahrbiicher, Abteilung fur Anatomie und Ontogenie der Tiere* 22:331–346.

Volz W. 1905c. Sur l'oeil de *Periophthalmus* et *Boleophthalmus. Archives des Sciences Physiques et Naturelles (Geneve)* 4:579–580.

Weber H. 1827. Uber das Geschmacksorgan der Karpfen und den Ursprung seiner Nerven. *Archive of Anatomy and Physiology* 1827:309–315.

Welsch U. and Storch V. 1969. Die Feinstruktur der Geschmacksknospen von Welsen (*Clarias batrachus (L.)* und *Kryptopterus bicirrhis* (Cuvier et Valenciennes). *Zeitschrift fur Zellforschung und Mikroskopische Anatomie* 100:552–559.

Whitear M. 1971. Cell specialization and sensory function on fish epidermis. *Journal of Zoology London* 163:552–559.

Whitear M. 1992. Solitary chemoreceptor cells. In *Chemoreception in Fishes* (ed. T.J. Hara) pp. 103–125. London: Chapman and Hall.

Whitear M. and Moate R.M. 1994. Microanatomy of taste buds in the dogfish *Scyliorhinus canicula. Journal of Submicroscopic Cytology and Pathology* 26:357–367.

Yamamoto S. 1931. Die statische Refraktion des Auge verchiedener Tiere. *Journal of Okayama Medical Society* 43:1407–1425.

Yamamoto M. and Ueda K. 1978. Comparative morphology of fish olfactory epithelium. V. Gasterosteiformes, Channiformes and Synbranchiformes. *Bulletin of the Japanese Society of Scientific Fisheries* 44:1309–1314.

Yamamoto M. and Ueda K. 1979. Comparative morphology of fish olfactory epithelium. IX. Tetraodontiformes. *Zoological Magazine (Tokyo)* 88:210–218..

Yashpal M., Kumari U., Mittal S. and Mittal A.K. 2006. Surface architecture of the mouth cavity of a carnivorous fish *Rita rita* (Hamilton 1822) (Siluriformes, Bagridae). *Belgian Journal of Zoology* 136:155–162.

Yew D.T. and Wu M.F. 1979. Ratio of rods and different cones in fishes of shallow sea. *Anatomische Anzeiger* 145:155–160.

Yoshizawa M., Jeffery W.R., van Netten S.M. and McHenry M.J. 2014. The sensitivity of lateral line receptors and their role in the behavior of Mexican blind cavefish (*Astyanax mexicanus*). *Journal of Experimental Biology* 217:886–95.

Zachar P.C. and Jonz M.G. 2012. Confocal imaging of Merkel-like basal cells in the taste buds of zebrafish. *Acta Histochemica* 114(2):101–115.

Zander C.D. 1974. Beziehungen zwischen korperbau und Lebensweise bei Blenniidae (Pisces) aus dem Roten Meer. III. *Morphologie des Auges Marine Biology* 28:61–71.

Zeiske E., Thiesen B. and Breucker H. 1992. Structure, development, and evolutionary aspects of the peripheral olfactory system. In *Fish Chemoreception* (ed. T.J. Hara) pp. 13–39. London: Chapman and Hall.

Zeyl J.N., Malavasi S., Holt D.E., Noel P., Lugli M. and Johnston C.E. 2016. Convergent aspects of acoustic communication in darters, sculpins and gobies. *Advances in Experimental Medicine and Biology* 877:93–120.

Zippel H.P., Hansen A. and Caprio J. 1997. Renewing olfactory receptor neurons do not require contact with the olfactory bulb to develop normal responsiveness. *Journal of Comparative Physiology A-Sensory Neural and Behavioral Physiology* 181:425–437.

Żuwała K. and Jakubowski M. 1993. Light and Electron (SEM, TEM) Microscopy of buds in the tench *Tinca tinca* (Pisces: Cyprinidae). *Acta Zoologica (Stockholm)* 74:277–282.

CHAPTER 7

Nitrogen Metabolism and Nitrogenous Waste Excretion

Shit F. Chew and Yuen K. Ip

CONTENTS

7.1 INTRODUCTION

Mudskippers (Perciformes: Gobiidae) are euryhaline, amphibious, and air-breathing fishes commonly found in mangrove areas throughout the Indo-Pacific. They construct burrows in the soft substrate, and store air in these burrows (Toba and Ishimatsu 2014; Ishimatsu et al. 2009, 2007, 1998; Lee et al. 2005). The stored air is a crucial source of oxygen for developing embryos, and may also serve as a subsidiary oxygen source for adult fishes during burrow confinement (Toba and Ishimatsu 2014). As adaptations to their natural habitats, mudskippers generally display high tolerance for terrestrial exposure (Chew et al. 2007; Ip et al. 2006, 2004a, 2001b, 1993; Lim et al. 2001), environmental hypoxia (Chew and Ip 1992a, b; Ip et al. 1991a; Chew et al. 1990; Gordon et al. 1978; Bandurski et al. 1968), environmental ammonia (Ip et al. 2005, 2004b,e; Randall et al. 2004, 1999; Peng et al. 1998), alkaline pH (Chew et al. 2003a), cyanide (Chew and Ip 1992c; Chew et al. 1998), and sulfide (Ip et al. 2004c). The degree of tolerance to various environmental parameters, however, varies between species.

Gills of fishes are important for respiration, ion transport, osmoregulation, acid–base balance and nitrogenous waste excretion, the latter predominantly in the form of ammonia. In mudskippers, gills are modified structurally, with different species displaying different branchial adaptation for air-breathing and nitrogenous waste excretion (Low et al. 1988, 1990). Published studies on nitrogen metabolism and nitrogenous waste excretion in mudskippers centered around two mudskipper species: *Periophthalmodon schlosseri* and *Boleophthalmus boddarti*. These mudskipper species display different behaviors despite sharing the same macrohabitat. In comparison to *Pn. schlosseri*, *B. boddarti* is less well adapted to land, and its burrows are found on the lower (more seaward) regions of mudflats (Low et al. 1988). Individuals of *B. boddarti* are usually seen on mudflats at low tide; as the tide rises, they retreat into their burrows and remain submerged until the tide ebbs. As for *Pn. schlosseri*, their burrows are found more landward, further away from the water's edge (Low et al. 1988). At high tide, individuals of *Pn. schlosseri* typically swim along the water's edge and continue to be active in the supralittoral zone. These behaviors mirror observations on gill structures in these two species. The branchial morphology and morphometry in *B. boddarti* are comparable to those of other water-breathing teleosts (Low et al. 1990, 1988). By contrast, *Pn. schlosseri* exhibits unusual branchial modifications for air breathing, including branched filaments with intrafilamentous interlamellar fusions whereby adjacent secondary lamellae of the same filament are fused to form numerous fenestrae (Low et al. 1990, 1988), which affect nitrogen metabolism and nitrogenous waste excretion when out of water. Furthermore, dietary protein composition directly influence excretory nitrogen metabolism. Differences in foraging modes between the carnivorous *Pn. schlosseri* and the herbivorous *B. boddarti* thus affect adaptations to excretory nitrogen metabolism.

7.2 AMMONIA PRODUCTION IN THE LIVER

The major source of amino acids for animals is dietary proteins. In the digestive tract, proteins are hydrolyzed to amino acids for absorption. Some absorbed amino acids can be metabolized in the intestine before reaching the liver (Tng et al. 2008; Karlsson et al. 2006). Whereas carbohydrates and lipids can be stored as glycogen and triglycerides, respectively, there is no known storage of amino acids. Therefore, dietary amino acids in excess of the amounts needed for growth and maintenance of protein turnover must be broken down into carbon dioxide, water, and ammonia (Campbell 1991). In fishes, approximately 40–60% of the dietary nitrogen intake is excreted within 24 hours (Ip et al. 2004d; Lim et al. 2004). During fasting bouts, fishes utilize muscle proteins as a source of amino acids for adenosine triphosphate (ATP) or carbohydrate production (Houlihan et al. 1995).

In fishes, dietary amino acid catabolism occurs mainly in hepatocytes (Campbell et al. 1983), and ammonia is produced mainly from the α-amino group of amino acids that are catabolized in the liver (Van den Thillart and van Raaji 1995). In fish hepatocytes, certain amino acids can be catabolized by specific deaminases (such as histidase, asparaginase, serine dehydratase, and threonine dehydratase) in the cytosol (Youngson et al. 1982), but the majority of amino acids are catabolized through the combined actions (transdeamination; Figure 7.1) of cytosolic aminotransferases and mitochondrial

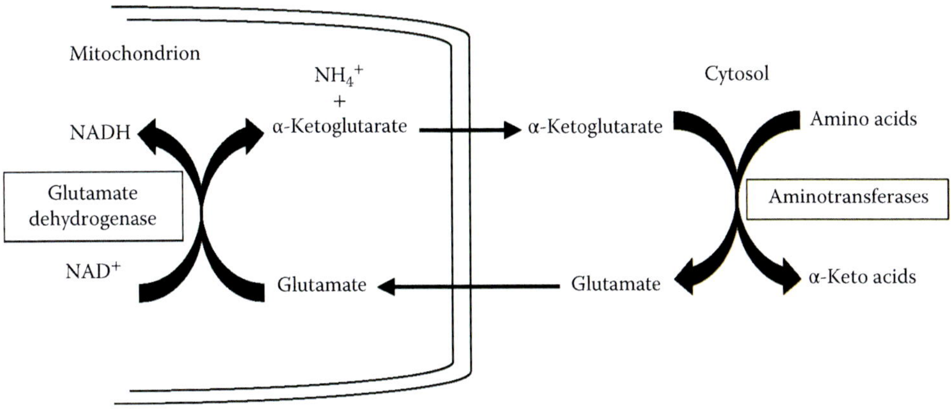

Figure 7.1 The release of the α-amino group of amino acids as NH_4^+ via transdeamination, involving cytosolic aminotransferases and the mitochondrial glutamate dehydrogenase. (Courtesy of Shit F. Chew and Yuen K. Ip.)

glutamate dehydrogenase (Ballantyne 2001; French et al. 1981; Walton and Cowey 1977). Studies on the enzymes associated with nitrogen metabolism in *Pn. schlosseri*, *B. boddarti*, and *Periophthalmus chrysospilos* reveal that ammoniagenesis occurs mainly through transdeamination with glutamate dehydrogenase operating in conjunction with aspartate aminotransferase and alanine aminotransferase in the liver (Chew and Ip 1993; Ip et al. 1990a; Chew and Ip 1987). As glutamate dehydrogenase is localized exclusively in the mitochondrial matrix, NH_4^+ is released through the deamination of glutamate in liver mitochondria. Subsequently, ammonia produced in the matrix of liver mitochondria has to permeate the mitochondrial and plasma membranes before entering the blood. Once in the blood, ammonia is delivered to every part of the body, including the brain. As a result, carnivorous fishes with a high protein diet, like *Pn. schlosseri*, may have to defend against ammonia toxicity after feeding.

7.3 AMMONIA IS TOXIC

Ammonia is toxic and has multiple deleterious effects on many cellular processes (Ip and Chew 2010; Chew et al. 2006; Ip et al. 2001a). It stimulates glycolysis in fishes by activating the cytosolic phosphofructokinase-1. It also impairs the tricarboxylic acid cycle, and interferes with energy metabolism in the mitochondria (Arillo et al. 1981). The most acute effects of ammonia probably arise from the ability of NH_4^+ to substitute for K^+ to activate ion channels/transporters (Binstock and Lecar 1969) and disrupt the electrochemical potential gradient of K^+ in the central nervous system (Cooper and Plum 1987). Several theories (glutamatergic dysfunction, glutamine accumulation leading to astrocyte swelling, activation of N-methyl-D-aspartate [NMDA] receptors) have been proposed to address the mechanisms of acute ammonia toxicity in mammalian brains (Albrecht et al. 2010; Albrecht and Norenberg 2006; Brusilow 2002; Felipo and Butterworth 2002; Rose 2002). The excessive activation of NMDA receptors by ammonia (Kosenko et al. 1999; Hermenegildo et al. 1996) is neurotoxic, leading to oxidative damages, neuronal degeneration, and death of neurons (Miñana et al. 1996). The mechanisms of ammonia neurotoxicity may also involve nitrosative/oxidative stress and mitochondrial permeability transition (Görg et al. 2013; Bemeur et al. 2010; Görg et al. 2010; Häussinger and Görg 2010; Reddy et al. 2009). The mitochondrial permeability transition leads to the opening of a pore in the inner mitochondrial membrane, resulting in a collapse of ionic gradients and mitochondrial dysfunction. Additionally, the mitochondrial permeability transition can enhance the permeation of glutamine through the inner mitochondrial membrane, and cause an increase in the production of ammonia catalyzed by glutaminase in the mitochondrial matrix of astrocytes (Albrecht and Norenberg 2006). Although it has been reported that ammonia exposure can also lead to oxidative stress in the brain of *B. boddarti* (Ching et al. 2009), many mechanisms proposed for ammonia toxicity have yet to be validated in brains of fishes (Ip and Chew 2010; Wee et al. 2007; Ip et al. 2005). Due to its toxicity, excess ammonia produced endogenously must be effectively excreted, and the uptake of exogenous ammonia from the environment must be avoided. Under conditions where the excretion of endogenous ammonia is impeded, the accumulation of ammonia to toxic levels must be averted through a variety of defense mechanisms, including a reduction in endogenous ammonia production and/or the conversion of ammonia into less toxic compounds such as glutamine or urea.

7.4 AMMONIA EXCRETION THROUGH GILLS

In water, ammonia exists as molecular NH_3 and cationic NH_4^+, the ratio of which is pH dependent. In this chapter, ammonia stands for both NH_3 and NH_4^+, while NH_3 and NH_4^+ represent specifically unionized molecular ammonia and ammonium ion, respectively. The p*K* of the equilibrium reaction, $NH_3 + H_3O^+ \Leftrightarrow NH_4^+ + H_2O$, is 9.0–9.5. As ammoniagenesis occurs in the mitochondrial matrix which is more alkaline (pH ~7.8) than the cytosol (pH ~7.2), the P_{NH_3} gradient would drive NH_3 to exit the mitochondrion. The exit of NH_3 through the inner mitochondrial membrane would theoretically uncouple oxidative phosphorylation as it would combine with H^+ outside the mitochondrion and disrupt the proton gradient (Ip and Chew 2010). How ammonia exits the liver mitochondrion of fishes, including mudskippers, which excrete ammonia as the major nitrogenous waste but without uncoupling oxidative phosphorylation is an enigma at present. As the pH of fish blood is approximately 2 units below the p*K* of ammonia, ammonia is circulated in the plasma to other parts of the body mainly (>95%) in the form of NH_4^+. In animals, whether the ammonia produced endogenously is excreted as NH_3/NH_4^+ or converted to less toxic compounds like urea or uric acid before excretion is dependent on the availability of water. Fully aquatic fishes are generally ammonotelic, excreting more than 50% of the nitrogenous waste as ammonia into the surrounding water.

Fish gills have large surface area, high perfusion by the total cardiac output, large ventilation rates, small diffusion distances, and constant contact with a voluminous external medium (Evans et al. 2005), and therefore act as the primary site of ammonia excretion (Weihrauch et al. 2009; Wilkie 2002, 1997; Evans and Cameron 1986). In water, the majority of ammonia is excreted as NH_3 down a favorable blood-to-water P_{NH_3} gradient across the branchial epithelium (Evans et al. 2005; Wilkie 2002, 1997). Permeation of ammonia through fish gills can occur across the phospholipid bilayer of the branchial epithelium or involve various channels and transporters (Weihrauch et al. 2009; Wright and Wood 2009; Huang and Peng 2005). Members of the Rhesus glycoprotein (RhGP) family may play an important role in transmembrane ammonia transport (Bakouh et al. 2006). Among the six clusters of the RhGPs, RhAG, RhBG, and RhCG are present in most vertebrates and are responsible for transporting ammonia and methylammonium (Planelles 2007). In ammonotelic fishes, Rhag, Rhbg, and Rhcg are vital ammonia conduits in gills (Nawata et al. 2010, 2007; Braun et al. 2009; Nakada et al. 2007a, b). The aquaporin family generally functions for the selective passage of water, but particular aquaporin isoforms can conduct unconventional permeants, including ammonia (Saparov et al. 2007; Wu and Beitz 2007; Nakhoul et al. 2001). Indeed, there are indications that aquaporin can be involved in branchial ammonia excretion in fishes (Ip et al. 2013b). As NH_4^+ has some ionic properties similar to those of K^+, it can compete with K^+ and permeate through the K^+ channels (Marcaggi and Coles 2001; Thomas 1984). NH_4^+ can also replace K^+ to be transported through Na^+/K^+-ATPase (Ip et al. 2012a; Wall and Koger 1994) and Na^+:K^+:$2Cl^-$ cotransporter (Loong et al. 2012; Kinne et al. 1986). Additionally, NH_4^+ can be transported through Na^+/H^+ exchangers with NH_4^+ substituting for H^+ (Nagami 1988).

7.4.1 Mudskipper Gill Morphologies Affect Branchial Ammonia Excretion

As a facultative air breather, the gills of *B. boddarti* appear better adapted for aquatic than aerial respiration. This species has larger gill surface area, mass-specific gill area, and gill area to skin area ratio compared to *Pn. schlosseri* (Low et al. 1990); the gill surface area of a 10 g *B. boddarti* is twofold greater than that of *Pn. schlosseri* with the same body mass (Low et al. 1990). In *B. boddarti*, gill rakers are broad and leaf-like, acting as filters to separate algae from mud during feeding (Low et al. 1988). Its gill filaments are long, arranged in a sheet-like fashion, with the secondary lamellae being aligned parallel to the respiratory water current to facilitate efficient gaseous exchange through the counter-current system (Figure 7.2). While on land and without the buoyancy of water, these filaments collapse and the secondary lamellae coalesce, which would impede branchial oxygen uptake and ammonia excretion, and consequently impose a limit on the terrestrial affinity of *B. boddarti*. In contrast, *Pn. schlosseri* has knob-like gill rakers, reflecting its carnivorous diet (Low et al. 1988). It has fewer and shorter filaments, and smaller gill surface area for gaseous exchange than *B. boddarti*. The short gill filaments are heavily branched, deriving good skeletal support from the gill rods. Extensive intrafilamentous secondary lamellar fusions between adjacent lamellae prevent the filaments from coalescing upon emersion (Figure 7.3). Water can be trapped within the fenestrae formed by these fusions, thereby reducing the risk of desiccation during terrestrial excursions. These branchial modifications provide *Pn. schlosseri* with the unique opportunity to continuously excrete ammonia into the fenestral water while on land. However, the ammonia concentration would quickly build up to high levels in such minute volumes of water, and therefore *Pn. schlosseri* must have the ability to actively excrete ammonia against inwardly driven P_{NH_3} and NH_4^+ concentration gradients. Despite these advantages, the intrafilamentous interlamellar fusion would hinder the flow of water between lamellae, rendering the counter-current mechanisms for gaseous exchange ineffective in water. Consequently, *Pn. schlosseri* is an obligatory air breather that can repay oxygen debts in air but not in water (Takeda et al. 1999). It can survive terrestrial

Figure 7.2 Gill filaments of *Periophthalmodon schlosseri*, showing structural complexity in the branched and thickened gill rods, with extensive intra-filamentous inter-lamellar fusions. (Courtesy of Shit F. Chew and Yuen K. Ip.)

Figure 7.3 Gill filaments of *Boleophthalmus boddarti*, in sheet-like fashion with the arrangement of secondary lamellae to facilitate counter-current distribution. (Courtesy of Shit F. Chew and Yuen K. Ip.)

exposure better than other mudskipper species, but would drown upon forced submergence for approximately 24 hours, even in well-aerated water (Kok et al. 1998; Ip et al. 1993).

The gills of *Pn. schlosseri* have a very high density of mitochondrion-rich cells, or ionocytes (Wilson et al. 1999), and are therefore metabolically active. They appear to be an oxygen-demanding organ rather than a site of oxygen uptake (Wilson et al. 1999). The abundance of ionocytes in the lamellar epithelium of the gills of *Pn. schlosseri* is highly unusual, even among marine fishes that have to actively excrete Na^+ and Cl^- through their gills (Wilson et al. 1999). In addition, most of the ionocytes in the gills of *Pn. schlosseri* are not associated with accessory cells with typical shallow tight junctions, which are required for NaCl excretion during hypoosmotic osmoregulation in seawater. The branchial Na^+/K^+-ATPase activity of *Pn. schlosseri* is probably the highest among teleosts; it is threefold higher than that of *B. boddarti* and approximately 10-fold higher than those of other fishes (Randall et al. 1999). In comparison with those of *B. boddarti*, the gills of *Pn. schlosseri* have a greater potential for glycolytic flux (fourfold greater phosphofructosekinase-1 activity), a higher aerobic capacity (fourfold greater citrate synthase activity), a greater capacity for utilizing exogenous glucose (fourfold higher hexokinase activity), a greater capacity for lactate production (high pyruvate kinase and lactate dehydrogenase activities), and a higher ratio of pyruvate reductive to lactate oxidative capacities (Low et al. 1993). Unlike the gills of other teleosts that are lactate-consuming organs, the gills of *Pn. schlosseri* are adapted to produce lactate during hypoxia (Low et al. 1992; Ip and Low 1990; Ip et al. 1990b). As *Pn. schlosseri* has high terrestrial affinity, its gills are also well adapted for ammonia excretion and acid–base regulation during emersion.

7.4.2 Mudskipper Skins Can Facilitate Ammonia Excretion or Uptake

The buccal and pharyngeal epithelia of mudskippers are well vascularized to support buccopharyngeal respiration (Aguilar et al. 2000; Gonzales et al. 2011). During air breathing, the buccal cavity is filled with air; consequently, the gills are no longer irrigated with water and cannot excrete ammonia effectively. The only known exception is *Pn. schlosseri*, which has unique structural and biochemical branchial adaptations to facilitate continuous ammonia excretion during emersion.

Mudskippers can also respire through their highly vascularized skin (Beon et al. 2013; Park et al. 2006), and the contribution of the skin to gaseous exchange is dependent on its degree of vascularization (Clayton 1993; Schöttle 1931). In *B. boddarti*, the cutaneous vascular system includes large button-shaped papillae on the dorsal surface of the head and on the sides of the body behind the dorsal fin. Each scale has a greatly differentiated respiratory papilla associated with it. As the gills of *B. boddarti* are ill adapted for air breathing, it has a high dependency on cutaneous respiration that involves the turning behavior to moisten the dorsal cutaneous surfaces while out of water (Ip et al. 1991b). The epidermis of *Pn. schlosseri* is thicker and the papillae structures are more complicated. The blood vessels in the finger-like projections of the dermis rise to the epidermal layer and branch out into several intraepithelial capillaries situated very close to the surface (Schöttle 1931). With cutaneous vascularization, the skin can augment branchial respiration, and also act as an auxiliary route for ammonia excretion during emersion (Morii et al. 1978). Despite that, highly vascularized skin has an undesirable effect of facilitating ammonia uptake during exposure to high concentrations of environmental ammonia.

7.5 DEFENSE AGAINST UNDESIRABLE ACCUMULATION AND/OR UPTAKE OF AMMONIA

When an amphibious air-breathing fish is out of water, it is confronted with problems of endogenous ammonia excretion because there is no water to flush the branchial and cutaneous surfaces. As many mudskippers are amphibious, they are adapted to ameliorate ammonia toxicity during emersion (Chew and Ip 2014; Chew et al. 2006; Ip et al. 2004a, 2001a). Besides, mudskippers build burrows in the mud and some of them (such as *B. boddarti*) presumably stay therein during high tides. During the breeding season which last 1–2 months, males move in and out of the burrow frequently to care for the developing embryos (Ishimatsu et al. 2007). As the burrow water is not well flushed, the ammonia concentrations therein can be high (Ip et al. 2004b). Even low concentrations of environmental ammonia would impede the diffusive excretion of ammonia. Mudskippers are further confronted with the retention of endogenous ammonia and the uptake of exogenous ammonia simultaneously when exposed to high concentrations of environmental ammonia. They must therefore have special adaptations to defend against environmental ammonia toxicity (Chew and Ip 2014; Ip and Chew 2010; Chew et al. 2006; Ip et al. 2004b, 2001a). As in the case of *Pn. schlosseri*, there could be a correlation between high terrestrial affinity and high environmental ammonia tolerance.

7.5.1 Mudskippers are Non-Ureogenic and Non-Ureotelic

In fully aquatic fishes, urea is an appreciable component of the excretory nitrogen wastes (Campbell and Anderson 1991), but it does not necessarily imply that ammonia is detoxified to urea. Urea can be produced from ammonia or glutamine via the ornithine-urea cycle (Anderson 2001; Campbell and Anderson 1991), the breakdown of arginine by argininolysis, or the degradation of uric acid through uricolysis. The ornithine-urea cycle represents a synthetic (anabolic) pathway, but argininolysis and uricolysis produce urea through catabolism instead. Fishes with a functional ornithine-urea cycle that sustains at least a low rate of urea synthesis are described as ureogenic, whereas fishes that excrete more than half of the total nitrogenous waste as urea-N are termed ureotelic. The majority of teleosts are non-ureogenic and non-ureotelic, although some normally ammonotelic fishes may increase urea excretion after feeding (Ip et al. 2004d; Lim et al. 2004). Urea excretion in fishes occurs mainly through the gills with the involvement of urea transporters (McDonald et al. 2006).

Earlier studies on excretory nitrogen metabolism in mudskippers focused on comparing the rates of urea and ammonia excretion in fishes that were first held in water for a control period, followed with exposure to air for a set time, and then returned to water (Morii 1979; Morii et al. 1979, 1978; Gordon et al. 1978). The idea behind doing so was that the post-emersion excretion pattern would indicate changes in nitrogen metabolism taking place in the fish during terrestrial exposure. The purpose of such experiments was to determine whether the rate of urea excretion increased upon reimmersion. The more logical objective, however, should have been whether mudskippers possess a functional ornithine-urea cycle, and whether urea synthesis played a major role in detoxifying endogenous ammonia during terrestrial exposure, which requires a quantitative analysis of the nitrogen budget of the experimental fish.

Gordon et al. (1969) reported that when *Periophthalmus argentilineatus* (reported as *Ps. sobrinus*) was exposed to terrestrial conditions for 12 hours, urea production increased more than threefold. However, Gregory (1977) could not detect activities of carbamoyl phosphate synthetase, argininosuccinate synthetase, and argininosuccinate lyase from the liver of *Periophthalmus novaeguineaensis* (reported as *Periophthalmus expeditionium*), *Periophthalmus gracilis*, and *Scartelaos histophorus*, and concluded that these three species of mudskipper could not synthesize urea through the ornithine-urea cycle. As the liver of *Ps. novaeguineaensis* and *Ps. gracilis* displayed urate oxidase, allantoinase, and allantoicase activities (Gregory 1977), urea was probably produced via uricolysis. Working on *Periophthalmus modestus* (reported as *Periophthalmus cantonensis*) and *Boleophthalmus pectinitrostris*, Morii (1979) and Morii et al. (1979, 1978) observed no change in the ratio of urea to ammonia excreted during the first 6 hours after a transfer from terrestrial conditions to water, indicating that ammonia accumulated during terrestrial exposure was not converted to urea. Iwata et al. (1981) and Iwata (1988) reported that urea production in *Ps. modestus* remained unchanged after exposure to ammonia or terrestrial conditions. When *Ps. modestus* was exposed to ^{15}N-labeled ammonia, urea-N was only weakly labeled (Iwata and Deguchi 1995). Subsequently, Lim et al. (2001) confirmed that no activity of carbamoyl phosphate synthetase I or III (a crucial enzyme in the ornithine-urea cycle) could be detected from the liver mitochondria of *B. boddarti*. Taken altogether, it can be concluded that urea synthesis *de novo* does not occur in species of *Periophthalmus*, *Scartelaos*, or *Boleophthalmus*; these mudskippers probably produce urea through argininolysis and uricolysis.

Although *Pn. schlosseri* possesses a full complement of ornithine-urea cycle enzymes, the activities of some of these enzymes, especially the mitochondrial carbamoyl phosphate synthetase, are too low to render the cycle functional (Lim et al. 2001). Endogenously produced ammonia is not detoxified to urea through the ornithine-urea cycle in *Pn. schlosseri* exposed to terrestrial conditions (Chew et al. 2007; Ip et al. 2004a; Lim et al. 2001) or environmental ammonia (Ip et al. 2004b; Randall et al. 1999; Peng et al. 1998). The synthesis of urea *de novo* in fishes is highly energy intensive; a total of 5 mol of ATP are hydrolyzed to ADP for each mole of urea synthesized, corresponding to 2.5 mol of ATP used for each mole of nitrogen assimilated. In comparison, 4 mol of ATP are hydrolyzed for each mole of urea synthesized in tetrapods. That may be the main reason why urea synthesis via the ornithine-urea cycle is rare in adult air-breathing teleosts (Chew and Ip 2014; Chew et al. 2006; Ip et al. 2001a). Furthermore, detoxifying a net influx of ammonia to urea during environmental exposure, and subsequently excreting it, requires not only a major expenditure of energy but also maintains an inwardly driven P_{NH_3} gradient by keeping low concentrations of internal ammonia. This could be the reason why urea synthesis does not play a major role in ammonia detoxification in air-breathing teleosts (Chew et al. 2014; Ip et al. 2001a, 2004b), including mudskippers (Randall et al. 1999; Peng et al. 1998; Iwata and Deguchi 1995; Iwata 1988), during ammonia exposure.

7.5.2 Avoidance of Postprandial Ammonia Toxicity in *Periophthalmodon schlosseri*

Tropical regions are characterized by high temperature, which often stays above 30°C. The dietary protein requirement of fishes increases with temperature (Delong et al. 1958), and an increase in temperature leads to a marked increase in food intake and ammonia excretion in tropical fishes (Leung et al. 1999). Notably, as air temperature fluctuates between day and night, and also throughout the day, the temperature on the mudflat can be higher than the temperature of the running water nearby. Ip et al. (2004d) postulated that the presence of a normally non-functional ornithine-urea cycle in the carnivorous *Pn. schlosseri* could be attributed to its high protein diet, and hence the need to defend against postprandial ammonia toxicity. They demonstrated that the ammonia and urea excretion rates increased 1.7-fold and 1.9-fold, respectively 3 hours post-feeding (Ip et al. 2004d). Simultaneously, there were significant decreases in ammonia concentrations in the plasma and the brain, and significant decreases in urea concentrations in the muscle and the liver. It is apparent that *Pn. schlosseri* could unload ammonia originally present in some of its tissues in anticipation of the ammonia released from the catabolism of excess amino acids obtained through the diet, although the mechanisms involved remain enigmatic. At hour 6 post-feeding, there were significant increases in urea concentrations in the muscle (1.39-fold), liver (2.17-fold), and plasma (1.62-fold). Between hour 3 and hour 6, the estimated rate of urea synthesis increased 5.8-fold (Ip et al. 2004d). The excess urea accumulated in the body at hour 6 was completely excreted in the following 6 hours, resulting in a significant increase in the percentage of urea-N in the total nitrogenous waste to 26% between hour 6 and hour 12; but, it never turned ureotelic (Ip et al. 2004d). Increased urea synthesis probably occurred in the liver of *Pn. schlosseri* as the liver manifested the greatest increase in urea content after feeding. From these results, it can be deduced that an ample supply of energy resources, for example, after feeding, is a prerequisite for the possible induction of urea synthesis, which is an energy-intensive process, through the feeble ornithine-urea cycle in the liver of *Pn. schlosseri*. Together, increases in nitrogenous waste excretion and urea synthesis after feeding effectively avert a postprandial surge of plasma ammonia in *Pn. schlosseri*, which is uncommon among fishes. Thus, in contrast to other fish species (Wicks and Randall 2002), the ammonia concentration in the brain of *Pn. schlosseri* decreases significantly throughout the 24-hour post-feeding period, while the brain glutamine concentration decreases significantly between hour 12 and hour 24 post-feeding (Ip et al. 2004d).

7.5.3 Defense against Endogenous Ammonia Toxicity during Terrestrial Exposure

In general, mudskippers have high tolerance to terrestrial exposure. In the laboratory, they can be kept out of water for up to 4–5 days (Morii 1979; Morii et al. 1978). In nature, different species have different terrestrial affinities. When out of water, *Pn. schlosseri* can survive better than *B. boddarti* (Kok et al. 1998; Ip et al. 1993). *Pn. schlosseri* is the only known teleost that exhibits bradycardia in water (Kok et al. 1998; Garey 1962); during emersion, it has a heart rate of 55 beats per minute, but the heart rate decreases to 11 beats per minute within 5 minutes of submergence (Kok et al. 1998). Concomitantly, *Pn. schlosseri* has a higher metabolic rate on land than in water, and the oxygen consumption rate increases significantly upon terrestrial exposure (Kok et al. 1998). In contrast, there are no significant differences between the aerial and aquatic respiratory rates in *Ps. modestus* (reported as *Ps. cantonensis*, Gordon et al. 1978; Tamura et al. 1976), *Ps. argentilineatus* (reported as *Periophthalmus vulgaris*, Milward 1974), *Boleophthalmus pectinirostris* (reported as *Boleophthalmus chinensis*, Tamura et al. 1976), and *B. boddarti* (Kok et al. 1998). Terrestrial exposure also significantly increases the blood P_{O_2} level and pH of *Pn. schlosseri* (Kok et al. 1998). The increase in blood pH, uncommon among fishes, could be attributed to an increase in the production and retention of bicarbonate (Kok et al. 1998).

In the muscle of *Pn. schlosseri*, terrestrial exposure leads to a significant increase in the glycogen content, with the lactate content remaining relatively unchanged (Ip et al. 2006), indicating either a decrease in glycogen degradation or an increase in glycogen synthesis. As there is no change in the muscle ATP content and energy charge, *Pn. schlosseri* probably uses stored energy other than glycogen during terrestrial exposure. Similar phenomena have not been observed in *B. boddarti* (Ip et al. 2006) or reported in other mudskipper species. Exposure to terrestrial conditions for

6 hours also leads to a decrease in phosphofructokinase-1 activity, which becomes more sensitive to ATP inhibition, from the muscle of *Pn. schlosseri*, indicating that the glycolytic flux can be impeded transiently at this locus (Ip et al. 2006). During terrestrial exposure, an increase in fructose-2,6-bisphosphate concentration in the muscle (Ip et al. 2006) may lead to an increase in the utilization of amino acids for energy production to support locomotor activities on land (Ip et al. 2001b, 1993). As follows, *Pn. schlosseri* would have to defend against endogenous ammonia toxicity when active on land. By the same reason, when *Pn. schlosseri* is quiescent on land, different mechanisms may be involved in dealing with endogenous ammonia toxicity.

7.5.3.1 Reduction in Proteolysis and/or Amino Acid Catabolism

Iwata et al. (1981) observed a marked increase in concentrations of free amino acids (FAAs), particularly the non-essential ones, in tissues of *Ps. modestus* (reported as *Ps. cantonensis*) kept out of water for 48 hours. Terrestrial exposure also led to significant increases in the activities of glutamate dehydrogenase in the liver and muscle of *Ps. modestus*. It was concluded that the incorporation of ammonia into non-essential amino acids was one of the mechanisms adopted by *Ps. modestus* to maintain a low plasma ammonia concentration during emersion (Iwata et al. 1981). However, low plasma or internal ammonia concentration can also be achieved through a reduction in the production of endogenous ammonia. If the rate of proteolysis remains the same while the rate of amino acid catabolism decreases, ammonia production will decrease (even though ammonia may accumulate in the tissues if its excretion is impeded), but there would be increases in concentrations of FAAs. Increases in concentrations of FAAs under experimental conditions in which fishes may have difficulty in excreting ammonia is therefore not indicative of detoxification of ammonia to amino acids (Iwata 1988; Iwata et al. 1981). Instead, it may simply denote a decrease in the rate of amino acid catabolism or a change in the catabolic pathways involved.

Lim et al. (2001) also reported very little accumulation of ammonia in the tissues of *Pn. schlosseri* and *B. boddarti* during exposure to terrestrial conditions under constant darkness, in spite of decreases in ammonia and urea excretion rates. In complete darkness, *Pn. schlosseri* remained quiescent on land, and the total FAA concentrations in the liver and plasma decreased significantly (Ip et al. 2001b; Lim et al. 2001). For *B. boddarti* under similar experimental conditions, there was a significant decrease in the total FAA concentration in the muscle. As the experimental fishes were fasted, the rate of protein degradation should be higher than the rate of protein synthesis, leading to a net proteolysis. If the rate of proteolysis remained relatively constant and was unaffected by terrestrial exposure, there could be an accumulation of FAAs due to a suppression of amino acid catabolism to reduce ammonia production. Instead, the total FAA concentration decreased in some tissues of these two species of mudskipper (Lim et al. 2001). It can be thus postulated that simultaneous decreases in rates of proteolysis and amino acid catabolism would have occurred to reduce the production of ammonia. In addition, the decrease in proteolytic rate was probably greater than the decrease in the rate of amino acid catabolism, leading to decreases in the steady-state concentrations of various FAAs and consequently lowering the total FAA concentration (Figure 7.4). Indeed, an analysis of the balance between nitrogen excretion and nitrogen accumulation in *Pn. schlosseri* and *B. boddarti* supports such a supposition (Lim et al. 2001). Despite a decrease in ammonia production, the decrease in the rate of nitrogenous waste excretion is apparently greater than the reduction in the rate of ammonia production, leading to small increases in the ammonia concentrations in tissues of these two species of mudskipper.

Reductions in rates of protein and amino acid catabolism constitute an effective strategy to slow down the internal build up of ammonia. However, it prevents the utilization of amino acids as an energy source. Operating by itself, it may not be a useful mechanism for fishes that are active on land (e.g., *Pn. schlosseri*, Kok et al. 1998). To overcome this, *Pn. schlosseri* undergoes partial

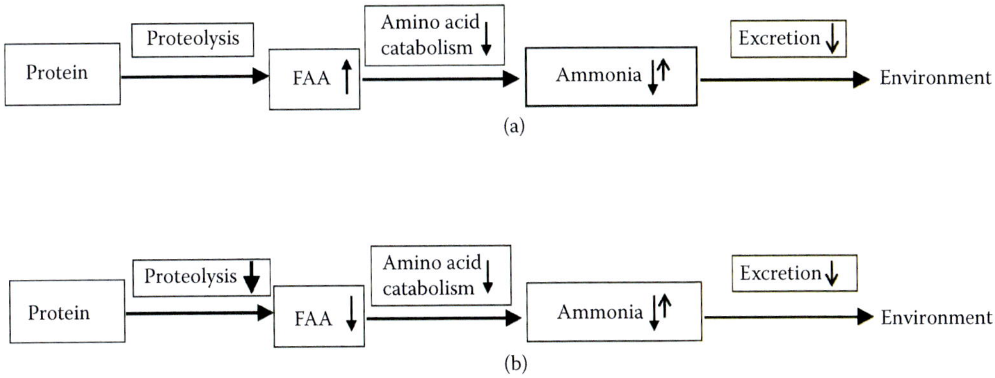

Figure 7.4 Possible changes in the steady state concentrations of protein, free amino acids (FAAs) and ammonia during terrestrial exposure due to (a) a decrease in the rate of ammonia excretion with a responding decrease in the rate of amino acid catabolism, or (b) a decrease in the rate of ammonia excretion with responding decreases in the rate of amino acid catabolism and the rate of proteolysis. (Courtesy of Shit F. Chew and Yuen K. Ip.)

amino acid catabolism in conjunction with a suppression of ammonia production. This allows *Pn. schlosseri* to use protein and amino acids as energy sources to support locomotor activities on land without releasing ammonia.

7.5.3.2 Partial Amino Acid Catabolism with the Formation of Alanine

Unlike fishes exposed to terrestrial conditions in constant darkness, the concentrations of some essential amino acids (e.g., isoleucine, leucine, proline, serine, lysine, and valine) increase in the liver and muscle tissues of *Pn. schlosseri* that performs normal activity on land in a dark and light regime, indicating the mobilization of amino acids through increased proteolysis (Ip et al. 2001b). More importantly, the concentration of alanine increases significantly in the muscle, liver and plasma, accompanied by an elevation of the total FAA concentrations in the muscle and plasma (Ip et al. 2001b, 1993).

Certain amino acids (e.g., arginine, glutamine, histidine, and proline) can be converted to glutamate, which can undergo deamination catalyzed by glutamate dehydrogenase, producing NH_4^+ and α-ketoglutarate (Campbell, 1991), and α-ketoglutarate can be channeled into the tricarboxylic acid cycle. Alternatively, glutamate can undergo transamination with pyruvate, catalyzed by alanine aminotransferase, producing α-ketoglutarate and alanine without the release of ammonia (Chew et al. 2003b; Ip et al. 2001b). If there were a continuous supply of pyruvate, transamination would facilitate the oxidation of carbon chains of some amino acids as α-ketoglutarate without polluting the internal environment with ammonia (Figure 7.5). Under normal circumstances, α-ketoglutarate can be oxidized completely to carbon dioxide and water through the tricarboxylic acid cycle and the electron transport chain, producing ATP and/or its equivalent. For *Pn. schlosseri* exposed to terrestrial conditions under a dark and light regime, α-ketoglutarate may undergo partial oxidation only through portions of the tricarboxylic acid cycle to malate, which can then be turned into pyruvate in the presence of malic enzyme to sustain transamination with glutamate to form alanine (Figure 7.5). This would cause a reduction in the efficiency of ATP production, because α-ketoglutarate is not completely oxidized. Nonetheless, partial amino acid catabolism leading to the formation of alanine would provide energy in the form of ATP without NH_4^+ release, which would compensate for the impediment of ammonia excretion when *Pn. schlosseri* is out of the water (Ip et al. 2001b). This would facilitate the utilization of certain amino acids as energy sources and,

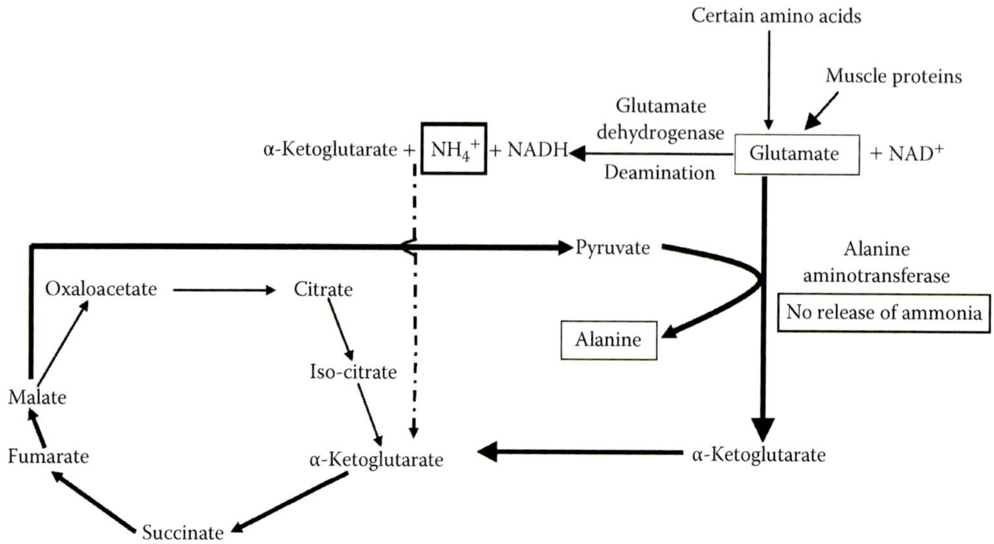

Figure 7.5 Partial amino acid catabolism of certain amino acids leading to the formation of alanine without producing ammonia, with the participation of alanine aminotransferase, glutamate dehydrogenase, and tricarboxylic acid cycle. (Courtesy of Shit F. Chew and Yuen K. Ip.)

at the same time, minimize ammonia accumulation and avert endogenous ammonia toxicity during terrestrial exposure.

When *Pn. schlosseri* is forced to exercise for 3 minutes on land after 3 hours of exposure to terrestrial conditions in light, there is no change in the glycogen content despite an increase in the lactate concentration in the muscle. Muscle ammonia and alanine concentrations also increase, with the ammonia level being twice that of the fish exercised in water (Ip et al. 2001b). The efficiency of alanine production through partial amino acid catabolism is apparently dependent on the period of terrestrial exposure. For *Pn. schlosseri* forced to exercise for 3 minutes after exposure to terrestrial conditions in a dark and light regime for 24 hours, there is no change in the glycogen level and no further accumulation of ammonia. However, the muscle alanine concentration of these experimental fish increases fourfold, and the amount of alanine accumulated is much higher than that in the fish exercised in water (Ip et al. 2001b). Such an adaptation reduces the dependency of *Pn. schlosseri* on carbohydrate as an energy source, sparing the glycogen store and allowing the fish to sustain a high metabolic rate for an extended period on land (Kok et al. 1998). As exercise in water leads to a decrease in glycogen content and increases in concentrations of lactate, ammonia, and alanine in the muscle of *Pn. schlosseri*, it can be deduced that both glycogen and amino acids are mobilized in this situation (Ip et al. 2001b).

By contrast, alanine concentration remains unchanged in the muscle of *B. boddarti* exposed to terrestrial conditions in a dark and light regime, indicating that *B. boddarti* does not rely on amino acids as an energy source during terrestrial exposure (Ip et al. 2001b). Unlike *Pn. schlosseri*, *B. boddarti* uses glycogen as a metabolic fuel to support locomotor activity on land. This strategy offers a limited amount of energy for a short period on land, and, as a consequence, the muscle ATP content decreased to a 10th of the control value after a short period of locomotor activity (Ip et al. 2001b). Together with the high concentration of ammonia being accumulated in the muscle of the exercised fish, these results suggest that it is highly unfavorable for *B. boddarti* to stay out of water for long durations.

7.5.3.3 Active Ammonia Excretion in Connection with H^+ Excretion during Emersion

Active excretion of ammonia would facilitate the maintenance of low internal ammonia concentrations, thereby preventing the brain from ammonia intoxication. Only a few teleost fishes are capable of active ammonia excretion (Chew et al. 2007, 2003a; Tay et al. 2006; Ip et al. 2004e,f; Randall et al. 1999). In the case of *Pn. schlosseri*, it can uniquely excrete ammonia continuously into the water trapped in the fenestrae of the branchial interlamellar fusions while out of water. Chew et al. (2007) demonstrated that ammonia concentrations in water samples (pH 6.3) collected from the branchial surface of *Pn. schlosseri* exposed to terrestrial conditions increased continuously, reaching 25–32 $mmol{\cdot}L^{-1}$ after 24 hours. Ammonia was apparently excreted into the water inside the fenestrae and then diffused or dispersed to water covering the branchial surface. As the plasma has an ammonia concentration of 0.6 $mmol{\cdot}L^{-1}$ at pH 7.6, an estimation based on an ammonia concentration of 32 $mmol{\cdot}L^{-1}$ at pH 6.3 in the branchial water reveals that both P_{NH_3} and NH_4^+ concentration gradients are acting inward across the branchial epithelium. It can therefore be concluded that *Pn. schlosseri* is capable of active ammonia excretion when exposed to terrestrial conditions.

Described in detail in Section 5.4.1, the putative mechanisms of active ammonia excretion in *Pn. schlosseri* may involve apical Na^+/H^+ (NH_4^+) exchangers (Wilson et al. 2000; Randall et al. 1999), and hence ammonia is actively excreted as NH_4^+ by *Pn. schlosseri*. While out of water, the rate of active NH_4^+ excretion must be greater than the rate of NH_3 back-flux; only then, would there be a continuous excretion of ammonia into the branchial water. Chew et al. (2007) discovered that the branchial water collected from *Pn. schlosseri* after 24 hours of emersion had a pH of 6.3, which was lower than the pH (7.8) of the water added to the bottom of the tank. These results indicate that H^+ is excreted into the water in the fenestrae, which corroborates reports of *Pn. schlosseri* having the ability to lower the pH of the external medium, especially in an alkaline medium and/or in the presence of environmental ammonia (Ip et al. 2004e; Chew et al. 2003a). Despite the high concentration of ammonia (32 $mmol{\cdot}L^{-1}$), the continuous excretion of acid to the fenestrae maintains a low pH in the water therein (Ip et al. 2004e), and prevents the excreted NH_4^+ from dissociating into NH_3 and H^+. This would avoid a back flux of NH_3, which Ip et al. (2004d) referred to as "NH_4^+ trapping." As *Pn. schlosseri* does not possess any accessory breathing organ and its gills are modified with intrafilamentous interlamellar fusions (Low et al. 1988), the excretion of H^+ to trap the actively excreted NH_4^+ is likely to be more effective than modifying the fluidity of the branchial epithelial surface to decrease NH_3 permeability. This is because a reduction in the fluidity of the branchial epithelium may also decrease the permeability of the epithelium to other gaseous molecules including oxygen. At present, the origin of the H^+ being excreted is unclear, but it may involve the hydration of carbon dioxide by carbonic anhydrase inside the ionocyte. The hydration of carbon dioxide produces H^+ and HCO_3^- (Figure 7.6). H^+ can be excreted through the apical membrane to trap the excreted NH_4^+. If HCO_3^- is transported through a basolateral Cl^-/HCO_3^- exchanger into the blood, it would explain why the blood pH of *Pn. schlosseri* exposed to terrestrial conditions would become uniquely more alkaline (Kok et al. 1998). Indeed, carbonic anhydrase is immunolocalized to the apical cytoplasmic region of branchial ionocytes of *Pn. schlosseri* (Wilson et al. 2000), and ammonia excretion can be reduced significantly by inhibiting carbonic anhydrase activity with acetazolamide (Wilson et al. 2000; Kok 2000) and Cl^-/HCO_3^- exchanger with 4-acetamido-4′-isothiocyanostilbene-2,2′-disulfonate (SITS; Kok 2000). In freshwater teleost fishes, the Cl^-/HCO_3^- exchanger is known to be present in the apical membrane of branchial ionocytes, participating in Cl^- absorption (Hwang and Lee 2007), but *Pn. schlosseri* is a euryhaline brackish water fish. Hence, efforts should be made in the future to determine the types of Cl^-/HCO_3^- exchangers present, and to elucidate their subcellular localization, in the gills of *Pn. schlosseri*.

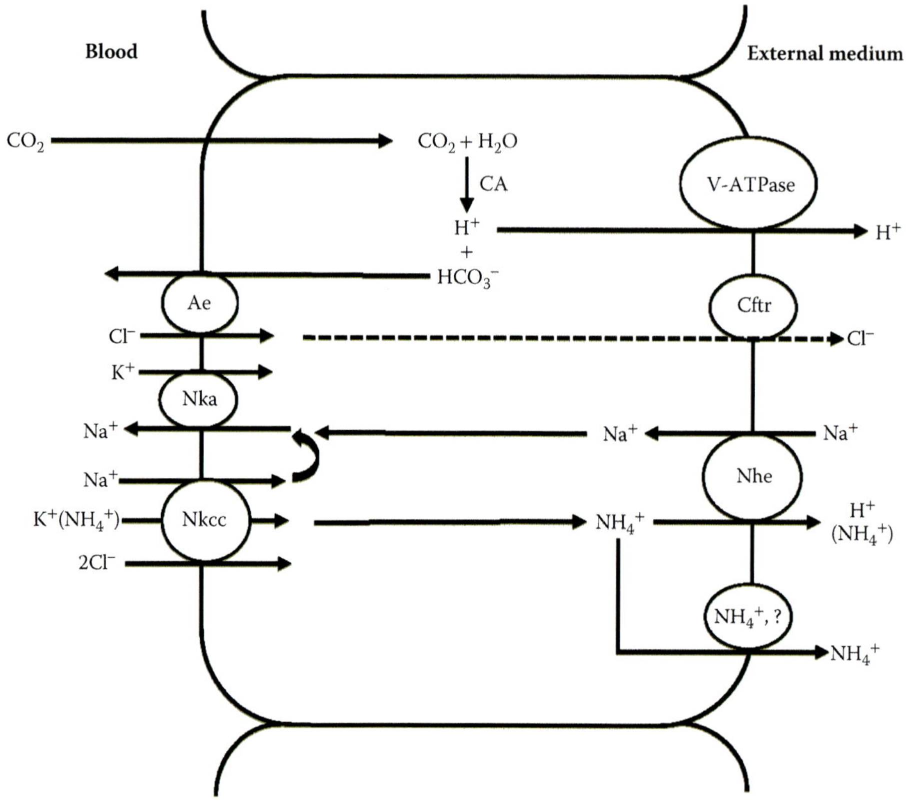

Figure 7.6 A proposed model for the active excretion of NH_4^+ through a branchial ionocyte in the gills of *Periophthalmodon schlosseri* during terrestrial or environmental ammonia exposure. Anion exchanger, Ae; carbonic anhydrase, CA; cystic fibrosis transmembrane conductance regulator-like Cl^- channel, Cftr; $Na^+/H^+(NH_4^+)$ exchanger, Nhe; $Na^+:K^+:2Cl^-$ cotransporter 1, Nkcc1; Na^+/K^+-ATPase, Nka; putative NH_4^+ transporter, ?. (Courtesy of Shit F. Chew and Yuen K. Ip.)

Chew et al. (2007) demonstrated that when ammonium acetate (8 $\mu mol \cdot g^{-1}$) was injected intraperitoneally into *Pn. schlosseri* and followed with 24 hours of terrestrial exposure, the cumulative ammonia excreted was significantly greater than the saline-injected control. At hour 2, the ammonia excretion rate in the experimental fish was transiently greater than that in fish kept in water. By hour 6, 33% of the injected ammonia was excreted by the experimental fish from the head region, probably through the gills (Chew et al. 2007). As ammonia could be excreted only into the small amount of water trapped in the fenestrae in the gills of *Pn. schlosseri* during emersion, ammonia concentration built up quickly therein and reached an extraordinarily high concentration of approximately 90 $mmol \cdot L^{-1}$. These results indicate that *Pn. schlosseri* can effectively excrete a high load of ammonia on land, indirectly corroborating the proposition that active NH_4^+ excretion contributes in part to its high terrestrial affinity and high tolerance of terrestrial exposure.

7.5.4 Defense against Endogenous and Exogenous Ammonia Toxicity during Exposure to Environmental Ammonia

As ammonia is toxic, many teleosts would succumb to 1 $mmol \cdot L^{-1}$ ammonium chloride in several hours. However, mudskippers have the extraordinary ability to tolerate very high concentrations of environmental ammonia. For instance, *Ps. modestus* (Iwata and Deguchi 1995; reported as *Ps. cantonensis* in Iwata 1988) can survive for 1 week in 15 mmol·L^{-1} ammonium chloride. *Boleophthalmus boddarti* has a 96-hour LC_{50} of 13.5 mmol·L^{-1} total ammonia (or 60.2 μmol·L^{-1} NH_3) at pH 7.0. In particular, *Pn. schlosseri* has the exceptional ability to survive in 100 mmol·L^{-1} ammonium chloride for more than 1 week; it has a 96-hour LC_{50} of 120 mmol·L^{-1} total ammonia at pH 7.0 (or 536 μmol·L^{-1} NH_3) (Peng et al. 1998). Evidently, the highly ammonia-tolerant *Pn. schlosseri* must have special adaptations, which are atypical among mudskippers and fishes in general, to avert or ameliorate environmental ammonia toxicity, and these adaptations must be in some way related to the unique structural and biochemical modifications of its gills.

7.5.4.1 Active Ammonia Excretion during Ammonia Exposure

The extraordinarily high tolerance of *Pn. schlosseri* to environmental ammonia is mainly attributable to its ability to actively excrete ammonia against inwardly directed P_{NH_3} and NH_4^+ concentration gradients (Chew et al. 2007; Ip et al. 2005; Randall et al. 1999; Peng et al. 1998), a feat which is shared by only two other teleost fishes, which both inhabit freshwater, *Anabas testudineus* (Ip et al. 2012a,b; Loong et al. 2012; Tay et al. 2006), and *Clarias gariepinus* (Ip et al. 2004d). As a result of active ammonia excretion, *Pn. schlosseri* can maintain plasma ammonia at low concentrations during exposure to high concentrations of environmental ammonia (Peng et al. 1998), up to at least 30 mmol·L^{-1} ammonium chloride at pH 7.2 (Randall et al. 1999), and sustain a normal rate of ammonia excretion when exposed to alkaline water (Chew et al. 2003a).

As active NH_4^+ excretion in *Pn. schlosseri* can be inhibited by ouabain (a Na^+/K^+-ATPase inhibitor) or amiloride (a Na^+/H^+ exchanger inhibitor), Randall et al. (1999) proposed that NH_4^+ was transported from the blood into a type of ionocyte by the basolateral Na^+/K^+-ATPase with NH_4^+ substituting for K^+, and from the ionocyte to the external medium by the apical Na^+/H^+ exchanger with NH_4^+ substituting for H^+. If NH_4^+ were to be actively transported into branchial ionocytes through basolateral Na^+/K^+-ATPase (Hwang et al. 2011; Weihrauch et al. 2009), the K_m values of branchial Na^+/K^+-ATPase for both K^+ and NH_4^+ must be comparable such that NH_4^+ can effectively replace K^+ to activate Na^+/K^+-ATPase. However, the ionocyte will face difficulty in maintaining a high intracellular K^+ concentration needed for K^+ homeostasis and a low intracellular NH_4^+ concentration to prevent ammonia cytotoxicity. As ammonia is toxic, it cannot be allowed to accumulate in the ionocyte (Ip and Chew 2010). Thus, active excretion of NH_4^+ against an electrochemical gradient should occur at the apical instead of the basolateral membrane of the ionocyte; only with active excretion of NH_4^+ through the apical membrane, would the intracellular ammonia concentration be maintained at low and non-toxic levels. Moreover, the basolateral membrane has an electrical potential of approximately 75 mV (inside negative; Wright 1991); unlike K^+ which has a very high intracellular concentration (~120 mmol·L^{-1}), NH_4^+ does not need to be actively transported across the basolateral membrane through Na^+/K^+-ATPase.

Recently, Chew et al. (2014) reported that the Na^+/K^+-ATPase from the gills of *Pn. schlosseri* preacclimated to slightly brackish water (salinity 3) for 2 weeks had substantially higher affinity to K^+ (K_m of 1.6 mmol·L^{-1}) than NH_4^+ (K_m of 15 mmol·L^{-1}). Assuming the plasma K^+ and NH_4^+ concentrations to be 10 and 1 mmol·L^{-1}, respectively, then branchial Na^+/K^+-ATPase would be saturated with K^+, and it is highly improbable that branchial Na^+/K^+-ATPase would transport NH_4^+ in replacement of K^+. In addition, after *Pn. schlosseri* was exposed to 75 mmol·L^{-1} ammonium chloride in slightly brackish water for 6 days, the K_m for NH_4^+ increased significantly to 23 mmol·L^{-1} with the K_m for K^+ remained relatively unchanged. Furthermore, the *Na^+/K^+-ATPase* α isoforms cloned and sequenced from the gills of *Pn. schlosseri* possessed K^+ binding sites that could apparently differentiate K^+ from NH_4^+ (Chew et al. 2014). Therefore, Na^+/K^+-ATPase cannot be the major mechanism involved in the transport of NH_4^+ from the blood into the ionocyte during active NH_4^+ excretion in *Pn. schlosseri*, as suggested previously (Hwang et al. 2011; Weihrauch et al. 2009; Randall et al. 1999).

Why then would there be an increase in Na^+/K^+-ATPase activity from the gills of *Pn. schlosseri* exposed to high concentrations of environmental ammonia (Chew et al. 2014; Randall et al. 1999; Peng et al. 1998)? Chew et al. (2015) hypothesized that it could be related to how ammonia entered the ionocyte from the blood; they made an effort to clone and sequence the *Na^+:K^+:2Cl^- cotransporter 1* from the gills of *Pn. schlosseri*, and to determine the effects of ammonia exposure on its mRNA expression and protein abundance after preacclimation to slightly brackish water for 2 weeks. They reported that exposure to 75 mmol·L^{-1} ammonium chloride in slightly brackish water had no significant effect on the mRNA expression of *Na^+:K^+:2Cl^- cotransporter 1a*. However, Western blotting revealed a significant increase in the protein abundance of multiple Na^+:K^+:2Cl^- cotransporter-immunoreactive bands of molecular mass 170–250 kDa in the gills of *Pn. schlosseri* exposed to ammonia (Chew et al. 2015). Furthermore, immunofluorescence microscopy revealed that exposure of *Pn. schlosseri* to 75 mmol·L^{-1} ammonium chloride in slightly brackish water led to increases in the expression of both basolateral Na^+:K^+:2Cl^- cotransporter 1a and basolateral Na^+/K^+-ATPase α-subunit, which were colocalized to certain ionocytes along the length of the branchial secondary lamellae (Chew et al. 2015). Hence, it can be concluded that ammonia exposure led to an increase in the protein expression and glycosylation of Na^+:K^+:2Cl^- cotransporter 1a in the gills of *Pn. schlosseri*. The dependency on translational and post-translational regulation of branchial Na^+:K^+:2Cl^- cotransporter 1a in *Pn. schlosseri* would facilitate prompt responses to changes in environmental condition. As NH_4^+ can replace K^+, NH_4^+ can probably enter ionocytes through the basolateral Na^+:K^+:2Cl^- cotransporter 1a during active ammonia excretion (Figure 7.6). However, increased influx of Na^+, NH_4^+, and 2Cl^- would alter the transmembrane Na^+ gradient. Consequently, exposure of *Pn. schlosseri* to ammonia would also result in an increase in branchial activity of Na^+/K^+-ATPase with decreased NH_4^+ affinity (Chew et al. 2014), so as to maintain intracellular Na^+ and K^+ homeostasis (Figure 7.6). The inhibitory effect of ouabain on the active excretion of NH_4^+ in *Pn. schlosseri* observed by Randall et al. (1999) probably resulted from the consequential decrease in the basolateral Na^+/K^+-ATPase activity leading to the dissipation of the Na^+ electrochemical potential gradient, which in turn affected the operation of Na^+:K^+:2Cl^- cotransporter 1a to transport NH_4^+ across the basolateral membrane. Furthermore, the transport of NH_4^+ by the basolateral Na^+:K^+:2Cl^- cotransporter 1a implies that NH_4^+ and HCO_3^- are not formed by the reaction of NH_3, CO_2 and H_2O catalyzed by carbonic anhydrase inside the branchial ionocytes of *Pn. schlosseri* as suggested by Wilson et al. (2000).

The NH_4^+ transported into the branchial ionocyte through Na^+:K^+:2Cl^- cotransporter 1a may be excreted through the apical membrane as NH_4^+ or as NH_3 and H^+. However, it is highly unlikely that active ammonia excretion in *Pn. schlosseri* is achieved through NH_3 trapping, as the elimination of the acidic conditions of the boundary water layer of the branchial surface by the addition of a buffer (HEPES or TRIS) to the external medium has no effect on the rate of active ammonia excretion (Chew et al. 2003a; Wilson et al. 2000). Also the rate of active ammonia excretion is unaffected by the V-type H^+-ATPase inhibitor, bafilomycin A1 (Ip et al. 2004e). The inhibition of active ammonia excretion in *Pn. schlosseri* by amiloride implies that NH_4^+ is excreted in substitution for H^+ across the apical membrane by a Na^+/H^+ exchanger, although the possible involvement of a novel active NH_4^+ transporter in the apical membrane of branchial ionocytes of *Pn. schlosseri* cannot be ignored (Figure 7.6).

Despite being immersed in an external medium containing ammonia, NH_4^+ is excreted into the water trapped in the fenestrae formed by the interlamellar fusion in the gills of *Pn. schlosseri*. How then can the excreted ammonia be released to the external medium? The ionocytes in the gills of *Pn. schlosseri* are uniquely isolated from one another by filament-rich cells (Wilson et al. 1999). It is possible that these filament-rich cells provide contractions to help release the fenestral water to the external medium for water renewal.

Boleophthalmus boddarti is also capable of excreting ammonia against inwardly directed P_{NH_3} and NH_4^+ concentration gradients but in much smaller magnitudes (8 mmol·L^{-1} NH_4Cl at pH 7.2;

Peng et al. 1998; Chew et al. 2003a). Active ammonia excretion in *B. boddarti* is also inhibited by ouabain and amiloride (Kok 2000). The inferior ability of *B. boddarti* to actively excrete NH_4^+ as compared with *Pn. schlosseri* could be due to a lack of interlamellar fusion and thus a low capacity to manipulate the branchial boundary water layer and/or a lack of ability to reduce a back flux of NH_3.

7.5.4.2 Lowering of Environmental pH

In water-breathing fishes at circumneutral and alkaline environmental pH, the water passing over the gills is acidified by the hydration of excreted carbon dioxide, giving rise to $HCO_3^- + H^+$, and to a lesser extent by direct H^+ excretion (Wright et al. 1989). As the gill functions as a counter-current exchanger and ventilation volumes are generally quite large, the acidification of expired water on ammonia speciation will only be relevant in the boundary layer close to the surface of the epithelium. This boundary layer acidification has been shown to be important in shifting the ammonia equilibrium toward NH_4^+ formation thereby lowering NH_3 to maintain the transbranchial P_{NH_3} gradient (Wilson et al. 1994; Wright et al. 1989). When fishes remain in stagnant water of finite volume (puddle, tidepool or water-filled burrow), it is possible for the acidifying effects of their excretions of carbon dioxide and H^+ to have a significant impact on the pH of the environment. The lowering of the pH of the environment has obvious advantages in dealing with the elevations in environmental ammonia levels by reducing the concentration of the more permeable species, NH_3, relative to the cationic NH_4^+.

When *Pn. schlosseri* is kept in an artificial burrow, the ambient ammonia increased to 10 mmol·L^{-1} within 5–6 days (Ip et al. 2004e). It is also capable of increasing the rate of net acid excretion when exposed to 20 or 30 mmol·L^{-1} ammonium chloride at pH 7.0 or 8.0 (Ip et al. 2004e). The net acid excretion can be inhibited by the specific V-type H^+-ATPase inhibitor, bafilomycin A1, and the V-type H^+-ATPase has been localized to the apical crypt of the branchial ionocyte of *Pn. schlosseri* (Figure 7.6; Wilson et al. 2000). The inhibition of the titratable acid flux by bafilomycin A1 has no effect on the ammonia flux, indicating that there is no dependence of ammonia excretion on acidification (as would be related to NH_3 trapping in the boundary layer). This is supported by the lack of an effect of boundary pH perturbation by buffers on the titratable acid flux (Wilson et al. 2000), which contrasts markedly with other teleost fishes (such as trout) that depend on the P_{NH_3} gradient for ammonia excretion (Wilson et al. 1994; Cameron and Heisler 1983). As mudskippers build burrows on the mudflat, and lay eggs therein during the breeding season, the secretion of acid into a finite volume of stagnant water would have a significant impact on environmental pH. The lowering of environmental pH represents an effective mechanism to deal with elevations in environmental ammonia by reducing the concentration of NH_3, which is the more permeant species of ammonia, constituting what Chew et al. (2003a) regarded as "environmental ammonia detoxification."

7.5.4.3 Reduction in the Cutaneous NH_3 Permeability

Many mudskipper species have highly vascularized skin (Clayton 1993), and therefore a certain amount of ammonia can be excreted through their cutaneous surfaces (Graham 1997; Morii et al. 1978). Passive NH_3 excretion can occur through Rhgps which are expressed in the skin of fishes in general (Weihrauch et al. 2009). For air-breathing fishes that are frequently confronted with environmental ammonia toxicity, it would be essential for their skin to have low permeability to NH_3 (Marcaggi and Coles 2001). In ammonia-loading conditions, a reduction in NH_3 permeability of the skin would help reduce the influx of NH_3. This is especially important for *Pn. schlosseri,* which actively excretes NH_4^+. In the event that the ambient ammonia builds up to high concentrations and the excreted NH_4^+ dissociates to NH_3, having a low NH_3 permeability will ensure that there is little or no back diffusion of NH_3 through the skin (Ip et al. 2004e; Randall et al. 1999).

Ip et al. (2004e) demonstrated that the skin of *Pn. schlosseri* mounted in an Ussing-type chamber bathed by 10 mmol·L^{-1} ammonium chloride with a 1 unit pH gradient (pH 8.0–7.0) had a small total ammonia flux of 0.010 μmol cm^{-2}·min^{-1}, which was lower than the ammonia flux through the abdominal skin of the frog, *Rana catesbiana*. Phosphatidylcholine is known to stabilize the lipid bilayer (Lande et al. 1995), and it happens to be the most abundant phospholipid (50%) in the skin of *Pn. schlosseri* (Ip et al. 2004e). By contrast, phosphatidylethanolamine, which is associated with high membrane fluidity, is present at low levels (<15%); hence, the phosphatidylcholine to phosphatidylethanolamine ratio, which acts as an indicator of membrane fluidity, is much higher in the skin of *Pn. schlosseri* (3.4) compared to values obtained in temperate teleosts (1.7; *Anguilla rostrata* and *Onchorhynchus mykiss*; Crockett 1999). The fraction of saturated (52%) to unsaturated fatty acids (28% mono- and 20% polyunsaturated) in the skin of *Pn. schlosseri* (Ip et al. 2004e; Barzaghi 2002) is also higher than that of the trout *O. mykiss* (Ghioni et al. 1997). Furthermore, the skin of *Pn. schlosseri* has a high content of cholesterol (4.5 μmol·g^{-1}) suggesting a low membrane fluidity. Teleosts with high NH_3 permeability have a cholesterol:phospholipid ratio of less than one. As the skin of *Pn. schlosseri* has a cholesterol:phospholipid ratio greater than six, its NH_3 permeability must be low (Ip et al. 2004e). Notably, after 6 days of exposure to 30 mmol·L^{-1} ammonium chloride at pH 7, the cutaneous cholesterol content of *Pn. schlosseri* increases significantly to 5.5 μmol·g^{-1}, accompanied with an increase in the total saturated fatty acid content in the skin (Ip et al. 2004e). Similar to cholesterol, a high level of sphingomyelin implies low membrane fluidity as it is structurally rigid and hydrophobic (Lande et al. 1995). The content of sphingomyelin in the skin of *Pn. schlosseri* also increases following ammonia exposure. These results confirm that, during environmental ammonia exposure, *Pn. schlosseri* can further reduce the membrane fluidity, and hence the NH_3 permeability, of its skin. As membrane fluidity is also temperature dependent, it is possible that *Pn. schlosseri* can only be found in the tropics because the low membrane fluidity of its skin restricts it to regions with high temperature.

7.5.4.4 Increased Production of Glutamine

Although *Pn. schlosseri* and *B. boddarti* do not accumulate glutamine when out of the water (Ip et al. 2001b; Lim et al. 2001), they do so when exposed to sublethal concentrations of environmental ammonia (Peng et al. 1998). Under an ammonia-loading condition of 100 mmol·L^{-1} and 8 mmol·L^{-1} ammonium chloride for *Pn. schlosseri* and *B. boddarti*, respectively, the total FAA concentrations in the brain, liver, and muscle of the experimental fishes increase significantly, largely due to increases in the glutamine concentration (Peng et al. 1998). In particular, the glutamine concentrations in the brain increase to 29.8 μmol·g^{-1} (12-fold) and 12.1 μmol·g^{-1} (4.3-fold) for *Pn. schlosseri* and *B. boddarti*, respectively. Glutamine accumulation occurs not only in the brain, but also to a lesser extent in the liver and muscle. As *Pn. schlosseri* is capable of partial amino acid catabolism, it would have to vary its dependence on alanine formation and glutamine synthesis to deal with terrestrial exposure and exposure to environmental ammonia, respectively. At present, it is unclear how these two biochemical pathways are regulated in *Pn. schlosseri* under these two different environmental conditions.

7.5.4.5 High Brain Ammonia Tolerance

Animals cannot maintain a brain ammonia concentration lower than that of the blood (Cooper and Plum 1987); when plasma ammonia concentration rises, the brains of most animals are affected. Although many animals can momentarily ameliorate ammonia toxicity in the brain by detoxifying ammonia into glutamine, increases in glutamine synthesis, and accumulation in the brain can lead to multiple problems (Albrecht and Norenberg 2006; Brusilow 2002). Encephalopathy usually develops in mammals when the brain ammonia concentration elevates beyond 1 mmol·g^{-1}.

However, some air-breathing fishes have developed extraordinarily high brain ammonia tolerance. Ip et al. (2005) reported that the ammonia concentration in the brains of *Pn. schlosseri* and *B. boddarti* increased drastically and reached 18 and 14.5 $\mu mol \cdot g^{-1}$, respectively, after 96 hours of exposure to sublethal ammonia concentrations (100 and 8 $mmol \cdot L^{-1}$ ammonium chloride, respectively, at pH 7.0). In addition, glutamine accumulated to exceptionally high levels in the brains of *Pn. schlosseri* (29.8 $\mu mol \cdot g^{-1}$) and *B. boddarti* (12.1 $\mu mol \cdot g^{-1}$) without causing death, indicating that these two species of mudskipper can ameliorate those problems associated with increased glutamine synthesis and accumulation as observed in patients suffering from hyperammonemia.

The administration of (5R,10S)-(+)-methyl-10,11-dihydro-5H-dibenzo[a,d]cyclohepten-5,10-imine hydrogen maleate (MK801), which is an antagonist of NMDA receptors, at a dosage of 2 $\mu g \cdot g^{-1}$ fish has no protective effect on *Pn. schlosseri* and *B. boddarti* injected with a lethal dose of ammonium acetate, indicating that an activation of NMDA receptors is not the major cause of death during acute ammonia intoxication (Ip et al. 2005). Hence, unlike mammals (Kosenko et al. 1999; Marcaida et al. 1992), activation of NMDA receptors may not be the correct explanation for acute ammonia toxicity in these two species of mudskipper. As membrane depolarization can lead to the removal of the Mg^{2+} block on NMDA receptors and result in their activation (Fan and Szerb 1993), it would appear that these two species of mudskipper can regulate the intracellular ammonia level in their brains despite drastic increases in the brain (intracellular + extracellular) ammonia concentration. NH_4^+ can replace K^+ in the facilitated diffusion of K^+ through K^+ channels and/or active transport of K^+ through Na^+/K^+-ATPase; both these processes have direct or indirect deleterious effects on the membrane potential of a cell. In view of the high concentrations of ammonia in the brains of *Pn. schlosseri* and *B. boddarti* exposed to chronic and acute ammonia toxicity, and the lack of protective effect of MK801, it can be deduced that either membrane depolarization has occurred but does not lead to activation of NMDA receptors, or membrane potentials are resilient to NH_4^+ interference due to the presence of K^+ channels and Na^+/K^+-ATPase with high substrate specificities for K^+, in the brains of these two species of mudskipper. It has been reported recently that the Na^+/K^+-ATPase from the brain of *Monopterus albus*, which also displays high brain ammonia tolerance, has a greater K^+ specificity, compared to NH_4^+, at physiological ionic concentrations, and the overall Na^+/K^+-ATPase protein abundance also decreases significantly after 6 days of ammonia exposure (Chen et al. 2013). Simultaneously, there is a downregulation of the mRNA and protein expression of Na^+:K^+:$2Cl^-$ cotransporter 1b, which is known to transport NH_4^+ in substitution for K^+ into cells, in the brain of *M. albus* confronted with high concentrations of ammonia (Ip et al. 2013a). Whether the brains of *Pn. schlosseri* and *B. boddarti* show similar adaptations must await future studies.

7.6 SUMMARY

Like all fishes, mudskippers obtain amino acids through the catabolism of ingested proteins. When they catabolize amino acids, ammonia is produced. In water, mudskippers are ammonotelic; endogenous ammonia is excreted mainly through their gills. However, no water is available to irrigate the gills during a terrestrial excursion, and this leads to the impediment of endogenous ammonia excretion. Furthermore, mudskippers can be confronted with environmental ammonia toxicity when they retreat into burrows, where high levels of ammonia may be present. Ammonia is toxic and cannot be accumulated in the body to high concentrations. Mudskippers have thus evolved multiple strategies and mechanisms to combat ammonia toxicity during terrestrial or ammonia exposure. Strategies involved in terrestrial exposure include the following: (1) reduction of ammonia production; (2) avoidance of ammonia accumulation; and (3) enhancement of ammonia excretion. Those involved in exposure to environmental ammonia comprise the following: (1) enhancement of ammonia excretion; (2) reduction of ammonia entry; (3) conversion of ammonia to a less toxic product for accumulation, and (4) tolerance of ammonia at cellular and tissue levels. These strategies support and complement each other; they are often operating in various combinations to combat ammonia toxicity. Among mudskippers, *Pn. schlosseri* exhibits unique gill structures and biochemistry in relation to terrestrial adaptations. Consequently, *Pn. schlosseri* has the highest terrestrial affinity and the most effective defense against ammonia toxicity during exposure to terrestrial conditions (Figure 7.7) or environmental ammonia (Figure 7.8).

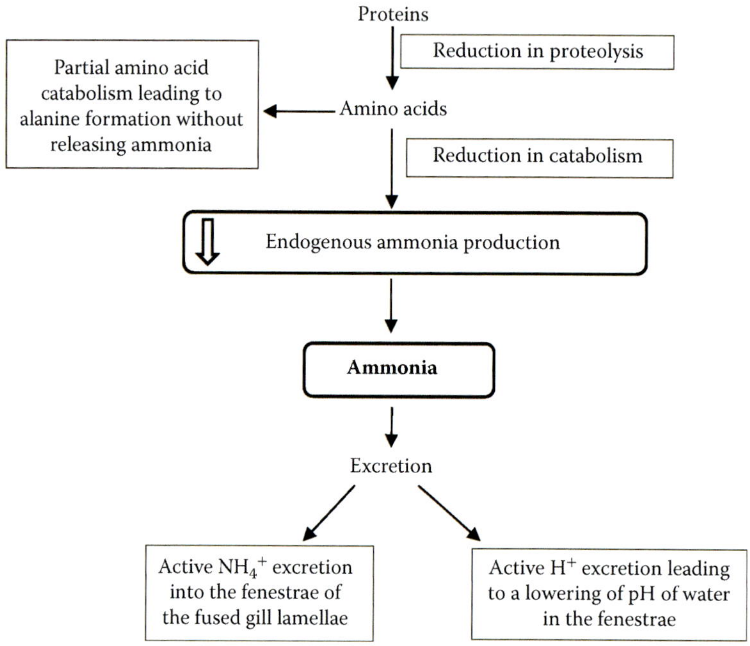

Figure 7.7 A variety of mechanisms employed by *Periophthalmodon schlosseri* to defend against endogenous ammonia toxicity during exposure to terrestrial conditions. (Courtesy of Shit F. Chew and Yuen K. Ip.)

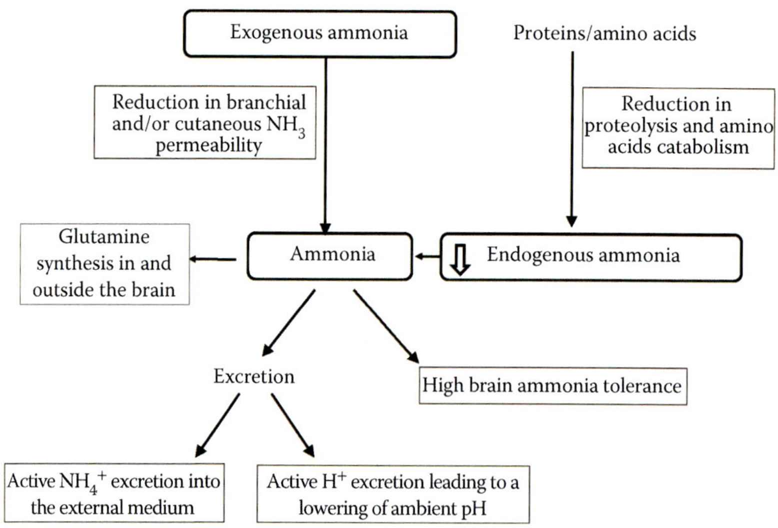

Figure 7.8 A variety of mechanisms employed by *Periophthalmodon schlosseri* to defend against exogenous and endogenous ammonia toxicity during exposure to high concentrations of environmental ammonia. (Courtesy of Shit F. Chew and Yuen K. Ip.)

REFERENCES

Aguilar N.M., Ishimatsu A., Ogawa K. and Khoo K.H. 2000. Aerial ventilatory responses of the mudskipper, *Periophthalmodon schlosseri*, to altered aerial and aquatic respiratory gas concentrations. *Comparative Biochemistry Physiology* 127A:285–292.

Albrecht J. and Norenberg M.D. 2006. Glutamine: A trojan horse in ammonia neurotoxicity. *Hepatology* 44:788–794.

Albrecht J., Zielinska M. and Norenbery M.D. 2010. Glutamine as a mediator of ammonia neurotoxicity: A critical appraisal. *Biochemical Pharmacology* 80:1303–1308.

Anderson P.M. 2001. Urea and glutamine synthesis: Environmental influences on nitrogen excretion. In *Fish Physiology Volume 20 Nitrogen Excretion* (eds. P.A. Wright and P.M. Anderson) pp. 239–277. New York, NY: Academic Press.

Arillo A., Margiocco C., Melodia F., Mensi P. and Schenone G. 1981. Ammonia toxicity mechanisms in fish: Studies on rainbow trout (*Salmo gairdneri* Rich). *Ecotoxicology and Environmental Safety* 5:316–325.

Bakouh N., Benielloun F., Cherif-Zahar B. and Planelles G. 2006. The challenge of understanding ammonium homeostasis and the role of Rh glycoproteins. *Transfusion Clinique et Biologique* 13:139–146.

Ballantyne J.S. 2001. Amino acid metabolism. In *Fish Physiology Vol. 19 Nitrogen Excretion* (eds. P.A. Wright and P.M. Anderson) pp. 77–107. New York, NY: Academic Press.

Bandurski R.S., Bradstreet E. and Scholander P.F. 1968. Metabolic changes in the mudskipper during asphyxia or exercise. *Comparative Biochemistry and Physiology* 24:271–274.

Barzaghi C. 2002. The mudskipper *Periophthalmodon schlosseri* increases acid (H^+) excretion and alters the skin lipid composition during short term and long term exposure to ammonia, respectively. MSc dissertation, National University Singapore.

Bemeur C., Desjardins P. and Butterworth R. F. 2010. Evidence for oxidative/nitrosative stress in the pathogenesis of hepatic encephalopathy. *Metabolism Brain Disease* 25:3–9.

Beon M.S., Oh M.K., Lee Y.J., Kim C.H. and Park J.Y. 2013. A comparative study on vascularization and the structure of the epidermis of an amphibious mudskipper fish, *Scartelaos gigas* (Gobiidae, Teleostei), on different parts of the body and the appendages. *Journal of Applied Ichthyology* 29:410–415.

Binstock L. and Lecar H. 1969. Ammonium ion currents in the squid giant axon. *Journal of General Physiology* 53:342–361.

Braun M.H., Steele S.L., Ekker M. and Perry S.F. 2009. Nitrogen excretion in developing zebrafish (*Danio rerio*): A role for Rh proteins and urea transporters. *American Journal of Physiology Renal Physiology* 296:F994–F1005.

Brusilow S.W. 2002. Hyperammoniemic encephalopathy. *Medicine* 81:240–249.

Cameron J.N. and Heisler N. 1983. Studies of ammonia in the rainbow trout: Physico-chemical parameters, acid–base behaviour and respiratory clearance. *Journal of Experimental Biology* 105:107–126.

Campbell J.W. 1991. Excretory nitrogen metabolism. In *Environmental and Metabolic Animal Physiology. Comparative Animal Physiology*, 4th ed. (ed. C.L. Prosser) pp. 277–324. New York, NY: Wiley-Interscience.

Campbell J.W. and Anderson P.M. 1991. Evolution of mitochondrial enzyme systems in fish: The mitochondrial synthesis of glutamine and citrulline. In *Biochemisty and Molecular Biology of Fishes Vol 1. Phylogenetic and Biochemical Perspectives* (eds. P.W. Hochachka and T.P. Mommsen) pp. 43–75. Amsterdam: Elsevier.

Campbell J.W., Aster P.A. and Vorhaben J.E. 1983. Mitochondrial ammoniagenesis in liver of the channel catfish *Ictalurus punctatus*. *American Journal of Physiology* 244:R709–R717.

Chen X.L., Wee N.L.J.E, Hiong K.C. et al. 2013. Properties and expression of Na^+/K^+-ATPase a-subunit isoforms in the brain of the swamp sel, *Monopterus albus*, which has unusually high brain ammonia tolerance. *PLoS ONE* 8(12):e84298.

Chew S.F., Goh E., Lim C.B. and Ip Y.K. 1998. Cyanide exposure affected the production and excretion of ammonia in the mudskipper, *Boleophthalmus boddaerti*. *Comparative Biochemistry and Physiology* 120C:441–448.

Chew S.F., Hiong K.C., Lam S.P., Chen X.L., Ching B. and Ip Y.K. 2015. Ammonia exposure increases the expression of Na^+:K^+:$2Cl^-$ cotransporter 1a in the gills of the giant mudskipper, *Periophthalmodon schlosseri*. *Journal of Comparative Physiology B* 185:57–72.

Chew S.F., Hiong K.C., Lam S.P. et al. 2014. Functional roles of Na^+/K^+-ATPase in active ammonia excretion and seawater acclimation in the giant mudskipper, *Periophthalmodon schlosseri*. *Frontiers in Physiology* 4:362.

Chew S.F., Hong L.N., Wilson J.M., Randall D.J. and Ip Y.K. 2003a. Alkaline environmental pH has no effect on the excretion of ammonia in the mudskipper *Periophthalmodon schlosseri* but inhibits ammonia excretion in the related species *Boleophthalmus boddaerti*. *Physiological and Biochemical Zoology* 76:204–214.

Chew S.F. and Ip Y.K. 1987. Ammoniagenesis in *Boleophthalmus boddaerti* and *Periophthalmodon schlosseri*. *Comparative Biochemistry and Physiology* 87B:941–948.

Chew S.F. and Ip Y.K. 1992a. Tolerance of the mudskipper, *Boleophthalmus boddaerti*, to a lack of oxygen. *Zoological Science* 9:227–230.

Chew S.F. and Ip Y.K. 1992b. Biochemical adaptations of the mudskipper, *Boleophthalmus boddaerti*, to a lack of oxygen. *Marine Biology* 112:567–572.

Chew S.F. and Ip Y.K. 1992c. Cyanide detoxification in the mudskipper, *Boleophthalmus boddaerti*. *Journal of Experimental Zoology* 261:1–8.

Chew S.F. and Ip Y.K. 1993. Respiration in the muscle mitochondria of the mudskipper, *Boleophthalmus boddaerti*. *Comparative Biochemistry and Physiology* 104B:681–688.

Chew S.F. and Ip Y.K. 2014. Excretory nitrogen metabolism and defence against ammonia toxicity in air-breathing fishes. *Journal of Fish Biology* 84:603–638.

Chew S.F., Lim A.L.L., Low W.P., Lee C.G.L., Chan K.M. and Ip Y.K. 1990. Can the mudskipper, *Periophthalmus chrysospilos*, tolerate acute environmental hypoxic exposure? *Fish Physiology and Biochemistry* 8:221–227.

Chew S.F., Sim M.Y., Phua Z.C., Wong W.P. and Ip Y.K. 2007. Active ammonia excretion in the giant mudskipper, *Periophthalmodon schlosseri* (Pallas), during emersion. *Journal of Experimental Zoology* 307A:357–369.

Chew S.F., Wilson J.M., Ip Y.K. and Randall D.J. 2006. Nitrogenous excretion and defense against ammonia toxicity. In *Fish Physiology Volume 23, The Physiology of Tropical Fishes* (eds. V. Val, V. Almedia-Val and D.J. Randall) pp. 307–395. New York, NY: Academic Press.

Chew S.F., Wong M.Y., Tam W.L. and Ip Y.K. 2003b. The snakehead *Channa asiatica* accumulates alanine during aerial exposure, but is incapable of sustaining locomotory activities on land through partial amino acid catabolism. *Journal of Experimental Biology* 2065:693–704.

Ching B., Chew S.F., Wong W.P. and Ip Y.K. 2009. Environmental ammonia exposure induces oxidative stress in gills and brain of *Boleophthalmus boddarti* (mudskipper). *Aquatic Toxicology* 95:203–212.

Clayton D.A. 1993. Mudskippers. *Oceanography Marine Biology—An Annual Review* 31:507–577.

Cooper J.L. and Plum F. 1987. Biochemistry and physiology of brain ammonia. *Physiological Review* 67:440–519.

Crockett E.L. 1999. Lipid restructuring does not contribute to elevated activities of Na^+/K^+-ATPase in basolateral membranes from the gill of seawater acclimated eel (*Anguilla rostrata*). *Journal of Experimental Biology* 202:2385–2392.

DeLong D.C., Halver J.E. and Mertz E.T. 1958. Nutrition of salmonid fishes. VI. Protein requirements of chinook salmon at two water temperatures. *Journal of Nutrition* 65:589–599.

Evans D.H. and Cameron J.N. 1986. Gill ammonia transport. *Journal of Experimental Zoology* 239:17–23.

Evans D.H., Piermarini P.M. and Choe K.P. 2005. The multifunctional fish gill: Dominant site of gas exchange, osmoregulation, acid–base regulation, and excretion of nitrogenous waste. *Physiological Review* 85:97–177.

Fan P. and Szerb J.C. 1993. Effects of ammonium ions on synaptic transmission and on responses to quisqualate and N-methyl-d-aspartate in hippocampal CA1 pyramidal neurons in vitro. *Brain Research* 632:225–231.

Felipo V. and Butterworth R.F. 2002. Neurobiology of ammonia. *Progress in Neurobiology* 67:259–279.

French C.J., Mommsen T.P. and Hochachka P.W. 1981. Amino acid utilization in isolated hepatocytes from rainbow trout. *European Journal of Biochemistry* 113:311–317.

Garey W.F. 1962. Cardiac responses of fishes in asphyxic environments. *Biological Bulletin* 122:362–368.

Ghioni C., Bell J.G., Bell M.V. and Sargent J.R. 1997. Fatty acid composition, eicosanoid production and permeability in skin tissues of rainbow trout (*Oncorhynchus mykiss*) fed control or an essential fatty acid deficient diet. *Prostoglandins, Leukotrienes and Essential Fatty Acids* 56:479–489.

Gonzales T.T., Katoh M., Ghaffar M.A. and Ishimatsu A. 2011. Gross and fine anatomy of the respiratory vasculature of the mudskipper, *Periophthalmodon schlosseri* (Gobiidae: Oxudercinae). *Journal of Morphology* 271:629–640.

Gordon M.S., Bofinus J., Evans D.H., Mccarthy R. and Oglesby L.C. 1969. Aspects of the physiology of terrestrial life in amphibious fishes. I. The mudskipper, *Periophthalmus sobrinus*. *Journal of Experimental Biology* 50:141–149.

Gordon M.S., Ng W.W.M. and Yip A.Y.W. 1978. Aspects of the physiology of terrestrial life in amphibious fishes. III. The Chinese mudskipper *Periophthalmus cantonensis*. *Journal of Experimental Biology* 72:57–75.

Görg B., Morwinsky A., Keitel V., Quartskhava N., Schrör K. and Häussinger D. 2010. Ammonia triggers exocytotic release of l-glutamate from cultured rat astrocytes. *Glia* 58:691–705.

Görg B., Schliess F. and Hausinger D. 2013. Osmotic and oxidative/nitrosative stress in ammonia toxicity and hepatic encephalopathy. *Archives of Biochemistry and Biophysics* 536:158–163.

Graham J.B. 1997. *Air-Breathing Fishes*. San Diego, CA: Academic Press.

Gregory R.B. 1977. Synthesis and total excretions of waste nitrogen by fish of the *Periophthalmus* (mudskipper) and *Scartelaos* families. *Comparative Biochemistry Physiology* 57A:33–36.

Häussinger D. and Görg B. 2010. Interaction of oxidative stress, astrocyte swelling and cerebral ammonia toxicity. *Current Opinion in Clinical Nutritional and Metabolic Care* 13:87–92.

Hermenegildo C., Marcaida G., Montoliu C., Grisolia S., Minana M. and Felipo V. 1996. NMDA receptor antagonists prevent acute ammonia toxicity in mice. *Neurochemical Research* 21:1237–1244.

Houlihan D.F., Carter C.G. and McCarthy I.D. 1995. Protein turnover in animals. In *Nitrogen Metabolism and Excretion* (eds. P.J. Walsh and P.A. Wright) pp. 307–395. Boca Raton, FL: CRC Press.

Huang C.H. and Peng J. 2005. Evolutionary conservation and diversification of Rh family genes and proteins. *Proceeding of National Academy of Science USA* 102:15512–15517.

Hwang P.P. and Lee T.H. 2007. New insights into fish ion regulation and mitochondrion-rich cells. *Comparative Biochemistry and Physiology* 148A:479–497.

Hwang P.P., Lee T.H. and Lin L.Y. 2011. Ion regulation in fish gills: Recent progress in the cellular and molecular mechanisms. *American Journal of Physiology Regulatory, Integrative and Comparative Physiology* 301:R28–R47.

Ip Y.K. and Chew S.F. 2010. Ammonia production, excretion, toxicity and defense in fish: A review. *Frontiers in Physiology* 1(134):1–20. doi:10.3389/fphy.2010.00134

Ip Y.K., Chew S.F. and Lim R.W.L. 1990a. Ammoniagenesis in the mudskipper, *Periophthalmus chrysospilos*. *Zoological Science* 7:187–194.

Ip Y.K., Chew S.F. and Low W.P. 1991a. Effects of hypoxia on the mudskipper *Periophthalmus chrysospilos* (Bleeker 1855). *Journal of Fish Biology* 38:621–624.

Ip Y.K., Chew S.F. and Randall D.J. 2001a. Ammonia toxicity, tolerance and excretion. In *Fish Physiology Vol. 19 Nitrogen Excretion* (eds. P.A. Wright and P.M. Anderson) pp. 109–148. New York, NY: Academic Press.

Ip Y.K., Chew S.F. and Randall D.J. 2004a. Five tropical fishes, six different strategies to defend against ammonia toxicity on land. *Physiological and Biochemical Zoology* 77:768–782.

Ip Y.K., Chew S.F. and Tang P.C. 1991b. Evaporation and the turning behavior of the mudskipper, *Boleophthalmus boddaerti*. *Zoological Science* 8:621–623.

Ip Y.K., Chew S.F., Wilson J.M. and Randall D.J. 2004b. Defences against ammonia toxicity in tropical fishes exposed to high concentrations of environmental ammonia: A review. *Journal of Comparative Physiology B* 174:565–575.

Ip Y.K., Hou Z., Chen X.L. et al. 2013a. High brain ammonia tolerance and down-regulation of Na^+:K^+:$2Cl^-$ cotransporter 1b mRNA and protein expression in the brain of the swamp eel, *Monopterus albus*, exposed to environmental ammonia or terrestrial conditions. *PLoS ONE* 8(9):e69512.

Ip Y.K., Kuah S.S.L. and Chew S.F. 2004c. Strategies adopted by the mudskipper *Boleophthalmus boddaerti* to survive sulfide exposure in normoxia or hypoxia. *Physiological and Biochemical Zoology* 77:824–837.

Ip Y.K., Lee C.Y., Chew S.F., Low W.P. and Peng K.W. 1993. Differences in the responses of two mudskippers to terrestrial exposure. *Zoological Science* 10:511–519.

Ip Y.K., Leong M.W.F., Sim M.Y., Goh G.S. and Chew S.F. 2005. Chronic and acute ammonia toxicity in mudskippers, *Periophthalmodon schlosseri* and *Boleophthalmus boddaerti*: Brain ammonia and glutamine contents, and effects of methionine sulfoximine and MK801. *Journal of Experimental Biology* 208:1993–2004.

Ip Y.K., Lim C.B. and Chew S.F. 2006. Intermediary metabolism in mudskipper, *Periophthalmodon schlosseri* and *Boleophthalmus boddaerti*, during immersion or emersion. *Canadian Journal of Zoology* 84:981–991.

Ip Y.K., Lim C.B., Chew S.F., Wilson J.M. and Randall D.J. 2001b. Partial amino acid catabolism leading to the formation of alanine in *Periophthalmodon schlosseri* (mudskipper): A strategy that facilitates the use of amino acids as an energy source during locomotory activity on land. *Journal of Experimental Biology* 204:1615–1624.

Ip Y.K., Lim C.K., Lee S.L.M., Wong W.P. and Chew S.F. 2004d. Postprandial increases in nitrogenous excretion and urea synthesis in the giant mudskipper *Periophthalmodon schlosseri. Journal of Experiment Biology* 207:3015–3023.

Ip Y.K., Loong A.M., Kuah J.S. et al. 2012a. The roles of three branchial Na^+/K^+-ATPase α-subunit isoforms in freshwater adaptation, seawater acclimation and active ammonia excretion in *Anabas testudineus. American Journal of Physiology Regulatory Integrative and Comparative Physiology* 303:R112–R125.

Ip Y.K. and Low W.P. 1990. Lactate production in the gills of the mudskipper *Periophthalmodon schlosseri* exposed to hypoxia. *Journal of Experimental Zoology* 253:99–101.

Ip Y.K., Low W.P., Lim A.L.L. and Chew S.F. 1990b. Changes in lactate content in the gills of the mudskippers *Periophthalmus chrysospilos* and *Boleophthalmus boddaerti* in response to environmental hypoxia. *Journal of Fish Biology* 36:481–487.

Ip Y.K., Randall D.J., Kok T.K.T. et al. 2004e. The mudskipper *Periophthalmodon schlosseri* facilitates active NH_4^+ excretion by increasing acid excretion and having a low NH_3 permeability in the skin. *Journal of Experimental Biology* 207:787–801.

Ip Y.K., Soh M.M.L., Chen X.L. et al. 2013b. Molecular characterization of branchial *aquaporin 1aa* and effects of seawater acclimation, emersion or ammonia exposure on its mRNA expression in the gills, gut, kidney and skin of the freshwater climbing perch, *Anabas testudineus. PLoS ONE* 8(4):e61163.

Ip Y.K., Wilson J.M., Loong A.M. et al. 2012b. Cystic fibrosis transmembrane conductance regulator-like Cl^- channel in the gills of the climbing perch, *Anabas testudineus,* is involved in both hypoosmotic regulation during seawater acclimation and active ammonia excretion during ammonia exposure. *Journal of Comparative Physiology B* 182:793–812.

Ip Y.K., Zubaidah R.M., Liew P.C. et al. 2004f. African Sharptooth Catfish *Clarias gariepinus* does not detoxify ammonia to urea or amino acids but actively excretes ammonia during exposure to environmental ammonia. *Physiological and Biochemical Zoology* 77:242–254.

Ishimatsu A., Hishida Y., Takita T. et al. 1998. Mudskippers store air in their burrows. *Nature* 391:237–238.

Ishimatsu A., Takeda T., Tsuhako Y., Gonzales T.T. and Khoo K.H. 2009. Direct evidence for aerial egg deposition in the burrows of the Malaysian mudskipper, *Periophthalmodon schlosseri. Ichthyological Research* 56:417–420.

Ishimatsu A., Yoshida Y., Itoki N., Takeda T., Lee H. J. and Graham J.B. 2007. Mudskippers brood their eggs in air but submerge them for hatching. *Journal of Experimental Biology* 210:3946–3954.

Iwata K. 1988. Nitrogen metabolism in the mudskipper, *Periophthalmus cantonensis*: Changes in free amino acids and related compounds in carious tissues under conditions of ammonia loading with reference to its high ammonia tolerance. *Comparative Biochemistry Physiology* 91A:499–508.

Iwata K. and Deguichi M. 1995. Metabolic fate and distribution of ^{15}N-Ammonia in an ammonotelic amphibious fish, *Periophthalmus modestus,* following immersion in ^{15}N-ammonium sulphate: A long term experiment. *Zoological Science* 12:175–184.

Iwata K., Kakuta M., Ikeda G., Kimoto S. and Wada N. 1981. Nitrogen metabolism in the mudskipper, *Periophthalmus cantonensis*: A role of free amino acids in detoxification of ammonia produced during its terrestrial life. *Comparative Biochemistry Physiology* 68A:589–596.

Karlsson A., Eliason E.J., Mydland L.T., Farrell A.P. and Kiessling A. 2006. Postprandial changes in plasma free amino acid levels obtained simultaneously from the hepatic portal vein and the dorsal aorta in rainbow trout (*Oncorhynchus mykiss*). *Journal of Experimental Biology* 209:4885–4894.

Kinne R., Kinne-Saffran E., Schutz, H. and Schölermann B. 1986. Ammonium transport in medullary thick ascending limb of rabbit kidney: Involvement of the Na^+,K^+,Cl^--cotransporter. *Journal of Membrane Biology* 94:279–284.

Kok T.W.K., Lim C.B., Lam T.J. and Ip Y.K. 1998. The mudskipper *Periophthalmodon schlosseri* respires more efficiently on land than in water and vice versa for *Boleophthalmus boddaerti. Journal of Experimental Zoology* 280:86–90.

Kok W.K. 2000. Can the mudskipper *Periophthalmodon schlosseri* excrete NH_4^+ against a concentration gradient? MSc. dissertation, National University Singapore.
Lim C.B., Anderson P.M., Chew S.F. and Ip Y.K. 2001. Reduction in the rates of protein and amino acid catabolism to slow down the accumulation of endogenous ammonia: A strategy potentially adopted by mudskippers (*Periophthalmodon schlosseri* and *Boleophthalmus boddaerti*) during aerial exposure in constant darkness. *Journal of Experimental Biology* 204:1605–1614.
Kosenko E., Kaminski Y., Lopata O., Muravyov N. and Felipo V. 1999. Blocking NMDA receptors prevents the oxidative stress induced by acute ammonia intoxication. *Free Radical Biology and Medicine* 26:1369–1374.
Lande M.B., Donovan J.M. and Zeidel M.L. 1995. The relationship between membrane fluidity and permeabilities to water, solute, ammonia, and protons. *Journal of General Physiology* 106:67–84.
Lee H.J., Martinez C.A., Hertzberg K.J., Hamilton A.L. and Graham J.B. 2005. Burrow air phase maintenance and respiration by the mudskipper *Scartelaos histophorus* (Gobiidae: Oxudercinae). *Journal of Experimental Biology* 208:169–177.
Leung K.M.Y., Chu J.C.W. and Wu R.S.S. 1999. Effects of body weight, water temperature and ration size on ammonia excretion by the areolated grouper (*Epinephelus areolatus*) and mangrove snapper (*Lutjanus argentimaculatus*). *Aquaculture* 170:215–227.
Lim C.K, Wong W.P., Lee S.M.L., Chew S.F. and Ip Y.K. 2004. The ammonotelic African lungfish *Protopterus dolloi* increases the rate of urea synthesis and becomes ureotelic after feeding. *Journal of Comparative Physiology B* 174:555–564.
Loong A.M., Chew S.F., Wong W.P., Lam S.H. and Ip Y.K. 2012. Both seawater acclimation and environmental ammonia exposure lead to increases in mRNA expression and protein abundance of Na^+:K^+:$2Cl^-$ cotransporter in the gills of the freshwater climbing perch, *Anabas testudineus*. *Journal of Comparative Physiology B* 182:491–506.
Low W.P., Chew S.F. and Ip Y.K. 1992. Differences in electrophoretic patterns of lactate dehydrogenases form the gills, heats, and muscles of three mudskippers. *Journal of Fish Biology* 40:975–977.
Low W.P., Ip Y.K. and Lane D.J.W. 1990. A comparative study of the gill morphometry in three mudskippers—*Periophthalmus chrysospilos*, *Boleophthalmus boddaerti* and *Periophthalmus schlosseri*. *Zoological Science* 7:29–38.
Low W.P., Lane D.J.W. and Ip Y.K. 1988. A comparative study of terrestrial adaptations in three mudskippers—*Periophthalmus chrysospilos*, *Boleophthalmus boddaerti* and *Periophthalmodon schlosseri*. *Biological Bulletin* 175:434–438.
Low W.P., Peng K.W., Phuan S.K., Lee C.Y. and Ip Y.K. 1993. A comparative study on the responses of the gills of two mudskippers to hypoxia and anoxia. *Journal of Comparative Physiology B* 163:487–494.
Marcaggi P. and Coles J.A. 2001. Ammonium in nervous tissue: Transport across cell membranes, fluxes from neurons to glial cells, and role in signaling. *Progress in Neurobiology* 64:157–183.
Marcaida G., Felipo V., Hermenegildo C., Minana M.D. and Grisolia S. 1992. Acute ammonia toxicity is mediated by NMDA type of glutamate receptors. *FEBS Letters* 296:67–68.
McDonald M.D., Smith C.P. and Walsh P.J. 2006. The physiological and evolution of urea transport in fishes. *Journal of Membrane Biology* 212:93–107.
Milward N.E. 1974. Studies on the taxonomy, ecology, and physiology of Queensland mudskippers. PhD dissertation, University Queensland.
Miñana M.D., Hermenegildo C., Llansola M., Montoliu C., Grisolia S. and Felipo V. 1996. Carnitine and choline derivatives containing a trimethylamine group prevent ammonia toxicity in mice and glutamate toxicity in primary cultures of neurons. *Journal of Pharmacology and Experimental Therapeutics* 279:194–199.
Morii H. 1979. Changes with time ammonia and urea concentrations in the blood and tissue of mudskipper fish, *Periophthalmus cantonensis* and *Boleophthalmus pectinirostris* kept in water and on land. *Comparative Biochemistry Physiology* 64A:235–243.
Morii H., Nishikata K. and Tamura O. 1978. Nitrogen excretion of mudskipper fish *Periophthalmus cantonensis* and *Boleophthalmus pectinirostris* in water and on land. *Comparative Biochemistry Physiology* 60A:189–193.
Morii H., Nishikata K. and Tamura O. 1979. Ammonia and urea excretion from mudskipper fishes, *Periophthalmus cantonensis* and *Boleophthalmus pectinirostris* transferred from land to water. *Comparative Biochemistry Physiology* 63A:23–28.

Nagami G.T. 1988. Liminal secretion of ammonia in the mouse proximal tubule perfused in vitro. *Journal of Clinical Investigation* 81:159–164.

Nakada T., Hoshjima K., Esakl M., Nagayoshi S., Kawakami K. and Hirose S. 2007a. Localization of ammonia transporter Rcgl in mitochondria-rich cells of yolk sac, gill, and kidney of zebrafish and its ionic strength-dependent expression. *American Journal of Physiology Regulatory Integrative and Comparative Physiology* 293:R1743–R1753.

Nakada T., Westhoff C.M., Kato A. and Hirose S. 2007b. Ammonia secretion from fish gills depends on a set of Rh proteins. *FASEB Journal* 21:1–8.

Nakhoul N.L., Hering-Smith K.S., Abdulnour-Nakhoul S.M. and Hamm L.L. 2001. Transport of NH_3/NH_4^+ in oocytes expressing aquaporin-1. *American Journal of Physiology* 281:F255–F263.

Nawata C.M., Hung C.C.Y., Tsui T.K.N., Wilson J.M., Wright P.A. and Wood C.M. 2007. Ammonia excretion in rainbow trout (*Oncorhynchus mykiss*): Evidence for Rh glycoprotein and H^+-ATPase involvement. *Physiological Genomics* 31:463–474.

Nawata C.M., Wood C.M. and O'Donnell M.J. 2010. Functional characterization of Rhesus glycoproteins from an ammoniotelic teleost, the rainbow trout, using oocyte expression and SIET analysis. *Journal of Experimental Biology* 213:1049–1059.

Park J.Y., Kim I.S. and Lee Y.J. 2006. A study on the vascularization and structure of the epidermis of the air-breathing mudskipper, *Periophthalmus magnuspinnatus* (Gobiidae, Teleostei), along different parts of the body. *Journal of Applied Ichthyology* 22:62–67.

Peng K.W., Chew S.F., Lim C.B., Kuah S.S.L., Kok T.W.K. and Ip Y.K. 1998. The mudskippers *Periophthalmodon schlosseri* and *Boleophthalmus boddaerti* can tolerate environmental NH_3 concentration of 446 and 36 μM, respectively. *Fish Physiology and Biochemistry* 19:59–69.

Planelles G. 2007. Ammonium homeostasis and human rhesus glycoproteins. *Nephron Physiology* 105:11–17.

Randall D.J., Ip Y.K., Chew S.F. and Wilson J.W. 2004. Air breathing and ammonia excretion in the mudskipper, *Periophthalmodon schlosseri. Physiological and Biochemical Zoology* 77:783–788.

Randall D.J., Wilson J.M., Peng K.W. et al. 1999. The mudskipper, *Periophthalmodon schlosseri*, actively transports NH_4^+ against a concentration gradient. *American Journal of Physiology Regulatory Integrative and Comparative Physiology* 46:R1562–R1567.

Reddy P.V.B., Rama Rao K.V. and Norenberg M.D. 2009. Inhibitors of the mitochondrial permeability transition reduce ammonia-induced cell swelling in cultured astrocytes. *Journal of Neuroscience Research* 87:2677–2685.

Rose C. 2002. Increased extracellular brain glutamate in acute liver failure: Decreased uptake or increased release? *Metabolic Brain Disease* 17:251–261.

Saparov S.M., Liu K., Agre P. and Pohl P. 2007. Fast and selective ammonia transport by aquaporin-8. *Journal of Biological Chemistry* 282:5296–5301.

Schöttle E. 1931. Morphologie und Physiologie der Atmung bei wasser-, schlamm- und landlebenden Gobiiformes. *Zeitschrift für wissenschaftliche Zoologie* 140:1–114.

Takeda T., Ishimatsu A., Oikawa S., Kanda T., Hishida Y. and Khoo K.H. 1999. Mudskipper *Periophthalmodon schlosseri* can repay oxygen debts in air but not in water. *Journal of Experimental Zoology* 284:265–270.

Tamura S.O., Morii H. and Yuzuriha M. 1976. Respiration of the amphibious fishes *Periophthalmus cantonensis* and *Boleophthalmus chinensis* in water and on land. *Journal of Experimental Biology* 65:97–107.

Tay Y.L., Loong A.M., Hiong K.C. et al. 2006. Active ammonia transport and excretory nitrogen metabolism in the climbing perch, *Anabas testudineus*, during 4 days of emersion or 10 min of forced exercise on land. *Journal of Experimental Biology* 209:4475–4489.

Thomas R.C. 1984. Experimental displacement of intracellular pH and the mechanism of its subsequent recovery. *Journal of Physiology (London)* 354:3–22.

Tng Y.Y.M., Wee N.L.J., Ip Y.K. and Chew S.F. 2008. Postprandial nitrogen metabolism and excretion in juvenile marble goby, *Oxyeleotris marmorata* (Bleeker, 1852). *Aquaculture* 284:260–267.

Toba A. and Ishimatsu A. 2014. Roles of air stored in burrows of the mudskipper *Boleophthalmus pectinirostris* for adult respiration and embryonic development. *Journal of Fish Biology* 84:774–793.

Van den Thillart G. and van Raaij M. 1995. Endogenous fuels; non-invasive versus invasive. In *Biochemistry and Molecular Biology of Fishes, Volume 4, Metabolic Biochemistry* (eds. P.W. Hochachka and T.P. Mommsen) pp. 33–63. Amsterdam: Elsevier.

Wall S.M. and Koger L.M. 1994. NH_4^+ transport mediated by Na^+-K^+-ATPase in rat inner medullary collecting duct. *American Journal of Renal Physiology* 267:F660–670.

Walton M.J. and Cowey C.B. 1977. Aspects of ammoniagenesis in rainbow trout, *Salmo gairdneri*. *Comparative Biochemistry and Physiology* 57:143–149.

Wee N.L.J., Tng Y.Y.M., Cheng H.T., Lee S.M.L., Chew S.F. and Ip Y.K. 2007. Ammonia toxicity and tolerance in the brain of the African sharptooth catfish, *Clarias gariepinus*. *Aquatic Toxicology* 82:204–213.

Weihrauch D., Wilkie M.P. and Walsh P.J. 2009. Ammonia and urea transporters in gills of fish and aquatic crustaceans. *Journal of Experimental Biology* 212:1716–1730.

Wicks B.J. and Randall D.J. 2002. The effect of feeding and fasting on ammonia toxicity in juvenile rainbow trout, *Oncorhynchus mykiss*. *Aquatic Toxicology* 59:71–82.

Wilkie M.P. 1997. Mechanisms of ammonia excretion across fish gills. *Comparative Biochemistry and Physiology* 118A:39–50.

Wilkie M.P. 2002. Ammonia excretion and urea handling by fish gills: Present understanding and future research challenges. *Journal of Experimental Zoology* 293:284–301.

Wilson J.M., Randall D.J., Donowitz M., Vogl W.A. and Ip Y.K. 2000. Immunolocalization of ion transport proteins to the mudskipper (*Periophthalmodon schlosseri*) branchial epithelium mitochondria-rich cells. *Journal of Experimental Biology* 203:2297–2310.

Wilson J.M., Randall D.J., Kok T.W.K., Vogl W.A. and Ip Y.K. 1999. Fine structure of the gill epithelium of the terrestrial mudskipper, *Periophthalmodon schlosseri*. *Cell Tissue Research* 298:345–356.

Wilson R.W., Wright P.A., Munger S. and Wood C.M. 1994. Ammonia excretion in freshwater rainbow trout (*Oncorhynchus mykiss*) and the importance of gill boundary layer acidification: Lack of evidence for Na^+–NH_4^+ exchange. *Journal of Experimental Biology* 191:37–58.

Wright P.A. and Wood C.M. 2009. A new paradigm for ammonia excretion in aquatic animals: Role of Rhesus (Rh) glycoproteins. *Journal of Experimental Biology* 212:2303–2312.

Wright P.A., Randall D.J. and Perry S.F. 1989. Fish gill boundary layer: A site of linkage between carbon dioxide and ammonia excretion. *Journal Comparative Physiology* 158:627–635.

Wright S.H. 1991. The interface of animal and aqueous environment: Strategies and constraints on the maintenance of solute balance. In *Biochemistry and Molecular Biology of Fishes 1, Phylogenetic and Biochemical Perspectives* (eds. P.W. Hochachka and T.P. Mommsen) pp. 165–180. Amsterdam: Elsevier.

Wu B. and Beitz E. 2007. Aquaporins with selectivity for unconventional permeants. *Cellular and Molecular Life Sciences* 64:2413–2421.

Youngson A., Cowey C.B. and Walton M.J. 1982. Some properties of serine pyruvate aminotransferase purified from liver of rainbow-trout *Salmo gairdneri*. *Comparative Biochemistry and Physiology* 73B:393–398.

CHAPTER 8

Aquatic and Terrestrial Locomotion

Cinnamon Pace

CONTENTS

8.1 INTRODUCTION

Mudskippers, like other fishes, must be able to successfully move to obtain food, avoid predation, defend territories, and reproduce (e.g., Clayton 1993). In contrast with the vast majority of fish species, mudskippers perform many of these behaviors both in terrestrial and aquatic environments (e.g., Ishimatsu and Gonzales 2011; Graham 1997; Martin 1995; Clayton 1993). Hence, mudskippers join an interesting cohort of organisms that must be able to cope with two worlds: water and land. For organisms that bridge aquatic and terrestrial habitats, a key locomotor challenge is how to reconcile moving through different physical regimes without sacrificing locomotor performance.

Land and water place drastically different demands on the musculoskeletal system (Ashley-Ross et al. 2013). Water is 800 times denser and 55 times more viscous than air, but the gravitational load to the organism is greatly reduced while in water (Biewener 2003; Graham 1997). The high density and viscosity of the aquatic medium mean that overcoming drag forces is the main challenge for locomotion. Water in the intertidal zone is in constant motion, generating complex and turbulent flow patterns, thus posing another challenge for the aquatic locomotion of mudskippers. In contrast, mass-related gravitational forces dominate while on land (Sayer 2005; Biewener 2003; Graham 1997). The increased influence of gravity also affects the orientation in which mudskippers generate propulsive thrusts. Rather than pushing on the aquatic medium surrounding the body using their fins, lateral body, and/or tail, mudskippers on land must use those same structures to generate thrust in a posterior and ventral direction in order to create ground reaction forces (GRFs) that propel them forward. In addition, when on land, the slope a mudskipper moves across and the nature of the substrate (e.g., fine sediment or rocks) can vary greatly and will likely dictate locomotor mode and effectiveness.

Because mudskippers use the same anatomical structures to move in such mechanically dissimilar environments, they represent an excellent model system in which to study how locomotion varies according to environment. In this chapter, I consider three areas: (1) anatomical modifications that facilitate movement on the land; (2) fundamental movement patterns generated on land and in water; and (3) locomotor strategies that allow successful movement in both aquatic and terrestrial habitats. Finally, I discuss the gaps in the current understanding of mudskipper locomotion and reflect on the broader conclusions to be drawn from studies of mudskipper anatomy and locomotor behaviors.

8.2 ANATOMICAL MODIFICATIONS OF LOCOMOTOR STRUCTURES

Mudskippers share general morphological features of other gobies that likely influence their locomotor capabilities. For example, most gobies are generally small, cylindrically shaped, and somewhat dorsoventrally flattened (Zander 2011). Most mudskippers also have fused pelvic fins, which are present in many gobioid fishes (Zander 2011; Murdy 1989; Harris 1960). However, the degree of pelvic fin fusion varies among and within the oxudercine genera. For example, the innermost rays of the pelvic fins of many species of *Periophthalmus* are completely separate or form a partial disc (Murdy 1989), whereas pelvic fins of *Boleophthalmus* spp. and *Ps. chrysospilos* are fused (Zander 2011). Similar to other gobies, the primary propulsive elements available to mudskippers in either habitat are the fins and axial musculature. Any modifications to these structures to facilitate life on land may have ramifications for aquatic locomotor performance.

Mudskippers have specialized morphological adaptations to their pectoral fins that enhance their ability to move on land and distinguish them from other gobies (Murdy 1989). Pectoral fins in a generalized ray-finned fish comprise two main elements: the proximal radials and the distal rays. The radials comprise a very small percentage of total-fin length and do not emerge from the body wall of the fish. However, in four mudskipper genera (*Boleophthalmus*, *Periophthalmodon*, *Periophthalmus*, and *Scartelaos*), the radials are substantially elongate and protrude from the body wall (Figure 8.1; Zander 2011; Pace and Gibb 2009; Murdy 1989; Harris 1960; Eggert 1929). Functionally, radial elongation creates a fin with two moveable joints—a shoulder joint where the radials abut the cleithrum of the pectoral girdle, and an intrafin joint where the radials support the rays (Pace and Gibb 2009; Harris 1960). This elongate pectoral fin may increase the range of motion of the pectoral fin relative to non-mudskipper gobioid fishes (Harris 1960). For example, the elongate radials have been suggested to "clear the fin from the body wall" (Harris 1960) and manipulations of cleared and stained specimens of *Periophthalmus argentilineatus* support this idea (Pace and Gibb 2009). The opercular bones of the head appear to occlude the radials from maximal anterior extension; however, the elongate radials extend the rays out and away from the obstruction, allowing them to be placed farther anteriorly (Pace and Gibb 2009). This substantially increases mudskippers terrestrial stride length (Figure 8.2), a key parameter for terrestrial locomotor performance. In *Periophthalmus*, the elongate radials also create an overall declination of the fin axis, which is observed in *Ps. argentilineatus* and *Ps. barbarus* (Pace and Gibb 2009; Harris 1960). This is achieved by the most dorsal radial being more elongate than the most ventral radial. The resultant change in the overall axis of the radial lobe probably facilitates the rays being positioned against the substrate when compared to other non-amphibious gobies (Harris 1960).

The degree of terrestriality in different mudskipper genera is reflected in their morphologies (Zander 2011; Murdy 1989). For example, the two most terrestrial genera (i.e., *Periophthalmus* and *Periophthalmodon*) have the longest pectoral fins relative to their body length (Murdy 1989).

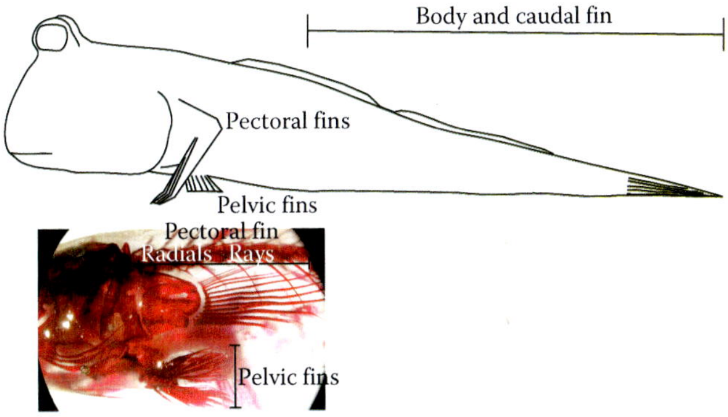

Figure 8.1 Depiction of the primary propulsive organs of mudskippers accompanied by a cleared and stained picture of the intact pectoral and pelvic fins of *Periophthalmus argentilineatus*. The picture is taken from a lateral and ventral angle so that both pectoral fin elements and pelvic fins can be visualized. (Modified from Pace, C.M. and Gibb, A.C., *Journal of Experimental Biology*, 212, 2279–2286, 2009.)

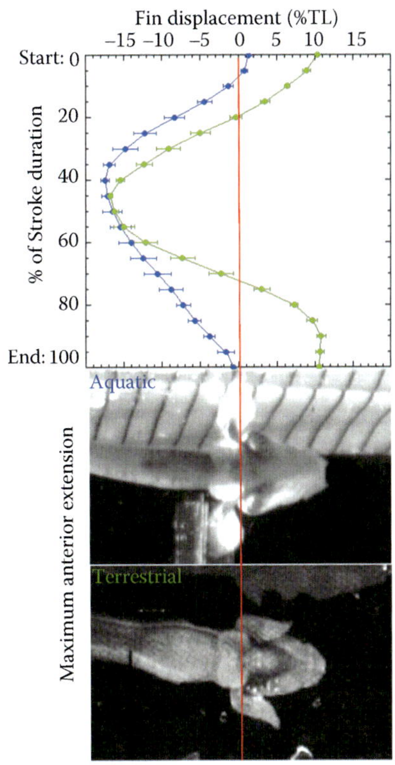

Figure 8.2 The top panel depicts the mean kinematic profile of the anterior and posterior extension of the pectoral fin of *Periophthalmus argentilineatus* (n = 3). The y-axis is the duration of a single stroke relativized to 100% and the x-axis is the anterior and posterior extension of the pectoral fin relativized by the total length (%TL). On the x-axis, a value of 0 occurs at the transverse plane through the shoulder joint. Negative values indicate that the tip of the fin is posterior to the shoulder joint and anterior values indicate that the fin tip is anterior to the shoulder joint. The middle and bottom panels display images of the maximum anterior extension of the pectoral fin of *Periophthalmus argentilineatus* while swimming versus crutching. The red line in all three panels demarcates the hypothetical transverse plane through the shoulder joint. Notice the increased anterior extension of the fin during terrestrial locomotion. (Modified from Pace, C.M. and Gibb, A.C., *Journal of Experimental Biology*, 212, 2279–2286, 2009.)

The elongation of the pectoral fin is accompanied by several other musculoskeletal adaptions, such as a more robust cleithrum, larger interradial muscles, and changes to the muscle architecture of the pectoral fin (Zander 2011; Harris 1960; Eggert 1929). The *abductor superficialis* muscle of *Periophthalmus*, *Periophthalmodon*, and *Boleophthalmus* is divided into two different muscle bands (*abductor superficialis I* and *abductor superficialis II*); whereas *Scartelaos* and the other oxudercine gobies that spend less time on land have the basal condition of a single fan-shaped pectoral fin abductor muscle (Zander 2011; Harris 1960; Eggert 1929). It is likely that the observed morphological variation of the pectoral fin in mudskippers has functional consequences. For example, the mudskipper genera with shorter pectoral fins probably also have correspondingly shorter pectoral fin stride lengths. However, explicit measures of movement that relate form to function among the mudskipper genera are currently lacking.

8.3 MUDSKIPPER LOCOMOTOR PATTERNS

In both aquatic and terrestrial environments, mudskippers are capable of multiple forms of locomotion that can be categorized into "sustained" and "burst" locomotor patterns (Table 8.1). During sustained locomotion in water, mudskippers have been reported to swim via axial undulations (Pace and Gibb 2009; Harris 1960; Ip et al. 1990), a form of body caudal fin (BCF) propulsion (Webb 1984). Although they have not been quantified these movements appeared to be similar to that of a generalized ray-finned fish (Stebbins and Kalk 1961). Mudskippers perform two distinct aquatic burst behaviors: (1) a fast-start behavior prevalent among ray-finned fishes (Swanson and Gibb 2004) and (2) they "skip" across the surface of the water in a series of jumps (De and Nandi 1984; Harris 1960). In all reported instances of aquatic locomotion in mudskippers, the axial musculature and caudal fin are pivotal in powering movements. This differs from terrestrial locomotion, where sustained locomotor bouts are powered by the pectoral fins and burst behaviors are powered by the axial musculature and caudal fin (Pace and Gibb 2009; Swanson and Gibb 2004; De and Nandi 1984; Dijk 1960; Harris 1960). Thus, aquatic locomotor behaviors are shared with other fishes whereas mudskipper terrestrial locomotor behaviors are more unusual, differ substantially from aquatic movements even when they appear superficially similar (Pace and Gibb 2009; Swanson and Gibb 2004), and are facilitated by specialized morphological adaptations (e.g., Zander 2011).

Table 8.1 Locomotor Behaviors in Aquatic and Terrestrial Environments

Habitat	Locomotor Category	Locomotor Pattern	Species in Which Locomotion Is Described by Detailed, Quantitative, Kinematic Data
Aquatic	Sustained	Slow swimming (axial + pectoral)	*Ps. argentilineatus* (Pace and Gibb 2009)
	Sustained	Fast swimming (axial only)	
	Burst	Fast-start	*Ps. argentilineatus* (Swanson and Gibb 2004)
	Burst	Skipping across water surface	*Ps. barbarus* (Harris 1960)
	Other	Maneuvering	
Terrestrial	Sustained	Crutching	*Ps. argentilineatus* (Pace and Gibb 2009); *Ps. barbarus* (Harris 1960; Kawano and Blob 2013)
	Burst	Jump: escape response	*Ps. argentilineatus* (Swanson and Gibb 2004)
	Burst	Jump: prey capture	
	Burst	Jump: courting	
	Other	Climbing	
	Other	Burrowing/digging	

8.3.1 Sustained Aquatic Locomotion

Many species of mudskippers are reported to retreat into burrows or move with the waterline up and down the intertidal zone rather than engage in active swimming during high tide (Polgar and Crosa 2009; Lee et al. 2005; Sayer 2005). There is a paucity of information on mudskipper aquatic locomotor behaviors, in part because field observations of aquatic behaviors are difficult. At high speeds, mudskippers swim using their axial musculature and caudal fin, but at slower speeds, they may also use "rowing" motions of the pectoral fins in combination with tail movements (Pace and Gibb 2009; Harris 1960). Similar to other fishes, it is likely that mudskippers also employ pectoral fins to aid in maneuvering, steering, and braking (Drucker et al. 2006). Mudskippers have also been observed to swim at the water surface with their head and eyes sticking out of the water (Ip et al. 1990; Harris 1960).

Paddling of pectoral fins during slow swimming is one of the few aquatic behaviors that had been quantified for mudskippers (Figure 8.3). During the pectoral fin paddling, the fins move either synchronously (both fins in the stroke phase at the same time) or asynchronously (one fin in stroke phase while the other is in recovery phase). At the initiation of a paddling stroke, the pectoral fin of *Ps. argentilineatus* is extended perpendicularly away from the body (Pace and Gibb 2009). In this position, the tip of the fin is approximately in the same transverse plane as the shoulder joint and the intrafin joint is slightly bent (~130°; Figure 8.2; Pace and Gibb 2009). From this position, the fin is retracted posteriorly until it meets the body wall; during retraction, the intrafin joint is straightened through 180° and the entire fin axis is slightly angled dorsally by the end of the stroke (Pace and Gibb 2009). During the recovery phase, when the fin is returned to its starting position, the fin is both protracted and rotated so that the fin is "feathered," which minimizes the surface area relative to flow (Figure 8.3), a fairly common practice for appendages utilizing drag-based aquatic locomotion (Pace and Gibb 2009).

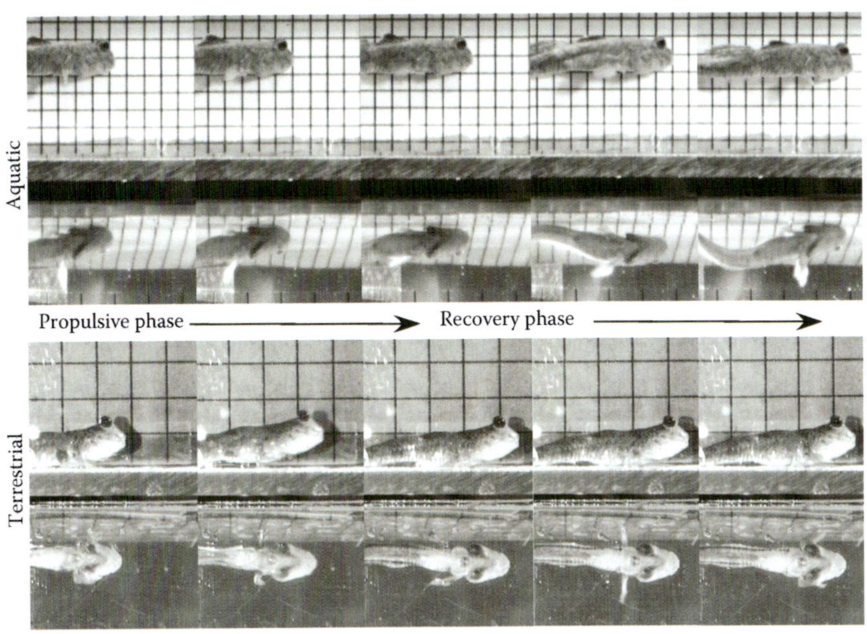

Figure 8.3 Image series of the pectoral fin movements of the mudskipper *Periophthalmus argentilineatus* in both lateral and ventral views while swimming and crutching. (Reproduced from Pace, C.M. and Gibb, A.C., *Journal of Experimental Biology*, 212, 2279–2286, 2009. With permission.)

8.3.2 Bursting Aquatic Locomotion

Aquatic fast-starts, or C-starts, are a common aquatic escape response for many fishes, and comprise two phases: a preparatory phase or Stage 1 and a propulsive phase or Stage 2 (Gibb et al. 2013; Swanson and Gibb 2004; Domenici and Blake 1997; Webb 1978). It is unknown how often mudskippers perform aquatic fast-start responses. *Periophthalmus argentilineatus* is the only mudskipper species in which aquatic escape responses had been quantified and the behavior was difficult to elicit (Swanson and Gibb 2004). However, the movements match the stereotypical kinematic pattern of a C-start (Swanson and Gibb 2004). The behavior begins when the fish curls its body into a "C" shape in response to a perturbation (Figure 8.4). During the preparatory stage, the body of *Ps. argentilineatus* forms a curve of fairly constant radius along the body where the curvature coefficient (i.e., the length of the curved vertebral span over the length of the straight vertebral span) is 0.55 ± 0.02 (Swanson and Gibb 2004). Similar curvature coefficients have been reported for other fishes that have approximately the same number of intervertebral joints (Swanson and Gibb 2004). During the propulsive phase, *Ps. argentilineatus* unfolds its body to rapidly accelerate away in a different direction (Swanson and Gibb 2004).

Mudskippers have also been reported to produce a second aquatic burst behavior. When startled, they may burst from the water and skip across the surface in a series of brief aerial bouts (Ip et al. 1990; De and Nandi 1984; Stebbins and Kalk 1961; Dijk 1960; Harris 1960). Harris (1960) divides a skip into three phases: active swimming, projection through the air, and landing on the water surface. During the active swimming phase, *Ps. barbarus* generates propulsive force by vigorous undulation of the tail, which continues to thrash until the entire tail has cleared the water (Harris 1960). During the aerial phase, *Ps. barbarus* holds its pectoral fins close to the side of the body (Harris 1960). However, De and Nandi (1984) observed that *Boleophthalmus boddarti* uses the pectoral fins to "maintain balance" while in the air, implying that the fins stick out in this species. Ip et al. (1990) stated that *Periophthalmodon* spp. and *Periophthalmus* spp. would skip across the water surface, while *Boleophthalmus* spp. and *Scartelaos* spp. would dive into the water and remain submerged when startled. Because there has been no comprehensive examination in the variation of locomotor patterns between species, and in locomotor variability within a species, these reports may not be contradictory. Rather, all of these anecdotes may be correct for the given authors' observations of that particular species under those specific circumstances.

Figure 8.4 Depiction of the changes in the axial patterning of *Periophthalmus argentilineatus* in four different locomotor behaviors; burst behavior images. (Adapted from Swanson, B.O. and Gibb, A.C., *Journal of Experimental Biology*, 207, 4037–4044, 2004.)

8.3.3 Sustained Terrestrial Locomotion

Sustained terrestrial locomotion of mudskippers is called "crutching" due to its superficial resemblance to a person walking on crutches (Harris 1960). During crutching, mudskippers synchronously use their pectoral fins to lift their body (and center of mass [COM]) off the substrate, swing it forward, and then lower it back down onto their pelvic fins (Figure 8.3). While resting on the substrate, mudskippers reposition their pectoral fins by lifting them off the substrate and moving them anteriorly to prepare for the next stride (Pace and Gibb 2009; De and Nandi 1984; Stebbins and Kalk 1961; Harris 1960). There is generally no axial or caudal fin propulsive contribution to crutching and the body and tail trail in a straight line behind the head and pectoral girdle (Figure 8.5; De and Nandi 1984; Harris 1960); although, it has been demonstrated that *Ps. barbarus* will assist with its tail when crutching up a 20° slope of loose granular media (McInroe et al. 2016). Among all amphibious fishes studied to date, crutching has only been documented in mudskippers; other fishes that use their pectoral fins for sustained terrestrial locomotion combine pectoral fin movements with movements of the body and caudal fin (Pace and Gibb 2014).

Crutching has been observed in four oxudercine genera (Clayton 1993; De and Nandi 1984; Harris 1960); however, kinematic and biomechanical details have only been quantified in two species of *Periophthalmus*: *Ps. argentilineatus* (Pace and Gibb 2009) and *Ps. barbarus* (Kawano and Blob 2013; Harris 1960, reported as *Ps. koelreuteri*). At the start of the pectoral fin stride, the entire fin is bent at the intrafin joint (in *Ps. argentilineatus* the intrafin angle ~100°) and the tip of the fin is in contact with the ground anterior to the shoulder joint (Figures 8.2 and 8.3; Pace and Gibb 2009; Harris 1960). During the propulsive phase (also known as the stance phase) of the stride, the mudskipper's body is elevated and launched over, and beyond, the fin tips. In *Ps. barbarus*, the peak GRF of a pectoral fin occurs during the middle of the stance phase (57% of stance; Kawano and Blob 2013). When the GRF is broken into vertical, anteroposterior, and mediolateral components, it is clear that mudskippers have an acceleratory anterior component and a strong medial component (which probably adds stability) during the propulsive phase of the stride, in addition to a large vertical component (Kawano and Blob 2013). By the end of the stance phase, the fin tips, which have been in contact with the ground for the entire stride, are now positioned posterior to the shoulder joint and the entire fin has been straightened so that it is almost a flat plane (Figure 8.3; in *Ps. argentilineatus* the intrafin joint is straightened to ~167°, see Pace and Gibb 2009). During the recovery phase (also known as the swing phase) of the stride, the fin tips are lifted off the substrate and the fins are flexed at the intrafin joint as they are drawn forward and repositioned anteriorly for the next stride (Figure 8.3; Pace and Gibb 2009; Harris 1960).

Figure 8.5 Dorsal view of *Periophthalmus argentilineatus* crutching over sandy-mud substrate at Paya mangroves, Tioman Island, Peninsular Malaysia. Note the track marks from the pectoral fins and the tail. (Courtesy of Heok Hui Tan.)

Table 8.2 Comparative Kinematic Data between *Periophthalmus argentilineatus* and *Periophthalmus barbarus*

	Periophthalmus argentilineatus	*Periophthalmus barbarus*	
Variable	**Pace and Gibb (2009)**	**Kawano and Blob (2013)**	**Harris (1960)**
TL (cm)	6.0 ± 0.5	13.7 ± 0.1	14
Mass (g)	2.3 ± 0.1	25.1 ± 0.5	NA
N	3	5	1
Average velocity (cm/s)	4.47 ± 0.27	7.6 ± 0.3	*10.75*
Average velocity (TL/s)	*0.75*	*0.55*	*0.77*
Distance per stride	2.13 ± 0.09	*4.48*	4.3
Distance per stride (%TL)	*0.36*	*0.33*	*0.31*
Stride duration (s)	0.49 ± 0.04	0.59 ± 0.02	0.4
Stance duration (s)	0.23 ± 0.02	0.39 ± 0.01	0.25
Swing duration (s)	0.26 ± 0.02	0.20 ± 0.01	0.15
Duty factor	0.47 ± 0.01[a]	0.66 ± 0.01	*0.63*
Stride frequency (Hz)	*2.04*	1.78 ± 0.6	*2.5*

Italicized values were not present in the original manuscripts, but were calculated from the available data.

[a] In Pace and Gibb (2009), this value is reported at 0.45 because it excluded a refractory period of the fin. In order to improve the comparison to the data for *Periophthalmus barbarus*, the refractory period has been included in the stance phase, making the duty factor 0.47.

It is possible to compare some basic kinematic parameters between *Ps. argentilineatus* and *Ps. barbarus* (Table 8.2). Two different studies on terrestrial crutching in *Ps. barbarus* reported a pectoral fin duty factor (i.e., the ratio of the stance phase duration to the total stride duration) of ~0.65 (Kawano and Blob 2013; Harris 1960). This contrasts with the reported 0.47 duty factor of *Ps. argentilineatus* indicating that the propulsive ground contact phase of the stride comprises a smaller percentage of the total stride duration in this species. In tetrapods, duty factor is affected by velocity, such that as velocity increases duty factor decreases (Biewener 2003). Size and velocity differences confound comparing the three studies, but when the data are relativized by total length the average velocity in *Ps. argentilineatus* is close to that reported for *Ps. barbarus* by Harris (1960), yet its duty factor is much lower (Table 8.2). *Periophthalmus argentilineatus* also appeared to move relatively further for each stride than *Ps. barbarus* (36%TL vs. 33–31%TL). Murdy (1989) reported an average relativized total pectoral-fin length for *Ps. argentilineatus* that is slightly longer than that for *Ps. barbarus* (26.9%SL vs. 26.4%SL), but the range of pectoral-fin lengths of the two species can overlap. From the available data, it cannot be ascertained if pectoral-fin kinematics differs, or if morphological differences in size and fin length account for the observed differences. A closer examination of the comparative crutching kinematics both within and among the mudskipper genera could be a fruitful avenue of further study.

Although they are not the primary propulsive organs, pelvic fins do have a role during terrestrial crutching. They interact with the substrate out of phase with the pectoral fins (i.e., when pectoral fins are contacting the ground the pelvic fins are not and vice versa; De and Nandi 1984; Harris 1960). This enables them to prop the fish off the substrate during the recovery phase, which may help reduce wet adhesion and/or friction with the surface (Pace and Gibb 2009; Harris 1960). However, whether they contribute to forward propulsion is uncertain. *Periophthalmus barbarus* continues to move forward after the pectoral fins have been lifted from the substrate, but this could be due to the forward momentum of the COM generated by the pectoral fins rather than pelvic-fin propulsion (Harris 1960). A pilot study using *Ps. argentilineatus* suggested that any contribution of the pelvic fins to forward movement was negligible based on the evidence that when the pectoral fins were in the recovery phase and only the pelvic fins were in contact with the substrate, forward movement was less than 7% of the total movement during the stride (Pace and Gibb 2005). Another potential function of the pelvic fins is to generate a forward thrust to slow the deceleration of the body (Harris 1960). Harris (1960) observed that *Ps. barbarus* decelerates in two phases. The reduced deceleration phase could stem from muscle activation of the pelvic fins that aids in braking; alternatively it could simply be a passive material property of the pectoral fins as they are compressed against the substrate by the body (Harris 1960). Finally, it is highly likely that aspects of pelvic-fin function are species specific due to the variation in pelvic-fin morphology, both in terms of the general shape and the degree of fusion (Zander 2011; Murdy 1989).

8.3.4 Bursting Terrestrial Locomotion

When on land, mudskippers jump as an escape response, while pursuing prey, during aggressive interactions with other mudskippers, and as part of the courtship behavior (e.g., R. Mleczko, personal communication, June 2016; Swanson and Gibb 2004; Clayton and Snowden 2000; Clayton 1993; De and Nandi 1984; Dijk 1960). In contrast to crutching, the terrestrial jump is primarily powered by the tail and the caudal fin and is a powerful ballistic movement that can propel mudskippers several body lengths through the air (Swanson and Gibb 2004; De and Nandi 1984; Stebbins and Kalk 1961; Harris 1960). Mudskippers primarily perform prone jumps—defined as jumps where the ventral surface of the fish is in contact with the substrate (Figure 8.6; Gibb et al. 2013; Swanson and Gibb 2004). However, mudskippers can also produce tail-flip jumps in which the mudskipper peels its head and body off the substrate and curls them over the tail such that the lateral side of the tail pushes into the substrate and the whole body forms a "C" shape perpendicular to the ground (Gibb et al. 2013). The mudskipper then jumps in the direction the tail was pointing (Gibb et al. 2013). The latter jumping technique is most likely a response to an anteriorly directed negative stimulus because it results in the fish moving in a direction opposite to the stimulus (Gibb et al. 2013).

The escape response jump of *Ps. argentilineatus* has been filmed and quantified using high-speed imaging (Swanson and Gibb 2004). Similar to an aquatic C-start, each terrestrial jump can be broken into a preparatory phase and a propulsive phase (Swanson and Gibb 2004). During the preparatory phase, the caudal fin is lifted off the substrate and placed anteriorly, such that the body of the fish is shaped like a "J" (Figure 8.4). The caudal fin is placed just behind the head (near the COM) and the maximum degree of bending occurs in the last third of the body with intervertebral

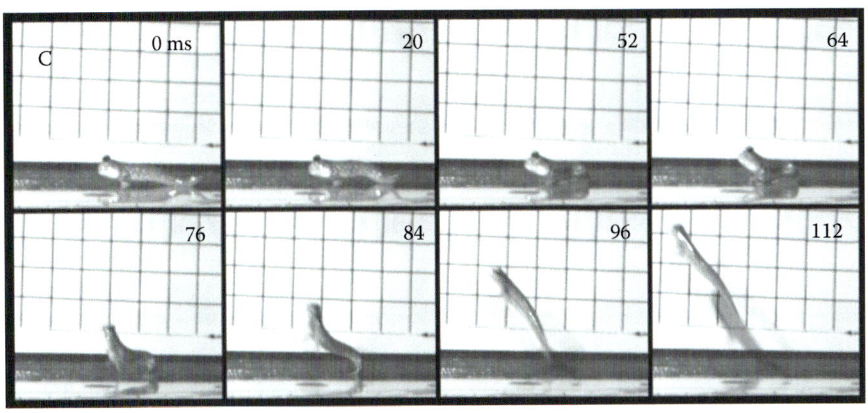

Figure 8.6 Sequence of movements from the lateral view of *Periophthalmus argentilineatus* executing a terrestrial escape response jump. Panels read from the top left down to the bottom right and time (ms) is printed on each panel. (Reproduced from Gibb, A.C. et al., *Integrative and Comparative Biology*, 53, 295–306, 2013. With permission.)

bending reaching approximately 20° (Swanson and Gibb 2004). The preparatory phase of the jump ends once maximum curvature has been achieved. At the start of the propulsive phase, the head is slightly elevated by the pelvic and pectoral fins, and this elevation is followed by a rapid unfolding of the body, a movement that straightens the tail (ending the propulsive phase; Figure 8.6). The fish is propelled vertically and anteriorly in the direction its head was initially oriented. This movement is powered by the tail, which pushes both laterally and ventrally against the substrate as the mudskipper's body is extended. *Periophthalmus argentilineatus* demonstrate take-off angles ranging from 27° to 59° (to maximize distance traveled take-off angle should be ~45°) and the preparatory stage has a longer duration than the propulsive stage (~88 ms vs. ~42 ms; Swanson and Gibb 2004). Mudskippers in this study were jumping in response to a negative stimulus. To date, there are no data of the jumping kinematics for pursuing prey or courting mates; the kinematics associated with these behaviors may differ from those observed during the escape response. Additionally, body size is a factor in jumping kinematics. For example, small (<2 cm) mudskippers have been reported to jump without using their tail (Harris 1960), but the behavior has not been confirmed with high-speed imaging and it is unclear that how these jumps would be powered by the musculoskeletal system. Mudskippers are an intriguing model system to explore comparative questions on relationships between size and jump performance and mechanics.

8.4 LOCOMOTOR STRATEGIES FOR GENERATING MOVEMENT IN AQUATIC AND TERRESTRIAL ENVIRONMENTS

Amphibious fish have been repeatedly shown to move differently while on land than in water (Perlman and Ashley-Ross 2016; Clardy 2012; Pace and Gibb 2011, 2009; Hsieh 2010; Swanson and Gibb 2004; Gillis 1998). Two strategies for generating effective locomotion in such disparate habitats are to utilize locomotor partitioning among structures or locomotor flexibility within a structure (Pace and Gibb 2014; Gibb et al. 2013; Biewener and Gillis 1999). For example, some amphibious fishes (e.g., juvenile *Sicyopterus stimpsoni*—Blob et al. 2006; *Clarias batrachus*—Pace and Gibb 2014) partition different locomotor tasks to different propulsive organs (Pace and Gibb 2014). In contrast, other amphibious fishes (e.g., *Erpetoichthyes calabaricus*—Pace and Gibb 2011; *Kryptolebias marmoratus*—Pronko et al. 2013) use the same propulsive organ in both habitats, but in different ways (locomotor flexibility; Pace and Gibb 2014; Gibb et al. 2013). Locomotor flexibility can occur through changes in muscle activation patterns, intensity, and/or duration (Gillis and Blob 2001; Biewener and Gillis 1999). However, it is possible that with locomotor flexibility, functional trade-offs exist such that locomotor function could be compromised for a given task. In contrast, locomotor partitioning takes advantage of utilizing different structures to achieve different outcomes.

Mudskippers utilize elements of both strategies. For sustained locomotion, they use a form of locomotor partitioning by shifting the primary propulsive system used in each habitat, that is, aquatic BCF swimming versus terrestrial pectoral-fin crutching (Pace and Gibb 2014). However, even when mudskippers use the same structure in water and on land, these structures move in kinematically distinct ways. For example, when aquatic pectoral-fin paddling is compared to terrestrial crutching, there are biologically relevant differences in the stride length (Figure 8.2), intrafin joint angle, recovery phase orientation of the fin rays, and timing of elevation/depression of the fin axis relative to the shoulder joint (Pace and Gibb 2009). Similarly, when aquatic and terrestrial escape responses are compared, the degree of curvature differs along the body (i.e., the aquatic "C" shape vs. the terrestrial "J" shape; Figure 8.4), and the duration of kinematic events differs (Swanson and Gibb 2004). These differences suggest that muscle activation patterns are varied for each behavior in each environment. In amphibious fishes, differences in muscle activity between aquatic and terrestrial environments have only been studied in the eel *Anguilla rostrata* (Gillis and Blob 2001; Gillis 2000), the lungfish *Protopterus annectens* (Horner and Jayne 2014), and the mangrove rivulus

Kryptolebias marmoratus (Perlman and Ashley-Ross 2016). However, these studies, and work done on non-piscine vertebrates, have found changes in intensity and duration of muscle activity, in addition to shifts in the timing of muscle activity, when comparing aquatic and terrestrial locomotion (Gillis and Blob 2001). Data from *K. marmoratus* in particular, suggest that muscle activation patterns for the terrestrial tail-flip jump evolved from the aquatic C-start (Perlman and Ashley-Ross 2016). Mudskippers are potentially useful model organisms to examine how plasticity of the neuromuscular system can generate different locomotor outputs and how new locomotor patterns evolve.

Mudskippers display kinematic patterns and morphology specially adapted for, and suited to, terrestrial locomotion. However, it is possible that functional trade-offs exist between aquatic and terrestrial locomotion such that these adaptations have occurred to the detriment of aquatic locomotion. Explicit tests of functional trade-offs are understudied in all amphibious fishes, including mudskippers (Gibb et al. 2013). While the mudskipper pectoral fins are the most obviously modified locomotor structure, because of locomotor partitioning it is possible that there are no major detrimental ramifications for aquatic locomotion. In contrast, the axial musculature of mudskippers generates several different types of movement in both environments (Figure 8.4). Intriguingly, there is some evidence that the axially powered aquatic escape response performance of *Ps. argentilineatus* may be impaired (Figure 8.7; Gibb et al. 2013). When comparing several similarly sized fishes that exhibit variation in terrestrial behavior, it was found that increasing terrestriality was

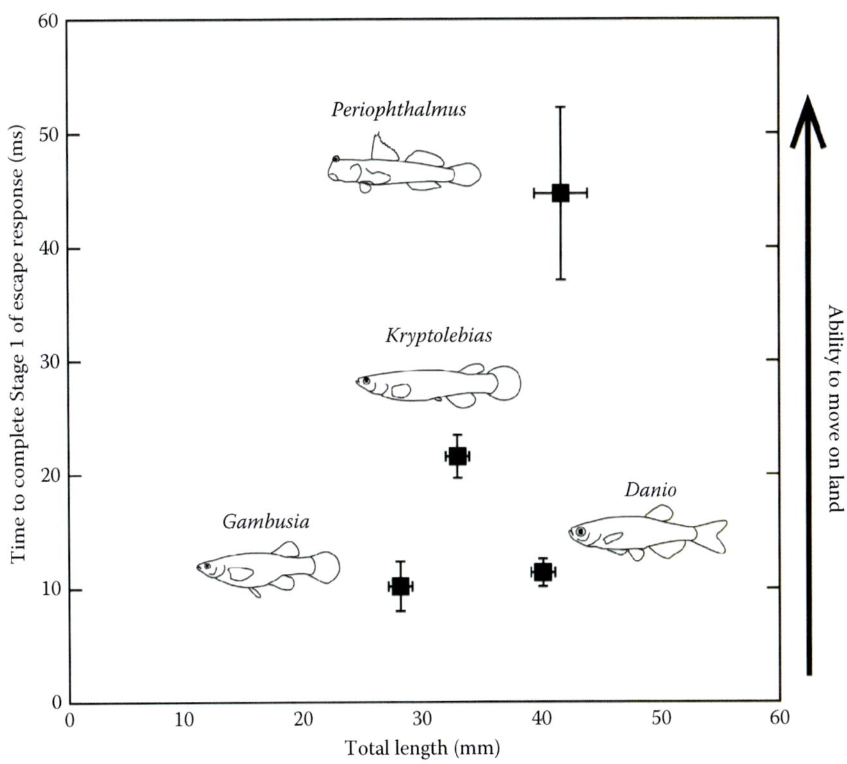

Figure 8.7 Variation in timing of the average preparatory phase (Stage 1) of the aquatic escape response of four similarly sized teleost fish. As fish species increase in terrestriality, the preparatory phase of the aquatic escape response lengthens, such that the highly amphibious *Periophthalmus argentilineatus* exhibits the longest preparatory phase. This increased delay in successfully executing an escape response could have negative consequences for individual fish. This data suggests that there may be a functional trade-offs for increased terrestrial locomotor performance. (Reproduced from Gibb, A.C. et al., *Integrative and Comparative Biology*, 53, 295–306, 2013. With permission.)

accompanied by a longer aquatic C-start preparatory phase (Gibb et al. 2013). The mudskipper genera—*Boleophthalmus*, *Periophthalmodon*, *Periophthalmus*, and *Scartelaos*—with their hypothesized gradient in terrestriality (Polgar and Crosa 2009; Murdy 1989) have the potential to be an excellent model system to test hypotheses on these functional trade-offs.

8.5 CONCLUSION

Mudskippers exhibit variation in the extent of terrestrial behavior among species. Among the four genera discussed above, *Periophthalmus* and *Periophthalmodon* are considered the most derived and the most terrestrial genera, *Scartelaos* is the most basal and the least terrestrial, and *Boleophthalmus* is intermediate in both evolutionary position and terrestriality (Polgar et al. 2010; Polgar and Crosa 2009; Murdy 1989). It has also been hypothesized that parallels of adaptive radiation for increasing terrestriality may have occurred within each genus (Polgar et al. 2010). Against this backdrop of variation in habitat use, our current understanding of corresponding locomotor variation is rather limited. Most of our understanding of mudskipper locomotion comes from two species, both in the genus *Periophthalmus*, and quantified kinematic information is missing entirely for several behaviors (Table 8.1). Expanding our knowledge of variation in locomotor performance, and how that variation connects to both habitat use and morphological variation among and within the mudskipper genera, could be very informative toward understanding both mudskipper evolutionary relationships and ecological patterns of habitat use.

Understanding mudskipper locomotion can also provide insights into tetrapod locomotor evolution. There are always caveats to utilizing modern analogs as a window to the past (Pierce et al. 2013). However, some questions about the evolutionary transitions of tetrapods can only be inferred from studying extant organisms (Graham and Lee 2004; Long and Gordon 2004; Graham 1997). There exists a continuum of locomotor stages in the water-to-land transition of tetrapods (Long and Gordon 2004) and while mudskippers are perhaps inappropriate functional analogs for crown-group tetrapods, they may prove informative in understanding the functional demands faced by some stem-group tetrapods (Pierce et al. 2013). It was recently posited that, based on three dimensional limb-joint modeling, the most likely locomotor mode for the pectoral appendages of *Ichthyostega* may be mudskipper-like crutching (Pierce et al. 2012). Another recent study demonstrated that locomotor endurance and postexercise recovery of *Ps. modestus* was enhanced by hyperoxia, suggesting that the rise of atmospheric oxygen in the middle and late Paleozoic could have increased the terrestrial performance of early tetrapods (Jew et al. 2013). Ground-reaction-force data from *Ps. barbarus* suggests that an evolutionary transition from mudskipper-like crutching to a sprawling limbed posture may have accompanied a change in the orientation of the medial GRF, which can impact weight bearing capacities of the appendicular system and associated stresses and safety factors of limb/fin elements (Kawano and Blob 2013). A recent study has even utilized a mudskipper-inspired robot to ask functional questions about tail movements assisting locomotion up slopes on granular media for early tetrapods (McInroe et al. 2016). While we have developed a basic understanding of mudskipper locomotion, it is clear that future mudskipper locomotor research could progress our understanding of how different locomotor patterns evolve, how afferent sensory information is integrated with our neuromuscular system to produce different locomotor outputs, and even inform us on the water-to-land transition undergone by extinct tetrapods.

ACKNOWLEDGMENTS

Thank you to Alice Gibb, Zeehan Jaafar, Jamie Lamit, Richard Mleczko, and Edward Murdy for improving the quality of this chapter by providing instructive feedback on the earlier drafts.

REFERENCES

Ashley-Ross M.A., Hsieh S.T., Gibb A.C. and Blob R.W. 2013. Vertebrate land invasions–past, present and future: An introduction to the symposium. *Integrative and Comparative Biology* 53:192–196.

Biewener A.A. 2003. *Animal Locomotion*. Oxford: Oxford University Press, 296 pp.

Biewener A.A. and Gillis G.B. 1999. Dynamics of muscle function during locomotion: Accommodating variable conditions. *Journal of Experimental Biology* 202:3387–3396.

Blob R.W., Rai R., Julius M.L. and Schoenfuss H.L. 2006. Functional diversity in extreme environments: Effects of locomotor style and substrate texture on the waterfall-climbing performance of Hawaiian gobiid fishes. *Journal of Zoology* 268:315–324.

Clardy T.R. 2012. Aquatic and terrestrial locomotion of the rock prickleback, *Xiphister mucosus* (Cottiformes: Zoarcoidei: Stichaeidae). *Northwestern Naturalist* 93:203–210.

Clayton D.A. 1993. Mudskippers. *Oceanographic Marine Biology Annual Review* 31:507–577.

Clayton D.A. and Snowden R. 2000. Surface activity in the mudskipper, *Periophthalmus waltoni* Koumans 1941 in relation to prey activity and environmental factors. *Tropical Zoology* 13:239–249.

De J.K. and Nandi N.C. 1984. A note on the locomotory behaviour of the mudskipper *Boleophthalmus boddarti*. *Indian Journal of Fisheries* 31:407–409.

Dijk D.E. 1960. Locomotion and attitudes of the mudskipper, *Periophthalmus*, a semi-terrestrial fish. *South African Journal of Science* 56:158–162.

Domenici P. and Blake R.W. 1997. The kinematics and performance of fish fast start swimming. *Journal of Experimental Biology* 200:1165–1178.

Drucker E.G., Walker J.A. and Westneat M.W. 2006. Mechanics of pectoral fin swimming in fishes. In *Fish Biomechanics, vol. 23* (eds. R.E. Shadwick and G.V. Lauder). San Diego, CA: Elsevier.

Eggert B. 1929. Die Gobiidenflosse und ihre Anpassung an das Landleben. *Zeitschrift für wissenshaftliche Zoologie* 133:411–440.

Gibb A.C., Ashley-Ross M.A. and Hsieh S.-T.T. 2013. Thrash, flip, or jump: How do new behaviours evolve when teleost fishes are confronted with a novel environment? *Integrative and Comparative Biology* 53:295–306.

Gillis G.B. 1998. Environmental effects on undulatory locomotion in the American eel *Anguilla rostrata*: Kinematics in water and on land. *Journal of Experimental Biology* 201:949–961.

Gillis G.B. 2000. Patterns of white muscle activity during terrestrial locomotion in the American eel (*Anguilla rostrata*). *Journal of Experimental Biology* 203:471–480.

Gillis G.B. and Blob R.W. 2001. How muscles accommodate movement in different physical environments: Aquatic vs. terrestrial locomotion in vertebrates. *Comparative Biochemistry and Physiology. Part A* 131:61–75.

Graham J.B. 1997. *Air-Breathing Fishes: Evolution, Diversity, and Adaptation*. San Diego: Academic Press, 324 pp.

Graham J.B. and Lee H.J. 2004. Breathing air in air: In what ways might extant amphibious fish biology relate to prevailing concepts about early tetrapods, the evolution of vertebrate air breathing, and the vertebrate land transition. *Physiological and Biochemical Zoology* 77:720–731.

Harris V.A. 1960. On the locomotion of the mud-skipper *Periophthalmus koelreuteri* (Pallas):(Gobiidae). *Proceedings of the Zoological Society of London* 134:107–135.

Horner A.M. and Jayne B.C. 2014. Lungfish axial muscle function and the vertebrate water to land transition. *PLoS ONE* 9:e96516. doi:10.1371/journal.pone.0096516

Hsieh S.-T.T. 2010. A locomotor innovation enables water-land transition in a marine fish. *PLoS ONE* 5:e11197. doi:10.1371/journal.pone.0011197

Ip Y.K., Chew S.F., Lim A.L.L. and Low W.P. 1990. The Mudskipper. In *Essay in Zoology: Papers Commemorating the 40th Anniversary of the Department of Zoology* (eds. L.M. Chou and P.K.L. Ng) pp. 85–95. Singapore: Department of Zoology, National University of Singapore.

Ishimatsu A. and Gonzales T.T. 2011. Mudskippers: Front runners in the modern invasion of land. In *The Biology of Gobies* (eds. R.A. Patzner, J.L. Van Tassell, M. Kovačić and B.G. Kapoor) pp. 609–638. Boca Raton: CRC Press/Taylor and Francis.

Jew C.J., Wegner N.C., Yanagitsuru Y., Tresguerres M. and Graham J.B. 2013. Atmospheric oxygen levels affect mudskipper terrestrial performance: Implications for early tetrapods. *Integrative and Comparative Biology* 53:248–257.

Kawano S.M. and Blob R.W. 2013. Propulsive forces of mudskipper fins and salamander limb during terrestrial locomotion: Implications for the invasion of land. *Integrative and Comparative Biology* 53:283–294.
Lee H.J., Martinez C.A., Hertzberg K.J., Hamilton A.L. and Graham J.B. 2005. Burrow air phase maintenance and respiration by the mudskipper *Scartelaos histophorus* (Gobiidae: Oxudercinae). *Journal of Experimental Biology* 208:169–177.
Long J.A. and Gordon M.S. 2004. The greatest step in vertebrate history: A paleobiological review of the fish-tetrapod transition. *Physiological and Biochemical Zoology* 77:700–719.
Martin K.L. 1995. Time and tide wait for no fish: Intertidal fishes out of water. *Environmental Biology of Fishes* 44:165–181.
McInroe B., Astley H.C., Gong C. et al. 2016. Tail use improves performance on soft substrates in models of early vertebrate land locomotors. *Science* 353:154–158.
Murdy E.O. 1989. A taxonomic revision and cladistic analysis of the oxudercine gobies (Gobiidae: Oxudercinae). *Records of the Australian Museum Supplement* 11:1–93.
Pace C.M. and Gibb A.C. 2005. Do mudskippers really "crutch"? Testing the inverted pendulum model in an amphibious fish. *Integrative and Comparative Biology* 44:P1.42.
Pace C.M. and Gibb A.C. 2009. Mudskipper pectoral fin kinematics in aquatic and terrestrial environments. *Journal of Experimental Biology* 212:2279–2286.
Pace C.M. and Gibb A.C. 2011. Locomotor behavior across an environmental transition in the ropefish, *Erpetoichthys calabaricus. Journal of Experimental Biology* 214:530–537.
Pace C.M. and Gibb A.C. 2014. Sustained periodic terrestrial locomotion in air-breathing fishes. *Journal of Fish Biology* 84:639–660.
Perlman B.M. and Ashley-Ross M.A. 2016. By land or by sea: A modified C-start motor pattern drives the terrestrial tail-flip. *Journal of Experimental Biology* 219:1860–1865. doi:10.1242/jeb.128744
Pierce S.E., Clack J.A. and Hutchinson J.R. 2012. Three-dimensional limb joint mobility in the early tetrapod *Ichthyostega. Nature* 486:523–526.
Pierce S.E., Hutchinson J.R. and Clack J.A. 2013. Historical perspectives on the evolution of tetrapodomorph movement. *Integrative and Comparative Biology* 53:209–223.
Polgar G. and Crosa G. 2009. Multivariate characterisation of the habitats of seven species of Malayan mudskippers (Gobiidae: Oxudercinae). *Marine Biology* 156:1475–1486.
Polgar G., Sacchetti A. and Galli P. 2010. Differentiation and adaptive radiation of amphibious gobies (Gobiidae: Oxudercinae) in semi-terrestrial habitats. *Journal of Fish Biology* 77:1645–1664.
Pronko A.J., Perlman B.M. and Ashley-Ross M.A. 2013. Launches, squiggles and pounces, oh my! The water-land transition in mangrove rivulus (*Kryptolebias marmoratus*). *Journal of Experimental Biology* 216:3988–3995.
Sayer M.D. 2005. Adaptations of amphibious fish for surviving life out of water. *Fish and Fisheries* 6:186–211.
Stebbins R.C. and Kalk M. 1961. Observations on the natural history of the mudskipper *Periophthalmus sobrinus. Copeia* 18–27.
Swanson B.O. and Gibb A.C. 2004. Kinematics of aquatic and terrestrial escape responses in mudskippers. *Journal of Experimental Biology* 207:4037–4044.
Webb P.W. 1978. Fast-start performance and body form in seven species of teleost fish. *Journal of Experimental Biology* 74:211–226.
Webb P.W. 1984. Body form, locomotion, and foraging in aquatic vertebrates. *American Zoologist* 24:107–120.
Zander C.D. 2011. Morphological adaptions to special environments of gobies. In *The Biology of Gobies* (eds. R.A. Patzner, J.L. Van Tassell, M. Kovačić and B.G. Kapoor) pp. 345–366. Boca Raton: CRC Press/Taylor and Francis.

CHAPTER 9

Review of Reproductive Strategies

Karen L. M. Martin and Atsushi Ishimatsu

CONTENTS

9.1 INTRODUCTION

Oxudercine mudskippers engage in many social behaviors out of water, including courtship and territorial defense (Graham 1997; Clayton 1993; Murdy 1989; Clayton and Vaughan 1986). These fishes are highly mobile and are capable navigators that show fidelity to burrows and habitats (Gibson 1999). While most intertidal fishes are exposed to terrestrial conditions only during low tides, some mudskipper species from the genus *Periophthalmus* remain emerged even during high tides (Takita et al. 2011; Ikebe and Oishi 1996). Different intertidal fish species have varying tolerances to emersion, thus influencing microhabitat choice (Martin 2014; Horn and Martin 2006).

Residence habitat of intertidal fishes may change in tidal height, with temperature, or by season, and may differ between sexes (Zander et al. 1999).

In many intertidal fish species, spawning occurs underwater and is synchronized around the time of the highest tides. This allows their clutches to be deposited in the higher intertidal zone to permit exposure of the developing embryos to air by an ebbing tide, either occasionally or continuously (Martin et al. 2004; Martin and Swiderski 2001; DeMartini 1999; Tewksbury and Conover 1987; Taylor et al. 1977; Jones 1972). Incubation in air has many advantages for embryo development, including high oxygen availability and temperature modification (Martin and Carter 2013; Martin and Strathmann 1999). This is exemplified in the location of nests at species-specific tidal heights in beach-spawning teleosts such as *Takifugu niphobles* (Yamahira 1996), *Clinocottus acuticeps* (Marliave 1981), *Leuresthes tenuis* (Martin et al. 2004), and many other species (Martin 2015).

In mudskippers, courtship occurs only during low tide (Figure 9.1a,b), and the mating pair enters a burrow while the mudflat is exposed (Matoba and Dotsu 1977). They remain inside the burrow for several hours, during which time the incoming tide may cover the mudflat surface. Spawning may take place within the nesting chamber above the water surface, as deduced from the presence of air in the spawning chamber of a burrow prior to spawning, and therefore may not be closely synchronized with tide (Ishimatsu et al. 2007). The female leaves the burrow after the spawning event and the male begins guarding the burrow (R. Ono and A. Ishimatsu, unpublished data). Embryonic development occurs in air stored in a spawning chamber within the burrow. Parental care comprises continuous addition of fresh air into the spawning chamber during low tides, but terminates with submersion of hatch-competent embryos to trigger hatching during high tide (see Section 9.7).

9.2 WATER AND AIR AS RESPIRATORY MEDIA

Air contains roughly 30 times more oxygen per unit volume than water, when air and water are equilibrated. The volumetric composition of the atmosphere is at present about 79% nitrogen, 20.95% oxygen, 0.04% carbon dioxide, and small volumes of several other biologically inert gases such as argon, helium, and neon. When in equilibrium with the atmosphere, the partial pressure of dissolved oxygen in water is by definition identical to the partial pressure of atmospheric oxygen. However, oxygen concentration is much lower in water because of the lower solubility of oxygen in water than in air (Dejours 1981). The gas composition of air is usually highly stable due to a much higher diffusion coefficient, which will quickly dissipate any local concentration gradient. In addition, air is 800 times less dense and 60 times less viscous than water (Dejours 1981). In contrast, oxygen and carbon dioxide concentrations in water are highly variable. Aquatic hypoxia and hypercapnia characterize many aquatic habitats, for example, tropical freshwater swamps and ponds, creeks through mangrove forests, and bottom layers of enclosed bays in summer (see below). This is mainly due to the lower solubility and diffusion coefficient of oxygen in water, which retards oxygen supply from higher oxygen zones unless convective water movements by waves, tides, and currents replenish oxygen to hypoxic or anoxic water bodies. Salinity reduces solubility of oxygen in water as it increases, and temperature also has an inverse relationship with oxygen solubility.

Fish eggs are highly metabolically active with a continuous, increasing demand for oxygen (Rambough 1988). Small fish eggs typically have a relatively large surface area per unit volume, which facilitates gas diffusion to the metabolizing cells of embryos. As long as dehydration can be avoided, air offers an excellent environment for embryonic development (Strathmann and Hess 1999). Thus, the use of air as a respiratory medium is advantageous not only for adults but also for embryos of the species that inhabit waters prone to develop hypoxic conditions, such as mudflats. The novel reproductive strategy of filling an egg chamber with air, instead of water, as employed by mudskippers (see Section 9.7), ensures 30 times more oxygen available to embryos in the same

(a)

(b)

Figure 9.1 Courtship displays of oxudercine gobies on mudflats in the Ariake Sea, Japan. (a) A male *Boleophthalmus pectinirostris* jumps to attract females. (Courtesy of Kenshin Nakahama.) (b) A male *Periophthalmus modestus* (foreground) with dorsal fins erect and pale body color, displaying to female (background). (Courtesy of Mizuri Murata.)

volume, and oxygen rapidly diffuses to the surface of egg membranes. Dehydration is avoided because egg chambers are excavated in water-saturated, fine mud sediment (often <10 μm in diameter, Ishimatsu et al. 2007), which is usually regularly inundated by tidal water. Air exposure for egg-enclosed embryos is probably the major selective advantage of beach spawning in fishes, and oviposition that facilitates this is a type of parental care (Martin 2015).

9.3 MARINE HYPOXIA

Evolutionary biologists in the past did not consider hypoxia tolerance for most marine organisms (Romer 1967), because dissolved oxygen in seawater is typically high and constant (Graham et al. 1978). However, hypoxic conditions do occur in the oceans on both small and large scales, and this environmental factor is now considered important in evolution of amphibious fishes (Graham and Lee 2004).

Oceanic hypoxia has occurred through geologic time irrespective of recent human impacts. Hypoxia occurs in huge volumes at the intermediate (several hundred to 1000 m) depths of the oceans (Stramma et al. 2008). Organisms in these areas, known as "oxygen minimum zones," have evolved various adaptations to living in oxygen-poor environment (Childress and Seibel 1998). Moreover, oceanic anoxia is considered to be one potential cause of mass extinctions in the past (Harnik et al. 2012). Coastal hypoxia has been reported from bays and continental shelves in both tropical and temperate zones mainly as a result of eutrophication and other human activities (Diaz 2001).

During the flood tide, the intertidal area may be well oxygenated because of wave action and the connection with the wider ocean. During the ebb tide, oxygen levels in intertidal water pools may still be high during daytime because oxygen production by photosynthesis often exceeds oxygen consumption of resident animals and plants, whereas hypoxia may develop in those pools during night time (Truchot and Duhamel-Jouve 1980). Water inundating mudflats, nearby creeks, and swamps of the tropical regions can also be extremely hypoxic. For example, Okamura et al. (2010) reported concentrations of dissolved oxygen lower than 1 mg/L (oxygen saturation <15%) in the creeks and mudflats in the Matang mangrove estuary in Malaysia. Mazda et al. (1990) also reported the development of severe hypoxia and anoxia in a mangrove area in Iriomote Island, a tropical island in Japan. Wolanski et al. (1992) discussed the hydrodynamics of mangrove forests, and listed vertical salinity gradient, continued input of detritus, and decomposition as the main causes for the development of hypoxic conditions in mangrove systems. These environmental conditions are important factors contributing to the biology and structure of fish communities inhabiting coastal habitats.

9.4 DEVELOPMENT OF FISH EGGS IN AIR

To avoid hypoxic aquatic conditions, developing eggs are exposed to air periodically or permanently. The location for oviposition is in itself parental care as substrate and conditions must be suitable. Some species oviposit at a vertical height on shore, such that the eggs emerge during the ebb tides, coinciding with the times when hypoxia and hypercapnia are most likely to occur (Martin 2015). Some estuarine beach-spawning fishes attach eggs to vegetation such as surfgrass (Middaugh et al. 1983) or deposit eggs in empty mussel shells (Taylor 1999). Not all estuarine fishes spawn on beaches; some release their eggs into the plankton and others attach their eggs onto vegetation in deeper waters.

Placing anamniotic vertebrate embryos in air is not the norm, but it has many benefits, along with some obvious potential disadvantages. Advantages include increased potential for high rates of oxygen diffusion, especially with the reduction of a boundary layer around the eggs, and the possibility of higher incubation temperatures during air exposure than in water (Martin and Strathmann 1999). Both of these physical advantages may increase the rate of embryonic development. Placing the eggs out of water may also help to avoid many aquatic predators (Martin 2015; Martin et al. 2009; Middaugh et al. 1983).

Oviposition in the high intertidal zone presents some obvious disadvantages for both developing embryos and spawning adults, including higher risk of embryo desiccation and exposure to novel terrestrial predators and pathogens (Martin 1999; Strathmann and Hess 1999). The risk of hatching into a hostile terrestrial environment is also a concern (Martin and Carter 2013; Martin 1999). The advantages of spawning on shore must sufficiently outweigh the disadvantages as some species of teleost fishes that typically live in deeper waters spawn on the beach (Martin 2015; Martin et al. 2004; DeMartini 1999).

Developing embryos are protected and self-contained within a zona pellucida or chorion. In many ways, these eggs are "helpless"; they are unable to move the layer of stagnant and oxygen-impoverished water that collects around them, or escape from hypoxic conditions. Embryonic development requires aerobic metabolism, and in general fishes develop rapidly. Placing the eggs so that they incubate in air may be considered a form of parental care that helps to avoid aquatic hypoxia for the embryos and improve their survival.

Avoiding aquatic hypoxia may help to explain the propensity for adhesive eggs that attach on the walls and ceiling of the incubation chamber in mudskippers and other beach-spawning fishes such as blennies and midshipmen. Gaseous exchange in air is necessary and important for the egg incubation, especially at low tide. For example, fishes in the rocky intertidal may nest beneath a boulder that shelters a shallow pool at its base, and attach the eggs to the bottom of the rock, the ceiling of the incubation chamber (Shimizu et al. 2006; Coleman 1999). As the tide recedes, a small amount of water remains that provides humidity. Because it is not in direct contact with the developing embryos or their eggs, oxygen availability is not a problem. The guarding adult may be tolerant of hypoxia (Craig et al. 2014) or may be able to tolerate emersion during low tides (Martin 1996).

Mudskippers avoid some of the risks of air exposure for embryos by laying eggs in burrows (see Section 9.7). Desiccation is less of a problem when the egg chamber is enclosed and surrounded by water-laden mud. In addition, males guard embryos from predators, and this guarding may also protect from pathogens as the eggs may otherwise be subject to fungal infestation (Martin 2015). Mudskippers monitor and maintain the oxygen concentration of the air within the burrow by repeatedly renewing it during incubation. Males actively trigger hatching by submerging the eggs in hypoxic water at an appropriate time to coincide with the rising tides at night. The disadvantage of this extensive parental care of the embryos is its high energetic cost to adults (Jones and Reynolds 1999).

9.5 MUDFLATS AS SPAWNING GROUNDS

Teleost fishes spawn on many different substrates, including gravel beaches, sandy shores, estuaries, tide pools, and mudflats (Martin 2015). Each substrate type has its own characteristics that affect its use by fishes for oviposition. Sand and gravel are porous and hold humidity well while allowing the rapid diffusion of oxygen through air spaces (McLachlan and Brown 2006). Mud becomes anoxic within a few centimeters of the surface and its tiny interstitial spaces allow little diffusion of oxygen from surrounding waters, but its lack of porosity allows animals to construct burrows that are more stable than burrows built in sand (Little 2000).

Mudflats are perceived as a challenging habitat for organisms to inhabit. Organisms must tolerate and adapt to daily cycles of emersion and submersion, large fluctuations in temperature and salinity, sedimentation, high current velocity, ultraviolet irradiation, and extreme oxygen shortages within the substrate (Little 2000). Mudflats are also a formidable place for reproduction; open mudflats extending offshore in estuaries or along the seaward edges of mangrove forests lack cover to protect embryos and egg-guarding parent fishes from direct sunlight, desiccation, and terrestrial predators during low tide (Figure 9.2a) as well as from siltation and displacement caused by strong current during high tide (Figure 9.2b). Tide pools and burrows provide necessary aquatic refuge to marine organisms on mudflats. Even though mudskippers tolerate exposure to air for hours during low tide, they will desiccate without frequent access to water.

9.6 CHANGES IN SPAWNING BEHAVIOR AS GOBIES COME ASHORE

Species within the highly diverse Gobiidae, to which mudskippers belong, often associate with a particular substrate while spawning. Many of these species also show parental care, typically by the male, who may attend to multiple clutches at any one time (Breder and Rosen 1966). The male

Figure 9.2 Mudflat in the Ariake Sea, Japan (a) exposed during low tide and (b) inundated during high tide. (Reproduced from Kaiyusha, Tokyo. With permission.)

typically excavates a space or takes advantage of an existing crevice, to create an area suitable for nesting (Breder and Rosen 1966). He then engages in complex courtship behaviors to attract gravid females (Mazzoldi et al. 2005). In some cases, a different male may attempt to take possession of a nest if the resident male is absent, suggesting that males guarding egg clutches are more attractive to females than are males without active nests (Hastings and Petersen 2010). In some dimorphic gobiid species, alternate male strategies such as "sneaker males" have evolved (Mazzoldi et al. 2005; Immler et al. 2004), wherein a male appears similar to the female. The guarding male may not behave aggressively toward such an innocuous appearing fish, and consequently the sneaker male may be able to surreptitiously fertilize some of the eggs spawned in the nest.

Many gobiid fishes are hermaphroditic (Cole 2010), with reproductive organs capable of becoming either ovaries or testes sequentially, and in some cases both simultaneously (Rodgers et al. 2007). In some species of *Coryphopterus*, individuals mature first as females, then become males later, but contain the precursor of both ovarian and testicular tissues from the start (Cole 2010). Four species of the genus *Eviota* are hermaphrodites, as is the Atlantic *Elacatinus multifasciatus* (Cole 2010). In all, 14 gobiid genera have been shown to have functional hermaphroditism, all within the subfamily Gobiinae. These include *Bryaninops*, *Fusigobius*, *Gobiodon*, *Lophiogobius*, *Luposicya*, *Lythrypnus*, *Paragobiodon*, *Pleurosicya*, *Priolepis*, *Rhinogobiops*, and *Trimma* (Cole 2010). Hermaphroditism, however, has not been observed in mudskippers.

In general, gobies produce many small, demersal eggs that may be attached to a substrate by a short peduncle or filament. Typically, gobies spawn in an inverted position, attaching eggs to the ceiling of a nest chamber, in empty shells of bivalves, under rocks, on the walls of shrimp burrows, on sponges, on algae and seagrass, on root filaments, on decaying wood, and even in old tin cans, bottles, and shoes (Breder and Rosen 1966). If under rocks, the eggs are often placed on the ceiling of a chamber formed by a space between a rock and the floor of the river, stream, or ocean bottom. In some gobiids, accessory male reproductive structures allow the application of a "sperm trail" on a spawning surface from which the spermatozoa can slowly emerge to fertilize eggs oviposited subsequently on that location (Scaggiante et al. 1999; Marconato et al. 1996). Egg incubation periods in gobies range from 1 to 20 days (Borges et al. 2011).

The habit of burrow spawning in some subtidal gobies suggests that ancestral species of mudskippers had probably developed this behavior before their evolution onto land. The behavior was further adapted so that larvae emerge at the appropriate time during the tidal cycle (see Section 9.7.3). It should be noted that all oxudercine goby species studied to date spawn in burrows. Following are some examples of the varieties of reproduction starting from aquatic gobies through amphibious oxudercine gobies.

9.6.1 Non-Oxudercine Gobies

9.6.1.1 Nesting on the Sea Bottom

Similar to many goby species, *Sicyopterus micrurus* lays eggs on the underside of rocks (Manacop 1953). Even though spawning in flowing freshwater streams, this species follows a periodicity related to the full moon and the tidal cycles of the ocean.

A male *Pomatoschistus minutus* constructs a nest by using an old mussel shell to excavate a hollow in a sandy substrate. He then places the shell over the hollow and covers it with sand to create a protected egg chamber (Breder and Rosen 1966, as *Gobius*). This effort may generate long channels in the substrate radiating out from the nest. The male guards the eggs and aerates them by fanning them. Male *Pomatoschistus microps* builds a marine nest similarly by covering an empty shell with sand and hollowing out the area underneath the shell. Males supply oxygen to incubating embryos by fanning with their fins, increasing the frequency to offset the severity of hypoxia (Jones and Reynolds 1999). The increased fanning effort appears to compromise energy allocation to subsequent reproductive activities (Jones and Reynolds 1999).

9.6.1.2 Burrowing

Several subtidal gobies are known to spawn in burrows in shallow sandy or muddy bottoms, such as *Acanthogobius flavimanus* (Dôtu and Mito 1955), *Acanthogobius hasta* (Kuno and Takita 1997), *Gymnogobius breunigii* (reported as *Chaenogobius castanea*, Dôtu 1954), and *Gymnogobius uchidai* (reported as *Paleatogobius uchidai*, Dôtu 1957a). Their behavior may provide clues for the evolution of the more specialized reproductive behavior of mudskippers in mudflats.

Typhlogobius californiensis lives in pairs in the burrows of a commensal shrimp, *Callianasa affinis* (MacGinitie 1939). The burrows may have many openings and branches that facilitate water flow over the eggs (Dumbauld et al. 1996). Pairs spend a few days in preparation for spawning by cleaning and smoothing out the walls of the burrow, and then eggs are laid by the female individually in a single layer, fertilized by the male, and guarded by both parents (Whiteman and Côté 2004). The developing embryos are fully aquatic, fanned intermittently by both parents throughout their incubation period of 10–12 days. During hatching, both parents attend to the eggs and move over the clutch continuously, prompting MacGinitie (1939) to speculate that this action aids in the process of hatching or facilitates the escape of the embryos from their chorions.

Eucyclogobius newberryi also spawn in burrows (Swenson 1997), but their burrows do not include air chambers (see Section 9.7). These eggs hang from the ceiling of the burrow attached by short filaments, allowing them to move with water flow. Male parental care in this species involves fanning and nudging the eggs; if the male leaves the nest, fungus quickly infects and destroys the eggs (Andrew Kinziger, personal communication, February 2013).

Males and females of the monogamous *Valenciennea longipinnis* work together to excavate a burrow with several openings in the sandy substrate, and only one opening has a conical mound, several centimeters above the substrate. While males provide parental care (fanning and egg cleaning with the mouth) to the eggs attached to the ceiling of the burrow, females maintain the mound to prevent the burrow from collapsing by wave action (Takegaki and Nakazono 1999). As the burrow opening surrounded by the mound is located higher than the others, it sucks burrow water out by the Bernoulli effect, thus accelerating burrow water renewal (Takegaki et al. 1999), as is also known for some shrimp burrows (Allanson et al. 1992).

9.6.2 Oxudercine Gobies

Parapocryptes serperaster builds burrows in mudflats for protection from predators and refuge during low tides, yet does not use these burrows for reproduction (Dinh et al. 2014a). Their U- or W-shaped burrows do not change form between seasons, as do those of mudskippers that reproduce in burrows.

Pseudapocryptes elongatus may be semi-catadromous in Vietnam (Bucholtz et al. 2009). Based on analysis of ovarian development, female adults living in canals and estuarine areas migrate to mangrove areas with higher water salinities to build nesting burrows. Post-larvae, juveniles, and adults are found in estuarine areas although eggs and larvae have not been discovered in the same area. The fecundity of *Pd. elongatus*, as indicated by oocyte numbers, was potentially higher than other mudskipper species (Bucholtz et al. 2009). In Vietnam, the aquaculture of this species involves the growing out of wild-caught juveniles (Minh et al. 2010).

Of the four genera of amphibious oxudercine gobies, *Boleophthalmus* and *Scartelaos* are considered more aquatic, whereas *Periophthalmodon* and *Periophthalmus* are regarded more terrestrial, and therefore we present them in this order. Of these four genera, all species studied to date spawn inside their burrows (see Figures 9.3, 9.4a–k; Table 9.1).

9.6.2.1 Species of Boleophthalmus

Breder and Rosen (1966) stated that Harms (1935) thought that *Boleophthalmus boddarti* was viviparous, but neither viviparity nor internal fertilization has been confirmed for this species. Jones (1937) was able to strip eggs from *B. boddarti*, fertilize them, and initiate development. In the Mekong Delta, Vietnam (A. Ishimatsu, personal observation, February 2016) and Bangladesh (Mst. Kaniz Fatema, personal communication, December 2015), *B. boddarti* is consumed by local communities.

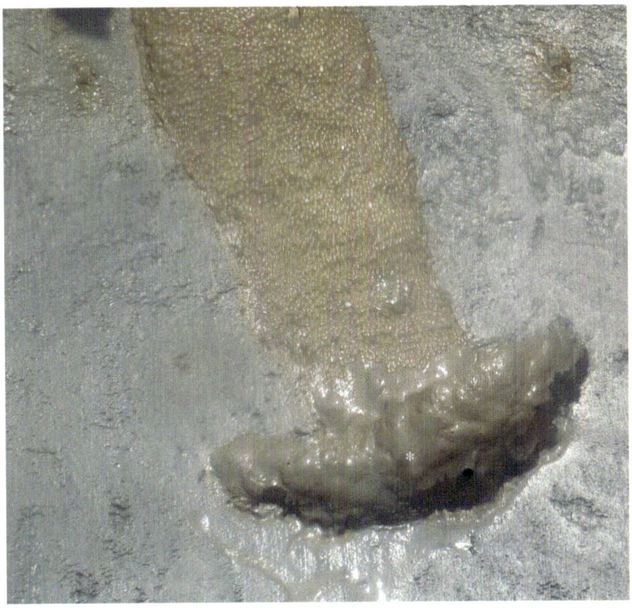

Figure 9.3 Eggs of *Periophthalmus modestus* on the wall of a spawning chamber excavated and longitudinally halved, Ariake Sea, Japan. Asterisk (*) indicates the bottom-most enlarged cavity. (Reproduced from Kaiyusha, Tokyo. With permission.)

Boleophthalmus dussumieri from the western coast of India spawn only once per year (Chung et al. 1991), but on the coast of Pakistan may spawn twice in a year (Hoda 1986). In Kuwait, *B. dussumieri* constructs mud-walled territories and defends them vigorously (Clayton and Vaughan 1986, reported as *B. boddarti*). Fecundity, as indicated by oocyte numbers, is 1,028–7,197 as estimated by Mutsaddi and Bal (1970) or 970–4,113 by Hoda and Akhtar (1985).

In China, *B. pectinirostris* can be induced to spawn by exogenous injection of hormones (Hong and Zhang 2004). Under natural conditions, the reproductive hormones estrogen (E_2), 17-alpha hydroxyprogesterone (17-αOHP), testosterone (T), and 11-ketotestosterone (11-KT) follow a semi-lunar cycle during the spawning season of May and June (Wang et al. 2008), possibly synchronized by the moon phase and/or the tidal regime. Males spawn repeatedly during the season but females spawn only once (Xie and Zhang 1990). In contrast, Washio et al. (1993) concluded that *B. pectinirostris* in the Ariake Sea, Japan, spawn several times in a single reproductive season based on histological examination of the ovaries. Oocyte numbers were 10,000–23,000 for *B. pectinirostris* (Zhang et al. 1989). For aquaculture, wild-caught fry are reared in captivity. With human chorionic gonadotropin (HCG) and luteinizing hormone releasing hormone (LHRH), captive adults were induced to spawn in artificial nests within a few days. After nest spawning and artificial fertilization, eggs were incubated on substrates to which they adhered in aerated currents of seawater.

Eggs of *B. pectinirostris* hatched between 96 and 168 hours after fertilization at 23–28°C, but hatching rates were relatively low (4–21% in artificial nests, up to 76% in buckets) even though the embryos appeared to be fully developed (Hong and Zhang 2004). The authors suggest that increased aeration, or mechanical stimulation by stirring the water, could have induced or assisted hatching for some embryos (Hong and Zhang 2004). Perhaps some form of parental behavior may also be a cue for hatching.

The semi-lunar rhythm of the spawning cycle may be related to the expression of melatonin receptors in the ovaries of *B. pectinirostris* (Hong et al. 2014). The source is the hypothalamic-pituitary axis, acting on receptors that peak during the part of the lunar cycle associated with

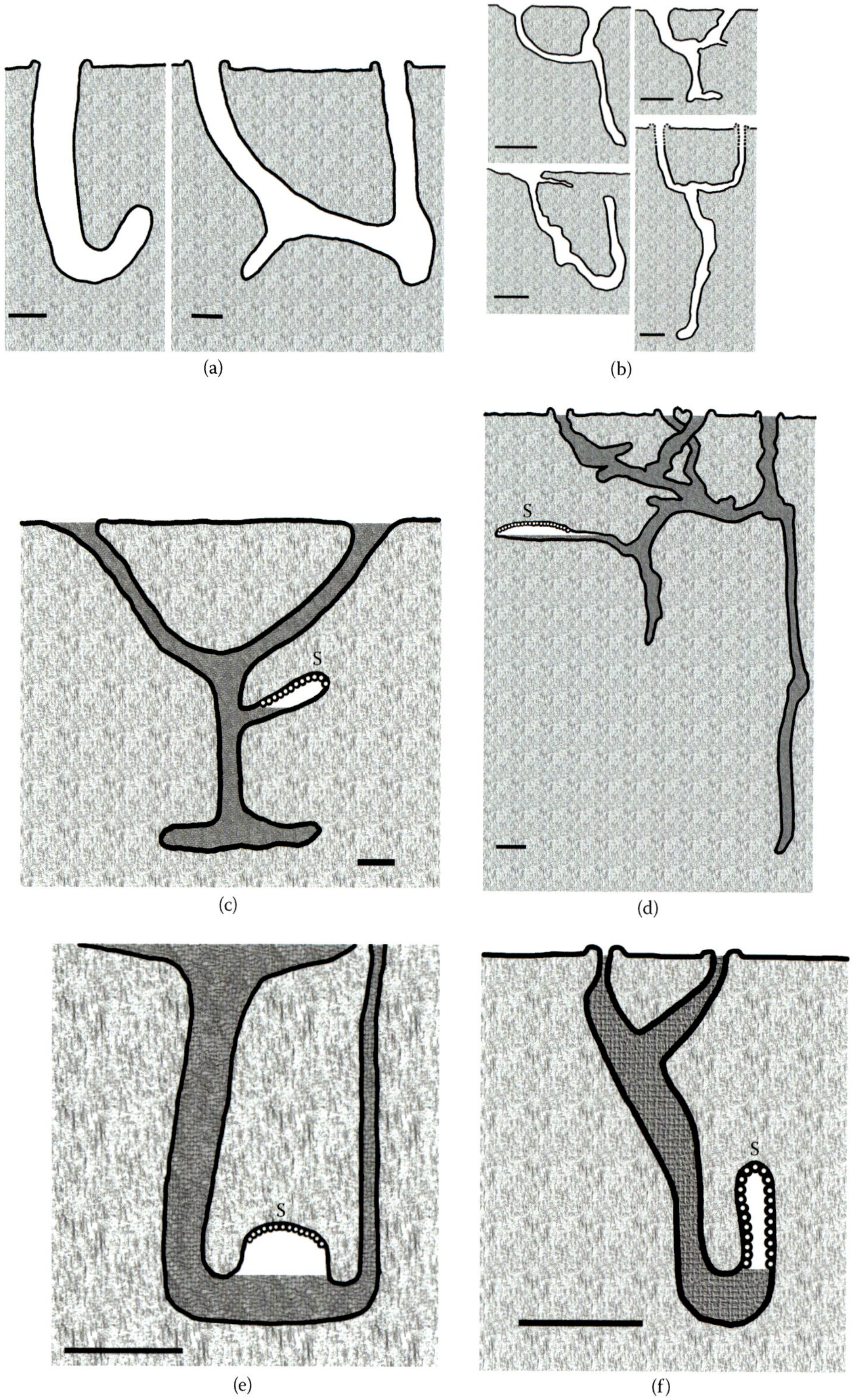

Figure 9.4 Variation in burrows constructed by oxudercine gobies. *(Continued)*

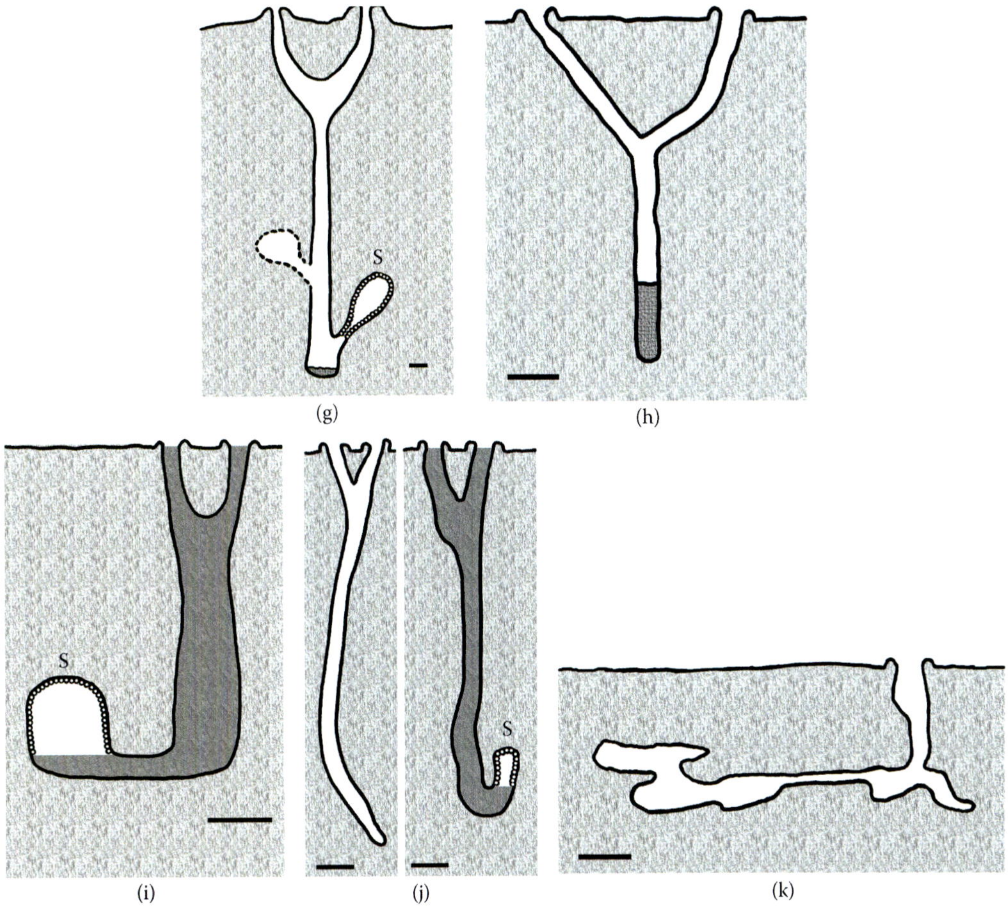

Figure 9.4 (*Continued*) Variation in burrows constructed by oxudercine gobies. (a) Two burrow types of *Boleophthalmus boddarti*, Soc Trang Province, Vietnam, scales = 5.5 cm (left) and 6.0 cm (right). (Modified from Dinh Q.M. et al., *Kasetsart University Fisheries Research Bulletin*, 38, 1–9, 2014b. With permission.) (b) Four burrow types of *Boleophthalmus dussumieri*, Sulaibikhat Bay, Kuwait, scale = 20 cm. (Modified from Clayton D.A. and Vaughan T.C., *Journal of Zoology*, 209, 501–519, 1986. With permission.) (c) Burrow of *Boleophthalmus pectinirostris* during reproductive season, Funing Bay, China, scale = 2 cm. (Modified from Hong W. et al., *Acta Oceanologica Sinica*, 26, 72–81, 2007. With permission.) (d) Burrow of *Boleophthalmus pectinirostris* during the reproductive season, Ariake Sea, Japan, scale = 10 cm. (Modified from Toba A. and Ishimatsu A., *Journal of Fish Biology*, 84, 774–793, 2014. With permission.) (e) Burrow of *Periophthalmodon schlosseri*, Penang, Malaysia, scale = 30 cm. (Modified from Ishimatsu A. et al., *Ichthyological Research*, 56, 417–420, 2009. With permission.) (f) Burrow of *Periophthalmus modestus*, Ariake Sea, Japan, scale = 10 cm. (Modified from Matoba M. and Dotsu Y., *Bulletin of the Faculty of Fisheries, Nagasaki University*, 43, 23–33, 1977. With permission.) (g) Burrow of *Periophthalmus argentilineatus*, western coast of Madagascar, scale = 2 cm (Modified from Brillet C., *Revue d'Ecologie: La Terre et la Vie*, 30, 465–483, 1976. With permission.) (h) Burrow of *Periophthalmus chrysospilos*, eastern coastline of the Kuwait Bay, scale = 12 cm. (Modified from El-Ziady S. et al., *Bulletin of Faculty of Science, Cairo University*, 48, 113–133, 1979. With permission.) (i) Burrow of *Periophthalmus magnuspinnatus*, Suncheonman Bay, Korea, scale = 5 cm. (Modified from Baeck G.W. et al., *Ichthyological Research*, 55, 43–52, 2008. With permission.) (j) Two burrow types of *Periophthalmus minutus* during reproductive (right) and non-reproductive (left) seasons, scale = 5 cm. (Modified from Takeda T. et al., *Australian Journal of Zoology*, 59, 312–320, 2011. With permission.) (k) Burrow of *Scartelaos gigas* is more extensive horizontally than vertically, scale = 10 cm. (Modified from Kim J.K. et al., *Fisheries and Aquatic Sciences*, 14, 403–410, 2011. With permission.) The letter "S" denotes the spawning chamber in c through g and i through j; open circles denote eggs in c through g and i through j; darkened area within burrow indicates presence of water in c through j, water level in j (left) varies in the duration of mudflat immersion during the dry season, the water levels in a and b were not stated in the original publications.

Table 9.1 Burrow Excavation, Air Deposition, and Spawning in Oxudercine Gobies

Species	Burrow Excavation		Burrow Air Deposition		Burrow Spawning	Note	References
	RS	*Non-RS*	*RS*	Non-RS			
Boleophthalmus birdsongi	Yes*		Unknown	Unknown	Unknown		Takita et al. (2011)
Boleophthalmus boddarti	Yes*		Unknown	Unknown	Unknown	Fish with stage V gonads were caught by hand (from within a burrow?).	Dinh et al. (2014b)
Boleophthalmus caeruleomaculatus	Yes*		Unknown	Unknown	Unknown		Takita et al. (2011)
Boleophthalmus dussumieri	Yes*		Unknown	Unknown	Unknown		Clayton and Vaughan (1986)
Boleophthalmus pectinirostris	Yes	Yes	Yes	Yes	Yes	Both males and females have burrows throughout the year. Air volume is low in winter.	Toba and Ishimatsu (2014), Hong et al. (2007)
Periophthalmodon freycineti	Yes*		Inferred from behavior	Unknown	Probably yes	We observed at Site C, that pairs of *Pn. freycineti* alternately gulped air and dove into their burrows, suggesting air storage in egg chambers and biparental embryonic care in both species.	Takita et al. (2011)
Periophthalmodon schlosseri	Yes	Yes	Yes	Yes	Yes	Gas was collected in November and December (non-breeding season) in Ishimatsu et al. 1998.	Ishimatsu et al. (1998, 2009)
Periophthalmus argentilineatus	Yes*		Unknown	Unknown	Unknown		Takita et al. (2011), Brillet (1976)
Periophthalmus chrysospilos	Yes*		Unknown	Unknown	Unknown		El-Ziady et al. (1979)
Periophthalmus darwini	Yes*		Unknown	Unknown	Unknown		Takita et al. (2011)
Periophthalmus minutus	Yes	Yes	Stated but no evidence	Probably no	Yes		Takeda et al (2011)
Periophthalmus magnuspinnatus	Yes	Yes	Stated but no evidence	Unknown	Yes	The air stored by the male was confirmed in all spawning chambers with eggs laid	Baeck et al. (2008)

(Continued)

Table 9.1 (*Continued*) **Burrow Excavation, Air Deposition, and Spawning in Oxudercine Gobies**

	Burrow Excavation		Burrow Air Deposition				
Species	***RS***	***Non-RS***	***RS***	***Non-RS***	**Burrow Spawning**	**Note**	**References**
Periophthalmus modestus	Yes	Yes	Yes	Unknown	Yes		Ishimatsu et al. (2007), Matoba and Dotsu (1977), Kobayashi et al. (1971)
Periophthalmus novaeguineaensis	Yes*		Unknown	Unknown	Unknown		Takita et al. (2011)
Scartelaos histophorus	Yes	Unknown	Stated but no evidence	Unknown	Yes	Larger males were territorial, but paired females were not observed to defend territories (see Takita et al. 2011).	Okinawa Prefecture (2000)
Scartelaos gigas	Yes	Unknown	Stated but no evidence	Unknown	Yes	Although Kim et al. (2011: fig. 7c) stated that air in the spawning burrow provides enough oxygen to the eggs, they provided no evidence to support this claim	Kim et al. (2011)

RS, reproductive season.
*Burrow excavation was observed but the distinction between RS and non-RS was not stated.

spawning behavior. Melatonin triggers production of 17α,20β-dihydroxy-4-pregnen-3-one (DHP), a maturation hormone, in the ovaries and may help to synchronize the reproductive rhythms. In Japan, China, Korea, and Taiwan, *B. pectinirostris* is exploited for food; catch levels have decreased and its habitat in some areas underwent degradation due to seawater pollution (Hong and Zhang 2004). This species can live up to 7 years (Nanami and Takegaki 2005).

9.6.2.2 *Species of* Scartelaos

Scartelaos gigas in Korea spawns between May and August (Kim et al. 2011), as its gonadosomatic index (GSI) is highest in May and summer is when growth rings are laid down in its bones (Park et al. 2008). In cold winter periods below 10°C, this species is not visible in the habitat and may hide underground (Park et al. 2008). *Scartelaos gigas* builds L-shaped burrows with a spawning chamber in the mud (Figure 9.4k), and the male guards the eggs, sometimes in conjunction with the female (Kim et al. 2011). Although the authors stated, "We newly found that air exists in the spawning burrow (Kim et al. 2011: fig. 7C) and may supply enough oxygen to the eggs," they gave no evidence to support this observation, as the figure being referred to is a photograph of newly hatched larvae. Eggs, numbering 1,300–18,300, eggs adhere to the ceiling of the nest chamber with fine filaments. This species is considered "at risk" because of over-exploitation and habitat loss. A closed season for fishing during the reproductive season is recommended (Kim et al. 2011).

Scartelaos histophorus in Queensland, Australia, shows sexual dimorphism, with males being larger with white-eye borders and smaller females showing black-eye borders (Townsend and Tibbetts 2005). Females are gravid from February to April (Townsend and Tibbetts 2005). The first juveniles appear during March and occupy the pneumatophore zone of the mudflat, higher toward shore than the adults (Townsend and Tibbetts 1995). Males and females communicate via dorsal-fin display behavior, and males also do the tail-standing behavior, lifting the anterior portion of the body off the substrate during courtship displays (Townsend and Tibbetts 2005). Between August and October, no *S. histophorus* is visible in the habitat (Townsend and Tibbets 1995).

9.6.2.3 *Species of* Periophthalmodon

In Malaysian mangrove areas and mudflats, *Periophthalmodon schlosseri* exhibits a long spawning season (Mazlan and Rohaya 2008), during which it reproduces repeatedly. Males build and repair burrows for the nests and to house the slightly larger females. The sex ratio is approximately 1:1. Peak GSI occurs in June and October, and spawning was inferred to occur from June to July and again from October to November (Mazlan and Rohaya 2008). Males are monogamous and aggressively defend their mate and the burrow. The female places her eggs on the ceiling and the wall of the spawning chamber that is guarded by the male (Ishimatsu et al. 2009; Tsuhako et al. 2003). Mazlan and Rohaya (2008: 295) suggest that "viviparity in giant mudskippers could be lecithotrophic" but they do not provide any observations to support this idea. Within its burrows, this species stores air (see Section 9.7.1; Ishimatsu et al. 1998, 2009). This species is consumed and used in traditional medicines in Bangladesh (Mst. Kaniz Fatema, personal communication, December 2015), Malaysia and the Mekong Delta, Vietnam (A. Ishimatsu, personal observation, March 2016), impacting wild populations.

9.6.2.4 *Species of* Periophthalmus

Male and female species of *Periophthalmus* can be differentiated by the morphology of the genital papilla (Udo 2002a) as is typical for gobies. The sex ratio of males to females was 1:1.3 for *Periophthlmus barbarus* (Udo et al. 2016). This may indicate that females live longer than males, or that males spend more time in burrows and thus are less likely to be captured. In Nigeria,

Ps. barbarus faces high fishing pressure and was initially thought to spawn only once per year between February and May (Etim et al. 2002). However, in the estuary of the Imo River, Nigeria, this species may breed more than once during the spawning season (Udo et al. 2016), with the GSI peaking from August to December (Etim et al. 2010) at 3.23% of body weight for males and 4.01% for females (Udo et al. 2016). Egg numbers are correlated with body size, and fecundity was between 900 and 24,000 eggs (Udo 2002b). For *Ps. barbarus* (reported as *Periophthalmus papilio*) in Lagos Lagoon, Nigeria, fecundity was between 508 and 15,700 ripe eggs, with a mean of 3,482 ± 102 (Lawson 2011).

Breder and Rosen (1966) cited literature sources, which described *Ps. barbarus* as having internal fertilization and giving birth to live young in burrows during the monsoon season (Harms 1934; Mayer 1929); these claims have not been corroborated. For *Ps. argentilineatus* (reported as *Periophthalmus vulgaris*), Harms (1934: 75, fig. 26) stated "I could observe in the ovary . . . already a young fish with optic vesicles, heart, spinal cord, notochord and somites. These animals already moved weakly within the egg membrane. . ." (translated from German by Ishimatsu). Data and observations on congeners (Section 9.7.1) support the hypothesis of egg laying in burrows followed by external fertilization for *Periophthalmus*.

Periophthalmus magnuspinnatus builds J-shaped burrows during the reproductive season of May through July (Figure 9.4i; Baeck et al. 2008), similar to those seen in *Ps. modestus*. Baeck et al. (2008) suggested that *Ps. magnuspinnatus* burrows are present year-round; the burrows are excavated high in the intertidal zone, and the water within them serves as a refuge after extensive periods of emersion. In preparation for spawning, the male expands the burrow by removing the mud using his mouth and depositing the pellets on the mudflat. He displays for the female by jumping nearly a body length in distance, up to half a body length in height, and quivering the body horizontally while holding the caudal fin erect (Baeck et al. 2008). When she responds, the male reduces the frequency of jumping and quivering as he leads her to his burrow entrance (Baeck et al. 2008). The female pauses at the burrow entrance for some time as the male continues to display. Once the pair bonds, she assists the male in completing the construction of the burrow, which is made longer and more complex. A spawning chamber is constructed within the burrow for egg deposition on the ceiling and upper walls of this structure. Once the eggs are present, the male guards the entrance of the burrow and also takes mouthfuls of air into the spawning chamber repeatedly to displace water and allow the embryos to incubate surrounded by air (Baeck et al. 2008). Juveniles appear on the mudflat in early summer and form loose aggregations while foraging and interacting.

Based on observations from a mudflat surrounded by mangroves near Darwin, Australia, *Ps. minutus* builds J-shaped burrows during the spawning season but straight burrows with no upturn at other times of year (Takeda et al. 2011). This extremely amphibious species takes refuge in its burrow during tidal inundation, but the mudflat in which they live may be exposed to air for more than 20 days during some neap tides (Itoki et al. 2012). The GSI peaked in January and February (the rainy season) for both males and females (Takeda et al. 2011). Fertilized eggs collected by Takeda et al. (2011) developed out of water and some hatched upon immersion in seawater after the incubation period was completed.

Periophthalmus modestus is the best-studied mudskipper with regard to reproductive behavior. *Periophthalmus modestus* builds J-shaped burrows during the spawning season in mid-May to late August in the Ariake Sea, Japan, and eggs are housed within the upturned end in a chamber filled with air (Figures 9.3 and 9.4f; Ishimatsu et al. 2007). The GSI peaks with a semi-lunar rhythm in May and June in the Ariake Sea, Japan, suggesting tidal cues for spawning (Soyano et al. 2008). Kobayashi et al. (1972) was able to raise artificially fertilized eggs (reported as *Ps. cantonensis*) which hatched 7 days after the fertilization at 19–20°C. The yolk is completely absorbed within 4 days after hatching (Kobayashi et al. 1972). The larvae spent 30 days in plankton, after which they metamorphosed to juveniles and became demersal. The fish started to emerge from water at about 50 days after hatching. Detailed accounts of the egg-care and hatch-induction behavior by *Ps. modestus* are given below.

Baeck et al. (2008) speculated that *Ps. modestus* does not maintain burrows year-round because they take refuge in the burrows of other species, including crabs, when threatened or stressed. However, burrows of *Ps. modestus* have been found throughout the year in breeding, non-breeding, and overwintering seasons in both the Ariake Sea (Kobayashi et al. 1971; Ishimatsu, personal observation, 2016) and Tokyo Bay, Japan (Shingo Tanabe, unpublished data). Peak GSI occurs in May and spawning is inferred from April to September in the southern coast of Korea, with fecundity of 1,664–13,428 (Kim et al. 2007). *Periophthalmus modestus* is considered "near threatened" by the Japanese government because of habitat loss due to human activities such as construction of dams and artificial structures near estuaries (Koyama et al. 2016).

In an estuary in the northwestern Arabian Gulf, *Ps. waltoni* spawns twice a year between December and March (Al-Daham and Al-Noor 2000). Fecundity was 11,560–35,005 eggs per female, considered relatively high for gobies, and sex ratios did not differ from 1:1. *Periophthalmus waltoni* builds shelters or burrows with two entrances, but it is unknown if these are used in reproduction (Al-Daham and Al-Noor 2000).

9.7 MUDSKIPPER REPRODUCTION ON THE MUDFLATS

9.7.1 Spawning in Burrows

A large variation in burrow shape and form exists among the genera of amphibious oxudercine gobies (Figure 9.4a through k). Whereas burrows in Figure 9.4 represent the typical shape observed for the species in question, burrow shape can be highly variable between individuals of the same species (Figure 9.4b) due to obstacles such as shells or roots of mangrove trees buried in the mud or the depth of the mud layer. For example, Brillet (1976) showed 10 different burrow shapes of *Ps. argentilineatus* (reported as *Periophthalmus sobrinus*) in mud or muddy sand. Whether these reported differences in burrow morphology reflect real variability between species or different sediment properties, or are possibly due to insufficient filling of burrows by a casting material, is currently unknown.

Burrows of different species of *Periophthalmus* are similar in shape (Figures 9.4f through j); typically two openings on the mudflat surface lead to a vertical shaft at a depth of 5–10 cm. During the reproductive season, a spawning chamber is added so that the burrow resembles the letter "J." In some cases, a spawning chamber may also be formed along the vertical shaft (Figure 9.4g). The shape of *Periophthalmodon* burrows is known only for *Pn. schlosseri*, which is U-shaped with a dome-shaped spawning chamber formed between the two vertical shafts (Figure 9.4e). In *Boleophthalmus*, the burrows differ between congeners. Burrows of *B. boddarti* (Figure 9.4a), *B. dussumieri* (Figure 9.4b), and *B. pectinirostris* in Funing Bay, China (Figure 9.4c) have one or two openings on the mudflat surface that lead to the vertical shaft, as in *Periophthalmus* burrows (Figure 9.4b,c), or have a horizontal tunnel, somewhat similar to *Periophthalmodon* burrows (Figure 9.4a right). In contrast, the burrows of *B. pectinirostris* in the Ariake Sea, Japan have two to seven openings and are more complex in the spatial disposition of shafts and tunnels (Figure 9.4d).

Burrows of *Scartelaos* mudskippers are not well studied, but they seem to extend more horizontally than vertically (Kim et al. 2011). Burrows of *Boleophthalmus, Periophthalmodon,* and *Periophthalmus*, on the other hand, are excavated deeper rather than over a wider area. Burrows with a spawning chamber were excavated by the male in *B. pectinirostris* (Toba and Ishimatsu 2014) and *Ps. modestus* (Ishimatsu et al. 2007), but perhaps by both sexes in *Pn. freycineti* (Takita et al. 2011), *Ps. magnuspinnatus* (Baeck et al. 2008), and *Pn. schlosseri* (Ishimatsu et al. 1998).

Burrows must contain sufficient oxygen supply for the successful development of embryos as they are unable to survive prolonged hypoxia. By bringing mouthfuls of air to expel underwater in a J-shaped nursery burrow, mudskippers fill the spawning chamber with a pocket of air, thus causing the eggs to avoid being surrounded by hypoxic water that would otherwise fill the burrow during incubation (Ishimatsu and Graham 2011; Ishimatsu et al. 2000, 2007). Air deposition in burrows for embryonic development is currently confirmed only in three species of mudskippers: *B. pectinirostris*, *Pn. schlosseri*, and *Ps. modestus*.

The strongest evidence for aerial incubation of mudskipper embryos comes from the direct observation of intact spawning chambers of *Pn. schlosseri* burrows using an endoscope (Ishimatsu et al. 2009). By inserting a 10-m-long endoscope into burrows of *Pn. schlosseri* in a mudflat in Malaysia, these authors confirmed that the spawning chambers were filled with air. Of the five burrows in which they observed the spawning chambers, only two had laid eggs in a monolayer on the ceiling of the chamber; the other three did not have eggs. These burrows without eggs were either in a pre-spawning condition or abandoned by the parent fish for some reason, assuming that *Pn. schlosseri* induces embryonic hatching by removing the air and thereby submerging the eggs (as was confirmed for *Ps. modestus*; Section 9.7.2).

There is strong evidence for aerial embryonic incubation in *Ps. modestus* with detailed information on air deposition behavior by a burrow-guarding male (Ishimatsu et al. 2007). Although evidence for *B. pectinirostris* is circumstantial with no direct observations, available data are strong enough to affirm it. Hong et al. (2007) demonstrated that a dye infused into burrows of male *B. pectinirostris* did not stain eggs laid on the upper wall of chambers, and suspected that the chambers were filled with air. Toba and Ishimatsu (2014) collected more than 700 mL of gas from a burrow, which they confirmed contained eggs, and also found the redox potential determined on the upper wall was significantly higher than the value recorded from the base, which indicates that the upper wall had been exposed to a higher level of oxygen than the bottom.

Aerial incubation of embryos can be inferred for other species by the following additional methods. Behavioral observation indicated air deposition in *Pn. freycineti* (Takita et al. 2011), and an upturned terminal chamber suggestive of air deposition was confirmed by either digging (Figure 9.4g, for *Ps. argentilineatus* reported as *Ps. sobrinus*, Brillet 1976 and Figure 9.4i, *Ps. magnuspinnatus*, Baeck et al. 2008) or inferred from burrow casting (Figure 9.4j; *Ps. minutus*, Takeda et al. 2011). Fertilized eggs were found in the chambers of all three species. These eggs have adhesive filaments and attach to the inner surface in a monolayer. Kim et al. (2011) stated that for *S. gigas* air occurred in an egg chamber where eggs were laid on the underside of the upper surface, but they did not offer evidence.

Data on burrow excavation, air deposition, and spawning are summarized for 16 mudskipper species (Table 9.1). All the mudskippers studied to date excavate burrows, often as the center of their territory. Five species maintain burrows in both breeding and non-breeding seasons, but this distinction is not clear for the other species. Eggs were recovered from burrows in seven species: *B. pectinirostris*, *Pn. schlosseri*, *Ps. modestus*, *Ps. minutus*, *Ps. magnuspinnatus*, *S. gigas*, and *S. histophorus*. The data are relatively rich for *B. pectinirostris*, *Ps. modestus*, and *Pn. schlosseri*, whereas data for the other mudskipper species are fragmentary.

9.7.2 Egg Care and Parental Behavior of Mudskippers

Spawning in a burrow surrounded by anoxic intertidal mud requires sophisticated parental care for developing embryos. These burrows must be excavated below the highest high spring tide such that the burrow openings will be inundated by seawater at the appropriate time. This is crucial for larvae dispersing from within the burrows. Larvae of mudskippers spend 30–40 days in the plankton (*B. pectinirostris*, Koga et al. 1989; Zhang et al. 1989; *P. modestus*, Kobayashi et al. 1972). The spawning chamber must be large enough to ensure that the amount of oxygen in the burrow is sufficient for embryonic respiration during high tide when the burrow-guarding male fish does not replenish the stored air. This means that at the end of a low tide, the male must sense the oxygen concentration of the stored air and raise it high enough to satisfy the oxygen demand of the embryos throughout the subsequent high tide. In addition, when the time is right, the male must be able to perceive when the embryos are able to hatch, and then initiate the hatching process while the burrow opening is covered by a high tide. Mudskippers have evolved an extraordinary suite of reproductive tactics to satisfy all these requirements (Ishimatsu and Graham 2011).

In the Ariake Sea, Japan, males of *Ps. modestus* deposit mouthfuls of fresh air into the spawning chamber at low tide (Ishimatsu et al. 2007; Figure 9.5a). This action raises the oxygen concentration of the stored air in the burrow (Figure 9.6a; Ishimatsu et al. 2007). The male ceases air deposition when the burrow openings are covered by the rising tide. The oxygen level within the egg-chamber declines during high tide, mainly due to metabolism of developing embryos, given that the male is not able to replenish the air within the burrow during high tide (Etou et al. 2007). During high tide, burrow-guarding males likely remain inside or near a burrow opening, inferred by comparing the low male/female ratio of *Ps. modestus* collected at high tide during the reproductive season (0.04–0.48) to that of the non-reproductive season (0.89–1.54, Matoba and Dotsu 1977). A motion detector at the bottom of the burrow shaft recorded few hits during high tide demonstrating that the male rarely swam through the burrow shaft (Figure 9.6a). This probably indicates that the male respired in the open water overlying the mudflat surface (oxygen saturation ≈60%, Gonzales et al. 2006) during high tide. Our preliminary laboratory study showed that *Ps. modestus* could survive submersion in normoxic water for several months (unpublished). Toba and Ishimatsu (2014) also demonstrated that *B. pectinirostris* did not incur lactate buildup in the muscles after an 8-hour submersion in normoxic water.

Air deposition during low tide appears to be governed by sensing the oxygen concentration of the stored air within the burrow by the male, and also probably by unknown mechanism(s) with

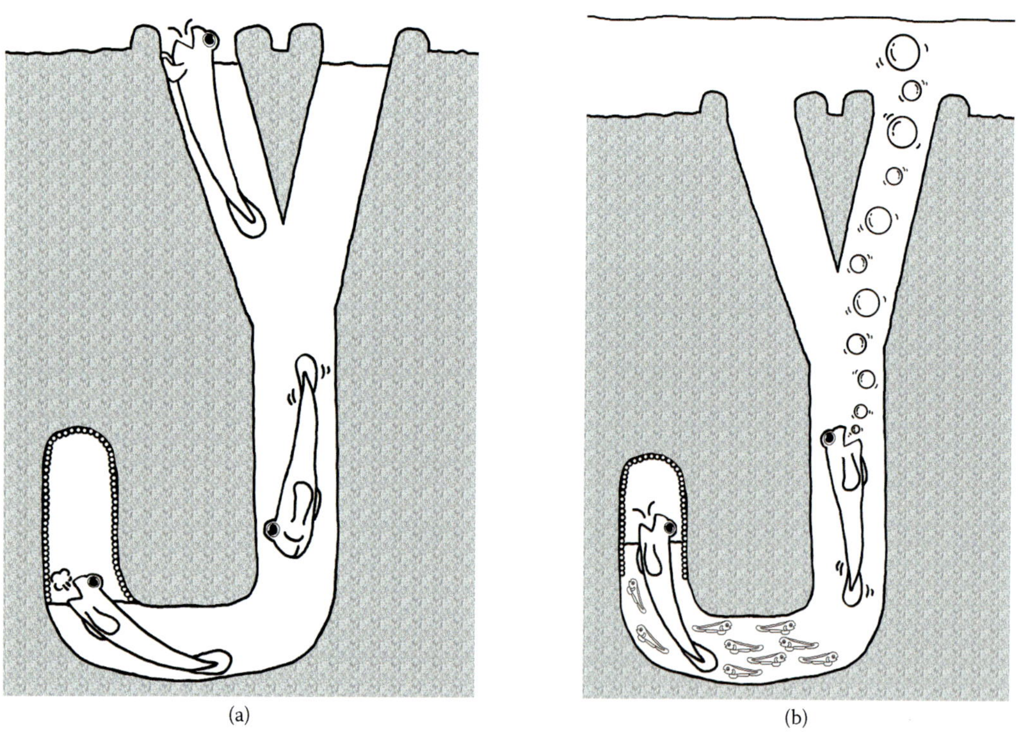

Figure 9.5 (a) Burrow-guarding male fish gulps air at the burrow opening and expels in the egg chamber, observed when mudflat surface is exposed during low tide. (b) Male fish gulps air in the egg chamber and expels it in the burrow shaft at the end of an egg-incubation period, during rising tide in the evening or at night. (Courtesy by Mizuri Murata.)

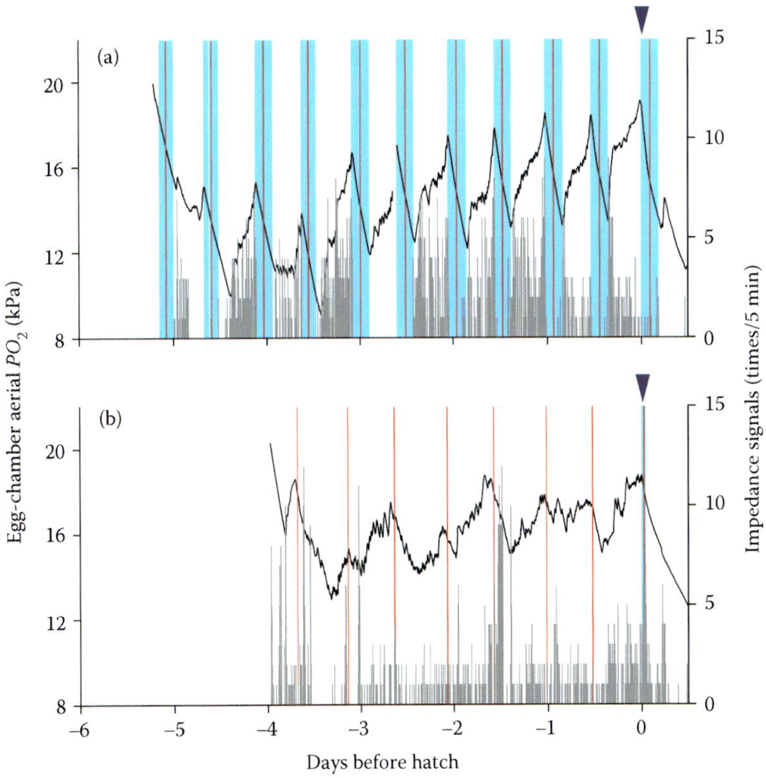

Figure 9.6 Continuous PO_2 readings (denoted by black line) and visits of the guarding male (denoted by impedance signals, gray spikes) to the egg chamber of *Periophthalmus modestus*; data acquisition started on the day of instrumentation introduction and ended when eggs hatched (denoted by solid triangle); day 0 is set to when a burrow was covered by the incoming tide preceding eggs hatching. The initial high PO_2 reflects the exposure of the egg chamber to air during instrumentation introduction. (a) Data for a burrow inundated with seawater (blue bars, red lines show the time of highest tidal height) and exposed to natural tidal regime; impedance data absent for day 3 due to instrument malfunction; (b) Data for a burrow located high on the mudflat, and continually exposed to air until day 0, when rising tide (blue bar) covered the burrow and eggs hatched (solid triangle). (Reproduced from Ishimatsu A. et al., *Journal of Experimental Biology*, 210, 3946–3954, 2007. With permission.)

which the male can recognize or predict the time left before the next submersion of burrow openings by the rising tide (Ishimatsu et al. 2007). The data from those burrows excavated high in the intertidal zone suggest that the male modulates air deposition according to the tidal rhythm at the specific location of his burrow (Figure 9.6b). Although these burrows remain uncovered by the highest neap tides, the burrow-gas oxygen concentration still reflected the tidal cycle.

Eggs of *Ps. argentilineatus* will die, if they were left underwater throughout the development, but they must be submerged to hatch (Brillet 1976). All mudskipper eggs must be immersed in water to hatch, similar to eggs of many other fishes that spawn on littoral substrates (Martin 2015). As the final step of egg tending, the male removes the egg-chamber air mouthful by mouthful, and releases it outside the burrow during a rising tide at night (Ishimatsu et al. 2007). Removal of the air pocket from the spawning chamber floods the eggs and induces immediate hatching (Figures 9.5b and 9.7a through h). Approximately 80 minutes after the burrow openings are inundated by the rising tide, the egg-guarding male starts to expel air from the spawning chamber by transporting mouthfuls of air into the main burrow shaft leading to the burrow openings and releasing them. As this was

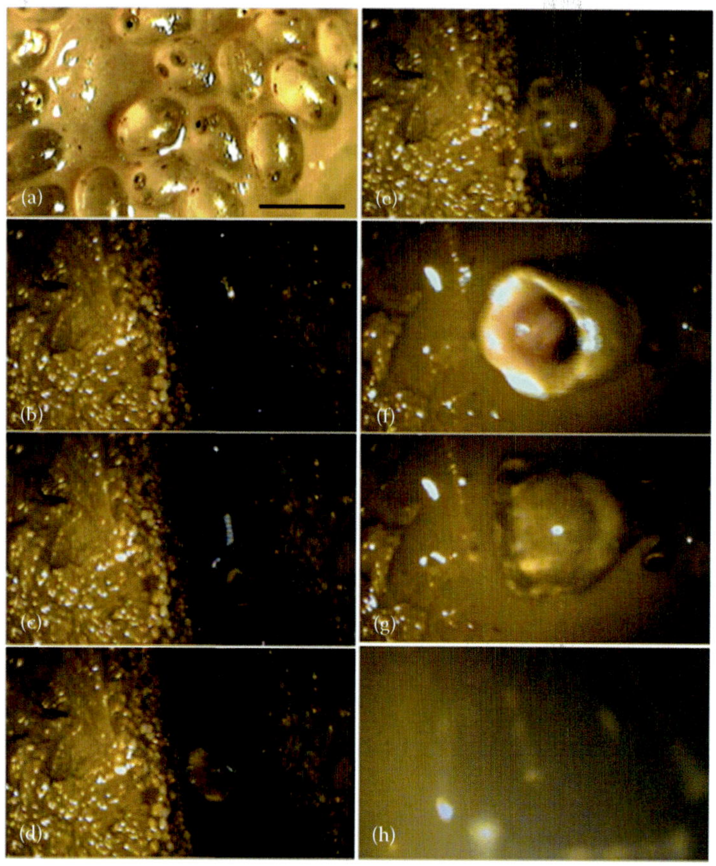

Figure 9.7 View of the egg chamber interior within the burrow of *Periophthalmus modestus*. (a) Eggs attached to the wall of an egg chamber, laboratory-incubated, scale bar =1 mm; (b)–(g) video frames captured by an endoscope that is positioned through the top of the egg chamber, June 9, 2002; endoscope camera. (b) Note the single layer of eggs on mud walls, water surface visible at the bottom of the egg chamber. (c)–(g) Series of frames in which male enters the egg chamber and becomes progressively closer to the camera with mouth open in (d) and (f). Removal of air raises the water level within the burrow, causing the eggs to be immersed and subsequently hatched. (h) Larvae, average 2.84 mm total length (Kobayashi et al. 1972), swimming in a flooded egg chamber 48 minutes after frame (g). (Reproduced from Ishimatsu A. et al., *Journal of Experimental Biology*, 210, 3946–3954, 2007. With permission.)

occurring, bubbles of air were observed exiting from the burrow openings. Video records showed that flooding of the chamber required the removal of approximately 100 air gulps, which gave an almost exact volume of total spawning-chamber air (47 mL) based on the mouth capacity of egg-guarding *Ps. modestus* (0.46 mL). Hatching occurred during rising tides in the evening or at night, but was not correlated with the lunar cycle.

9.7.3 Evolution of Air Deposition in Mudskippers

We hypothesize that air-breathing habits had been acquired among gobies prior to their invasion of mudflats, and that the behavior of air deposition into burrows evolved in mudskippers as an auxiliary oxygen source for the eggs in embryonic development and also for fishes confined in burrows during high tide. This supposition is based on the observation of *Odontamblyopus lacepedii* of the subfamily Amblyopinae, which is closely related to the Oxudercinae (Thacker 2003). This species is a facultative air-breathing fish but does not emerge from water (Gonzales et al. 2006). *O. lacepedii* is not amphibious and is restricted to its extensive mudflat burrow where, during low tide, it breathes air to sustain metabolism, by extending its head out of water. The reproductive behavior of *O. lacepedii* is not known, however, fish that spawned in captivity attached their eggs along the upper, inner side of the submerged clay pipes in their holding tank (Dotsu and Takita 1967), implying that burrow spawning may occur. Eggs of *O. lacepedii* have not been collected in the field despite almost 60 years of searching (Dôtu 1957b). Our hypothesis is also supported by the observation that both males and females of *B. pectinirostris* store air not only during the breeding season but also throughout the year (Toba and Ishimatsu 2014). Further, as mudskippers became more terrestrial and avoided submersion during high tide, as seen in some species of *Periophthalmus*, they may have abandoned the habit of air storage during non-breeding season, but retained it only during the breeding season. Burrows of *Ps. modestus* (Kobayashi et al. 1971) and *Ps. minutus* (Figure 9.4j; Takeda et al. 2011), for example, are J-shaped only during the breeding season and lacked the upturned portion in the non-breeding season. The behavior of maintaining an air phase within the burrow is also seen in *S. histophorus* under laboratory conditions without the presence of eggs (Lee et al. 2005). More investigation is needed to test the hypothesis.

9.8 FUTURE STUDIES

The mangroves and mudflat habitats where most mudskippers live are challenging to access and therefore poorly investigated. Additionally, these habitats are rapidly disappearing due to coastal development. While progress has been made in the past several decades to obtain insights to mudskipper reproduction, we list several major unknowns that await clarification.

1. Although there is an indication that spawning chambers are already filled with air before spawning and therefore both spawning and fertilization occur in air, we lack evidence. The presence of air in burrows without eggs for *Pn. schlosseri* (Ishimatsu et al. 2009) seems to support the idea, assuming that spawning chambers are flooded for hatching. Hence, burrows with no eggs must have been unused. Shimizu et al. (2006) observed aerial egg deposition and fertilization in the amphibious blenny, *Andamia tetradactyla*, which also suggests that aerial spawning by mudskippers is a possibility. However, it is difficult for us to envisage how a relatively larger mudskipper is able to oviposit without collapsing the ceiling and side walls of a spawning chamber. In particular, the giant mudskipper, *Pn. schlosseri*, can weigh over 200 g and grow more than 30 cm long, and it seems to us almost impossible for the fish to spawn on the ceiling of an air-filled chamber.
2. Some mudskipper species inhabit freshwater areas far from coastlines. For example, Khaironizam and Norma-Rashid (2003) reported that three species of mudskippers (*Periophthalmodon septemradiatus, Periophthalmus novemradiatus*, and *Periophthalmus gracilis*) were found on the mud banks and in small tributaries along the upper reaches of an estuary, nearly 15 km (in linear distance) from the river mouth. The water was fresh (0 ppt) at low tide. During extreme high tides, the salinities recorded in the river ranged 1–3 ppt. Recently, Ishimatsu et al. have found *Pn. septemradiatus* in several sites that are more than 80 km (linear distance) from the mouth of the Mekong River, in Vietnam. The salinity at the time of sampling was 0 ppt at all these sites. There were juveniles and local people said that they reproduced in holes. Questions abound, including whether these mudskippers are able to complete their life history in freshwater, and how far upstream can these mudskippers infiltrate and reproduce successfully? The reproductive behavior of mudskippers living on intertidal mudflats appears to need tidal fluctuation to cue air replenishment and hatch induction.
3. The possibility of live birth in mudskippers remains an enigma. Breder and Rosen (1966) suggested that *Ps. barbarus* has internal fertilization and possibly gives birth to live young. However, this has never been confirmed. They cite Harms (1935), but he does not report any observations of *Ps. barbarus* in that publication. Harms (1935) reported observations of copulation in *Periophthalmus* and

Boleophthalmus species and suggested that both *Ps. barbarus* and *B. boddarti* may reproduce via internal fertilization with live birth. Neither has been confirmed by others, although Harms (1934: fig. 26) provided a drawing of an advanced embryo taken from the ovary of *Ps. argentilineatus.* Harms (1934) described in German ". . . The oldest stage, which I could observe in the ovary, was already a young fish with optic vesicles, heart, spinal cord, notochord and somites. These animals already moved weakly within the egg membrane. . . ." Could some mudskippers have internal fertilization, or asexual reproduction by parthenogenesis, or be hermaphrodites?

4. Studies of the role of the parent fishes in hatching, and the environmental cues that initiate hatching behavior in both the adult and the embryos, would be useful both within and between species. Whether these cues are mechanical, circadian, chemical, or some combination of the above is not known.
5. In some mudflats, several species of oxudercine gobies live sympatrically. For example, five species (*B. boddarti, Oxuderces nexipinnis, Pn. schlosseri, Periophthalmus chrysospilos*, and *S. histophorus*) live in one mudflat in the Soc Trang Province, Vietnam (A. Ishimatsu, personal observation). Comparison of reproductive ecology and physiology of these sympatric species is of great biological interest, particularly as they differ greatly in amphibious tendencies.

ACKNOWLEDGMENT

The authors thank Mizuri Murata of Nagasaki University for her help in preparing figures.

REFERENCES

Al-Daham N.K. and Al-Noor S.S. 2000. Reproductive biology of the spotted mudskipper, *Periophthalmus waltoni* Koumans, 1941 (Pisces, Gobiidae) in Khor Al-Zubair, Northwest Arabian Gulf. *Marina Mesopotamica* 15:425–437.

Allanson B.R., Skinner D. and Imberger J. 1992. Flow in shrimp burrows. *Estuarine, Coastal and Shelf Science* 35:253–266.

Baeck G.W., Takita T. and Yoon Y.H. 2008. Lifestyle of Korean mudskipper *Periophthalmus magnuspinnatus* with reference to a congeneric species *Periophthalmus modestus. Ichthyological Research* 55:43–52.

Borges R., Faria C., Gil F. and Gonçalves E.J. 2011. Early development of gobies. In *The Biology of Gobies* (eds. R.A. Patzner, J.L. Van Tassell, M. Kovačić and B.G. Kapoor) pp. 403–462. Enfield: Science Publishers.

Breder C.M. and Rosen D.E. 1966. *Modes of Reproduction in Fishes.* Garden City: Natural History Press, 941 pp.

Brillet C. 1976. Struture du terrier, reproduction et comportment des jeunes chez le poisson amphibie *Periophthalmus sobrinus* Eggert. *Revue d'Ecologie: La Terre et la Vie* 30:465–483.

Bucholtz R.H., Meilvang A.S., Cedhagen T., Christensen J.T. and Macintosh D.J. 2009. Biological observations on the mudskipper *Pseudapocryptes elongatus* in the Mekong Delta, Vietnam. *Journal of World Aquaculture Society* 40:711–723.

Childress J.J. and Seibel B.A. 1998. Life at stable low oxygen levels: Adaptations of animals to oceanic oxygen minimal layers. *Journal of Experimental Biology* 201:1223–1232.

Chung E.Y., An C.M. and Lee T.Y. 1991. Sexual maturation of the bluespotted mud hopper, *Boleophthalmus pectinirostris* (Linnaeus). Korean Journal of Fisheries and Aquatic Sciences 24:167–176 (in Korean with English abstract).

Clayton D.A. 1993. Mudskippers. *Oceanography and Marine Biology: An Annual Review* 31:507–577.

Clayton D.A. and Vaughan T.C. 1986. Territorial acquisition in the mudskipper *Boleophthalmus boddarti* (Teleostei, Gobiidae) on the mudflats of Kuwait. *Journal of Zoology* 209:501–519.

Cole K.S. 2010. *Reproduction and Sexuality in Marine Fishes: Patterns and Processes.* Berkeley, CA: University of California Press, 408 pp.

Coleman R.M. 1999. Parental care in intertidal fishes. In *Intertidal Fishes: Life in Two Worlds* (eds. M.H. Horn, K.L.M. Martin and M.A. Chotkowski) pp. 165–180. San Diego, CA: Academic Press.

Craig P.M., Fitzpatrick J.L., Walsh P.J., Wood C.M. and McClelland G.B. 2014. Coping with aquatic hypoxia: How the plainfin midshipman (*Porichthys notatus*) tolerates the intertidal zone. *Environmental Biology of Fishes* 97:163–172.

Dejours P. 1981. *Principles of Comparative Respiratory Physiology*. Amsterdam: Elsevier, 282 pp.

DeMartini E.E. 1999. Intertidal spawning. In *Intertidal Fishes: Life in Two Worlds* (eds. M.H. Horn, K.L.M. Martin and M.A. Chotkowski) pp. 143–164. San Diego, CA: Academic Press.

Diaz R.J. 2001. Overview of hypoxia around the world. *Journal of Environmental Quality* 30:275–281.

Dinh Q.M., Qin J.G., Dittmann S. and Tran D.D. 2014a. Burrow morphology and utilization of the goby (*Parapocryptes serperaster*) in the Mekong Delta, Vietnam. *Ichthyological Research* 61:332–340.

Dinh Q.M., Nguyen T.T.G., Ngo N.L.D., Dang H.D. and Lam T.H. 2014b. Burrow configuration and utilization of the blue-spotted mudskipper, *Boleophthalmus boddarti* caught in Soc Trang, Vietnam. *Kasetsart University Fisheries Research Bulletin* 38:1–9.

Dôtu Y. 1954. On the life history of a goby, *Chaenogobius castanea* O'Shaughnessy. *Japanese Journal of Ichthyology* 3:133–138 (in Japanese with English abstract).

Dôtu Y. 1957a. The bionomics and life history of a gobioid fish, *Paleatogobius uchidai* Takagi. *Japanese Journal of Ichthyology* 6:97–104 (in Japanese with English abstract).

Dôtu Y. 1957b. On the bionomics and history of the eel-like goby, *Odontamblyopus rubicundus* (Hamilton). *Science Bulletin of the Faculty of Agriculture, Kyushu University* 16:101–110 (in Japanese with English abstract).

Dôtu Y. and Mito S. 1955. On the breeding-habits, larvae and young of a goby, *Acanthogobius flavimanus* (Temminck et Schlegel). *Japanese Journal of Ichthyology* 4:153–161 (in Japanese with English abstract).

Dotsu Y. and Takita T. 1967. Induced spawning by hormone operation, egg development and larva of blind gobioid fish, *Odontamblyopus rubicundus*. *Bulletin of the Faculty of Fisheries, Nagasaki University* 23:135–144 (in Japanese with English abstract).

Dumbauld B.R., Armstrong D.A. and Feldman K.L. 1996. Life-history characteristics of two sympatric thalassinidean shrimps, *Neotrypaea californiensis* and *Upogebia pugettensis*, with implications for oyster culture. *Journal of Crustacean Biology* 16:689–708.

El-Ziady S., Eissa S.M. and Al-Naqi Z. 1979. Ecological studies on the mud-skipper fish, *Periophthalmus chrysospilos* (Bleeker) in Kuwait. *Bulletin of Faculty of Science, Cairo University* 48:113–133.

Etim L., King R.P. and Udo M.T. 2002. Breeding, growth, mortality and yield of the mudskipper *Periophthalmus barbarus* (Linneaus 1766) (Teleostei: Gobiidae) in the Imo River estuary, Nigeria. *Fisheries Research* 56:227–238.

Etim L., Olele F.N., Udo M.T. and Brownson I.I. 2010. The reproductive characteristics of mudskipper *Periophthalmus barbarus* (Linnaeus 1760) (Teleostei, Gobiidae) in the estuarine swamps of the Imo river, Nigeria. *Global Journal of Pure and Applied Sciences* 16:395–400.

Etou A., Takeda T., Yoshida Y. and Ishimatsu A. 2007. Oxygen consumption during embryonic development of the mudskipper (*Periophthalmus modestus*): Implication for the aerial development in burrows. In *Fish Respiration and Environment* (eds. M.N. Fernandes, F.T. Rantin, M.L. Glass and B.G. Kapoor) pp. 83–91. Enfield, NH: Science Publishers.

Gibson R.N. 1999. Movement and homing in intertidal fishes. In *Intertidal Fishes: Life in Two Worlds* (eds. M.H. Horn, K.L.M. Martin and M.A. Chotkowski) pp. 97–125. San Diego, CA: Academic Press.

Gonzales T.T., Katoh M. and Ishimatsu A. 2006. Air breathing of aquatic burrow-dwelling eel goby, *Odontamblyopus lacepedii* (Gobiidae: Amblyopinae). *Journal of Experimental Biology* 209:1085–1092.

Graham J.B. 1997. *Air-Breathing Fishes: Evolution, Diversity and Adaptation*. San Diego, CA: Academic Press, 299 pp.

Graham J.B. and Lee H.J. 2004. Breathing air in air: In what ways might extant amphibious fish biology relate to prevailing concepts about early tetrapods, the evolution of vertebrate air-breathing, and the vertebrate land transition? *Physiological and Biochemical Zoology* 77:720–731.

Graham J.B., Rosenblatt R.H. and Gans C. 1978. Vertebrate air breathing arose in fresh waters and not in the oceans. *Evolution* 32:459–463.

Harms J.W. 1934. *Wandlungen des Artgefüges unter natürlichen und künstlichen Umweltbedingungen*. Leipzig: Johann Ambrosius Barth, 212 pp.

Harms J.W. 1935. Die Realisation von Genen und die consecutive Adaption. 4. Mitteilung. Experimentell hervorgerufener Medianwechsel: Wasser zu Feuchtluft vzw. Zu Trockenluft bei Gobiformes (*Gobius*, *Boleophthalmus* und *Periophthalmus*). *Zeitschrift fur wissenschaftliche Zoologie, Leipzig* 146:417–462.

Harnik P.G., Lotze H.K., Anderson S.C. et al. 2012. Extinctions in ancient and modern seas. *Trends in Ecology & Evolution* 27:608–617.

Hastings P.A. and Petersen C.W. 2010. Parental care, oviposition sites, and mating systems of Blennioid fishes. In *Reproduction and Sexuality in Marine Fishes* (ed. K.S. Cole) pp. 91–116. Berkeley, CA: University of California Press.

Hoda S.M.S. 1986. Maturation and fecundity of the mudskipper *Boleophthalmus dussumeri* Cuv. & Val. from the Karachi Coast. Mahasager. *Mahasagar-Bulletin of the National Institute of Oceanography* 19:73–78.

Hoda S.M.S. and Akhtar Y. 1985. Maturation and fecundity of the mudskipper *Boleophthalmus dussumeri* in the northern Arabian Sea. *Indian Journal of Fisheries* 32:64–67.

Hong L.Y., Hong W.S., Zhu W.B., Shi Q., You X.X. and Chen S.X. 2014. Cloning and expression of melatonin receptors in the mudskipper *Boleophthalmus pectinirostris*: Their role in synchronizing its semilunar spawning rhythm. *General and Comparative Endocrinology* 195:138–50.

Hong W. and Zhang Q. 2004. Induced nest spawning and artificial hatching of the fertilized eggs of mudskipper, *Boleophthalmus pectinirostris. Chinese Journal of Oceanology and Limnology* 22:408–413.

Hong W., Chen S., Zhang Q. and Wang Q. 2007. Reproductive ecology of the mudskipper *Boleophthalmus pectinirostris. Acta Oceanologica Sinica* 26:72–81.

Horn M.H. and Martin K.L. 2006. Rocky intertidal zones. In *Ecology of Marine Fishes: California and Adjacent Waters* (eds. L.A. Allen, M.H. Horn and D. Pondella) pp. 205–226. Berkeley, CA: University of California Press.

Ikebe Y. and Oishi T. 1996. Correlation between environmental parameters and behaviour during high tides in *Periophthalmus modestus. Journal of Fish Biology* 49:139–147.

Immler S., Mazzoldi C. and Rassoto M.B. 2004. From sneaker to parental male: Change of reproductive traits in the black goby, *Gobius niger* (Teleostei: Gobiidae). *Journal of Experimental Zoology* 301A:177–185.

Ishimatsu A. and Graham J.B. 2011. Roles of environmental cues for embryonic incubation and hatching in mudskippers. *Integrative and Comparative Biology* 51:38–48.

Ishimatsu A., Hishida Y., Takita T. et al. 1998. Mudskippers store air in their burrows. *Nature* 391:237–238.

Ishimatsu A., Takeda T., Kanda T., Oikawa S. and Khoo K.H. 2000. Burrow environment of mudskippers in Malaysia. *Journal of Bioscience* 11:17–28.

Ishimatsu A., Takeda T., Tsuhako Y., Gonzales T.T. and Khoo K.H. 2009. Direct evidence for aerial egg deposition in the burrows of the Malaysian mudskipper, *Periophthalmodon schlosseri. Ichthyological Research* 56:417–420.

Ishimatsu A., Yoshida Y., Itoki N., Takeda T., Lee H.J. and Graham J.B. 2007. Mudskippers brood their eggs in air but submerge them for hatching. *Journal of Experimental Biology* 210:3946–3954.

Itoki N., Sakamoto T., Hayashi M., Takeda T. and Ishimatsu A. 2012. Morphological responses of mitochondria-rich cells to hypersaline environment in the Australian mudskipper, *Periophthalmus minutus. Zoological Science* 29:444–449.

Jones B.C. 1972. Effect of intertidal exposure on survival and embryonic development of Pacific herring spawn. *Journal of the Fisheries Research Board of Canada* 29:1119–1124.

Jones S. 1937. Observations on the breeding habits and development of certain brackish water fishes of Adyar, Madras. *Proceedings of the Indian Academy of Sciences Section B* 5:261–289.

Jones J.C. and Reynolds J.D. 1999. Cost of egg ventilation for male common gobies breeding in conditions of low dissolved oxygen. *Animal Behaviour* 57:181–188.

Khaironizam M.Z. and Norma-Rashid Y. 2003. First record of the mudskipper, *Periophthalmodon septemradiatus* (Hamilton) (Teleostei: Gobiidae) from Peninsular Malaysia. *Raffles Bulletin of Zoology* 51:97–100.

Kim J.K., Baeck H.J., Kim J.-W., Chang D.-S. and Kim J.-I. 2011. Sexual maturity and early life history of the mudskipper *Scartelaos gigas* (Pisces, Gobiidae): Implications for conservation. *Fisheries and Aquatic Sciences* 14:403–410.

Kim J.W., Yoon Y.H., Shin H.C. et al. 2007. Maturation and spawning of shuttles hoppfish *Periophthalmus modestus* in the mud flat of Suncheon Bay, Korea. *Journal of Korean Fisheries Society* 40:79–83 (in Korean with English abstract).

Kobayashi T., Dotsu Y. and Takita T. 1971. Nest and nesting behavior of the mud skipper, *Periophthalmus cantonensis* in Ariake Sound. *Bulletin of the Faculty of Fisheries, Nagasaki University* 32:27–40 (in Japanese with English abstract).

Kobayashi T., Dotsu Y. and Miura N. 1972. Egg development and rearing experiments of the larvae of the mud skipper, *Periophthalmus cantonensis*. *Bulletin of the Faculty of Fisheries, Nagasaki University* 33:49–62 (in Japanese with English abstract).

Koga H., Noda S., Noguchi T. and Aoto I. 1989. Studies on artificial propagation of mud skipper *Boleophthalmus pectinirostris* (Linnaeus)-III Hatching and larval rearing. *Bulletin of Saga Prefectural Ariake Fisheries Research and Development Center* 11:17–28 (in Japanese).

Koyama A., Inui R., Iyooka H., Akamatsu Y. and Onikura N. 2016. Habitat suitability of eight threatened gobies inhabiting tidal flats in temperate estuaries: Model developments in the estuary of the Kuma River in Kyushu Island, Japan. *Ichthyological Research* 63:307–314.

Kuno Y. and Takita T. 1997. The growth, maturation and feeding habits of the gobiid fish *Acanthogobius hasta* distributed in Ariake Sound, Kyushu, Japan. *Fisheries Science* 63:242–248.

Lawson E.O. 2011. Length-weight relationships and fecundity estimates in mudskipper, *Periophthalmus papilio* (Bloch and Schneider 1801) caught from the mangrove swamps of Lagos Lagoon, Nigeria. *Journal of Fisheries and Aquatic Sc*iences 6:264–271.

Lee H.J., Martinez C.A., Hertzberg K.J., Hamilton A.L. and Graham J.B. 2005. Burrow air phase maintenance and respiration by the mudskipper *Scartelaos histophorus* (Gobiidae: Oxudercinae). *Journal of Experimental Biology* 208:160–177.

Little C. 2000. *The Biology of Soft Shores and Estuaries*. New York, NY: Oxford University Press, 252 pp.

MacGinitie G.E. 1939. The natural history of the blind goby, *Typhlogobius californiensis* Steindachner. *American Midland Naturalist* 21:489–505.

Manacop P.R. 1953. The life history and habits of the goby *Sicyopterus extraneus* Herre (anga) Gobiidae with an account of the goby fry fishery of Cagayan River, Oriental Misamis. *Philippine Journal of Fisheries* 2:1–58.

Marconato A., Rasotto M.B. and Mazzoldi C. 1996. On the mechanism of sperm release in three gobiid fishes (Teleostei: Gobiidae). *Environmental Biology of Fishes* 46:321–327.

Marliave J.B. 1981. High intertidal spawning under rockweed, *Fucus distichus*, by the sharpnose sculpin *Clinocottus acuticeps*. *Canadian Journal of Zoology* 59:1122–1125.

Martin K.L.M. 1996. An ecological gradient in air-breathing ability among marine cottid fishes. *Physiological Zoology* 69:1096–1113.

Martin K.L.M. 1999. Ready and waiting: Delayed hatching and extended incubation of anamniotic vertebrate terrestrial eggs. *American Zoologist* 39:279–288.

Martin K.L.M. 2014. Theme and variations: Amphibious air-breathing intertidal fishes. *Journal of Fish Biology* 84:577–602.

Martin K.L.M. 2015. *Beach-spawning Fishes: Reproduction in an Endangered Ecosystem*. Boca Raton, FL: CRC Press, 219 pp.

Martin K.L.M. and Carter A.L. 2013. Brave new propagules: Terrestrial embryos in anamniotic eggs. *Integrative and Comparative Biology* 53:233–247.

Martin K.L.M. and Strathmann R.A. 1999. Aquatic organisms, terrestrial eggs: Early development at the water's edge. Introduction to the symposium. *American Zoologist* 39:215–217.

Martin K.L.M. and Swiderski D. 2001. Beach spawning in fishes: A phylogenetic approach. *American Zoologist* 41:526–237.

Martin K.L., Moravek C.L. and Flannery J.A. 2009. Embryonic staging series for the beach spawning, terrestrially incubating California grunion *Leuresthes tenuis* (Ayres 1860) with comparisons to other Atherinomorpha. *Journal of Fish Biology* 75:17–38.

Martin K.L.M., Van Winkle R.C., Drais J.E. and Lakisic H. 2004. Beach spawning fishes, terrestrial eggs, and air breathing. *Physiological and Biochemical Zoology* 77:750–759.

Matoba M. and Dotsu Y. 1977. Prespawning behavior of the mud skipper *Periophthalmus cantonensis* in Ariake Sound. *Bulletin of the Faculty of Fisheries, Nagasaki University* 43:23–33 (in Japanese with English abstract).

Mayer F. 1929. The *Periophthalmus koelreuteri* or mud leaper. *Aquatic Life, Baltimore* 11:212–213.

Mazda Y., Sato Y., Sawamoto S., Yokochi H. and Wolanski E. 1990. Links between physical, chemical and biological processes in Bashita-minato, a mangrove swamp in Japan. *Estuarine, Coastal and Shelf Science* 31:817–833.

Mazlan A.G. and Rohaya M. 2008. Size, growth and reproductive biology of the giant mudskipper, *Periophthalmodon schlosseri* (Pallas, 1770), in Malaysian waters. *Journal of Applied Ichthyology* 24:290–296.

Mazzoldi C., Petersen C. and Rasotto M.B. 2005. The influence of mating system on seminal vesicle variability among gobies (Teleostei: Gobiidae). *Journal of Zoological Systematics and Evolutionary Research* 43:307–314.

McLachlan A. and Brown A.C. 2006. *The Ecology of Sandy Shores*, 2nd Edition. San Diego, CA: Academic Press, 373 pp.

Middaugh D.P., Kohl H.W. and Burnett L.E. 1983. Concurrent measurement of intertidal environmental variables and embryo survival for the California grunion, *Leuresthes tenuis*, and Atlantic silverside, *Menidia menidia* (Pisces: Atherinidae). *California Fish & Game* 69:89–96.

Minh T.H., Gallardo W.G. and Nguyen T.P. 2010. Fishery and aquaculture of juvenile mudskipper *Pseudapocryptes elongatus* (Cuvier, 1816) in the coastal zone of Mekong Delta, Vietnam. *Asian Fisheries Science* 23:224–239.

Murdy E.O. 1989. A taxonomic revision and cladistic analysis of the Oxudercinae gobies (Gobiidae Oxudercinae). Records of the Australian Museum (Suppl. 11):1–93.

Mutsaddi K.K. and Bal D.V. 1970. Maturation and spawning of *Boleophthalmus dussumeri. Journal of the University of Bombay* 39:58–76.

Nanami A. and Takegaki T. 2005. Age and growth of the mudskipper *Boleophthalmus pectinirostris* in Ariake Bay, Kyushu, Japan. *Fisheries Research* 74:24–34.

Okamura K., Tanaka K., Siow R., Man A., Kodama M. and Ichikawa T. 2010. Spring tide hypoxia with relation to chemical properties of the sediments in the Matang mangrove estuary, Malaysia. *Japan Agricultural Research Quarterly* 44:325–333.

Okinawa Prefecture. 2000. *Tokage-haze no Hanashi* (An account of *Scartelaos histophorus*). Okinawa: Okinawa Sekkei Service, 15 pp. (in Japanese).

Park K-D., Kim J-K., Chang D-S., Kim, J-I. and Oh C-W. 2008. Age and growth of the mudskipper, *Scartelaos gigas* (Perciformes, Gobiidae) from Korea. *Animal Cells and Systems* 12:305–311.

Rambough P.J. 1988. Respiratory gas exchange, aerobic metabolism, and effects of hypoxia during early life. In *Fish Physiology XI, The Physiology of Developing Fish, Part A Eggs and Larvae* (eds. W.S. Hoar and D.J. Randall) pp. 59–161. San Diego, CA: Academic Press.

Rodgers E.W., Earley R.I. and Grober M.S. 2007. Social status determines sexual phenotype in the bi-directional sex changing blue banded goby *Lythrypnus dalli. Journal of Fish Biology* 70:1660–1668.

Romer A.S. 1967. Major steps in vertebrate evolution. *Science* 158:1629–1637.

Scaggiante M., Mazzoldi C., Petersen C.W. and Rasotto M.B. 1999. Sperm competition and mode of fertilization in the grass goby *Zosterisessor ophiocephalus* (Teleostei: Gobiidae). *Journal of Experimental Zoology* 283:81–90.

Shimizu N., Sakai Y., Hashimoto H. and Gushima K. 2006. Terrestrial reproduction by the air-breathing fish *Andamia tetradactyla* (Pisces: Blenniidae) on supralittoral reefs. *Journal of Zoology* 269:357–364.

Soyano K., Higuchi T., Kobayashi M., Takushima M., Aoki J.-Y. and Nagae M. 2008. Gonadal development and spawning rhythm in mudskipper (*Periophthalmus modestus*). *Cybium* 32(2):234.

Stramma L., Johnson G.C., Springtall J. and Mohrholz V. 2008. Expanding oxygen-minimum zones in the tropical oceans. *Science* 320:655–658.

Strathmann R.R. and Hess H.C. 1999. Two designs of marine egg masses and their divergent consequences for oxygen supply and desiccation in air. *American Zoologist* 39:253–260.

Swenson R.O. 1997. Sex-role reversal in the tidewater goby, *Eucyclogobius newberryi. Environmental Biology of Fishes* 50:27–40.

Takeda T., Hayashi M., Toba A., Soyano K. and Ishimatsu A. 2011. Ecology of the Australian mudskipper *Periophthalmus minutus,* an amphibious fish inhabiting a mudflat in the highest intertidal zone. *Australian Journal of Zoology* 59:312–320.

Takegaki T. and Nakazono A. 1999. Reproductive behavior and mate fidelity in the monogamous goby, *Valenciennea longipinnis. Ichthyological Research* 46:115–123.

Takegaki T., Mori K., Shikasho S. and Nakazono A. 1999. Effect of mound structure on water exchange in *Valenciennea longipinnis* burrow. *Fisheries Engineering* 36:87–90 (in Japanese).

Takita T., Larson H.K. and Ishimatsu A. 2011. The natural history of mudskippers in northern Australia, with field identification characters. *The Beagle, Record of the Museums and Art Galleries of the Northern Territory* 27:189–204.

Taylor M.H. 1999. A suite of adaptations for intertidal spawning. *American Zoologist* 39:313–320.

Taylor M.H., DiMichele L. and Leach G.J. 1977. Egg stranding in the life cycle of the mummichog, *Fundulus heteroclitus*. *Copeia* 1977:397–399.
Tewksbury H.T. and Conover D.O. 1987. Adaptive significance of intertidal egg deposition in the Atlantic silverside *Menidia menidia*. *Copeia* 1987:76–83.
Thacker C.E. 2003. Molecular phylogeny of gobioid fishes (Teleostei: Perciformes: Gobioidei). *Molecular Phylogenetics and Evolution* 26:354–368.
Toba A. and Ishimatsu A. 2014. Roles of air stored in burrows of the mudskipper *Boleophthalmus pectinirostris* for adult respiration and embryonic development. *Journal of Fish Biology* 84:774–793.
Townsend K.A. and Tibbetts I.R. 1995. Re-appearance of the blue mudskipper, *Scartelaos histophorus* (Pisces: Gobiidae) in the greater Brisbane area. *Memoirs of the Queensland Museum* 38:671–676.
Townsend K.A. and Tibbetts I.R. 2005. Behaviour and sexual dimorphism of the blue mudskipper, *Scartelaos histophorus* (Pisces: Gobiidae). *Proceedings of the Royal Society of Queensland* 112:53–62.
Truchot J.P. and Duhamel-Jouve A. 1980. Oxygen and carbon dioxide in the marine intertidal environment: Diurnal and tidal changes in rockpools. *Respiration Physiology* 39:241–254.
Tsuhako Y., Ishimatsu A., Takeda T., Khoo K.H. and Tachihara K. 2003. The eggs and larvae of the giant mudskipper, *Periophthalmodon schlosseri*, collected from a mudflat in Penang, Malaysia. *Ichthyological Research* 50:178–181.
Udo M.T. 2002a. Intersexual plasticity in aspects of the biology of the mudskipper *Periophthalmus barbarus* (Gobiidae) in the mangrove swamps of IMO Estuary, Nigeria. *Journal of Environmental Science* 14:95–101.
Udo M.T. 2002b. Morphometric relationships and reproductive maturation of the mudskipper, *Periophthalmus barbarus* from subsistence catches in the mangrove swamps of IMO estuary, Nigeria. *Journal of Environmental Science* 14:221–226.
Udo M.T., Abiaobo N. O. and Asuquo I. E. 2016. Aspects of the reproductive biology in mudskippers *Periophthalmus barbarus* (Gobiidae) (Linneaus 1766) in mangrove swamps of Iko Estuary, Southeast, Nigeria. *Continental Journal of Biological Sciences* 9:1–14.
Wang Q., Hong W., Chen S. and Zhang Q. 2008. Variation with semilunar periodicity of plasma steroid hormone production in the mudskipper *Boleophthalmus pectinirostris*. *General and Comparative Endocrinology* 155:821–826.
Washio M., Komiya S. and Takita T. 1993. Maturation of the mudskipper *Boleophthalmus pectinirostris* distributed in the mud flats of the Midori River, Kumamoto Prefecture. *Nippon Suisan Gakkaishi* 59:575–580.
Whiteman E.A. and Côté I.M. 2004. Monogamy in marine fishes. *Biological Reviews* 79:351–375.
Wolanski E., Mazda Y. and Ridd P. 1992. Mangrove hydrodynamics. In *Tropical Mangrove Ecosystems* (eds. A.I. Robertson and D.M. Alongi) pp. 43–62. Washington, DC: American Geophysical Union.
Xie, X.Y. and Zhang, Q.Y. 1990. Studies on the gonadal development in female mudskipper. *Journal of Oceanography of Taiwan Strait* 9:217–221 (in Chinese with English abstract).
Yamahira K. 1996. The role of intertidal egg deposition on survival of the puffer, *Takifugu niphobles* (Jordan et Snyder), embryos. *Journal of Experimental Marine Biology and Ecology* 198:291–306.
Zander C.D., Nieder J. and Martin K.L.M. 1999. Vertical distribution patterns. In *Intertidal Fishes: Life in Two Worlds* (eds. M.H. Horn, K.L.M. Martin and M.A. Chotkowski) pp. 26–53. San Diego, CA: Academic Press.
Zhang Q., Hong W., Dai Q., Zhang J. and Cai, Y-Y. 1989. Studies on induced ovulation, embryonic development and larval rearing of the mudskipper (*Boleophthalmus pectinirostris*). *Aquaculture* 83:375–386.

CHAPTER 10

Feeding Behavior: A Review

David Clayton

CONTENTS

10.1 INTRODUCTION

This review of feeding behavior in oxudercine gobies is centered on studies of their diet, but presents them in the context of relevant biology and ecology of this fish group, and draws on research from other fish groups to expand the theoretical base under which future study could be conducted. The review begins with a survey of studies using gut content analyses to understand seasonal, ontogenetic, diel, and sex-related patterns in dietary intakc. Using examples from Africa and India, in situ benthic fauna are compared to gut contents of *Periophthalmus* spp. in order to determine prey species availability and ascertain if mudskippers are selective, or opportunistic predators, subsequently framing the information in the wider context of the piscivorous community of which they are part. Stable isotope studies are also reviewed to gain further understanding of trophic relationships of mudskippers to the environment. Mechanisms of sensory detection, prey capture, and prey processing and digestion are discussed. Details of pharyngeal dentition in some species are provided. The roles of pharyngeal teeth, oral teeth, and gill rakers in food selection are reevaluated. The length and structure of the alimentary tract, and the enzymatic and bacterial floral distributions are reviewed. The transition from larval to adult feeding is briefly discussed. Finally, the few behavioral studies that look at feeding-time budgets, predator–prey interactions, and diatom feeding are reviewed before suggestions on the direction of future studies are presented.

10.2 FEEDING BEHAVIOR

Diets of some oxudercine goby taxa in their natural habitat have been reported throughout their natural range (Figure 10.1 and Table 10.1), whereas fish enthusiasts have described the feeding of captive mudskippers (see Chapter 13). Details of prey items are summarized in Tables 10.2 and 10.3; the level of taxonomic identification of prey items varied from general categories (e.g., Rathod and Patel 2009; Hajisamae and Ibrahim 2008) to detailed lists of prey species (e.g., Pankow and Hug 1979). Sampling duration varied from a single day collection (Sarker et al. 1980) to monthly collections over an annual cycle (Mutsaddi and Bal 1969). There is little information on the dietary inclination of some oxudercine genera such as *Oxuderces* and *Parapocryptes* (henceforth *O.* and *Pa.*, respectively), but with considerable simplification, *Periophthalmus* and *Periophthalmodon* (henceforth *Ps.* and *Pn.*, respectively) are carnivores, *Boleophthalmus* (henceforth *B.*) is an herbivore, and *Scartelaos* and *Pseudapocryptes* (henceforth *S.* and *Pd.*, respectively) are omnivores.

Species of *Periophthalmus* have been considered opportunistic predators, and their dietary intake of plant material is widely assumed to be accidental. Several studies suggest otherwise, especially in which plant material was the main component of the diet (Bob-Manuel 2011; Udoh et al. 2013). The diet of *Ps. waltoni* from Khor Al-Zubair, Iraq, for example, was found to consist predominantly of crustaceans (67.2% occurrence) and molluscs (34.5%) in the winter months from November to May (Mhaisen and Al Maliki 1996; Barak et al. 1994). Plant material was found in the diet of *Ps. waltoni* also collected by Al-Noor (1994) during the winter months. Interestingly, Mhaisen and Al Maliki (1996) attributed the presence of the plant material to the winter shortage of animal material. The seasonal, diel, and ontogenetic changes in diet form the bases of the following sections.

Figure 10.1 *Boleophthalmus boddarti* on an exposed mud bank, feeding during ebb tide at Ubin Island, Singapore. (Courtesy of Keren Sadanandan.)

Table 10.1 Mudskipper Diets Arranged by Study Location in a North to South Sequence

	Authority	Location	Coordinates	Species Code	Species
1	Ryu et al. (1995)	Korea: Obong-ri, Okgu-gun	(35°50′ N 126°30′ E)	Bpe Psca	*B. pectinirostris* *Ps. cantonensis* (=*Ps. modestus*)
2	Baeck et al. (2008)	Korea: Sangnim, Suncheonman Bay	35°02′ N 127°37′ E	Psms	*Ps. magnuspinnatus*
		Korea: Sangbong, Suncheomann Bay	34°57′ N 127°42′ E		
3	Baeck et al. (2013)	Korea: Suncheon Bay	34°50′ N 127°44′ E	Psms Psmo	*Ps. magnuspinnatus* *Ps. modestus*
4	Taher (2010)	Iraq: Shatt Al-Basrah Canal	30°22′ N 47°47′ E	Bd	*B. dussumieri*
5	Abidizadegan et al. (2015)	Iran: Bander Deylam	(30°02′ N 50°08′ E)	Pswl St	*Ps. waltoni* *S. tenuis*
6	Barak et al. (1994)	Iraq: Khor Al-Zubair	(30°11′ N 47°53′ E)	Pswl	*Ps. waltoni*
7	Sarker et al. (1980)	Iraq: Shatt Al-Arab, Fao Estuary	(29°58′ N 48°29′ E)	Bd	*B. dussumieri*
8	Abdoli et al. (2012)	Iran: coast of Baucher Province	27°50′ N 51°56′ E	St	*S. tenuis*
9	Fang (1996)	China: Cangnan Longgang	(27°34′ N 120°34′ E)	Bpe	*B. pectinirostris*
10	Afshar et al. (2012)	Iran: coast of Hormozgan Province	27°11′ N 56°24′ E	Pswl	*Ps. waltoni*
11	Feulner and Roobas (2013)	United Arab Emirates: location undisclosed	(24° N 53° E)	Pswl	*Ps. waltoni*
12	Huang (2013)	Taiwan: Tainan Coastal Wetlands, Keliao	23°04′ N 120°02′ E	Psmi	*Ps. minutus*
		Taiwan: Tainan Coastal Wetlands, Spoonbills area	23°3′ N 120°02′ E		
		Taiwan: Tainan Coastal Wetlands	23° N 120°05′ E		
		Taiwan: Tainan Coastal Wetlands, Anping	23° N 120°8′ E		
13	Bhatt et al. (2009)	India: Navinal Coast, Kachchh	22°47′ N 69°36′ E	Pnsh	*Pn. schlosseri*
14	Chan (1989)	China: Mai Po Marshes Nature Reserve, Hong Kong	22°30′ N 114°02′ E	Psca Bpe Sh	*Ps. cantonensis* (=*Ps. modestus*) *B. pectinirostris* *S. viridis* (=*S. histophorus*)
15	Yang et al. (2003)	China: Deep Bay, Hong Kong	(22°29′ N 114°01′ E)	Bpe	*B. pectinirostris*
16	Chaudhuri et al. (2012)	India: Matla and Bidya Rivers, Sundarbans	22°10′ N 88°40′ E	Pde	*Pd. elongatus*
17	Chaudhuri et al. (2014)	India: Matla and Bidya Rivers, Sundarbans	22°10′ N 88°40′ E	Pde Psnm Bbo	*Pd. elongatus* *Ps. novemradiatus* *B. boddarti*
18	Rathod and Patil (2009)	India: Thane City, Maharashtra State	(19°12′ N–19°19′ N 72°51′ E–72°59′ E)	Bd	*B. dussumieri*

(Continued)

Table 10.1 (*Continued*) Mudskipper Diets Arranged by Study Location in a North to South Sequence

	Authority	Location	Coordinates	Species Code	Species
19	Mutsaddi and Bal (1969)	India: Thana Creek, Mumbai	(19°21′ N 72°58′ E)	Bbo	*B. boddarti*
		India: Mahul Creek, Mumbai	(18°59′ N 72°51′ E)		
20	Ravi (2013)	India: Pichavaram Mangroves, Muzhukkuthvrai	11°27′ N 79°47′ E	Bbo	*B. boddarti*
21	Bucholtz et al. (2009)	Vietnam: Canal in Bac Lieu Province	9°15′ N 105°44′ E	Pde	*Pd. elongatus*
22	Turay et al. (2006)	Sierra Leone: Congotown	8°27′ N 15°8′ W	Psb	*Ps. barbarus*
23	Swennen et al. (1995)	Thailand: Pattani	6°55′ N 101°20′ E	Bbo	*B. boddarti*
				Pde	*Pd. lanceolatus(=Pd. elongatus)*
				Pnsh	*Pn. schlosseri*
24	Hajisamae and Ibrahim (2008)	Thailand: Pattani	6°55′ N 101°20′ E	Pde	*Pd. lanceolatus(=Pd. elongatus)*
25	Norma-Rashid and Khaironizam (2005)	Malaysia: Langkawi Island	(6°26′ N 99°50′ E)	Psxx	*Periophthalmus spp.*
				Bbo	*B. boddarti*
26	Bob-Manuel (2011)	Nigeria: Rumuolumeni Creek, Niger River Delta	(4°47′ N 6°58′ E)	Psb	*Ps. koelreuteri(=Ps. barbarus)*
27	Udo (2002b)	Nigeria: Imo Estuary, Ikot Abasi	4°34′ N 7°48′ E	Psb	*Ps. barbarus*
28	Udo et al. (2008)	Nigeria: Ikot Abasi, Akwa Ibon State	4°25′ N–6°25′N 5° E–7°40′ E	Psb	*Ps. barbarus*
29	Chukwu and Deekae (2013)	Nigeria: New Calabar River	4°15′ N–4°45′N 6°57′ E–7°04′ E	Psb	*Ps. barbarus*
30	Zulkifli et al. (2012)	Malaysia: Klang Strait	3°04′ N 101°23′ E	Pnsh	*Pn. schlosseri*
31	Mazlan et al. (2006a)	Malaysia: Kuala Gula, Matang Mangrove Reserve, Perak	(4°50′ N 100°38′ E)	Pnsh	*Pn. schlosseri*
32	Khaironizam and Norma-Rashid (2004)	Malaysia: Klang Strait, Jeram	3°13′ N 101°18′ E	Psxx	*Ps. argentilineatus; Ps. chrysospilos; Ps. gracilis; Ps. spilotus; Ps. walailakae; Ps. novemradiatus;*
		Malaysia: Klang Strait, Sementa	3°5′ N 101°21′ E		
		Malaysia: Klang Strait, Morib	2°46′ N 101°26′ E		
33	Colombini et al. (1996)	Kenya: Tana River Delta, Shekiko Mangroves	2°35′ S 40°02′ E	Pss	*Ps. sobrinus(=Ps. argentilineatus)*
34	Kruitwagen et al. (2007)	Zanzibar: Chwaka Bay	6°08′ S 39°48′ E	Psa	*Ps. argentilineatus*
		Tanzania: Mtoni Estuary	6°52′ S 39°17′ E		
35	Takeda et al. (2011)	Australia: Darwin	12°34′ S 130 °53′ E	Psmi	*Ps. minutus*
36	Milward (1974)	Australia: Townsville	(19°15′ S 146°49′ E)	Sh	*S. histophorus*
37	Stebbins and Kalk (1961)	Mozambique: Saco da Inhaca	(26°01′ S 32°55′ E)	Psa	*Ps. argentilineatus*

Note: Location names and coordinates are provided by the authors and standardized to degrees and minutes. For studies that do not specify the precise location, or which appear incorrect, a "guesstimate" is provided in parentheses. Most of these studies are also cited in the tabulated prey of the mudskippers (Tables 10.2 and 10.3) in which the same species codes are used.

10.2.1 Seasonal Variation

Seasonal, size-related changes in the diet of *Ps. barbarus* from the mangrove forests of the Imo River estuary, Nigeria, are well studied (Udo 2002b). During the year-long single-monthly sampling from April 1992 to March 1993, 734 specimens were collected by setting traps at low tide and retrieving them just before high tide (King and Udo 1998). As a side issue, this study and others (Bob-Manuel 2011; Chukwu et al. 2010; Turay et al. 2006) used fish provided by local artisanal fishermen who typically baited their traps with crabs that had been ground-up. If baiting could be shown to enhance capture rates, it would be the first evidence that olfactory or gustatory cues are used during foraging in mudskippers.

The major food items of *Ps. barbarus* from the Imo River were coarse and fine particulate detritus (50.64% expressed by the Modified Food Object Number [MFON]) followed by algae (32.5%) that was predominantly composed of the diatoms *Coscinodiscus* (15.4%), *Navicula* (8.2%), and *Pleurosigma* (4.8%) (Udo 2002b). Detritus (55.1%) was a significantly more important food item as was the diatom *Coscinodiscus* (17.8%) in the wet season (May to October). In the dry season (November to April), because consumption of *Navicula* (9.3%) and *Pleurosigma* (9.1%) increased significantly, overall, algae (38.6%) were the most important food item. There was also a marked increase in diet breadth as measured by Simpson's diversity index (D) in the dry season (0.9) when compared with the wet season (0.8).

In the mangroves of the New Calabar River, a tributary river in the adjacent Niger delta, Chukwu and Deekae (2013) also reported differences in prey items of *Ps. barbarus* between the wet and dry seasons as measured by the Index of Food Significance (IFS). The diet is, however, very different to that of this species from Imo River. Sampling bimonthly from February 2008 to January 2009, gut analyses showed fish scales and a variety of crabs and insects being dominant to the diet, with only little contribution from plant material, and the absence of diatoms. Whereas fish scales had the overall highest individual component IFS index (23.3% and 25.6% wet and dry seasons, respectively), it is likely that crabs were the most significant dietary component if the separate categories (*Ocypode* sp., parts of crabs, unidentified crabs, and *Sesarma* sp.) were pooled. Significant increases in gut content indices between the wet and dry seasons, as indicated by IFS values, were of unidentified crabs (7.5 and 13.5, respectively), annelids and nematodes (1.0 and 3.7), and *Penaeus* sp. (0.1 and 3.1). The mudskippers were more active during the dry season and possibly foraged more in the water as suggested by the increased shrimp consumption and the consistently, but not significantly, increased indices for the four species of molluscs (Chukwu and Deekae 2013) which again may have been significant if considered as a single category.

A second study in the Niger Delta from another mangrove area in Rumuolumeni Creek in Port Harcourt collected *Ps. barbarus* (reported as *Ps. koelreuteri*) on a weekly basis during the wet season from March 2009 to August 2009 (Bob-Manuel 2011). The analyses centered on a comparison of the diets of juveniles (46–63 mm SL) and adults (63–150 mm SL). The overall diet was different to that of the population from New Calabar River (Chukwu and Deekae 2013) and was more similar to the populations from the Imo River due to the presence of significant levels of diatoms (Udo 2002b). The points method of gut analysis showed that diatoms (29.8%), plant fragments (38.2%), and algal filaments (23.6%), accounted for the diet of juveniles which in adults, were merely found as traces. In adults, crustacean appendages (36.4%), insect parts (22.7%), and polychaetes (20.7%) were the main dietary elements, indicating that there was an ontogenetic shift from a predominantly herbivorous diet to a carnivorous one.

Rathod and Patil (2009) divided the food categories of *B. dussumieri* from the Ulhas River estuary in northwestern India into six food categories. Whereas "diatoms," "nematodes," and "harpacticoid copepods" were individual categories, the remaining categories were broad: "other algae," "plant materials," and "others," the latter included detritus, sand, mud, fish scales, ova, and unidentified food items. Based on the percentage frequency of occurrence, the categories were divided into main (>50%), compensatory (10–49%), and accidental/emergency (<10%) food items. These designations are somewhat misleading when, for example, the harpacticoid copepods during the monsoon period were compensatory, but became a main item in early post-monsoon (October–December) and thereafter accidental (January–May). During the 2-year study, the only main item during the pre-, post-, and monsoon periods were diatoms. Nematodes were always considered as compensatory.

Working in the same estuarine system as Rathod and Patil (2009), Mutsaddi and Bal (1969) showed seasonal changes in the diet of *B. dussumieri*. During the monsoon period, the percentage occurrence of algae, nematodes, and diatoms was reduced. Fish eggs (60% occurrence) were found only in June, and the highest intake of nematodes was post-monsoon (Mutsaddi and Bal 1969). Both studies reported the highest intake of algae premonsoon although this was not the case for *B. boddarti* from southeastern India. Ravi (2013) reported that algal intake by *B. boddarti* was higher after the monsoon, and confirmed that there was increased nematode intake post-monsoon. Diatoms formed the main part of the diet, but due to terminology inherent in the study, these became compensatory food items post-monsoon (Rathod and Patil 2009).

Ravi (2013) identified nine genera of diatoms in descending overall percentage occurrence: *Pleurosigma directum*, *Coscinodiscus centralis*, *Nitzschia granulata*, and *Navicula gracilis*. The percentage occurrences of these diatoms varied greatly between seasons. During the premonsoon months, *P. directum* was the most significant diatom (34%) followed by *C. centralis* (7.7%), *N. granulata* (2.8%), and *Na. gracilis* (1.3%) but during the monsoon months, the order was reversed: *Na. gracilis* (13.8%), *N. granulata* (13.5%), *P. directum* (10.6%), and *C. centralis* (8.3%). Diatoms were the predominant food items of *B. dussumieri* (reported as *B. boddarti*) in Khor Al-Zubair in southern Iraq, but during the summer (June and July) diatoms were superseded by blue-green algae (Al-Daham and Al-Noor 2000). While polychaetes were included as a food item between July and September, their low percentage occurrence was attributed to accidental ingestion (Al-Daham and Al-Noor 2000).

10.2.2 Ontogenetic Changes

Udo (2002c) considered size-related changes in the diet of *Ps. barbarus* and divided fishes he studied into small and large size classes based on total length attained at maturity. For comparability with Bob-Manuel (2011), the equivalent standard lengths were calculated using the formula for the relationship between the two measures (Turay et al. 2006). The small (31–73 mm SL) and large (81–113.5 mm SL) fish categories clearly overlap between the two studies. In agreement with Bob-Manuel's (2011) findings, larger fish became more carnivorous. The MFON values of the crustacean and fish prey categories of large fish (10.4 and 0.2, respectively) were significantly greater than that of the small fish (1.1 and 1.8). In contrast, large mudskippers took significantly fewer insects (1.8 vs. 2.9 for small fish) and while the diatoms were reduced in large mudskippers (MFON values: 25.7 vs. 35.6), they still formed a major part of their diet (Udo 2002b).

In the absence of any algae or diatoms in its diet, *Ps. waltoni* from Khor Al Zubair in Iraq was characterized as a carnivore; its diet also changed with size (Barak et al. 1994). Smaller individuals (<70 mm TL) took mainly copepods (83.3% occurrence), whereas larger (>90 mm TL) ones became piscivorous (80% occurrence). The source or species of fish eaten was not specified so it is unclear if cannibalism (see Zulkifli et al. 2012) was involved. A similar size-related change of diet was found in *Ps. argentilineatus* from East Africa (Kruitwagen et al. 2007). In Zanzibar, amphipods and copepods were the main (>50% gut content) dietary items in small (<50 mm TL) fishes.

In larger fishes (>70 mm TL) decapod crustaceans, mostly species of *Uca*, increased in the diet composition and dominated it (>60% gut content) in the largest (100–130 mm TL) individuals. In contrast, on mainland Tanzania, the fishes were smaller (40–80 mm TL) and their diet did not show any clear size-related changes except that crabs only appeared in the diet of the larger (>70 mm TL) fish. Polychaetes, amphipods, fishes, and insects were the main prey items of this population (Kruitwagen et al. 2007).

Juvenile *Pn. schlosseri* (95–107 mm SL: identified by the absence of gonads) and adults (155–222 mm SL) from Malaysia take the same prey groups but adults took more fiddler crabs (53.6%) and fewer fishes (38.1%) than did juveniles (26% and 65%, respectively). When the adult data are separated by sex, prey items of females are similar (fiddler crabs 27% and fishes 57%) to those of juveniles (Zulkifli et al. 2012). The fish prey, predominantly the killifish *Oryzias* sp., are relatively small fishes (up to 33 mm TL).

Using the frequency of occurrence, Ravi (2013) showed that ontogenetic diet changes also occurred in *B. boddarti*. Although the basis on which the fish were divided into juveniles and adults was not specified, they had a similar diet. However, there was a reduction in diatoms (56%) and an increase in nematodes (9.5%) and polychaetes (8.5%) in adults over the values found in juveniles (71.5%, 4% and 3.6%, respectively). This supports the earlier finding of Mutsaddi and Bal (1969) where the overall diet of *B. dussumieri* juveniles (as identified by undeveloped gonads: 16–95 mm SL) was predominantly of animal origin, with copepods, crustacean larvae, nematodes, and polychaetes having a higher combined percentage occurrence than that of the diatoms and algae. The same prey items were also found in adults (96–185 mm SL), but the relationship was reversed such that the diatom percentage in particular outweighed that of the animal component of the diet. Splitting the data into 10 mm increments of SL revealed that the smallest juveniles fed exclusively on animal material but, with increasing fish size, the plant percentage occurrence increased, exceeding that of animal occurrences in fishes measuring ≥60 mm SL. In the largest juveniles, the proportion becomes 80% plant and 20% animal. In all adult sizes, the ratio remains essentially the same at about 70% plant and 30% animal. Adults also consumed fish eggs, but nematodes (20% occurrence) accounted for the largest proportion of their animal intake, the latter only appearing in the diet of juvenile *B. boddarti* larger than 66 mm SL (Ravi 2013).

Nanjo et al. (2008) studied food habits of 67 fish species from the mangrove estuary of the Urauchi River on Iriomote Island in southern Japan. Seventeen of them, including *Ps. argentilineatus*, were identified as exhibiting an ontogenetic dietary shift. The mudskipper was among the largest fish guild in the estuary feeding on large benthic crustaceans. The adult (61–81 mm SL) diet predominantly consisted of crabs, and also included isopods, shrimp, gastropods, detritus, gammarid amphipods, and sedentary polychaetes. The latter two items predominated in the diet of smaller (24–60 mm SL) fish decreasing in importance with increasing fish size.

10.2.3 Diel Variation

Studies on the diel changes in the diet of *Ps. barbarus* were also carried out from the mangrove forests of the Imo River estuary in Nigeria, from 734 specimens collected between March 2004 and April 2005 (Udo et al. 2008). The sampling method follows Udo (2002b) detailed in Section 10.2.1. The collection technique poses uncertainties for a diel study. With several hours between setting the traps and collection, the dawn, day, dusk, and night categories are inappropriate if only because fish may be in the traps for the duration of the flood tide (Udo et al. 2008). This has significance for the temporal analyses since it is likely that gut retention times of prey vary considerably. Udo et al. (2008) considered *Ps. barbarus* to be crepuscular, but given the concerns about the short dawn and dusk periods in relation to sampling frequency, only their day/night data will be discussed here. Aside from the absence of detritus, the diet remained much the same as previously reported

(Udo 2002b). Basically, visually hunted prey was reduced, whereas diatoms were increased in the nocturnal diet. The MFON values indicated reduced nocturnal consumption of crustaceans, particularly *Sesarma* and shrimps (0.5% from 6.1%), fishes (0.5% from 1.5%), and molluscs (0.02% from 0.8%) but not insects (5.4% from 2.6%). Algae values increased (39.5% from 27.0% in the day) because of the increased consumption of *Coscinodiscus*, *Navicula*, and *Pleurosigma* diatoms.

Like Udo et al. (2008), Colombini et al. (1996) demonstrated clear differences between day and night foraging. However, even though Colombini et al. (1996) were careful to maximize the counting of prey by only considering stomach contents and largely intact specimens, they also included polychaete mandible pairs as well as arthropod heads. The latter items may well have been caught considerably earlier than the 2-hour fish capture interval so that their capture times in relation to the tidal cycle may be earlier than reported.

Baeck et al. (2013, 2008) also confirmed that the diet of *Periophthalmus* spp. is subject to diel and ontogenetic changes and further added details of resource partitioning between two sympatric species: *Ps. modestus* and *Ps. magnuspinnatus*. For the diel study, a separate collection of mudskippers for the size-related diet changes was made every 2 hours over 24 hours of a full-moon day during May (premonsoon) and August (monsoon). Both species foraged throughout the 24-hour period and whereas polychaetes again dominated the diet of *Ps. modestus*, the difference between day (May 61.4% and August 58.3% by weight) and night (May 52.4% and August 64.7%) or seasons is relatively minor. More amphipods (Amphipoda), however, were taken at night (May and August 33.4%) than during the day (May 10.2% and August 16.1%) (Baeck et al. 2013). The dominant prey item of *Ps. magnuspinnatus* is the crab *Ilyoplax pusilla*, other prey items included other crab species, amphipods, polychaetes, and insects. In August, crabs were taken more frequently during the day than night (50.2% and 32.2% by weight, respectively) and amphipods less so (13.4% and 27.9%, respectively). In May, however, the diurnal differences for these prey items were marginal (crabs: day 34.1% and night 31.1%; amphipods: day 28.7% and night 25.8%), but polychaete intake increased at night (day 15.2% and night 24%) (Baeck et al. 2013).

The 2-hourly sampling study of *Ps. argentilineatus* (Colombini et al. 1996) is discussed in Section 10.3.1, but feeding studies in general, and diel and gut fullness ones in particular, would be improved if more was known of gut evacuation rates (GERs) of mudskippers. Based on other piscivorous studies, temperature, fish size, and food composition and size may be factors that affect the GER, but none were considered in reports dealing only with mudskippers. Feces of *Pn. schlosseri* (reported as *Ps. barbarus*) and *B. boddarti* were collected at intervals after they had been fed charcoal particles and an algal–charcoal mixture, respectively (Lim 1967). Complete passage, as measured in charcoal output for both species, and additionally for undigested, broken or empty algal cells for *B. boddarti*, was about 35 hours. Reanalyses of the data, however, showed that 50% of the charcoal was evacuated within 15 hours for *Pn. schlosseri* and 16 hours for *B. boddarti*. For the combined algal categories, 50% was evacuated within 8 hours (Lim 1967). In contrast, Mazlan et al. (2004) reported that the gut evacuation time for *Pn. schlosseri* was 18 hours before feeding recommenced.

10.2.4 Sex Differences

Considering the number of fishes collected for gut analysis, it is surprising that few studies compare the diets between male and female fishes, preferring instead to report on size or adult diet. Perhaps, as Ravi (2013) found in both the overall or seasonal diet of *B. boddarti*, there is little difference in gut content and prey items between male and female mudskippers. Similarly, Udo (2002a) reported that there was little difference in the diets of male and female *Ps. barbarus* from the Imo estuary, but the index of food dominance showed that males took significantly more cyanobacteria, *Coscinodiscus* diatoms, and *Sesarma alberti* crabs than did the females. However, Zulkifli et al. (2012) demonstrated a difference in the diet of male and female *Pn. schlosseri* on the intertidal mudflats of the Puloh River that flows into the Klang Strait, Peninsular Malaysia. Leaving aside the text and figure discrepancies, it is clear that the subadult and adult diets were different. Whereas the adult male diet was predominantly composed of *Uca* spp., these featured less importantly in the diet of females in which *Oryzias* sp. was the main dietary component.

10.3 RELATIONSHIP BETWEEN DIET AND PREY AVAILABILITY

Following Mutsaddi and Bal's (1969) paper on the feeding behavior of the herbivorous mudskipper *B. dussumieri* in India, other studies on *Boleophthalmus* spp. (Zhang et al. 2006; Yang et al. 2003; Sponder and Lauder 1981; Pankow and Hug 1979) provided details of gut contents and compared it with prey species availability. Zhang et al. (2006), for example, investigated the nutritional value of particulate organic matter (POM) on which larval and young adult *B. pectinirostris* feed. They quantified both the community of organisms that included protozoa, bacteria, diatoms, and nematodes, as well as the free amino acid composition of the POM. This has not been attempted with any of the carnivorous species and it is therefore constructive to indirectly attempt such a comparison.

10.3.1 Benthic Surveys

It is possible to combine the details of faunal distribution in the mangroves at Saco da Inhaca, Mozambique (MacNae and Kalk 1962), with the prey items of *Ps. argentilineatus* identified by Stebbins and Kalk (1961), the details of which are provided in Table 10.2. The mudskippers were common among the pneumatophores of the fringing *Avicennia* mangrove trees where the ocypodid crab *Uca annulipes* was abundant and also among the buttressed roots of *Rhizophora* where *Uca urvillei* and *Uca chlorophthalmus* were also abundant. The *Rhizophora* mangrove trees line tidal channels along which the mudskippers migrate with the tide. Given their common location, it is not surprising that the "shade-loving" *Uca chlorophthalmus* is included in the fish's diet. However, why there are no similarly "shade-loving" *U. urvillei*, *U. annulipes* or the other widely distributed fiddler crab, *Uca inversa*, found in the diet? It simply may be that insufficient sampling was undertaken or that the fiddler crabs were too large, this latter contention being supported by the presence of the small sand crab, *Dotilla fenestrata*, and not the sympatric *U. annulipes* in the diet. These considerations do not explain why polychaetes, which Macnae and Kalk (1962) reported as being rare in the mangrove mud, are found in the diet. Most of the prey items are taken terrestrially from the mud, but the presence of crustacean larvae (megalopae and schizopods) suggests that *Ps. argentilineatus* may also feed aquatically using suction feeding (K.B. Michel, personal communication, 2016) as do other fishes.

These questions highlight the need for a more considered approach to the study of mudskipper predation and answers to these and other questions obviously require more than just gut content analyses. Colombini et al. went part way to providing some of the answers. Detailed observations of the movement of *Ps. argentilineatus* throughout the day and night of spring and neap tides (Colombini et al. 1995), together with gut content analysis of the mudskippers collected at 2-hour intervals, suggested that this movement was related to foraging (Colombini et al. 1996). The fish were observed in a muddy *Avicennia* mangrove forest on the Tana River delta in Kenya where the fiddler crabs *U. urvillei*, *U. inversa*, and *U. annulipes* were found, which is a habitat broadly similar to where Stebbins and Kalk (1961) worked.

Similar numbers of *Ps. argentilineatus* were seen during spring (314) and neap (315) tides. During spring tides, all fish were equally active between day (152) and night (162), but during neap tides only about 25% (43/121) were active during the day whereas considerably more (119/194) were active at night with the remainder resting on their turreted burrow entrances. The active *Ps. argentilineatus* foraged in different zones depending on the tidal cycle. During spring tides, they

Table 10.2 Prey Items Consumed by Species of *Periophthalmus*, Summarized to Phylum, Class, or Order

Genus		*Periophthalmus*																			
Species Code		Psca	Psmg				Psmo	Psms	Pswl			Psca	Psb					Psnm	Psa	Psxx	
Higher Classification	Lower Classification	Ryu et al. (1995)	Baeck et al. (2008)	Huang (2013)	Chaudhuri et al. (2014)	Takeda et al. (2011)	Baeck et al. (2013)	Baeck et al. (2013)	Barak et al. (1994)	Afshar et al. (2012)	Feulner and Roobas (2013)	Chan (1989)	Turay et al. (2006)	Bob-Manuel (2011)	Udo (2002b)	Udo et al. (2008)	Chukwu and Deekae (2013)	Chaudhuri et al. (2014)	Stebbins and Kalk (1961)	Norma-Rashid and Khaironizam (2005)	Khaironizam and Norma-Rashid (2004)
Total number of fish examined		NA	13	Obs.	NA	60	494	455	76	267	Obs.	488	364	405	734	734	450	NA		NA	120
Annelida																	Psb		Psa		
	Polychaeta			Psmg			**Psmo**	Psms		Pswl	Pswl	Psca	**Psb**	**Psb**					Psa	**Psxx**	Psxx
	Oligochaeta											Psca									
Nematoda							Psmo	Psms				Psca			Psb	Psb	Psb		Psa	Psxx	Psxx
Arthropoda	Crustacea	**Psca**		Psmg										**Psb**	Psb	Psb					**Psxx**
	Amphipoda		Psmg		Psmg	**Psmg**	Psmo	**Psms**				Psca							Psa		Psxx
	Tanaidacea							Psms				Psca									
	Ostracoda						Psmo														
	Copepoda				Psmg	Psmg	**Psmo**	Psms	**Pswl**			Psca						Psnm		Psxx	Psxx
	Isopoda				**Psmg**			Psms							Psb	Psb		Psnm	Psa	Psxx	Psxx
	Decapoda		**Psmg**		**Psmg**	**Psmg**	Psmo	Psms	Pswl	Pswl	**Pswl**				Psb		**Psb**		Psa	**Psxx**	Psxx
	Brachyura						Psmo	**Psms**				Psca	**Psb**						Psa		
	Cladocera				Psmg				Pswl									Psnm			Psxx
	Insecta	Psca		Psmg	Psmg	Psmg			Pswl	Pswl			Psb	**Psb**	Psb	Psb	**Psb**	**Psnm**	Psa	Psxx	
	Coleoptera																Psb				
	Diptera						Psmo	Psms			Pswl	Psca			Psb						Psxx
	Hemiptera											Psca									
	Hymenoptera					Psmg						Psca							Psa		Psxx
	Myriapoda															Psb					
	Chilopoda														Psb						
	Arachnida						Psmo	Psms				Psca				Psb					Psxx
Mollusca													Psb		Psb	Psb					
	Gastropoda					Psmg	Psmo	Psms		**Pswl**		Psca			Psb		Psb			Psxx	Psxx
	Bivalvia					Psmg	Psmo					Psca					Psb			Psxx	Psxx
Chordata	Teleost			Psmg	**Psmg**				**Pswl**	Pswl				Psb		Psb	**Psb**				Psxx
	Teleost eggs						Psmo	Psms				**Psca**									
Cyanobacteria															Psb	Psb					
Chrysophyta																	Psb				
	Diatoms	Psca			Psmg									Psb	**Psb**			**Psnm**		Psxx	Psxx
	Microalgae																	Psnm			Psxx
Chlorophyta														Psb	Psb	**Psb**	Psb				
Rhodophyta															Psb	**Psb**					
Detritus/sediment				Psmg	Psmg	Psmg							Psb	Psb	Psb	Psb	Psb	Psnm		Psxx	Psxx
Fungi															Psb	Psb					

Note: Species codes in red indicate main food item(s) from each study. The left to right sequence in the table reflects the north to south location of the fish for both the individual species and within species comparisons. Species codes follow Table 10.1.

Table 10.3 Prey Items Consumed by Species of *Boleophthalmus*, *Pseudapocryptes*, *Scartelaos*, and *Periophthalmodon*, Summarized to Phylum, Class, or Order

Genus		*Boleophthalmus*								*Pseudapocryptes*			*Scartelaos*				*Periophthalmodon*		
Species Code		Bpe		Bd				Bbo		Pde	Pdl		St		Sv	Sh	Pnsh		
Higher Classification	Lower Classification	Ryu et al. (1995)	Fang (1996)	Taher (2010)	Sarker et al. (1980)	Chaudhuri et al. (2014)	Mutsaddi and Bal (1969)	Chaudhuri et al. (2014)	Ravi (2013)	Chaudhuri et al. (2012)	Swennen et al. (1995)	Hajisamae and Ibrahim (2008)	Abidzadegan et al. (2015)	Abdoli et al. (2012)	Chan (1989)	Milward (1974)	Swennen et al. (1995)	Mazlan et al. (2006a)	Colombini et al. (1996)
Total number of fish examined		NA	24	40	159	NA	940	NA	156	25	36	34	50	150	24	NA	11	Obs.	96
Annelida										Pde								Pnsh	Pnsh
	Polychaeta						Bd		Bbo						**Sv**				Pnsh
	Oligochaeta																		**Pnsh**
	Hirudinea																		Pnsh
Nematoda			Bpe				Bd		Bbo						Sv	**Sh**			**Pnsh**
Khinorhyncha																Sh			
Arthropoda	Crustacea	Bpe									Pdl					Sh			Pnsh
	Amphipoda									Pde									Pnsh
	Tanaidacea																		**Pnsh**
	Ostracoda															Sh			Pnsh
	Copepoda		Bpe				Bd			Pde				St		Sh	Pnsh		Pnsh
	Isopoda									Pde									Pnsh
	Decapoda									**Pde**		Pdl	St	St			Pnsh		Pnsh
	Brachyura																	**Pnsh**	
	Cladocera									Pde									Pnsh
	Insecta												St					Pnsh	Pnsh
	Diptera														Sv	Sh			
	Myriapoda																		
	Acari																		Pnsh
Mollusca																			Pnsh
	Gastropoda													St					
	Bivalvia												**St**	St					
Cnidaria				Bd															
Chordata	Teleost									**Pde**							Pnsh		
	Teleost eggs		Bpe	Bd			Bd		Bbo						Sv				
Cyanobacteria			Bpe		Bd														
	Diatoms	**Bpe**	**Bpe**	Bd	**Bd**		**Bd**	**Bbo**	**Bbo**		Pdl	**Pdl**			**Sv**	**Sh**			
	Microalgae					**Bd**	Bd	Bbo				Pdl			**Sv**	Sh			
Chlorophyta				**Bd**	Bd				Bbo		**Pdl**								
Detritus/ sediment				Bd		Bd			Bbo	Pde	Pdl								
Plankton										Pde									
Foraminifera														St					

Note: Species codes in red indicate main food item(s) from each study. The left to right sequence in the table reflects the north to south location of the fish for both the individual species and within species comparisons. Species codes follow Table 10.1

followed the tide but remained in the mangroves at about the same mean distance from the extreme high water mark during the day (22.4 ± 2.5 m) and night (26.1 ± 4.2 m). During neap tides when the upper and midtidal areas were dry, the fish foraged close to the mean neap low-water mark, moving out from under the mangroves onto the exposed mudflat further during the night (62.3 ± 0.9 m) than during the day (59.3 ± 3.3 m) (Colombini et al. 1995).

Based on stomach analyses of 96 specimens of *Ps. argentilineatus*, 46 of which were caught during the day, more than five times the number of prey was taken during spring tides (1006) as opposed to neap tides (178) and almost double the number of prey items was taken during the day (763/spring 616) than night (421/spring 147). During spring tides, tanaid crustaceans accounted for more than 40% (407/1006) of the total number of prey, with nematodes, copepods, oligochaetes, and insects in descending order of numerical order accounting for the next 43%. During neap tides, nematodes were the most common prey (20.2%: 36/178) with the next 42% made up of polychaetes, oligochaetes, and insects. More nematodes, tanaids, insects, amphipods, and archianelid polychaetes were eaten during the day, whereas more polychaetes and oligochaetes were consumed at night (Colombini et al. 1996). As determined by the time of prey capture in relation to the tidal cycle, most feeding activity of *Ps. argentilineatus* occurred soon after spring low tides or neap high tides. These periods accounted for the majority of prey groups in the diet with the exceptions that insects were caught much closer to high spring tides and tanaids just before neap low tides (Colombini et al. 1996).

Diel or tidal differences in prey capture may be related to spatial or temporal distributions of the prey species (Colombini et al. 1996). Vertical (Nagelkerken et al. 2008; Kim et al. 2000; Vijayakumar et al. 1991) and horizontal (Shabdin and Othman 2008; Somerfield et al. 1998) distributions of prey, their association with mangroves (Ansari et al. 2014; Vijayakumar et al. 1991) or even upstream location in the estuarine habitat (Yeager et al. 2014; Kondalarao 1983) will all influence the availability of potential prey. Thus to answer the simple question as to whether *Periophthalmus* spp. are opportunistic, taking items in similar proportions to that found in the environment requires an approach that includes an equally detailed study of their macro- and meiofaunal prey.

Other than diatoms (see Section 10.2.1 for detailed discussion), nematodes varied seasonally in the diet of *B. boddarti* (Ravi 2013). It was lowest in the monsoon period (3.9% in October for males and 2% in December for females) and thereafter increased through the post-monsoon to a maximum in summer (19.3% in June for males and 15.1% in June for females). Similarly, the contribution of polychaetes as prey items was lowest in the monsoon period (3.5% October for males and 3.2% November for females) and increased in post-monsoon and early summer (15.8% March for males and 14.7% June for females) (Ravi 2013).

In a survey of meiofuna assemblages (retained on a 63 μm sieve) in the Prichavaram mangrove area, the same locality as the study of Ravi (2013), Chinnadurai and Fernando (2007) reported maximum densities of 827 and 890 individuals per 10 cm^2. The study was carried out in January, at the start of the post-monsoon season, over 2 years. The study found nematodes to be the most dominant (>85%) of the seven taxa obtained; other taxa in descending order of abundance were copepods, polychaetes, foraminifera, kinorhynchs, and reported separately, the foraminiferan allogromiids and ostracods. A similarity matrix and multidimensional scaling revealed that 37 nematode species have a "clear relationship to the type of mangrove cover" such that 16 of these species, including those taken as prey items by *B. boddarti*, occurred in association with both *Avicennia marina* and *Rhizophora apiculata* dominated mangrove systems. Nematode assemblages within the Rokkaku River estuary in Saga, Japan, a habitat of *B. pectinirostris*, showed similar results in depth distribution and seasonal variation (Matsumoto et al. 1992). The adjacent coastal area in the Ariake Sea, however, houses different nematode species composition, but mudskipper predation of nematodes was not discussed (Matsumoto et al. 1992).

Pravinkumar et al. (2013) also carried out a survey of meiofuna assemblages in the Prichavaram mangrove area but this study spanned a full year. They identified 16 species of polychaetes (retained on a 500 μm sieve) as the dominant group (47.5%) followed by crustaceans (41.7%), gastropods, and bivalves. During the monsoon season, the macrobenthic population densities were low and thereafter increased to reach a maximum in May (Pravinkumar et al. 2013). Thilagavathi et al. (2013) reported similar macrobenthos distribution patterns from Pichavaram, as well as from a riverine and an island mangrove ecosystem further south along the Tamil Nadu coast.

Changes in polychaete abundance are reflected in the diet of *B. boddarti* such that the percentage occurrence in the monsoon period was low, increasing in the post-monsoon and summer periods (Ravi 2013). Similarly, the same seasonal variation in nematode abundance (Kathiresan 2000) was also reflected in the diet of *B. boddarti*. However, the second most common taxon in the diet of these fishes, the copepods, was absent in the benthic studies.

As these benthic surveys were not conducted at the same time as the diet study, they can only be indicative of prey availability. Additionally, the community may change dramatically over time, as exemplified in Kakinada Bay, within the Godavari delta, where *B. boddarti* (reported as *B. sculptus*) is also found (Raut et al. 2005). Nevertheless, such comparisons suggest that there may be some form of prey selection being made by *B. boddarti*. Ideally, investigations of the diet and benthic surveys should be conducted concurrently, as was done for the estuarine goby *Pomatoschistus microps* (Jackson and Rundle 2008) and in the analysis of the feeding guild structure of fishes in the Matla and Bidya Rivers of the Indian Sundarbans (Chaudhuri et al. 2014).

10.3.2 Community Structure

Non-metric multidimensional scaling based on the percentage abundance of prey items for 31 fish species investigated in the Sundarbans, divided them into five guilds, of which the only herbivore was *B. boddarti*. *Periophthalmus novemradiatus* was considered as an insectivore within the larger plankti-benthivore guild while *Pd. elongatus* was a "cranci-piscivore" within the carnivore guild (Chaudhuri et al. 2014). Conventional percentage of occurrence data for the diet of *Pd. elongatus* included shrimps (17%), phytoplankton (14%), fishes (14%), cladocerans (11%), aquatic insects (10%), copepods (10%), isopods (5%), and amphipods (3%). However, as the abundance of prey items in the water column and mudflat was measured, it was possible to identify the prey preference of the fish by comparing Chesson's α index with random feeding values. Food items with an index greater than the random feeding value are considered as positively selected so that fishes ($\alpha > 0.24$), isopods ($\alpha > 0.21$), and shrimps ($\alpha > 0.09$) were preferred food items. The other items with an index less than the random value were considered as being avoided despite in some cases (e.g., fishes and aquatic insects) having a higher percentage occurrence (Chaudhuri et al. 2014).

In the past, to understand the context in which mudskippers forage, habitat information was gleaned from individual studies (Clayton 1993). Increasingly, mudskipper habitat descriptions are becoming more precise (Polgar and Bartolino 2010; Polgar and Crosa 2009; Norma-Rashid and Khaironizam 2005; Takita et al. 1999; Swennen et al. 1995), and combined with studies considering mudskippers in the wider context of the fish community to which they belong (Abdul-Razak et al. 2015; Jin et al. 2010, 2007; Quan et al. 2009) will aid greatly in understanding the environmental framework in which mudskippers compete and forage.

10.3.3 Stable Isotope Studies

Primary production sources, whether the producer is benthic or pelagic, and trophic levels of organisms can be determined by stable isotope values of carbon, sulfur, and nitrogen, respectively. The method is well established (DeNiro and Epstein 1978), but its application for mudskippers has been limited. Despite the early study of Rodelli et al. (1984) using $\delta^{13}C$ tracing mangrove carbon in a Malaysian ecosystem that included mudskippers, it is only in the last decade that this has been extended by dual isotope ($\delta^{13}C$ and $\delta^{15}N$) studies from East Africa (Kruitwagen et al. 2010, 2007),

China (Shang et al. 2008), Pacific islands (Kawamura et al. 2011), and Peninsular Malaysia (Zulkifli et al. 2014, 2012), the latter two studies being in the same general location as Rodelli et al. (1984).

Through $\delta^{13}C$ and $\delta^{15}N$ analyses, the diet of *Ps. argentilineatus* was placed in the wider context of the mangrove-mudflat community from the Mtoni estuary in East Africa (Kruitwagen et al. 2010). For mudflat and mangrove habitats, carbon stable isotope signatures were obtained from primary producers through macroinvertebrates to 30 fish species. The mean $\delta^{13}C$ signature for mangrove zoobenthivores was about −15.5‰, but *Ps. argentilineatus*, classified in this group, showed a more depleted carbon signature (−18.9‰). As an animal shows a $\delta^{13}C$ enrichment of about 1‰ relative to its diet, this value suggests that the mudskipper fed on mangrove-associated crabs of the genus *Uca* rather than *Sesarma*. *Uca* had $\delta^{13}C$ signatures (approximated mean values for species: −20.7 to −18.7‰) closer to that of the mudskippers than did the mangrove-leaf eating sesarmid crabs (−23.5 to −21.5‰). With a $\delta^{13}C$ signature of about −13.5‰ (Kruitwagen et al. 2007), *D. fenestrata* is unlikely to be part of the mudskipper's diet.

Stable isotope data were collected for a study of *Pn. schlosseri* on the mudflats of the Puloh River estuary within the Klang Strait (Zulkifli et al. 2012). Stable isotope data between this study and that from East Africa (Kruitwagen et al. 2010) are comparable to an extent; similar $\delta^{13}C$ values were observed for mangrove leaves (−26.5‰ and −26‰, respectively) and *Uca* spp. (−21.1‰ and −20.7‰ to −18.7‰). However, $\delta^{13}C$ values for the two species of mudskipper were different between the two studies. The mean $\delta^{13}C$ values for adult male and female *Pn. schlosseri* were −20.3‰ and −20.4‰, respectively (Zulkifli et al. 2012), although Rodelli et al. (1984) gave a value of −19.5‰ for the same species without specifying either the sex or life stage of the individuals.

The $\delta^{13}C$ isotopic composition of an animal reflects its diet with an average enrichment of about 1‰ relative to its diet (DeNiro and Epstein 1978). Accordingly, large *Uca* spp. (−16.4‰) were omitted as a possible prey item for *Pn. schlosseri* (Zulkifli et al. 2012). Females fed on *Oryzias* spp. (−21.4‰) and small *Uca* spp. (−21.1‰), whereas males fed only on small *Uca* spp. Given that the isotope signatures of subadult and juvenile *Pn. schlosseri* were similar to that of adults, it was suggested that the adults might be cannibalistic. Of equal possibility is that the adult males were taking other prey items with similar $\delta^{13}C$ values with obvious candidates being polychaetes (about −18.5‰: Kruitwagen et al. 2010) or *Nereis* sp. (−18.9‰: Rodelli et al. 1984) from the same location sampled by Zulkifli et al. (2012).

In a preliminary study some 15 km northwest off the coast on the Sungai Janggut intertidal mudflats, Zulkifli et al. (2014) obtained 10 stable isotope values for some organisms including algae, sediment, several crustacean and mollusc species, as well as two mudskipper species: *Ps. variabilis* (reported as *Ps. novemradiatus*) ($\delta^{13}C$ −21.5 ± 0.2‰; $\delta^{15}N$ 9.5 ± 0.3‰) and *B. boddarti* ($\delta^{13}C$ −16.2 ± 0.1‰; $\delta^{15}N$ 12.8 ± 0.4‰). Comparison of these values with those of the other organisms sampled led to the conclusion that *Ps. variabilis* is a molluscivore as the mean $\delta^{13}C$ values for five molluscan species ranged between −19.7‰ and −15.6‰. However, *Uca vocans* ($\delta^{13}C$ −21.4 ± 0.1‰) would also seem to be a likely prey item since there was also a large difference between the $\delta^{15}N$ values of the mudskipper and the crab. Algae ($\delta^{13}C$ −18.7 ± 0.7‰) rather than detritus ($\delta^{13}C$ −24.4 ± 0.9‰) were considered to be the major food source for *B. boddarti*. The methods employed by Zulkifli et al. (2014) to collect algal samples adhering to submerged vegetation, that is, by shaking it in algal free water and by taking 2 cm depth of surface sediment, would be unlikely to sample the same species that *B. boddarti* grazes on.

Rodelli et al. (1984) distinguished between the predatory *Periophthalmus* spp. ($\delta^{13}C$ −22.9–22.3‰) and the algae-eating *Boleophthalmus* spp. ($\delta^{13}C$ −15.7‰) without commenting on their prey items. From the $\delta^{13}C$ values for other sampled animals, it was likely that *Uca* spp. (−23.6‰ to −19.3‰), rather than *Sesarma* spp. (−27.2‰–22.4‰) or molluscs such as *Nerita* (−22.5‰ to −19.8‰) and *Cerithidea* (−22.8‰ to −20.5‰), that were eaten by *Periophthalmus* spp. Except for *Navicula* (−14.8‰), all the other diatoms and filamentous chlorophytes had $\delta^{13}C$ values that were more depleted than *Boleophthalmus*, thus making it an unlikely food source.

Working on the same Yangtzee River island as Jin et al. (2010), Shang et al. (2008) collected stable isotope ratios of the primary producers and common consumers including *B. pectinirostris* ($\delta^{13}C$ −17.7 ± 0.2‰; $\delta^{15}N$ 10.7 ± 0.1‰) and the ocypode crab *Ilyoplax deschampsi* ($\delta^{13}C$ −19.0 ± 0.2‰; $\delta^{15}N$ 10.0 ± 0.3‰). From the ratios of all the primary producers, microphytobenthos ($\delta^{13}C$ −19.3 ± 0.3‰; $\delta^{15}N$ 7.5 ± 0.3‰) were considered the primary probable contribution (>60%) to their diets. The secondary source (>18%) was C_3 (relatively $\delta^{13}C$ depleted) and C_4 (relatively enriched) plants for the crab and fish, respectively.

Using muscle stable isotope differences, the report of Kawamura et al. (2011) clearly demonstrated that there were significant differences in the prey of *Ps. argentilineatus* from Iriomote Island ($\delta^{13}C$ −17.4‰; $\delta^{15}N$ 7.4‰) and Okinawa Island ($\delta^{13}C$ −21.3‰; $\delta^{15}N$ 11.8‰) some 450 km to the northwest. Without isotope values for the prey items, the specific diets cannot be identified especially as *Ps. argentilineatus* from Iriomote Island exhibits an ontogenetic shift in their diet (Nanjo et al. 2008).

Stable isotope studies can provide useful information on the trophic relationships of mudskippers, especially as mudskipper stomachs are commonly empty when sampled (see Ekpo et al. 2014; Afshar et al. 2012). However, detailed information on other syntopic organisms will improve the accuracy of such information. Additionally, isotope studies could provide information on mudskipper movement patterns, for example, of migratory *Pseudapocryptes* or, of site fidelity in *Periophthalmus* (see Bouillon et al. 2011).

10.4 PREY CAPTURE AND PROCESSING

Ingestion of prey items involves prey capture and its separation from extraneous sedimentary material before it is passed into the alimentary canal. This section reviews the behavioral and structural features involved in this process.

10.4.1 Jaw and Pharyngeal Dentition

Initial prey capture by *Ps. barbarus* was analyzed using film and involves the reorientation of the oral jaws of the terminal mouth by pivoting forward on the pectoral fins (Sponder and Lauder 1981). The premaxilla slides anteroventrally and, with the mandible depressed against the substrate, the jaws close on the prey (Sponder and Lauder 1981). Using high-speed x-ray video, Michel et al. (2012) supplemented this observation with additional details; the premaxilla of *Ps. barbarus* can rotate dorsally during protrusion and the dentary has significant posterior rotation to maximize the gape, which is thus positioned above the prey parallel to the substrate. Mouth closure over the prey was due to the rapid movement of the lower jaw achieved through a combination of the unique structure of the suspensorium, shortening of the adductor mandibulae muscles and possibly, the elastic energy stored by the rotation of the intermandibular joint (Michel et al. 2012).

Both these studies used inert, non-moving prey, but another laboratory study on the escape response of *Ps. argentilineatus* is relevant to the capture of live, moving prey. Following the description by Harris (1961) of the escape response of *Ps. barbarus* (reported as *Ps. koelreuteri*), Swanson and Gibb (2004) compared this terrestrial response to its escape in water. On land, the fish adopts a J-shape posture with the caudal fin lying next to the body just behind the head. By straightening the axial musculature, the mudskipper propels itself aerially forward in the direction it faces. Because of the ability to bend the intervertebral joint (up to 20° instead of the usual 8°) two-thirds of the way down the body, the caudal fin is placed close to center of the fish's body mass thus providing both forward and vertical thrust. The action described by Swanson and Gibb (2004) precisely describes the catching phase when *Ps. waltoni* hunts ocypodid crab prey (Clayton and Snowden 2000). Significantly, Swanson and Gibb (2004) noted that in the preparatory phase of aligning the tail toward the head, the anterior portion of the body, including the head, does not move. This enables the fish to remain "locked onto" its prey.

Other ambulatory modes were described by Harris (1961), including crutching on the pectoral fins, which provides slower and shorter forward movement (Pace and Gibb 2009) rather than an escape response. When crutching, *B. boddarti* can advance 0.3–1.5 cm per pectoral-fin stroke, but following the J-shape posture the fish advances 8–30 cm (De and Nandi 1984). In this species, the latter action is a true escape response since crutching is sufficient to catch diatoms. In *Ps. argentilineatus*, a smaller species than *Ps. barbarus* (93 vs. 147 mm, respectively, Murdy 1989), the escape response achieves distances of 60–120 cm and heights of 15–30 cm in a single leap. Prey catching distances are markedly shorter, and, given the power suggested by this leap, it is not surprising that when this mudskipper lands "the snout usually disappears in the mud, the tail tipping up in the air" (Stebbins and Kalk 1961).

Having secured the prey, oral teeth appear to have little further involvement in the feeding process of *Ps. barbarus*. The pharyngeal plates are responsible for the posterior movement of the prey to the esophagus (Sponder and Lauder 1981). Most of the movement is generated by the retraction stroke of the independently articulated dorsal plates with their posteriorly pointing teeth. During this raking action, the fused ventral plates with their anteriorly pointing teeth remain stationary, and move significantly less than the dorsal ones, serving only as a support for the prey item. The pharyngeal plates of *Ps. barbarus* were seen as relatively generalized ones that function primarily as a transport system with no food preparation, other than some possible crushing and puncturing of the prey.

In the absence of any water in the opercular/buccopharyngeal cavity, the movement of the pharyngeal plates is obligatory for prey positioning during deglutination (Sponder and Lauder 1981). This conclusion was based on x-ray imagery showing that the buccal cavity was filled with air, and whatever water present in the mouth was lost when the fish opened its mouth over the prey. Examining this capture phase in more detail, Michel et al. (2015) showed that the "lost" water is better termed "protrusion" water as the buccal water is then "retracted" as in a form of a hydrodynamic tongue, so named as it functionally resembles the tongue movement in the feeding of lower tetrapods. The water movement is controlled by buccal and opercular actions and retraction was sufficient to move the prey to the pharyngeal jaws. Because of the hydrodynamic tongue, *Ps. barbarus* was able to swallow several prey without returning to water, a finding also reported by Sponder and Lauder (1981). Additionally, when Michel et al. (2015) placed prey on an absorbent surface that prevented water retraction, there was a marked drop in the ability to move the prey posteriorly to the pharyngeal jaws or esophagus. If the rock on which Sponder and Lauder (1981) placed their prey was porous, it would have had the same effect and thus, account for the loss of buccal water in their study. This raises the question of how effective the hydrodynamic tongue is on a mudflat where the hydration of the surface mud will vary considerably depending on the tide and time of day.

The view of the unimportance of the oral jaws and pharyngeal plates toward food processing (rather than mainly capture and transport, respectively) among the carnivorous species of mudskippers probably needs to be revisited. Further details of pharyngeal tooth plate architecture together with an analysis of the degree of damage suffered by prey prior to digestion may well change this interpretation (see Table 10.4 for summary on dentition characteristics of oxudercine gobies). Additionally, while not doubting the validity of the oral jaw actions during prey capture as described above, the studies were of necessity carried out in a laboratory setting with inert soft-bodied prey on a hard surface. Once captured the prey passes directly to the pharyngeal transport region. For hard-shelled prey is it the oral or pharyngeal jaws that break the shell to allow digestive juice access to the prey's soft parts?

The situation in the field may lead to additional considerations for the initial capture and subsequent deglutination processes for all mudskippers irrespective of their diet. For the herbivorous species feeding predominantly on benthic diatoms, the selectivity of prey is eloquently demonstrated in *B. pectinirostris* (Yang et al. 2003; Yang 1996). Of more than 23 species of diatoms identified and counted from the mud and gut contents, four species were considered to be positively selected

Table 10.4 Dentition Characteristics of Oxudercine Gobies

	Jaws									
	Dentary					Premaxilla				
				Symphysis					Symphysis	
Genus	Rows	No.	Form	No.	Form	Rows	No.	Form	No.	Form
Apocryptes	1	18–36	Bifid, caninoid	2	Canine	1	14–44	Bifid, caninoid	–	–
Apocryptodon	1	22–49	Bifid	2	Canine	1	15–30	Caninoid, blunt rounded tip	–	–
Boleophthalmus	1	31–81	Notched, compressed, horizontally orientated	2	Recurved canine	1	31–79	Caninoid, blunt rounded tip	–	–
Oxuderces	1	12–49	Caninoid, larger than in upper jaw	2	Recurved canine	1	15–40	Not specified	2	Enlarged canine
Parapocryptes	1	23–44	Subequal caninoid	2	Recurved canine	1	23–56	Caninoid, blunt rounded tip	–	–
Periophthalmodon	1	16–26	Stout caninoid larger than in upper jaw	–	–	2	16–35 3–11	Caninoid Caninoid	–	–
Periophthalmus	1	11–39	Stout caninoid larger than in upper jaw	–	–	1	13–39	Caninoid, anterior ones pointed	–	–
Pseudapocryptes	1	12–26	Caninoid, blunt rounded tip, horizontally orientated	2	Recurved canine	1	17–45	Caninoid, blunt rounded tip	–	–
Scartelaos	1	17–47	Subequal caninoid	2	Recurved canine	1	19–27	Caninoid, blunt rounded tip		
Zappa	1	19–22	Caninoid, compressed, blunt rounded tip	2	Recurved canine	1	35–48	Caninoid, blunt rounded tip		

Source: Adapted from Murdy E.O., *Records of the Australian Museum,* 11, 1–93, 1989.

and ingested, while five others were negatively selected. The positively selected diatoms, *Nitzschia palea*, *Pleurosigma* sp., and *Gyrosigma acuminatum*, are all larger than 50 μm, whereas the fourth, *Skeletonema* sp., is much smaller, but it is a chaining species of diatom. Diatom species less than 50 μm in size were negatively selected.

After the characteristic side-to-side head movement while grazing in *B. dussumieri*, buccal and opercular movements set up concentric water vibrations in the water around the mudskipper's head and eventually plumes of cloudy, lighter colored water are ejected from the opercular opening (Clayton and Vaughan 1988). This was also observed in *B. pectinirostris* (Yang 1996).

10.4.1.1 Sensory Detection of Prey

The description of prey capture while on terrestrial excursions by *Periophthalmus* spp. implies that prey items are visually detected and the anatomical features of the visual system as related to prey detection (see Clayton 1993 for review). One relevant example is the lowered threshold for movement detection, especially in the horizontal plane that results from the high density of visual cells in the horizontal band shaped "areae" (Munk 1970). However, only Kutschera et al. (2008) have begun to identify prey features that elicit hunting in mudskippers. Using a horizontally moving black rectangle (1 × 6 mm) oriented either horizontally (worm) or vertically (antiworm), *Ps. barbarus* preferred the worm configuration in 42 of 48 trials.

Tanaids and nematodes, the predominant prey items for *Ps. argentilineatus*, are common infauna of mangrove forest substrata; they are found in the greatest abundance within 4 mm of the surface (Somerfield et al. 1998). If *Ps. argentilineatus* is only a visual predator then, as Stebbins and Kalk (1961) suggested, it is probably detecting slight surface mud movements that are undetectable to observers. However, as well as tanaids and nematodes, the major infaunal nocturnal prey was polychaetes and oligochaetes (Colombini et al. 1996). During the dates of the study, the hunting reported was under a new moon on the neap tides and a first quarter moon on the spring tides, suggesting that sensory cues other than visual ones may be involved in prey detection.

When mudskippers forage in water, olfaction and gustation are likely to play a role in prey detection. This is especially likely in the more "aquatic" genera of mudskippers. This sensory modality has been somewhat neglected possibly because, in contrast to most fishes, mudskippers have an apparently impoverished olfactory apparatus. In describing the brain structure and gross external morphology of the olfactory system of *Ps. novemradiatus* (reported as *Ps. koelreuteri*), *B. boddarti*, and *Pd. lanceolatus*, Datta and Das (1980) reported that these species had no anterior olfactory bulb/rosette associated with the olfactory sacs in the nasal cavity. It is generally accepted that olfactory sensitivity in fishes is positively correlated with the number and size of the lamellae in the rosette (Fishelson and Gon 2009), but as Kuciel et al. (2011) demonstrated in *Ps. barbarus*, the olfactory epithelium is well developed. Rather than being aggregated into a rosette, islets of sensory epithelia are scattered along the tubular part of the nasal cavity and not within the sacs (Kuciel et al. 2011). Instead, the nasal sacs are central to the mechanism of ventilation of the sensory nasal tube (Kuciel 2013) with "sniffing" behavior both on land and in water. The "sniffing" behavior is often accompanied by eye retraction due to muscular contractions responsible for the water movement within the nasal cavity. The histological structure of the olfactory apparatus is now known for *Ps. barbarus* (Kuciel 2013), *S. gigas* (Kim et al. 2014), and *Pd. lanceolatus* (Sarkar et al. 2014; Sarkar and De 2011a,b).

Detection of chemical signatures of aquatic prey by mudskippers is one area for future research, as for example in the ability of *Periophthalmus* spp. to distinguish between burrows with and without eggs or for *Scartelaos* spp. and *Boleophthalmus* spp. to assess nematode or diatom richness in ephemeral mudflat pools at low tide. The techniques used by Gordon et al. (1985) and Chen et al. (2008) on microhabitat selection by adult *Ps. modestus* (reported as *Ps. cantonensis*) and juvenile *B. pectinirostris*, respectively, could be employed for this purpose. Parallel studies on the composition and distribution of the taste buds in the oral cavity and lips should also be conducted. Following the distinction made between those of blennies and gobies (Fishelson and Delarea 2004), the prediction would be that the visually hunting species would have fewer taste buds in both number and type than the grazing species.

10.4.2 Filtering Mechanism

The current consensus expressed in several studies is that gill rakers of herbivorous mudskipper species perform as a filtering mechanism (Yang et al. 2003; Low et al. 1990; Low 1988; Milward 1974; Khoo 1966). These studies, as well as those dealing with gill morphometrics (Gonzales et al. 2011; Pan et al. 2010; Low et al. 1990; Yadav and Singh 1989; Niva et al. 1981), were mainly concerned with the gills' adaptations to an amphibious life style. None include the size of gill rakers or distances between adjacent ones (see Abuzinadah 1995; MacNeill and Brandt 1990). Given that the details of the gill rakers have been investigated extensively in many pelagic planktivorous fishes, it is surprising that this aspect of oxudercine morphology has been neglected.

There are differences in gill raker architecture between carnivorous and herbivorous mudskipper species. In carnivorous *Periophthalmus* spp., gill rakers are short, knob-like projections that are widely separated. In herbivorous *Boleophthalmus* spp. and *Scartelaos* spp. gill rakers are closer together, and are better developed with a more flexible flap-like structure. Extrapolating from the scaled line drawings of Lim (1967), the mean distance between adjacent rakers of *B. boddarti* is about 122 μm (n = 21), whereas for *Pn. schlosseri* the gap mean is 311 μm (n = 10). Lim (1971) also prepared scaled drawings and for *Pn. schlosseri* the spacing averaged at 502 μm (n = 10) and apparently anomalously, that for *Ps. argentilineatus* was 79 μm (n = 7). In *Ps. modestus* (reported as *Ps. cantonensis*), the short knob-like rakers are spaced at about 170 μm, whereas the soft, leaf-like ones of *B. pectinirostris* are much closer together at about 60 μm (Yang 1996). The omnivorous *S. histophorus* (reported as *S. viridis*) has an intermediate spacing distance of 100–140 μm between its long flattened rakers (Yang 1996: plate 5.4: estimated measurements assuming similar sized individuals are illustrated). In the subsequent publication, Yang et al. (2003) gave the spacing of adjacent gill rakers in adult *B. pectinirostris* to be somewhere between 50 and 60 μm, sufficient to positively select the larger diatoms after taking into account the mucus covering of the gill membranes.

Support for differences in gill raker architecture between carnivorous and herbivorous species of mudskippers is indirectly confirmed by two other studies. In *Ps. chrysospilos*, the gill rakers are short and widely spaced at about 270 μm (Mazlan et al. 2006b: fig. 1). Although Ishimatsu and Gonzales (2011: fig. 4.5.2) confirm the close spacing of the gill rakers in *B. pectinirostris*, it is apparent that their architecture is more complex than suggested by Yang et al. (2003); there are differences in the spacing arrangement of the rakers on and between the four gill arches.

Size-dependent variation in the gaps between gill rakers could explain the small value for *Ps. argentilineatus* (Lim 1971). Barak et al. (1994) clearly established such a relationship for *Ps. waltoni* where the gaps were 86 μm for small (37–50 mm TL) and 290 μm for large (75–110 mm TL) fish, the latter value being very close to that just surmised for *Ps. chrysospilos*.

For herbivorous species, it would be useful to do interspecific comparisons, especially between large and small species such as *B. dussumieri* and *An. madurensis*. Additionally, ontogenetic studies within species take on greater significance when there appear to be diet shifts with maturation (Nanjo et al. 2008; Kruitwagen et al. 2007; Mutsaddi and Bal 1969). Yang et al. (2003) compared the species composition of diatom prey between large (fully grown >88 mm SL) and small (>28 mm SL) individuals of *B. pectinirostris*, and found that diatom prey were similar, although with different relative frequencies. The size and spacing of the gill rakers were not reported in the small fish, but it is unlikely that smaller gills, with presumably smaller gill raker spacing, would filter the same prey species. Given that prey species were similar, these results and future investigations should be viewed within the wider context of other piscivorous studies. As summarized by Smith and Sanderson (2013), gill raker filtration in planktivorous fish could involve several mechanisms including dead-end or cross-flow sieving with or without the assistance of mucus adhesion. Following Drenner et al. (1987), they showed that removal of the gill rakers in cichlids did not significantly affect the size of particles injected. For herbivorous mudskippers, the necessity to separate the food and sediment particles poses an additional element in prey ingestion.

10.4.3 Dentition Revisited

The oral dentition of oxudercine gobies is well documented and a summary for the genera is presented in Table 10.4 based on Murdy (1989). A clear relationship to diet is evidenced by the fact that the lower dentary teeth in the most obviously herbivorous genera, *Boleophthalmus* and *Pseudapocryptes*, are horizontally oriented in the jaw rather than vertically as in the other genera. Further, eight of the genera possess a pair of large teeth internal to the symphysis that Murdy (1989: 62) speculates may be related to prey capture (or burrow excavation) despite their absence in the most obviously carnivorous and terrestrial genera, *Periophthalmus* and *Periophthalmodon*.

Given the importance attributed to the pharyngeal jaw apparatus in increasing feeding versatility that leads to ecological diversification in other teleost groups (Burress 2015; Wainwright et al. 2012), it is surprising that this aspect of mudskipper study has been somewhat neglected (Burress 2015). Clayton (1993) suggested that in the absence of the details of diet items, the structure of the pharyngeal plates in other gobioid genera may clarify the issue. Murdy (1989) suggested that a broad survey of the pharyngeal jaw apparatus would be beneficial.

In many other fishes, the left and right fifth ceratobranchials are fused together so that the ventral or lower plate acts as one unit. The left and right second, third, and fourth pharyngobranchials that comprise the dorsal or upper plates are not fused so the two plates can move independently. The teeth on the dorsal and ventral plates point in opposite directions, posteriorly on the ventral plate and anteriorly on the dorsal one (Sponder and Lauder 1981). It is the arrangement and shape of the teeth on these plates that are considered indicative of diet.

In a study of the pharyngeal apparatus of mudflat gobies in Sulaibikhat Bay in Kuwait, carnivorous and herbivorous mudskippers are found to have different tooth architecture and shape (Figures 10.2a,b and 10.3a,b). In carnivorous mudskipper species, teeth on both plates are of a similar shape and are irregularly and widely distributed but increase in size posteromedially. The lateral edges of the ventral plate have an intermittently continuous row of teeth that is absent in the dorsal plate. In herbivorous mudskipper species, both plates have regular, closely spaced teeth arranged in rows that stretch across the plate with little or no size differentiation. The rows on the ventral plate are oriented in two directions, medially anteroposteriorly and laterally diagonally. Those on the dorsal plate have a common transverse direction. In contrast to the carnivorous species, the tooth shape differs between the plates: ventral plate teeth are hooked, while those on the dorsal plate are not (Figures 10.2a,b and 10.3a,b).

The pharyngeal plates of *Ps. waltoni* clearly identify this mudskipper as a carnivorous species as do the micrographs of the ventral and dorsal plates of *Ps. barbarus* (reported as *Ps. koelreuteri*, Sponder and Lauder 1981: pl. 1). The teeth in *Ps. waltoni* are thick, squat, pointed peg-like with some recurving tips, while those in *Ps. barbarus* are longer, and most with recurving tips. It is unclear if *Ps. waltoni* has depressible teeth as described by Sponder and Lauder (1981) for *Ps. barbarus*. Tooth wear is significant with some teeth broken off partially or completely, or split lengthwise but tooth replacement is continuous (see Figure 10.4a,b).

Acentrogobius cyanomos and *Istigobius ornatus* are not mudskippers, but are good examples of tooth morphology matching diet. Both species were found in alpheid shrimp burrows on the mudflats and, by their dentition, are clearly carnivores. Further, *I. ornatus* feeds on small crustaceans and thin-walled bivalve molluscs (reported as *Acentrogobius ornatus* in Clayton and Abouseedo 1986) and has large peg-like crushing teeth. *Periophthalmus variabilis* should have similar peg-like teeth as it is also considered to be a molluscivore (Zulkifli et al. 2014).

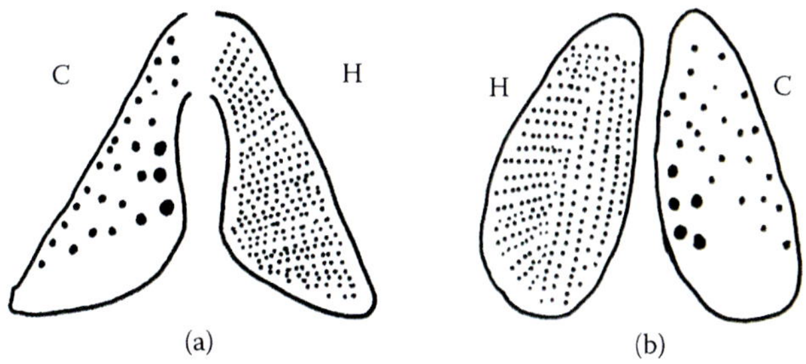

Figure 10.2 Generalized pharyngeal plate structure in oxudercine gobies (a) ceratobranchial and (b) pharyngobranchials; "C" denotes carnivores and "H" denotes herbivores. (Courtesy of David Clayton.)

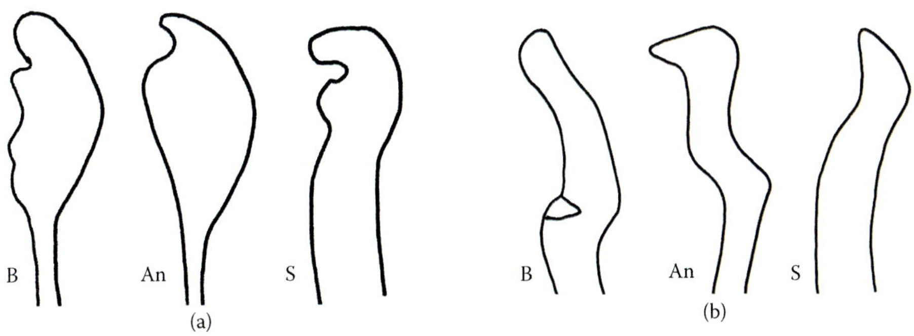

Figure 10.3 Generalized tooth structure, emphasizing the distal end of the teeth, in oxudercine gobies (a) on the ceratobranchial and (b) on the pharyngobranchials; "B" denotes *Boleophthalmus dussumieri*, "An" denotes *Apocryptodon madurensis*, and "S" denotes *Scartelaos tenuis*. Teeth of *Boleophthalmus dussumieri* (110 µm) are approximately twice the size of those in *Apocryptodon madurensis* (54 µm). (Courtesy of David Clayton.)

The herbivorous *B. dussumieri* has more delicate and clearly hooked teeth on the ventral jaw, and recurved blunt-tipped ones on the dorsal plate. *Apocryptodon madurensis* (reported as *Apocryptes madurensis*) feeds by sweeping its head from side to side in the same way that *B. dussumieri* grazes (Clayton 1986; Clayton and Abouseedo 1986). Given the size difference of the adults (*B. dussumieri* up to 260 mm SL, *An. madurensis* up to 50 mm SL), the pharyngeal plates and dentition are remarkably similar, indicating the herbivorous diet of both species. Broken teeth are also common in these species.

Scartelaos tenuis has dorsal hooked teeth and the ventral recurved ones are also arranged in rows, indicative of a herbivorous diet. Additionally, there appears to be a row of peg-like teeth along the lateral edges of the ventral plate. These are more clearly seen on the dorsal plate where there are also more strongly recurved peg-like teeth toward the middle of the plate, the latter resembling those of *Ps. waltoni*. The mixed dentition observed in *S. tenuis* suggests that this species is an omnivore (Clayton 1989, unpublished).

Parenti and Thomas (1998) studied phylogenetic relationships based on the ventral pharyngeal plates of several gobies including *Pd. elongatus* (reported as *Pd. lanceolatus*). The recurved blunt-tipped (Parenti and Thomas 1998: pl. 5C and D) and the hooked (Parenti and Thomas 1998: pl. 9B) teeth resemble those of *B. dussumieri* and *An. madurensis*. As the teeth are both from the same plate, it raises the possibility that the ventral plate of the latter two species may also have such tooth

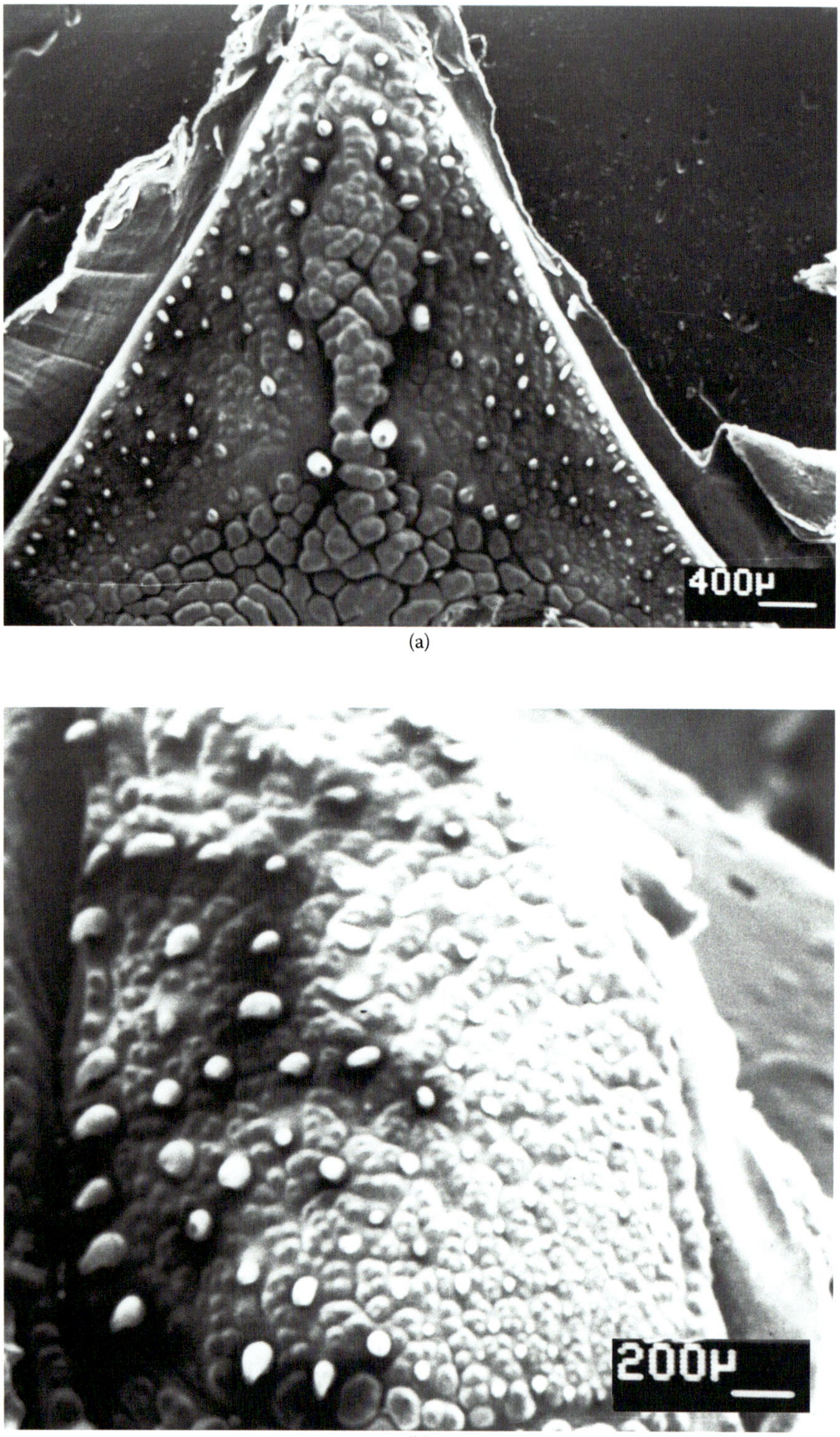

Figure 10.4 Scanning electron microscopy image of pharyngeal plates of *Periophthalmus waltoni* (a) ceratobranchial and (b) pharyngobranchials. (Courtesy of David Clayton.)

dimorphism, the hooked type being found in the transverse diagonal rows and the recurved ones in the anteroposterior rows. Whereas the dorsal plate of *Pd. elongatus* is not shown, whatever the tooth architecture of all three species demonstrates, these species are clearly herbivores. Additionally, it would be useful to identify the shape of the interacting ventral and dorsal teeth as this would help elucidate the sorting function of the pharyngeal jaws in the deglutination of food items.

Neglected areas of the adaptive strategies in oxudercine dentition are not restricted to interactive tooth architecture, but also include the size of the articulated surfaces and shape of the upper plate as well as that of the lower plate together with their associated musculature and basi-pharyngeal joint variation (see Liem 1974). Equally informative would be to study how ontogenetic changes in diet affect pharyngeal tooth plate dentition and architecture (Zulkifli et al. 2012; Kruitwagen et al. 2007). In *Neogobius melanostomus*, for example, pharyngeal teeth became more peg-like as the goby switched its diet from arthropods to bivalves (Andraso et al. 2011). Small *Acentrogobius viridipunctatus* are herbivorous and larger ones omnivorous taking crustaceans as well as plant material (Hoda 1991); a more extreme diet shift that might be expected to produce significant plate changes. As this suggestion echoes that for the gill rakers, the optimal research strategy would be to study all aspects in tandem (see Peretti and Andrian 2008).

Based on evidence presented above, the pharyngeal plates likely play a more central role in prey selection and sorting, possibly even greater than that of the gill rakers. It is also equally likely that herbivorous mudskipper species engage in winnowing behavior (see Drucker and Jensen 1991). Internal water currents between the pharyngeal jaws and the action of the pharyngeal teeth must serve as the filter, separating the particulate organic and inorganic material. Such a filtering mechanism would work regardless of the mode of feeding; skimming the surface film of mud as in *B. dussumieri* (Clayton and Vaughan 1988), scooping up as in *S. histophorus* (Townsend and Tibbetts 2005), or taking bites out of the sediment as in *Pd. elongatus* (reported as *Pd. lanceolatus*, Swennen et al. 1995). It may also be applicable to *Periophthalmus spp.* Stebbins and Kalk (1961) noted that after branchial water and air are expelled during prey capture, the mudskipper replenished both the air and water by rapid pumping at a pool of water. This action was attributed as part of the respiratory response, but it may be more parsimonious to attribute it to feeding, with the prey clamped between the pharyngeal jaws, the pumping action will remove the excess sediment.

10.4.4 Alimentary Tract

A shorter intestine is required to digest animal rather than plant material and the measurement of the relative gut length (RGL) values of the different genera support this generalization. Early studies showed that the RGL for *Periophthalmodon* ranged between 0.4–1.1, *Periophthalmus* 0.3–1.0, *Boleophthalmus* 1.5–2.6, and *Scartelaos* 0.6–2.0 (Clayton 1993). Table 10.5 shows recent RGL values for the latter three genera. Differences in which the ratios were calculated prevent direct cross-study comparisons, but still support the generalization. Additionally in two other omnivorous species, the RGL values differed; *Pa. serperaster* (gut content of 60% vegetable and 40% animal matter) had an RGL value of 1.8, whereas *Pd. elongatus* (reported as *Pd. lanceolatus*, gut content of 41% vegetable and 59% animal matter) had an RGL value of 0.7 (Dasgupta 2004). Despite some questionable percentage values (Dasgupta 2002), the buccal cavity and stomach accounted for a greater proportion of the alimentary canal in *Pd. elongatus* than in *Pa. serperaster.*

These RGL values are means calculated from pooled data sets rather than individual or size class groups as were the case for ontogenetic diet changes. It can be assumed that the gut length increases with increasing fish size and indeed Chan (1989) and Yang (1996) reported significant positive correlations between the two measures. Scatter plots of the data enabled calculation of the range of values such that for *Ps. modestus* (reported as *Ps. cantonensis*, mean 0.719) the range was 0.6–1.0, and for *B. pectinirostris* (mean 1.87) the range was 1.0–2.8 (Chan 1989). Similarly,

Table 10.5 Relative Gut Lengths of *Periophthalmus*, *Boleophthalmus*, and *Scartelaos*

Periophthalmus		*Boleophthalmus*		*Scartelaos*		
N	Mean ± SD	N	Mean ± SD	N	Mean ± SD	Reference
11	0.84 ± 0.02	8	2.61 ± 0.43	–	–	Yuan (1963)
244	0.719 ± 0.006	370	1.87 ± 0.018	23	0.97 ± 0.029	Chan (1989)
–	–	–	1.35–1.43	–	–	Zhu and Zhang (1993)
–	–	–	–	–	1.27	Abdoli et al. (1996)
–	–	283	3.02 ± 0.68	–	–	Yang et al. (2003)
101	1.14 ± 0.03	–	–	–	–	Kruitwagen et al. (2007)
85	0.96 ± 0.04	–	–	–	–	Mohammadpour et al. (2009)
–	0.47	–	–	–	–	Afshar et al. (2012)
–	–	–	–	–	1.4	Abdoli et al. (2012)
–	–	–	1.699	–	–	Abdoli et al. (2013)
–	0.88	–	1.43	–	–	Mitra and Homechaudhuri (2013)
52	1.02	–	–	50	1.02	Abidizadegan et al. (2015)

the mean value for *Pa. serperaster* from the Mekong River delta was 2.1 ± 0.6 but the range was 1.0–3.5 from fishes measuring 11.5–22.1 cm TL (Quang et al. 2013). A smaller specimen of *Pa. serperaster* (9.0 cm TL) from Selangor in Peninsular Malaysia fed predominantly on plant material (95.8% abundance, Khaironizam and Norma-Rashid 2000) so this individual might be expected to have a high RGL. Unfortunately, however, Quang et al. (2013) did not correlate the two measures.

The range of RGLs for *Ps. argentilineatus* (mean 1.04) was 0.4–2.0 demonstrating a major change in RGL with increasing fish size (Kruitwagen et al. 2007). It was hypothesized that elongation of the intestine for the Zanzibar population coincided with the size-related change in diet. Larger *Ps. argentilineatus* predominantly consumed crabs from the genus *Uca*; these prey items have a high chitin content that requires additional time to be digested and assimilated (Kruitwagen et al. 2007). If this was the case, then it might also be expected that male *Pn. schlosseri* would have a higher RGL than females that eat fewer crabs than do the males (Zulkifli et al. 2012). While a study of chitinolytic activity in either species might resolve the issue, it should be noted that Lindsay (1984) found no correlation between the level of chitinous intake and chitinase activity. Exoskeletal damage caused by the oral or pharyngeal teeth must also be considered since the activity of the enzyme was high in fishes that swallowed their prey whole, and low in those that broke the exoskeleton proir to deglutination (Lindsay 1984).

Whatever the explanation for the increased RGL in *Ps. argentilineatus*, it goes against the generalization for a carnivorous species. Details of the RGL for *Ps. barbarus* would be equally interesting because it has been found to have both a similar crab-based diet (Chukwu and Deekae 2013) or one with a considerable amount of plant material (Bob-Manuel 2011; Udo 2002b). Irrespective of the diet, RGL questions could be better answered if individual gut content analyses included gut length data as well as the size (TL) of the fish examined.

The intestinal section of the alimentary tract for *Periophthalmus*, *Periophthalmodon*, *Boleophthalmus*, and *Scartelaos* has been histologically described in several theses (Milward 1974; Lim 1971, 1967). Additionally, Suyehiro (1942) and later Yuan (1963) described the gross anatomy of the entire alimentary tract for *Ps. modestus* (reported as *Ps. cantonensis*) and *B. pectinirostris* (reported as *B. chinensis*), and Zhu and Zhang (1993) provided details of the histological structure of the intestine of *B. pectinirostris.* Yang and Zhang (1991) described the gross structural development, including the coiling of the intestines, for *B. pectinirostris* from the larval to juvenile/young adults (25 mm TL) stages. Cao et al. (2013) described the morphology and histology of the alimentary canal of *S. histophorus*, and noted differences in intestinal length, diameter, and gastric gland distribution from that found in *B. pectinirostris*. In relation to loss of gastric digestion in gobioids, the gastric gland distribution in *Periophthalmus* sp. was compared with that of other gobies by Kobegenova and Dzhumaliyev (1991).

The histological appearance and distribution of the gastrin-secreting G cells were compared for five marine fishes (Han et al. 2008). In adult *B. pectinirostris* (7–11 cm TL), they were widely distributed in the gastric cardia (esophageal entrance) and particularly in the pyloric stomach and in the intestine. In the carnivorous *Bostrychus sinensis*, G cells were absent from the stomach but found in the intestine and rectum (Han et al. 2008). It was suggested that in *B. pectinirostris*, gastrin might stimulate growth of the gastrointestinal mucosa to compensate for the damage to tissues caused by indigestible detrital material. Another possible explanation for the intestinal and rectal presence of G cells might be to facilitate gut evacuation through increased gut motility.

10.4.4.1 Enzymes and Bacterial Flora Distributions

The divisions of the digestive system from which the enzyme activity was recorded are not always comparable in published studies. Enzymes from the liver and stomach were common to most studies, but only Chaudhuri et al. (2012) and Wu et al. (2006) included enzymes from the intestine in its entirety with the latter authors dividing the intestine into three (Wu et al. 2009, 2006). Dhage and Mohamed (1977) excluded the intestinal bulb (possibly referring to the stomach) arbitrarily, dividing it in half, as did Mitra and Homechaudhuri (2013), but without stating if the stomach was excluded or not.

Mitra and Homechaudhuri (2013) compared the activity of five enzymes in *B. boddarti* and *Ps. novemradiatus*. The activities of four (Wu et al. 2010), and nine (Wu et al. 2006), enzymes in *B. pectinirostris* were compared with that of another goby, the carnivorous *Bostrychus sinensis*. Wu et al. (2006) assessed the changing activity of 12 enzymes in pre-larvae through juvenile to early adult *B. pectinirostris* and in a follow-up study (Wu et al. 2007), determined the activity of 15 enzymes in the stomachs and intestines of adult *B. pectinirostris*. The pre-larval to adult stages are based on fish sizes as described by Zhang et al. (2005). Additionally, protease, amylase, and cellulase were reported from 1- to 5-month-old *Pd. elongatus* (reported as *Pd. lanceolatus*, Thuy et al. 2006). Chaudhuri et al. (2012) compared the activity of six enzymes in *Pd. elongatus* as a carnivore in a guild fish from the Sundarbans mudflats.

The activity of the enzyme amylase, a catalyst for starch hydrolysis, was higher in the intestinal bulb (=stomach) and anterior intestine than in the esophagus, liver, or posterior intestine in a species reported as *Ps. koelreuteri* (Dhage and Mohamed 1977). However, the gut of this species has three coils (see Yang and Zhang 1991; Chan 1989) and is longer than its TL (RGL>1), suggesting that the mudskipper was more likely to be *B. dussumieri*, also studied by Rathod and Patil (2009) from the same locality.

Results from Dhage and Mohamed (1977) were corroborated in other studies; enzyme activity in the liver was low in both *Pd. elongatus* (Chaudhuri et al. 2014) and *B. pectinirostris* (Wu et al. 2007), higher in the stomach, and greatest in the anterior intestine (=Intestine I). Wu et al. (2010) reported that Intestine I α-amylase activity was greater (12.3× higher) than in the stomach. In *B. boddarti*, overall anterior gut α-amylase activity was higher than in the posterior gut, but almost equal to that of the liver. Enzyme activity in the liver was initially very low in smaller individuals (mean TL 4.75 ± 0.5 cm) but became dominant in the largest (10.70 ± 3.75 cm) fishes (Mitra and Homechaudhuri 2013).

In the comparative (herbivore/carnivore) studies between *B. boddarti* and *Ps. novemradiatus* (Mitra and Homechaudhuri 2013) and *B. pectinirostris* and *Bostrychus sinensis* (Wu et al. 2010), amylase activity was considerably greater in the herbivores. Although high amylase activity is usually associated with herbivorous or omnivorous rather than carnivorous fishes, Chaudhuri et al. (2014) found this value to be low in the herbivorous species they studied. *Pseudapocryptes elongatus*, although classed as a carnivore, had a high plant and detritus abundance (14.4%), and the carbohydrases, including α-amylase, was higher than in other carnivores (Chaudhuri et al. 2014).

For *B. pectinirostris*, the location of amylase indicates that starch breakdown occurs in the anterior intestine and other carbohydrases that break down disaccharides (maltase, sucrose, lactase, trehalase, and cellobiase) were found to be more active in the intestine II segment (Wu et al. 2007, 2006).

It is difficult to decide whether or not the two studies on lipid digestion are contradictory because intestinal activity is compared with the liver in one and with the stomach in the other. Ignoring size differences, the herbivorous *B. boddarti* has a higher esterase activity than *Ps. novemradiatus* with both species having higher enzyme activity in the liver (Mitra and Homechaudhuri 2013). Conversely, the activity of bile salt activated lipase (BAL) was lower in *B. pectinirostris* than in *Bostrychus sinensis* but both species had higher activity in the intestine than the stomach (Wu et al. 2010). The fate of lipids needs to be further studied, including the possible contribution that might arise from endosymbionts in the posterior intestine as described for other herbivorous fishes (see Clements et al. 1994).

While the activities of 12 proteolytic enzymes have been reported in the seven studies, only some will be discussed here. In agreement with the generalization that carnivorous fishes have higher protease activity than do omnivores or herbivores, pepsin activity in the liver and intestine of *Ps. novemradiatus* is about seven to eight times greater than in *B. boddarti* (Mitra and Homechaudhuri 2013). However, protease activity in the stomach (pepsin) of *B. pectinirostris* is less than half that of *Bostrychus sinensis*, whereas intestinal activity (alkaline protease) is similarly high at about between 2.3 and 2.7 U/g tissue in the two fishes (Wu et al. 2010). Other proteolytic enzymes (aminopeptidase, alkaline phosphatase, and γ-glutamyltranspeptidase) had significantly greater activity in intestine III than the other intestinal segments.

Additional digestive activity results from the presence of bacterial flora in the digestive tract: *Vibrio*, *Aeromonas*, and *Enterobacteriaceae*, for example, are major cellulose-digesting bacteria. These and others were identified in adult *B. pectinirostris* and *Ps. modestus* (reported as *Ps. cantonensis*) from the Honmyo estuary in Isahaya City, Nagasaki, Japan (Morii and Kasama 1989); *Vibrio* and coryneforms were the dominant floral bacteria in both species. *Boleophthalmus pectinirostris* was found to have considerably a higher number of isolates than did *Ps. modestus*. Whereas there were similar counts for the stomach and intestinal flora for *Ps. modestus*, those of the stomach were much higher than that of the intestine in *B. pectinirostris*. Primary factors controlling the distribution were the inhibitory effects of lower pH on coryneforms in the stomach of *Boleophthalmus*, and bile on *Vibrio* in the intestine of *Periophthalmus* (Morii and Kasama 1990).

Given that the bacteria could originate from the substrate, water, or food (Morii et al. 1998; Morii and Kasama 1989), samples were taken to complement the study of floral bacteria within mudskippers in Pulao [=Island in Indonesian] Bengkalis in Sumatra (Morii et al. 1998). With a preponderance of *Enterobacteriaceae* and coryneforms, the intestinal flora of mudskippers reflected those taken from the water and mud samples. These bacteria were common to both carnivorous (*Ps. chrysospilos* and *Pn. schlosseri*) and herbivorous (*B. boddarti*, *B. dussumieri*, and *Pd. lanceolatus*) mudskippers. The carnivores also possessed Vibrionacea, not present in the water or mud and presumably originated from their diet. *Scartelaos histophorus* from Sumatra was found to have all three bacterial groups, associated with an omnivorous diet (Morii et al.1998). Further support for the environmental origin of intestinal flora is that *Enterobacteriaceae* were considerably more prevalent in the Sumatran fishes than in the Japanese ones; *Enterobacteriaceae* were absent from mud samples in Japan (Morii and Kasama 1990). Of additional importance, especially for grazing mudskippers, are the demonstrable diel and seasonal changes in (aerobic) bacterial flora of mud (Morii and Kasama 1989). Whether these are reflected in changes in the composition of the transient or core gut flora (see Givens et al. 2015; Kim et al. 2007) remains to be determined.

10.4.4.2 Juvenile to Adult Transition

The development of the digestive tract, and the feeding and nutrition of many species of mudskipper fingerlings have been investigated due to their importance in artisanal fisheries and aquaculture both in Africa and in Asia (Udoh et al. 2013; Minh and Phuong 2011; Lawson 2010; Minh et al. 2010; Tran et al. 2009; Ip et al. 1990). Once hatched, the meroplanktonic larvae of *B. pectinirostris* have sufficient yolk to last 4–5 days; the mouth and anus open on day 1.9–2.5, whereas the gill arches require time to develop the mandibular and pharyngeal dentition (Hong and Zhang 2003; Yang and Zhang 1991). Gill rakers develop 15–17 days after hatching; at a maximum larval size of 7.7 mm SL, gill rakers are mere projections on the gill arches (Zhang and Zhang 1988). The rakers become fully developed when they are at least 22.1 mm SL (Yang and Zhang 1991). *Boleophthalmus pectinirostris* obtain nutrition from the yolk sac until this source runs out and supplements nutritional intake with plankton and particulate organic detritus (Owen 2009; Zhang and Zhang 1998, 1991).

The mouths of larval *Pn. schlosseri* (Tsuhako et al. 2003) and *S. gigas* (Kim et al. 2011) are open at hatching and the larval yolk sacs last 3 and 5 days, respectively. However, *Pn. schlosseri* was reported to begin feeding 1 day after hatching, whereas *S. gigas* switched from internal to external feeding 5 days after hatching.

The yolk sac of *Pa. rictuosus* is completely used up when the actively swimming post-larval is 7 mm TL (Venkataramanujam et al. 1993). While juveniles of 9.6 mm TL require 38 days to reach 11.5 mm TL, the full timeline for growth from the first post-larval stage (2.3 mm TL) is not reported. Post-larval and juvenile (6.3–15 mm TL) *Pa. rictuosus* fed on planktonic diatoms and crustacean larvae, whereas juveniles (TL > 9.6 mm) consumed larger plankton such as copepods and prawn mysid larvae. Adults also fed on mysids, but mainly on *Acetes indicus*, copepods and nereids (Venkataramanujam et al. 1993).

When mudskippers become amphibious, they begin feeding more on benthic organisms. A detailed transition study in *Ps. modestus* found that the change from pelagic to demersal living occurred approximately 30 days after hatching, but feeding on land only started 16 or 17 days after (Inaba et al. 2014). Demersal living was found to occur approximately 50 days after hatching for *Ps. modestus* and 40 days for *B. pectinirostris* (Kon and Yoshino 2002), the latter being very close to the 41-day duration reported by Ryu et al. (1995).

10.5 BEHAVIOR

Behavior influences diet, especially in prey-encounter opportunities. For example, mudskippers following the tide or remaining near their burrows will encounter different prey items (Baeck et al. 2008). Chua et al. (2011) showed that *Pn. schlosseri* moves with the tide. Dietary differences between male (mainly crabs) and female (killifish) *Pn. schlosseri* were attributed to females remaining near the edge of the water, whereas the males moved higher up the shore (Zulkifli et al. 2012). The latter study, however, was done during the breeding season, whereas the former one was conducted outside the season without the benefit of being able to separate the sexes by observation alone.

Periophthalmodon septemradiatus from tidal flats of the Navinal Coast, India engaged in grazing activity (Bhatt et al. 2009) that was very similar to that described for *Boleophthalmus* (Clayton and Vaughan 1988) except that the fish moved backward while the anterior body swept from side to side. This contrasts with the description of that of other members of this genus as well as *Periophthalmus* spp. that are considered as hunting carnivores and should be verified.

Interpretations based on dietary studies need to be augmented with direct behavioral observation of the predator–prey interactions. Initially, the behavior involved should be identified and described as was done for *B. dussumieri* in Sulaibikhat Bay, Kuwait (Clayton and Vaughan 1988) and for *S. histophorus* in the Brisbane River and Morton Bay in Queensland, Australia (Townsend and Tibbetts 2005) enabling time budgets to be calculated. While inactive for 85% of the time, feeding was the most common activity in *B. dussumieri*, accounting for almost two-thirds of its overall activity (Clayton and Wright 1989). *Scartelaos histophorus* was only inactive (resting or submerged) for about 40% of the time but feeding remained the most common activity, although males spent less time feeding than females because they were displaying (Townsend and Tibbetts 2005).

Using similarly broad categories of behavior, Muhammad and Norma-Rashid (2005) compared the behavioral time budgets of mudskipper species in the Morib estuary, Selangor, Peninsular Malaysia. *Boleophthalmus boddarti* spent more of its time feeding (49%) than did *Ps. gracilis* (34%), *Ps. chrysospilos* (22%), or *Pn. schlosseri* (10%). The proportion of time spent feeding was equal regardless of whether *B. boddarti* was feeding on the mudflats or riverbank. This was not the case for *Pn. schlosseri*, which spent more time on mudflats (15%) than on riverbanks (8%). *Periophthalmus chrysospilos* spent more time feeding in the mangroves (28%) than either the mudflat (20%) or riverbank (15%). Except for the case of juvenile *Ps. chrysospilos* that stayed on the forest floor, it is not clear if the observations were made on fishes that moved between the different habitats. For all species except *B. boddarti*, locomotion was the main activity (Muhammad and Norma-Rashid 2005).

Feeding behavior of *Pn. schlosseri* was studied at the Matang Mangrove Reserve in Perak, Peninsular Malaysia (Mazlan et al. 2006a). The behavior was divided into three main phases: searching (48.5% of observed time), capture (0.1%), and handling. The latter phase was further divided into tearing (1%), moving to a safer place (7.9%), and playing with prey (43%). These overall time budgets included low-tide data from catching prey, mostly crabs during the day and insects at night. Overall, feeding activities accounted for 9.7% of their time budget, a figure within the range calculated for *Pn. schlosseri* from the Morib estuary (Muhammad and Norma-Rashid 2005).

Up until the final "playing" behavioral category, which is not clear, the sequence of events parallels that of *Ps. waltoni* from Kuwait (Clayton and Snowden 2000). Excursions from burrows were significantly longer when the fish engaged in hunting than when they did not. After a hunting event (lunging at prey) that may or may not have been successful, *Ps. waltoni* rapidly returned to the burrow or pool of water within their territory. This study additionally provided data on the behavior of the prey that also impacts both encounter rates and reaction behavior of mudskippers. For undisturbed crabs, surface activity (median 299 seconds; range 178–584) was interspersed with short burrow sojourns (28.6 seconds), but in-burrow time significantly increased (480 seconds) after a hunt in their vicinity.

The size and abundance of crabs *Paracleistostoma* sp. and *Haberma* sp. from areas with and without *Pn. schlosseri* in Singapore indicated that the mudskipper preyed on smaller individuals from the population (Su and Lim 2011). Preference of *Paracleistostoma* sp. over *Haberma* sp. as prey was attributed to the different behavioral response to the presence of the mudskipper (Su and Lim 2014). Whereas both crabs displayed a variety of behavior (e.g., freezing, sitting [pressing the body to the sediment], and running) *Paracleistostoma* sp. relied primarily on crypsis (Su and Lim 2014). Using a variety of models of increasing likeness to *Pn. schlosseri*, the initial response of *Paracleistostoma* sp. was to sit while that of *Haberma* sp. was to run (Su and Lim 2015). *Haberma* sp. was also more sensitive to the presence of "eyes" on model predators (Su and Lim 2015).

Uca bengali and *Perisesarma indiarum* are common prey items of *Ps. argentilineatus* within the *Rhizophora* mangrove forest of Sikao creek in Trang, southern Thailand. Using predator exclusion cages, Kon et al. (2009) showed that the number of individuals of all epifauna, including the two crab species, was significantly higher inside the cages. A laboratory experiment mimicking the open mudflat (unplanted) and the forest root system (planted) showed that the crabs preferred the artificial root area irrespective of the presence or absence of the mudskipper. The presence of the "roots" was no hindrance to the ability of mudskippers to catch the crabs (Kon et al. 2009).

Field studies have already investigated the ability of mudskippers to orient themselves in the mangroves; *Periophthalmus argentilineatus* captured in the tidal creeks of the Tana River in Kenya and then released midstream invariable returned to the bank from which they were captured (Berti et al. 1992). The authors attributed the homing behavior to a sun compass mechanism of orientation, rather than those involving the direction of water or landscape recognition (Berti et al. 1994). The field observation that after a hunting event *Ps. waltoni* restarted the second foraging activity 2.3 m away from the first site (Clayton and Snowden 2000) suggests that mudskippers have a spatial awareness of their immediate environment.

It is assumed that the dietary selection and overall foraging of mudskippers is optimized (cf. Baeck et al. 2008; Mazlan et al. 2006a), however, this is yet to be demonstrated in mudskippers despite the suggestion by Clayton and Snowden (2000) that the predatory mudskippers could be good models with which to investigate aspects of foraging theory.

Laboratory setups can be easily adapted to investigate sensory detection of prey or even topographical spatial awareness. Mukuda et al. (2011) used a dry version of the Morris water maze to demonstrate spatial learning in mudskippers (Mukuda et al. 2013; Mukuda 2012; Furukawa et al. 2011), including the ability to distinguish right from left (Hayase et al. 2011). Additional experimental techniques investigating landmark spatial learning in other fishes (Fabre et al. 2014; White and Brown 2014; and other studies in this paper) also could be adapted for mudskippers. An added level of sophistication could be incorporated by also testing species that move between different habitats (see Takeda et al. 2011; Polgar and Bartolino 2010; Polgar and Crosa 2009).

The visual acuity of *Ps. novemradiatus* was also investigated in an operant conditioning situation, where the fish were rewarded for being able to discriminate vertical from horizontal striped patterns (Hanke et al. 2014).

Since the first description of polygonal mud-walled territories in *B. dussumieri* from Kuwait (Clayton and Vaughan 1982), polygonal territories have been reported elsewhere in the Gulf (Höpner and Maraschi 1999) and in other species. On the open mudflats of Cardwell in Queensland, Australia, Aguilar (2000) described how similar pentagonal territories constructed by *S. histophorus* retain pools of water at low tide with exposed mud slopes within the surrounding walls. In both species, construction of the walls is population density dependent. Feeding of the resident fish is restricted to the territory, including on its walls. Clayton and Wright (1989) reported that diatom densities in three areas of the territory were lowest on the boundary walls, highest on the mud slopes, and intermediate in the pools of water. The non-territorial open mudflat sites had similar, but more variable diatom densities. Their results raised the possibility, but did not prove, that the fish were "farming" the diatoms and the study is worth repeating with alternate measures of benthic microalgal biomass. Chen et al. (2007) did just that in their study on the overwintering behavior of *B. pectinirostris* in their walled territorial areas in non-tidal farming ponds in Funing Bay in Fujian, China. Chlorophyll-a and total chlorophyll-a measurements showed significantly increased biomass in the territorial areas compared to non-territorial areas.

How, and if, the behavior of mudskippers encourages benthic microalgal growth remains speculative. Chen et al. (2007) suggested that the elevated temperature of the pools facilitates diatom growth, whereas Clayton and Wright (1989) suggested that water movements might relieve diffusion constraints against growth in dense diatom populations. Another possibility is that the pool chemistry is altered to favor growth (see Shenoy et al. 2012).

10.6 RECOMMENDATIONS FOR FUTURE STUDIES

All aspects of mudskipper feeding studies would benefit from the wealth of information available in the wider piscivorous literature, and indeed a more holistic approach to the study of feeding in mudskippers is required.

Gut content analyses alone are insufficient to resolve basic questions such as if a species is a carnivore or herbivore, a selective or opportunistic predator, or if foraging techniques are optimal. It is also necessary to consider prey availability, predatory behavior, and prey selection processes. Additionally, detailed anatomy of the entire alimentary tract, including its enzyme composition, will prove valuable in our broad understanding of diet and foraging in mudskippers. While this ideal may be unrealistic, it argues for a multidisciplinary approach in which the dietary studies are but one element. Where such additional data is available, it gives greater validity to the conclusion. For example, while it is one of the least studied of the oxudercine genera, *Scartelaos* is considered to be an omnivore and this is confirmed by the details of its diet, pharyngeal tooth plate architecture and enzyme composition of the intestine.

Variations in mudskipper diet over different time scales are well established, but for *Periophthalmus* spp. especially, seasonal diet information should be augmented by concurrent benthic surveys. Similarly, for diel changes, gut retention times may be important. For ontogenetic changes, a study of pharyngeal tooth architecture would be a useful adjunct. For sex differences in diet, direct observations of movement and location of identified fish in relation to prey distribution could be attempted.

Studies on mudskipper behavior could benefit greatly by adopting some of the well-established methods used in similar studies on blennies or other gobioids (Fabre et al. 2014; White and Brown 2014). Tagging of individuals to understand homing patterns (Thyssena et al. 2014; White and Brown 2013; Santos et al. 1989; the former studies used Visible Implant Florescent Elastomer tags-VIE – Marine Technology, Inc. 2010) would enable sexually monomorphic species to be easily distinguished during observation.

Clear details of initial prey capture have been elucidated, but what happens to the prey in the buccopharyngeal cavity remains unclear. Is the body wall of larger animal prey scarified by the action of the pharyngeal plates of *Periophthalmus* spp. to allow digestive enzymes access to the soft tissue or are the plates a simple transport system from the oral teeth to the esophagus? Are the gill rakers the main selective filtration mechanism in *Boleophthalmus* spp. or do the pharyngeal plates also play a role in the process? In the general piscivorous literature, it is accepted that pharyngeal plate structure reflects the carnivorous, herbivorous, or omnivorous diets of the fish and those of mudskippers deserve more attention. Another neglected aspect of prey selection includes the role of sensory detection of microalgae or infaunal prey items.

Mudskipper habitat preferences, their trophic level, and guild role in their community all contribute toward the interpretation of the diet of a specific species. Additionally, behavioral studies that include interactions between sympatric species, as well as mudskippers and their prey, would further elucidate proximate mechanisms of food selection that determine diet.

REFERENCES

Abdoli L., Kamrani E., Abdoli A. and Kiabi B. 1996. Feeding habits of *Scartelaos tenuis* (Day, 1876) in Hormozgan Province, Iran. *Aquatic Ecology Quarterly* 1(4):43–53.

Abdoli L., Kamrani E., Abdoli A., Kiabi B. and Rezazadeh K.E. 2013. Comparison of some biological factors of *Boleophthalmus dussumieri* in Hormozgan and Bushehr Provinces. *Journal of Fisheries, Iranian Journal of Natural Resources* 65(4):421–428.

Abdoli L., Kiabi B., Kamrani E., Abdoli A., Rezazadeh K.E. and Keshavarz M. 2012. Feeding habits of *Scartelaos tenuis* in Bushehr Province, Iran. *Journal of Fisheries (Iranian Journal of Natural Resources)* 64(4):309–317.

Abdul-Razak M.M., Saddek A.H. and Laith F.L. 2015. Spatiotemporal variability of fish assemblage in the Shatt Al-Arab River, Iraq. *Journal of Coastal Life Medicine* 3(1):27–34.

Abidizadegan M., Esmaeilpoor P.S. and Rahmani H. 2015. Partial morphometrics and meristic evaluation of the two species mudskippers: *Scartelaos tenuis* (Day, 1876) and *Periophthalmus waltoni* (Koumans, 1941) from the Persian Gulf, Bushehr, Iran. *International Journal of Fisheries and Aquatic Studies* 2(5):353–358.

Abuzinadah O.A. 1995. Gill raker morphology in some Red Sea fishes of different feeding preferences. *Journal of King Abdul-Aziz University: Marine Science* 6:93–122.

Afshar T., Abdoli A. and Kiabi B. 2012. Food habits of *Periophthalmus waltoni* (Koumans, 1941) in Persian Gulf, Hormozgan Province, Iran. *Journal of Aquatic Ecology* 2(1):42–52.

Aguilar N.M. 2000. Comparative physiology of air-breathing gobies. PhD dissertation, University of California, San Diego.

Al-Daham N.K. and Al-Noor S.S. 2000. Food and feeding habits of the mudskipper *Boleophthalmus boddarti* (Pisces: Gobiidae) in Khor Al-Zubair, Southern Iraq. Basrah *Journal of Agricultural Science* 13(1):21–27.

Al-Noor S.H. 1994. The ecology and biology of *Periophthalmus waltoni* Koumans and *Boleophthalmus boddarti* (Pallas) in Kor Al-Zubair, southern Iraq. MSc thesis, University of Basrah.

Andraso G., Cowles J., Colt R., Patel J. and Campbell M. 2011. Ontogenetic changes in pharyngeal morphology correlate with a diet shift from arthropods to dreissenid mussels in round gobies *Neogobius melanostomus. Journal of Great Lakes Research* 27:738–743.

Ansari K.G., Manokaran S., Raja S., Lyla P.S. and Khan S.A. 2014. Interaction of free-living marine nematodes in the artificial mangrove environment (southeast coast of India). *Environment Monitoring Assessment* 186(1):293–305.

Baeck G.W., Takita T. and Yoon Y.H. 2008. Lifestyle of Korean mudskipper *Periophthalmus magnuspinnatus* with reference to a congeneric species *Periophthalmus modestus. Ichthyological Research* 55(1):43–52.

Baeck G.W., Yoon Y.H. and Park J.M. 2013. Ontogenetic and diel changes in diets of two sympatric mudskippers *Periophthalmus modestus* and *Periophthalmus magnuspinnatus* on the tidal flats of Suncheon Bay, Korea. *Fisheries Science (Tokyo)* 79(4):629–637.

Barak N.A.E., Salman N.A. and Ahmad S.M. 1994. The piscivorous feeding of mudskipper *Periophthalmus waltoni* Koumans from Khor Al-Zubair, northwest Arabian Gulf. *Pakistan Journal of Zoology* 26(3):280–283.

Berti R., Chelazzi L., Colombini I. and Ercolini A. 1992. Direction-finding ability in a mudskipper from the delta of the Tana River (Kenya). *Tropical Zoology* 5(2):219–228.

Berti R., Colombini I., Chelazzi L. and Ercolini A. 1994. Directional orientation in Kenyan populations of *Periophthalmus sobrinus* Eggert: Experimental analysis of the operating mechanisms. *Journal of Experimental Marine Biology and Ecology* 181(1):135–141.

Bhatt N.Y., Patel S.J., Patel D.A. and Patel H.P. 2009. Burrowing activities of goby fish in the recent intertidal mud flats along the Navinal coast, Kachchh, western India. *Journal of the Geological Society of India* 74(4):515–550.

Bob-Manuel F.G. 2011. Food and feeding ecology of the mudskipper, *Periophthalmus koelreuteri* (Pallas) Gobiidae at Rumuolumeni Creek, Niger Delta, Nigeria. *Agriculture and Biology Journal of North America* 2(6):897–901.

Bouillon S., Connolly R.M. and Gillikin D.P. 2011. Use of stable isotopes to understand food webs and ecosystem functioning in estuaries. In *Treatise on Estuarine and Coastal Science* (eds. E. Wolanski and D.S. McLusky) pp. 143–173. Waltham, MA: Academic Press.

Bucholtz R.H., Meilvang A.S., Cedhagen T., Christensen J.T. and Macintosh D.J. 2009. Biological observations on the mudskipper *Pseudapocryptes elongatus* in the Mekong Delta, Vietnam. *Journal of the World Aquaculture Society* 40(6):711–723.

Burress E.D. 2015. Cichlid fishes as models of ecological diversification: Patterns, mechanisms, and concequences. *Hydrobiologia* 748:7–27.

Cao F.-J., Ye N., Luo J. and Li X.-B. 2013. Histological observation of digestive system in Blue Mudhopper *Scartelaos viridis. Fisheries Science (Liaoning)* 32(2):63–67.

Chan K.-Y.P. 1989. The ecology of mudskippers (Pisces: Periophthalmidae) at the Mai Po Marshes Nature Reserve, Hong Kong. MSc thesis, University of Hong Kong.

Chaudhuri A., Mukherjee S. and Homechaudhuri S. 2012. Diet composition and digestive enzymes activity in carnivorous fishes inhabiting mudflats of Indian Sundarban estuaries. *Turkish Journal of Fisheries and Aquatic Sciences* 12(2):266–275.

Chaudhuri A., Mukherjee S. and Homechaudhuri S. 2014. Food partitioning among carnivores within feeding guild structure of fishes inhabiting a mudflat ecosystem of Indian Sundarban. *Aquatic Ecology* 48:35–51.

Chen S.X., Hong W.S., Su Y.Q. and Zhang Q.Y. 2008. Microhabitat selection in the early juvenile mudskipper *Boleophthalmus pectinirostris* (L.). *Journal of Fish Biology* 72(3):585–593.

Chen S.X., Hong W.S., Su Y.Q., Zhang Q.Y. and Su Y.Q. 2007. Why does the mudskipper *Boleophthalmus pectinirostris* form territories in farming ponds? *Journal of the Marine Biological Association of the United Kingdom* 87(2):615–619.

Chinnadurai G. and Fernando O.J. 2007. Meiofauna of mangroves of the southeast coast of India with special reference to the free-living marine nematode assemblage. *Estuarine, Coastal and Shelf Science* 72:329–336.

Chua J.L.T., Jaafar Z. and Sivasothi N. 2011. Tidal migration of the Giant Mudskipper, *Periophthalmodon schlosseri* (Pallas, 1770) in a Singapore mangrove. Undergraduate Research Opportunities Programme Report, National University of Singapore, 4 pp.

Chukwu K.O. and Deekae S.N. 2013. Foods of the mudskipper (*Periophthalmus barbarus*) from New Calabar River, Nigeria. *Journal of Agriculture and Vetinary Science* 5(6):45–48.

Chukwu K.O., Deekae S.N. and Gabriel U.U. 2010. Reproductive biology of *Periopthalmus barbarus* (Linneaus 1766) in new Calabar River, Nigeria. *Agriculture And Biology Journal of North America* 1(6):1158–1161.

Clayton D.A. 1986. Ecology of mudflats with particular reference to those of the northern Arabian Gulf. In *Marine Environment and Pollution: Proceedings of the Conference on the Marine Environment & Pollution* (eds. R. Halwagy, D.A. Clayton and M. Behbehani) pp. 83–96. Kuwait: Kuwait University, Kuwait Foundation for Advancement of Science & Environmental Protection Council.

Clayton D.A. 1993. Mudskippers. *Oceanography and Marine Biology an Annual Review* 31:507–577.

Clayton D.A. and Abouseedo F. 1986. Osteichthyes. In *A Field Guide to the Sea Shores of Kuwait and the Arabian Gulf* (ed. D.A. Jones) pp. 167–175. Kuwait: Blandford Press.

Clayton D.A. and Snowden R. 2000. Surface activity in the mudskipper *Periophthalmus waltoni* Koumans, 1941 in relation to prey activity and environmental factors. *Tropical Zoology* 13(2):239–249.

Clayton D.A. and Vaughan T.C. 1982. Pentagonal territories of the mudskipper *Boleophthalmus boddarti* (Pisces: Gobiidae). *Copeia* 1982(1):233–235.

Clayton D.A. and Vaughan T.C. 1988. Ethogram of *Boleophthalmus boddarti* (Teleostei: Gobiidae) a mudskipper found on the mudflats of Kuwait. *Journal of the University of Kuwait (Science)* 15:115–138.

Clayton D.A. and Wright J.M. 1989. Mud-walled territories and feeding behaviour of *Boleophthalmus boddarti* (Pisces: Gobiidae) on the mudflats of Kuwait. *Journal of Ethology* 7:91–95.

Clements K.D., Gleeson V.P. and Slaytor M. 1994. Short-chain fatty acid metabolism in temperate marine herbivorous fish. *Journal of Comparative Physiology B* 164(5):372–377.

Colombini I., Berti R., Ercolini A., Nocita A. and Chelazzi L. 1995. Environmental factors influencing the zonation and activity patterns of a population of *Periophthalmus sobrinus* Eggert in a Kenyan mangrove. *Journal of Experimental Marine Biology and Ecology* 190(1):135–149.

Colombini I., Berti R., Nocita A. and Chelazzi L. 1996. Foraging strategy of the mudskipper *Periophthalmus sobrinus* Eggert in a Kenyan mangrove. *Journal of Experimental Marine Biology and Ecology* 197(2):219–235.

Dasgupta M. 2002. Morphometrics of the alimentary canal of some freshwater fishes of West Bengal in relation to their food and feeding habits. *Indian Journal of Fisheries* 49(4):641–464.

Dasgupta M. 2004. Relative length of the gut of some freshwater fishes of West Bengal in relation to food and feeding habits. *Indian Journal of Fisheries* 51(3):381–384.

Datta N.C. and Das A. 1980. Anatomy of the olfactory apparatus of some Indian gobioids (Pisces: Perciformes). *Zoologischer Anzeiger* 205(3–4):241–252.

De J.K. and Nandi N.C. 1984. A note on the locomotory behaviour of the mudskipper *Boleophthalmus boddarti. Indian Journal of Fisheries* 31(3):407–409.

DeNiro M.J. and Epstein S. 1978. Influence of diet on the distribution of carbon isotopes in animals. *Geochimica et Cosmochimica Acta* 42(5):495–506.

Dhage K.P. and Mohamed O.M.M.J. 1977. Study of amylase activity in mudskipper *Periophthalmus koelreuteri* (Pallas). *Indian Journal of Fisheries* 24(1–2):129–134.

Drenner R.W., Hambright K.D., Vinyard G.L. and Gophen M. 1987. Particle ingestion by *Tilapia galilaea* is not affected by removal of gill rakers and microbranchiospines. *Transactions of the American Fisheries Society* 116(2):272–276.

Ekpo I.E., Obot O.I. and Essien-Ibok M.A. 2014. Studies on the food composition and feeding pattern of fish communities in Qua Iboe River, Niger Delta region of Nigeria. *American Journal of Biology and Life Sciences* 2(5):122–134.

Fabre N., García-Galea E. and Vinyoles D. 2014. Spatial learning based on visual landmarks in the freshwater blenny *Salarias fluviatilis* (Asso, 1801). *Learning and Motivation* 48:47–54.

Fang J.-H. 1996. Analysis of feeding habits of mudskipper *Boleophthalmus pectinirostris* (Linnaeus). *Marine Sciences* 106:26–28.

Feulner G.R. and Roobas B. 2013. Re-discovery of the mudskipper *Periophthalmus waltoni* Koumans, 1941 in the United Arab Emirates. *Tribulus* 21:42–77.

Fishelson L. and Delarea Y. 2004. Taste buds on the lips and mouth of some blenniid and gobiid fishes: Comparative distribution and morphology. *Journal of Fish Biology* 65(3):651–665.

Fishelson L. and Gon O. 2009. Nasal olfactory organs and olfactory bulbs in blennies. In *The Biology of Blennies* (eds. R.A. Patzner, E.J. Gonçalves, P.A. Hastings and B.G. Kapoor) pp. 187–213. Enfield, NH: CRC Press.

Furukawa Y., Hayase T. and Mukuda T. 2011. Spatial learning and object recognition task in mudskipper. Paper presented at 41st Annual Meeting of the Society for Neuroscience Washington, DC. 12–16 November.

Givens C.E., Ransom B., Bano N. and Hollibaugh J.T. 2015. Comparison of the gut microbiomes of 12 bony fish and 3 shark species. *Marine Ecology Progress Series* 518:209–223.

Gonzales T.T., Katoh M., Mazlan G.A. and Ishimatsu A. 2011. Gross and fine anatomy of the respiratory vasculature of the mudskipper, *Periophthalmodon schlosseri* (Gobiidae: Oxudercinae). *Journal of Morphology* 272(5):629–640.

Gordon M.S., Gabaldon D.J. and Yip A.Y.W. 1985. Exploratory observations on microhabitat selection within the intertidal zone by the Chinese mudskipper fish *Periophthalmus cantonensis*. *Marine Biology* 85(2):209–215.

Hajisamae S. and Ibranhim S. 2008. Seasonal and spatial variations of fish trophic guilds in a shallow, semi-enclosed tropical estuarine bay. *Environmental Biology of Fishes* 82(3):251–264.

Han S.-Z., Zhu A.-J., Ye H.-H. and Huang H.-Y. 2008. The localization of G cell in the digestive tract in five kinds of marine fish. *Marine Sciences* 32(1):52–55.

Hanke W., Bednarczyk S. and Brumme M. 2014. Visual acuity and learning in mudskippers (*Periophthalmus novemradiatus*). Paper presented at 107th Annual Meeting of the German Zoological Society, Georg-August-Univeritat, Gottingen. 11–14 September.

Harris V.A. 1961. On the locomotion of the mudskipper *Periophthalmus koelreuteri* (Pallas)(Gobiidae). *Proceedings of the Zoological Society of London* 134:107–135.

Hayase T., Mukuda T. and Furukawa Y. 2011. Do mudskippers tell their right from left? The 8th International Congress of Comparative Physiology and Biochemistry, Nagoya, May 2011.

Hoda S.M. 1991. Feeding habits and lenghth-weight relationship of *Acentrogobius viridipunctatus* (Cuv. & Val.) from Karachi coast. *Indian Journal of Fisheries* 38(3):177–182.

Hong W.-S. and Zhang Q.-Y. 2003. Review of captive bred species and fry production of marine fish in China. *Aquaculture* 227(1–4):305–318.

Höpner T. and Maraschi S.M.K. 1999. Tidal flats in Iran: Intertidal treasure Khowr-e Mussa—Unraised. *Wadden Sea Newsletter* 1:3–6.

Huang M.-C. 2013. Mudskippers in Tainan: The ecology and lifestyle of mudskipper *Periophthalmus modestus* in Tainan coastal wetland. *International Journal of Science and Engineering* 3(4):37–43.

Inaba S., Yokouchi K., Ishimatsu A. and Tagawa M. 2014. Ontogenetic transition from aquatic to amphibious life in the mudskipper, *Periophthalmus modestus*. Paper presented at 11th International Congress of the Biology of Fish, Heriot-Watt University, Edinburgh, August 3–7.

Ip Y.K., Chew S.F., Lim A.L.L. and Low W.P. 1990. The mudskipper. In *Papers Commemorating the 40th Anniversary of the Department of Zoology, National University of Singapore* (eds. L.M Chou and P.K.L Ng) pp. 85–93. Singapore: Singapore University Press.

Ishimatsu A. and Gonzales T.T. 2011. Mudskippers: Front runners in the modern invasion of land. In *The Biology of Gobies* (eds. R. Patzner, J.L. Van Tassell, M. Kovačić and B.G. Kapoor) pp. 609–638. Enfield: Science Publishers.

Jackson A. C. and Rundle S.D. 2008. Diet–Shifts by an estuarine goby (*Pomatoschistus microps*) in the face of variable prey availability. *Journal of Experimental Marine Biology and Ecology* 361(1):1–7.

Jin B., Fu C., Zhong J., Li B., Chen J. and Wu J. 2007. Fish utilization of a salt marsh intertidal creek in the Yangtze River estuary, China. *Estuarine, Coastal and Shelf Science* 73(3–4):844–852.

Jin B., Qin H., Xu W., et al. 2010. Nekton use of intertidal creek edges in low salinity salt marshes of the Yangtze River estuary along a stream-order gradient. *Estuarine Coastal and Shelf Science* 88(3):419–428.

Kathiresan K. 2000. A review of studies on Pichavaram mangrove, southeast India. *Hydrobiologia* 430:185–205.

Kawamura R., Fujimura H., Uemura R., Higuchi T., Casareto B.E. and Suzuki Y. 2011. Stable isotope approach for feeding structure of mudskipper *Periophthalmus argentilineatus* at different habitats in Okinawa Islands, Japan. *Mineralogical Magazine* 75(3):1160.

Khaironizam M.Z. and Norma-Rashid Y. 2000. A new record of the mudskipper *Parapocryptes serperaster* (Oxudercinae: Gobiidae) from Peninsular Malaysia. *Malaysian Journal of Science* 19:101–104.

Khaironizam M.Z. and Norma-Rashid Y. 2004. Diet of mudskippers of the genus *Periophthalmus* from the Klang Strait, Malaysia. In *Marine Science into the New Millennium: New Perspectives & Challenges* (eds. S.E. Phang, C.V. Ching, H.S. Chye, N. Mokhtar and J.O.L. Sim) pp. 223–228. Kuala Lumpur: University of Malaya Maritime Research Centre.

Khoo K.G. 1966. Studies on the biology of periophthalmid fishes in Singapore. BSc. Zoology Honours Thesis, National University of Singapore.

Kim D., Je J.G. and Lee J.H. 2000. The community structure and spatial distribution of meiobenthos in the Kanghwa tidal flat, west coast of Korea. *Ocean Research* 22(1):15–23.

Kim D.-H., Brunt J. and Austin B. 2007. Microbial diversity of intestinal contents and mucus in rainbow trout (*Oncorhynchus mykiss*). *Journal of Applied Microbiology* 102:1654–1664.

Kim H.T., Lee Y.J., Park J.S. and Park J.Y. 2014. A study on the structure of peripheral olfactory organ in the Korean mudskipper, *Scartelaos gigas* (Pisces, Gobiidae). *Korean Journal of Ichthyology* 26(4):281–287.

Kim J.K., Baek H.-J., Kim J.-W., Chang D.-S. and Kim J.-I. 2011. Sexual maturity and early life history of the mudskipper *Scartelaos gigas* (Pisces, Gobiidae): Implications for conservation. *Fisheries and Aquatic Sciences* 14(4):403–410.

King R.P. and Udo M.T. 1998. Seasonality in diet and foraging performance of the mudskipper, *Periophthalmus barbarus* (Goblidae) in Imo River estuary, Nigeria. In *Fish and Fisheries of Southeastern Nigeria* (eds. R.P. King and B.S. Moses) pp. 65–75. Nigeria: University of Imo.

Kobegenova S.S. and Dzhumaliyev M.K. 1991. Morpho-functional peculiarities of the digestive system in some gobioid fishes (Gobioidei). *Voprosyy Ikhtiologii* 31(6):965–973.

Kon K., Kurokura H. and Tongnunui P. 2009. Do mangrove root structures function to shelter benthic macrofauna from predators? *Journal of Experimental Marine Biology and Ecology* 370:1–8.

Kon T. and Yoshino Y. 2002. Diversity and evolution of life histories of gobioid fishes from the viewpoint of heterochrony. *Marine and Freshwater Research* 53:377–402.

Kondalarao B. 1983. Distribution of meiofauna in the Gautami–Godavari estuarine system. *Mahasagar* 16(4):453–457.

Kruitwagen G., Nagelkerken I., Lugendo B.R., Mgaya Y.D. and Bonga S.E.W. 2010. Importance of different carbon sources for macroinvertebrates and fishes of an interlinked mangrove-mudflat ecosystem (Tanzania). *Estuarine Coastal and Shelf Science* 88(4):464–472.

Kruitwagen G., Nagelkerken I., Lugendo B.R., Pratap H.B. and Bonga S.E.W. 2007. Influence of morphology and amphibious life-style on the feeding ecology of the mudskipper *Periophthalmus argentilineatus. Journal of Fish Biology* 71(1):39–52.

Kuciel M. 2013. The mechanism of olfactory organ ventilation in *Periophthalmus barbarus* (Gobiidae, Oxudercinae). *Zoomorphology* 132:81–85.

Kuciel M., Zuwala K. and Jakubowski M. 2011. A new type of fish olfactory organ structure in *Periophthalmus barbarus* (Oxudercinae). *Acta Zoologica (Copenhagen)* 92(3):276–280.

Kutschera U., Burghagen H. and Ewert J.-P. 2008. Prey-catching behaviour in mudskippers and toads: A comparative analysis. *OnLine Journal of Biological Sciences* 8(2):41–43.

Lawson E.O. 2010. Aspects of reproductive biology in mudskipper, *Periophthalmus papilio* from mangrove swamps of Lagos Lagoon, Lagos, Nigeria. *Journal of Fisheries International* 5(2):36–43.

Liem K.F. 1974. Evolutionary strategies and morphological innovations: Cichlid pharyngeal jaws. *Systematic Zoology* 22:425–441.

Lim G.S. 1967. Studies on the feeding biology of *Periophthalmus barbarus* (Linnaeus) and *Boleophthalmus boddaerti* (Pallas) Perciformes—Periophthalmidae. BSc honours thesis, University of Malaya.

Lim S.Y. 1971. The feeding biology of *Periophthalmus vulgaris* (Eggert) and *Periophthalmodon schlosseri*. MSc thesis, Department of Biological Sciences, University of Malaya, Kuala Lumpur.

Lindsay F.J.H. 1984. Distribution and function of digestive tract chitinolytic enzymes in fish. *Journal of Fish Biology* 24(5):529–536.

Low W.P. 1988. A comparative study of terrestrial adaptations of the gills in three mudskippers—*Periophthalmus chrysospilos, Boleophthalmus boddaerti*, and *Periophthalmodon schlosseri*. *Biological Bulletin* 175:434–438.

Low W.P., Ip Y.K. and Lane D.J.W. 1990. A comparative study of the gill morphometry in the mudskippers *Periophthalmus chrysospilos, Boleophthalmus boddaerti* and *Periophthalmodon schlosseri*. *Zoological Science* 7(1):29–38.

MacNae W. and Kalk M. 1962. The ecology of the mangrove swamps at Inhaca Island Moçambique. *Journal of Animal Ecology* 50:19–34.

MacNeill D.B. and Brandt S.B. 1990. Ontogenetic shifts in gill-raker morphology and predicted prey capture efficiency of the alewife, *Alosa pseudoharengus*. *Copeia* 1990:164–171.

Matsumoto M., Ishibashi N. and Kondo E. 1992. Nematode fauna in estuarine and coastal sediment of the Ariake Sea with reference to the habitat of mudskipper, *Beleophthalmus* [*Boleophthalmus*] *pectinirostris*. *Japanese Journal of Nematology* 22:37–42.

Mazlan A.G., Faridah Y., Shukor M.N. and Aziz A. 2006a. Foraging behaviour and food selection of giant mudskipper (*Periophthalmodon schlosseri*) at Kuala Gula, Matang Mangrove Reserve, Perak, Malaysia. *Coastal Marine Science* 30(1):263–267.

Mazlan A.G., Masitah A. and Mahani M.-C. 2006b. Fine structure of gills and skins of the amphibious mudskipper, *Periophthalmus chrysospilos* Bleeker, 1852, and a non-amphibious goby, *Favonigobius reichei* (Bleeker, 1853). *Acta Ichthyologica et Piscatoria* 36(2):127–133.

Mazlan A.G., Rohaya M., Faridah Y., Zaidi C.C., Abdullah S. and Othman B.H.R. 2004. Implementing the bio-mathematical concept of feeding modeling in fish toward enhancing aquaculture production in Malaysia. *Malaysian Applied Biology* 33(2):51–58.

Mhaisen F.T. and Al Maliki N.S. 1996. Parasites, diseases and food of the dark-blotched mudskipper *Periophthalmus waltoni* (Perciformes: Gobiidae) in the Khor Al-Zubair estuary (Iraq). *Zoology in the Middle East* 13:85–87.

Michel K.B., Heiss E., Aerts P. and Van Wassenbergh S. 2015. A fish that uses its hydrodynamic tongue to feed on land. *Proceedings of the Royal Society B* 282(1805).

Michel K.B., Van Wassenbergh S. and Aerts P. 2012. Comparison of Atlantic mudskippers (*Periophthalmus barbarus*) feeding in an aquatic and terrestrial environment: A detailed morphological and kinematical study. *Integrative and Comparative Biology* 52(Suppl. 1):E121.

Milward N.E. 1974. Studies on the taxonomy, ecology and physiology of Queensland mudskippers. PhD dissertation, University of Queensland.

Minh T.H., Gallardo W.G. and Phuong N.T. 2010. Fishery and aquaculture of juvenile mudskipper *Pseudapocryptes elongatus* (Cuvier, 1816) in the coastal zone of Mekong Delta, Vietnam. *Asian Fisheries Science* 23:224–239.

Minh T.H. and Phuong N.T. 2011. Study on mudskipper (*Pseudapocryptes elongatus* Cuvier, 1816) pond culture in the coastal area of Soc Trang and Bac Lieu Provinces. *Journal of Science* 18b:219–227.

Mitra A. and Homechaudhuri S. 2013. Trophic niche and nutritional profile of two mudskippers *Periophthalmus novemradiatus* and *Boleophthalmus boddarti* in their native estuarine habitat. *Science and Culture* 79(5/6):248–251.

Mohammadpour Z.-S., Nabavi S.M. and Madiseh S.D. 2009. *Periophthalmodon schlosseri* (Gobiidae), according to occurrence index in Samaeily Creek of Mahshahr. *Journal of Marine Biology* 1(2):92–102.

Morii H. and Kasama K. 1989. Bacterial flora in the digestive tracts of mudskipper fishes *Boleophthalmus pectinirostris* and *Periophthalmus cantonensis*. *Bulletin Japanese Society of Fisheries* 55(4):733–741.

Morii H. and Kasama K. 1990. The characteristics of bacterial flora in the digestive tracts of mudskipper fishes *Boleophthalmus pectinirostris* and *Periophthalmus cantonensis*. *Bulletin of the Facuty of Fisheries Nagasaki University-Chodai Suikenpo* 68:43–52.

Morii H., Takita T. and Dahril T. 1998. Bacterial flora of the intestine of mudskippers in Indonesia. *Fisheries Science* 64(5):842–843.

Muhammad Ali S.H. and Norma-Rashid Y. 2005. Size class distribution and behaviour budgets of sympatric mudskippers utilising mangrove microhabitats in Morib. In *Ecology of Klang Strait* (eds. A. Sasekumar and V.C Chung) pp. 115–126. Kuala Lumpur: Faculty of Science, University of Malaya.
Mukuda T. 2010. Immunohistochemial identification of telencephalic neurons expressing c-Fos protein during learning and memory in the mudskipper. 15th Biennial Scientific Meeting of the International Society for Comparative Psychology, Hiroshima University, Japan
Mukuda T. 2012. Behavioural and morphological identification of the teleost hippocampus using mudskipper. Grant summary Hiroshima University Graduate School.
Mukuda T., Koyama Y. and Furukawa Y. 2011. Brain loci involved in object discrimination learning in mudskipper. 8th International Congress of Comparative Physiology and Biochemistry, Nagoya.
Mukuda T., Koyama Y., Hamasaki S., Hayase T., Furukawa Y. and Tomikawa K. 2013. Spatial learning in a euryhaline teleost, mudskipper *Periophthalmus modestus*. Paper presented at The Chugoku-Shikoku Branch of The Zoological Society of Japan, Tokushima. 11–12 May.
Munk O. 1970. On the occurrence and significance of horizontal band-shaped retinal areae in teleosts. *Videnskabelige Meddelelser fra Dansk naturhistorisk Foening* 133:85–120.
Murdy E.O. 1989. A taxonomic revision and cladistic analysis of the Oxudercine gobies (Gobiidae: Oxudercinae). *Records of the Australian Museum* Supplement 11:1–93 pp.
Mutsaddi K.B. and Bal D.V. 1969. Food and feeding of *Boleophthalmus dussumieri* (Cuv. & Val.). *Journal of the University of Bombay* 38(65):42–55.
Nagelkerken I., Blaber S.J.M., Bouillon S. et al. 2008. The habitat function of mangroves for terrestrial and marine fauna: A review. *Aquatic Botany* 89(2):155–185.
Nanjo K., Kohno H. and Sano M. 2008. Food habits of fishes in the mangrove estuary of Urauchi River, Iriomote Island, southern Japan. *Fisheries Science* 74(5):1024–1033.
Niva B., Ojha J. and Munshi J.S.D. 1981. Morphometrics of the respiratory organs of an estuarine goby, *Boleophthalmus boddaerti*. *Japanese Journal of Ichthyology* 27(4):316–326.
Norma-Rashid Y. and Khaironizam M.Z. 2005. Some biological aspects of mudskippers (Gobiidae, Oxudercinae) from Langkawi Island. *Malaysian Journal of Science* 24:139–144.
Owen X. 2009. Study on ecological habits in mudskipper (*Boleophthalmus pectinirostris*). PhD dissertation, Xiamen University.
Pace C. M. and Gibb A.C. 2009. Mudskipper pectoral fin kinematics in aquatic and terrestrial environments. *Journal of Experimental Biology* 212(14):2279–2286.
Pan L.-L., Zhang G.-R., Wei K.-J. and Zhang J. 2010. A comparison of the gill morphometry of four mudskippers (Gobiidae: Oxudercinae) from Southeast Asia and Oceania: With comments on ecological adaptation. *Chinese Journal of Zoology* 45(4):1–10.
Pankow H. and Hug M.F. 1979. Diatoms in the stomach content of *Pseudapocryptes dentatus* a mudskipper from the Shatt al-Arab estuary (Iraq). *Wissenschaftliche Zeitschschrift der Wilhelm-Pieck-Universitat Rostock. Mathematisch-Naturwissenschaftliche Reiche* 6:547–554.
Parenti L.R. and Thomas K.R. 1998. Pharyngeal jaw morphology and homology in Sicydiine Gobies (Teleostei: Gobiidae) and allies. *Journal of Morphology* 237:257–274.
Peretti D. and Andrian I.F. 2008. Feeding and morphological analysis of the digestive tract of four species of fish (*Astyanax altiparanae, Parauchenipterus galeatus, Serrasalmus marginatus* and *Hoplias* aff. *malabaricus*) from the upper Paraná River floodplain, Brazil. *Brazilian Journal of Biology* 68(3):671–679.
Polgar G. and Bartolino V. 2010. Size variation of six species of oxudercine gobies along the intertidal zone in a Malayan coastal swamp. *Marine Ecology Progress Series* 409:199–212.
Polgar G. and Crosa G. 2009. Multivariate characterisation of the habitats of seven species of Malayan mudskippers (Gobiidae: Oxudercinae). *Marine Biology* 156(7):1475–1486.
Pravinkumar M., Murugesan P., Prakash R.K., Elumalai V., Viswanathan C. and Raffi S.M. 2013. Benthic biodiversity in the Pichavaram mangroves, southeast coast of India. *Journal of Oceanography and Marine Science* 4(1):1–11.
Quan W., Ni Y., Shi L. and Chen Y. 2009. Composition of fish communities in an intertidal salt marsh creek in the Changjiang River estuary, China. *Chinese Journal of Oceanology and Limnology* 27(4):806–815.
Quang D.M., Xuyen L.T.M, Thanh N.M, Lua T.T. and Vi D.H. 2013. Preliminary study on the food composition of the goby, *Parapocryptes serperaster*, at Bac Lieu. Paper presented at the 5th National Conference on Ecology and Biological Resources, Hanoi, 18 October 2013.

Rathod S.D. and Patil N.N. 2009. Feeding habits of *Boleophthalmus dussumieri* (Cuv. & Val.) from Ulhas River Estuary near Thane City, Maharashtra State. *Journal of Aquatic Biology* 24(2):1–7.

Raut D., Ganesh T., Murty N.V.S.S. and Raman A.V. 2005. Macrobenthos of Kakinada Bay in the Godavari delta, east coast of India: Comparing decadal changes. *Estuarine, Coastal and Shelf Science* 62(4):609–620.

Ravi V. 2013. Food and feeding habits of the mudskipper, *Boleophthalmus boddarti* (Pallas, 1770) from Pichavaram Mangroves, southeast coast of India. *International Journal of Marine Sciences* 3(12):98–104.

Rodelli M.R., Gearing J.N., Gearing P.J., Marshall N. and Sasekumar A. 1984. Stable isotope ratio as a tracer of mangrove carbon in Malaysian ecosystems. *Oecologia* 61(3):326–333.

Ryu B.S., Kim I.S. and Choi Y. 1995. Ecology and life history of *Boleophthalmus pectinirostris* in Korea. *Journal of Korean Fisheries Society* 28(3):316–324.

Santos R.S., Almada V.C. and Santos A.J. 1989. Field experiments and observations on homing and territoriality in intertidal blennies. In *EthoExperimental Approaches to the Study of Behavior* (eds. R.J. Blanchard, P.F. Brain, D.C. Blanchard and S. Parmigiani) pp. 623–632. New York, NY: Kluwer Academic Press.

Sarkar S.K., Acharya A., Jana S. and De S.K. 2014. Macro-anatomical variation of the olfactory apparatus in some Indian teleosts with special reference to their ecological habitat. *Folia Morphologica* 73(2):122–128.

Sarkar S.K. and De S.K. 2011a. Functional morphoanatomy of olfactory sensory epitheilial cells of *Pseudapocryptes lanceolatus* (Block and Schneider). *International Journal of Science and Nature* 2(1):1–6.

Sarkar S.K. and De S.K. 2011b. A study on olfactory neuroepthelium of *Pseudapocryptes lanceolatus* (Bloch and Schneider): Medical analog X-ray based anatomy and transmission electron microscopical interpretations. *International Journal of Science and Nature* 2(2):186–191.

Sarker A.L., Al-Daham N.K. and Bhatti M.N. 1980. Food habits of the mudskipper *Pseudapocryptes dentatus* (Val.). *Journal of Fish Biology* 17(6):635–639.

Shabdin M.L. and Othman H.R. 2008. Horizontal distribution of intertidal nematode from Sabah, Malaysia. *Journal of Tropical Biology and Conservation* 4(1):39–53.

Shang X., Zhang G.S. and Zhang J. 2008. Relative importance of vascular plants and algal production in the food web of a *Spartina*-invaded salt marsh in the Yangtze River estuary. *Marine Ecology Progress Series* 367:93–107.

Shenoy L., Patro V.K. and Ying X. 2012. Effects of mudskippers on the soil composition and texture in mangrove forests, and hence the growth of mangroves (*Avicennia alba*) (ed. Little Green Dot), Student Research Grant Project Report submitted to Nature Society (Singapore) Junior College Category Year 2012, Temasek Junior College, Singapore.

Smith J.C. and Sanderson S.L. 2013. Particle retention in suspension-feeding fish after removal of filtration structures. *Zoology* 116(6):348–55.

Somerfield P.J., Gee J.M. and Aryuthaka C. 1998. Meiofaunal communities in a Malaysian mangrove forest. *Journal of the Marine Biological Association of UK* 78:717–732.

Sponder D.L. and Lauder G.V. 1981. Terrestrial feeding in the mudskipper *Periophthalmus* (Pisces: Teleostei): A cineradiographic analysis. *Journal of Zoology* 193(4):517–530.

Stebbins R.C. and Kalk M. 1961. Observations on the natural history of the mud-skipper *Periophthalmus sobrinus*. *Copeia* 1961(1):18–27.

Su T.L. and Lim S.L. 2011. Effect of mudskipper predation on brachyuran abundance and biodiversity in a Singapore mangrove. The Crustacean Society Summer Meeting, Honolulu, HA.

Su T.L. and Lim S.L. 2014. Anti-predatory behaviours employed by two common genera of crabs in response to a mudskipper model. The XV Congress of the International Society for Behavioural Ecology, New York.

Su T.L. and Lim S.L. 2015. To flee or not to flee: Decisions of two genera of crabs in response to a series of predator model variants. The Crustacean Society and International Association of Astacology, Sydney, Australia.

Suyehiro Y. 1942. A study on the digestive system and feeding habits of fish. *Japanese Journal of Zoology* 10(1):1–303.

Swanson B.O. and Gibb A.C. 2004. Kinematics of aquatic and terrestrial escape responses in mudskippers. *Journal of Experimental Biology* 207(23):4037–4044.

Swennen C., Ruttanadakul N., Haver M. et al. 1995. The five sympatric mudskippers (Teleostei: Gobioidea) of Pattani area, southern Thailand. *Natural History Bulletin of the Siam Society* 43(1):109–129.

Taher M.M. 2010. Specialisation, trophic breadth and diet overlap of thirteen small marine fish species from Shayy Al-Basrah Canal, Southern Iraq. *Marsh Bulletin* 5(2):118–131.

Takeda T., Hayashi M., Toba A., Soyano K. and Ishimatsu A. 2011. Ecology of the Australian mudskipper *Periophthalmus minutus*, an amphibious fish inhabiting a mudflat in the highest intertidal zone. *Australian Journal of Zoology* 59(5):312–320.

Takita T., Agusnimar A. and Ahyaudin B. 1999. Distribution and habitat requirements of oxudercine gobies (Gobiidae: Oxudercinae) along the Straits of Malacca. *Ichthyological Research* 46(2):131–138.

Thilagavathi B., Varadharajan D., Babu A., Manoharan J., Vijayalakshmi S. and Balasubramanian T. 2013. Distribution and diversity of macrobenthos in different mangrove ecosystems of Tamil Nadu coast, India. *Journal of Aquactic Research and Development* 4(6):1–12.

Thuy H.T.L., Vinh P.K., Trang T.T.H. and Thuy N.T.B. 2006. Some characteristics on digestive physiology of *Pseudapocryptes lanceolatus* (Block & Schneider, 1801). *Journal of Agricultural Sciences and Technology* 1(3):117–121.

Thyssena L., Triay-Portellab R., Santana del Pinob A. and Castrob J.J. 2014. Homing behaviour of rock pool blenny *Parablennius parvicornis* (Pisces: Blenniidae). *Journal of Natural History* 48(19–20):1169–1179.

Townsend K.A. and Tibbetts I.R. 2005. Behaviour and sexual dimorphism of the blue mudskipper, *Scartelaos histophorus* (Pisces: Gobiidae). *Proceedings of the Royal Society of Queensland* 112:53–62.

Tran N.H., Nguyen T.P. and Nguyen T.N. 2009. Intensive farming of the goby fish in the Mekong Delta, Vietnam. *AQUA Culture AsiaPacific Magazine* pp. 39–40.

Tsuhako Y., Ishimatsu A., Takeda T., Khoo K.H. and Tachihara K. 2003. The eggs and larvae of the giant mudskipper, *Periophthalmodon schlosseri*, collected from a mudflat in Penang, Malaysia. *Ichthyological Research* 50(2):178–181.

Turay I., Vakily J.M., Palomares M.L.D. and Pauly D. 2006. Growth, reproduction and food of the mudskipper, *Periophthalmus barbarus* on mudflats of Freetown, Sierra Leone. In *Fishes in Databases and Ecosystems. Fisheries Centre Research Report* (eds. M.L.D. Palomares, K.I. Stergiou and D. Pauly) pp. 49–54. Canada: University of British Columbia, Fisheries Centre.

Udo M. T. 2002a. Intersexual plasticity in aspects of the biology of the mudskipper *Periophthalmus barbarus* (Gobiidae) in the mangrove swamps of Imo Estuary, Nigeria. *Journal of Environmental Sciences* 14(1):95–101.

Udo M.T. 2002b. Trophic attributes of the mudskipper, *Periophthalmus barbarus* (Gobiidae: Oxudercineae) in the mangrove swamps of Imo River Estuary, Nigeria. *Journal of Environmental Sciences* 14(4):508–517.

Udo M.T. 2002c. Morphometric relationships and reproductive maturation of the mudskipper, *Periophthalmus barbarus* from subsistence catches in the mangrove swamps of Imo Estuary, Nigeria. *Journal of Environmental Sciences* 14(2):221–226.

Udo M.T., Ekpo I.E. and Akpan B.E. 2008. Diel feeding periodicity of the intertidal mudskipper, *Periophthalmus barbarus* (Gobiidae) in Imo River Estuary, Nigeria. *Asian Journal of Microbiology, Biotechnology and Environmental Sciences* 10(3):469–476.

Udoh J.P., Brownson I., Udo M.T. and Ofor C. 2013. Population dynamics of mudskipper, *Periophthalmus barbarus* (Linneaus, 1766) (Teleostei: Gobiidae) in the artisanal fishery of Imo River Estuary, Southeast Nigeria. *Journal of Fisheries and Aquaculture* 4(3):148–153.

Venkataramanujam K., Ramanathan N. and Venkataramani V.K. 1993. Postlarval development and feeding habits of the gobiid, *Parapocryptes rictuosus* (Koumans) (Pisces: Gobiidae). *Matsya* 15–16:53–55.

Vijayakumar R., Ansari Z.A. and Parulekar A.H. 1991. Benthic fauna of Kakinada Bay and backwaters, east coast of India. *Indian Journal of Marine Sciences* 20:195–199.

Wainwright P.C., Smith W.L., Price S.A. et al. 2012. The evolution of pharyngognathy: A phylogenetic and functional appraisal of the pharyngeal jaw key innovation in labroid fishes and beyond. *Systematic Biology* 61(6):1001–27.

White G.E. and Brown C. 2013. Site fidelity and homing behaviour in intertidal fishes. *Marine Biology* 160(6):1365–1372.

White G.E. and Brown C. 2014. A comparison of spatial learning and memory capabilities in intertidal gobies. *Behavioral Ecology and Sociobiology* 68:1393–1401.

Wu R.X., Hong W.S., Zhang Q.Y. and Chen S.X. 2009. Comparative enzyme activities of the intestinal brush border membranes of the herbivorous mudskipper *Boleophthalmus pectinirostris* and the carnivorous Chinese black sleeper *Bostrichthys sinensis. Journal of Applied Ichthyology* 25(5):571–575.

Wu R-X, Ge W., Hong W.-S. and Zhang Q.Y. 2007. Digestive enzymes of adult mudskipper, *Boleophthalmus pectinirostris. Journal of Fishery Sciences of China* 14(1):99–105.

Wu R-X, Ge W., Hong W.-S. and Zhang Q.Y. 2010. Digestive enzyme activities in mudskipper *Boleophthalmus pectinirostris* and Chinese black sleeper *Bostrichthys sinensis. Chinese Journal of Oceanology and Limnology* 28(4):756–761.

Wu R-X, Hong W.-S., Zhang Q.Y. and Chen S.-X. 2006. Comparison of the activities of digestive enzymes in the intestinal brush border membrane between the mudskipper *Boleophthalmus pectinirostris* and Chinese black sleeper *Bostrichthys sinensis. Acta Zoologica Sinica* 52(6):1088–1095.

Wu R.-X., Hong W.-S., Zhang Q.Y., Ge W. and He Z.-B. 2007. Studies on the activities of digestive enzymes of Chinese black sleeper (*Bostrichthys sinensis*). *Journal of Xiamen University Natural Science* 46(1):118–122.

Wu R.-X., Hong W.-S., Zhang Q.Y, Chen S.-X. and Wang Q. 2006. Digestive enzyme activities in larval, juvenile and early young fish of *Boleophthalmus pectinirostris. Journal of Fisheries of China* 30(6):733–739.

Yadav A.N. and Singh B.R. 1989. Gross structure and dimensions of the gill in an air-breathing estuarine goby, *Pseudapocryptes lanceolatus. Japanese Journal of Ichthyology* 36(2):252–259.

Yang K.-Y. 1996. The feeding ecology of the mudskipper *Boleophthalmus pectinirostris* (Pisces: Periophthalmidae) at the Mai Po Marches nature reserve, Hong Kong. MSc thesis, University of Hong Kong.

Yang K.Y., Lee S.Y. and Williams G.A. 2003. Selective feeding by the mudskipper (*Boleophthalmus pectinirostris*) on the microalgal assemblage of a tropical mudflat. *Marine Biology* 143(2):245–256.

Yang K.Q. and Zhang Q.Y. 1991. Study of the development of the juvenile digestive system and feeding of *Boleophthalmus* larvae. *Acta Oceanologia Sinica* 13(2):240–246.

Yeager L.A., Layman C.A. and Hammerschlag-Peyer C.M. 2014. Diet variation of a generalist fish predator, grey snapper *Lutjanus griseus*, across an estuarine gradient: Trade-offs of quantity for quality? *Journal of Fish Biology* 85(2):264–77.

Yuan C. 1963. Comparative anatomy of the alimentary tracts of *Periophthalmus cantonensis* (Osbeck) and *Boleophthalmus chinensis* (Osbeck) in relation to their feeding habits. *Journal of Nanjing University (Biology)* 2:36–46.

Zhang Q.-Y., Hong W.-S, Chen S.-X., Wu R.-X. and Wang Q. 2006. Biological composition and FAA content of particulate organic detritus from mudflat of intertidal zone. *Marine Fisheries Research* 27(4):71–76.

Zhang Q., Hong W.S., Ye Q.Y., Chen S.X. and Jiang G.Q. 2005. Initial feeding of early larvae *Boleophthalmus pectinirostris* in earthen pond rearing. *Marine Fisheries Research,* 26:38–44.

Zhang Q.-Y. and Zhang J. 1988. On the feeding, growth and survival of larval mudskipper (*Boleophthalmus pectinirostris*). *Journal of Fisheries China* 12(3):203–212.

Zhang Q.-Y. and Zhang J. 1991. Initial feeding and survival of larval mudskipper (*Boleophthalmus pectinirostris*) under rearing condition. *Proceedings of the 4th China Oceanology and Limnology Conference.* Beijing: Science Press, pp. 206–211.

Zhu Y. and Zhang Q.-Y. 1993. On feeding habits and histological structure of digestive tract of the mudskipper, *Boleophthalmus pectinirostris*, in intertidal zone of Jiulong River estuary. *Journal of Oceanography Taiwan Strait* 12(3):225–232.

Zulkifli S.Z., Mohamat-Yusuff F., Ismail A. and Miyazaki N. 2012. Food preference of the giant mudskipper *Periophthalmodon schlosseri* (Teleostei: Gobiidae). *Knowledge and Management of Aquatic Ecosystems* 405:7–10.

Zulkifli S.Z., Mohamat-Yusuff F., Mukhtar A., Ismail A. and Miyazaki N. 2014. Determination of food web in intertidal mudflat of tropical mangrove ecosystem using stable isotope markers: A preliminary study. *Life Science Journal* 11(3):427–431.

CHAPTER 11

Territoriality and Courtship Behavior

David Clayton and Kathy Townsend

CONTENTS

11.1 INTRODUCTION

Mudskippers, particularly species of the genus *Periophthalmus*, are renowned for their inter- and intraspecific interactions while out of water. Displays include those for courtship (Townsend and Tibbets 2005; Brillet 1980a) and maintaining territorial boundaries (e.g., Clayton and Vaughan 1982). Although other areas of mudskipper research have progressed (e.g., Sakamoto et al. 2015; Quan et al. 2012; Kanemori et al. 2006) since the literature was reviewed (Clayton 1993), the understanding of social behavior has not. Indeed, with few exceptions, there is little available information to distinguish between the description of territorial and courtship displays.

For progress in understanding social behavior in mudskippers, more detailed descriptive ethological studies need to be undertaken (see Section 12.5). Seminal studies on aggression, territoriality, and courtship behaviors in *Periophthalmus argentilineatus* that incorporate film, field observations, and laboratory experiments serve as useful models to follow (Brillet 1984, 1980b, 1976, 1975, 1969a, reported as *Periophthalmus sobrinus*), especially given the advances in all forms of recording technology for field (Vichaidist et al. 2015) and laboratory (Abril-de-Abreu et al. 2015;

Hu et al. 2015; Polgar et al. 2011) investigations and in mudskipper husbandry (Rupp 2015a, b, 2014, see Chapter 13) in the last 50 years.

To encourage more studies on the social behavior of oxudercine gobies, this review considers the wider theoretical framework of territoriality, aggression, and courtship and the role of hormones for each category of behavior. Except for courtship displays, sexual dimorphism in other behaviors has not been considered. This review further includes the seasonal distribution of and attempts to identify the composite elements of commonly assumed display categories such as courtship, agonistic, and territorial behaviors.

11.2 CATEGORIES AND ELEMENTS OF BEHAVIOR

The ease in observing mudskippers in their natural habitats of fringing mangrove areas, riverbanks, and exposed mudflats prompted studies of their feeding activities as well as territorial and courtship behavior. General behaviors such as basking, splash zone shuttling, rolling, and locomotion in *Periophthalmus* spp. relate to thermoregulation and are easily recognizable (Ikebe and Oishi 1996; Ip et al. 1991; Tytler and Vaughan 1983). The side-to-side head movements in *Boleophthalmus* spp. and the lunges in *Periophthalmus*, associated with foraging (see Section 10.5), are easily quantifiable. However, the elements of social behavior in mudskippers are often labeled as "aggressive" or "sexual" by the season in which observations were made (Huang 2013; Baeck et al. 2008; Taru et al. 2006). In their paper on *Periophthalmodon schlosseri*, Mazlan et al. (2004) included "aggression" and "mating" as two of the six named behavioral categories, the others being "feeding," "locomotion," "resting," and "other." Description of the behavior of *Boleophthalmus boddarti* used the aforementioned categories except that "resting" was replaced with "burrowing" (Vichaidist et al. 2015). Similarly, behavioral budgets of four Malaysian species were analyzed in six similar-grouped behavior categories: aggression, mating, feeding, locomotion, burrowing, and grooming (Muhammad and Norma-Rashid 2005). In the early stages of ethological study, described in these three papers, it may have been useful to categorize behavior in "catch-all" categories, but such presumptions may lead to confusion about the motivation of a particular behavioral element in later and more detailed studies. For example, biting was considered as "courtship" in females of *Pn. schlosseri* by Mazlan et al. (2004) but as "aggression" in other studies of various mudskippers species (e.g., Muhammad and Norma-Rashid 2005; Milward 1974).

Part of the difficulty in assigning motivation may be that, although the overall displays appear familiar, the composite elements may differ. This is true especially if interactions are fast and for this reason, Brillet (1975) considered "head butting" and "biting" as one category. In contrast, the different modes of locomotion are well known (Harris 1961) and easily distinguished. Assigning motivation to a particular behavior is further compounded by the fact that there are similarities between agonistic and pair bonding behaviors (Townsend and Tibbetts 2005).

The ethological descriptive approach for mudskipper behavior has been neglected in recent years despite the foundation laid by Brillet (1986, 1984) for *Ps. argentilineatus* (reported as *Ps. sobrinus*) and *Periophthalmus kalolo* (reported as *Periophthalmus koelreuteri*). Aside from Townsend and Tibbetts (2005), no recent papers have provided ethological-style descriptions of observed and categorized behaviors. In an attempt to rectify this situation, Table 11.1 uses stylized images of the behavioral elements of *Boleophthalmus dussumieri* described by Clayton and Vaughan (1988) to identify the elements and their presumed motivation in the general literature. Although other genera have different body outlines, it is not difficult to identify their behavioral elements equivalent to those shown.

The descriptions of the construction of walls by *Scartelaos histophorus* concur with those in Table 11.1. Mud was moved by either being carried in the mouth or pushed to the edge of the territory with head and body movements (Aguilar 2000). Townsend and Tibbetts (2005) named this

Table 11.1 **Summary of Generalized Social Behaviors of Mudskippers**

Fin Displays	Proposed Name	Name	Species	Context	Authority
	Rest	Horizontal	*B. dussumieri*	Inactive	Clayton and Vaughan (1988)
		Immobile	*Pn. schlosseri*	Feeding strategy	Swennen et al. (1995)
	Flag (further defined as being fully or partially erected)	D1 erection	*B. dussumieri*	Agonistic	Clayton and Vaughan (1988)
			B. boddarti	Courtship	Vichaidist et al. (2015)
				Territorial	Swennen et al. (1995)
			S. histophorus	Pair formation	Townsend and Tibbetts (2005)
		Dorsal fin flagging		Aggression	Aguilar (2000)
			Ps. modestus		Takita et al. (2011)
					Baeck et al. (2008)
	Flick	D2 erection	*B. dussumieri*	Agonistic	Clayton and Vaughan (1988)
			B. boddarti	Territorial	Swennen et al. (1995)
			B. birdsongi		Takita et al. (2011)
	Flash	D1 + D2 erection	*B. dussumieri*	Agonistic	Clayton and Vaughan (1988)
			B. boddarti		Vichaidist et al. (2015)
			Ps. novaeguineaensis	Courtship	Takita et al. (2011)
			Ps. argentilineatus		Brillet (1980a)
					Stebbins and Kalk (1961)
			Ps. modestus		Matoba and Dotsu (1977)
			Ps. modestus		Taru et al. (2006)
			S. gigas		Kim et al. (2011)
	Flare	Dorsal fins + caudal fin erection	*B. dussumieri*	Agonistic	Clayton and Vaughan (1988)
			B. boddarti	Aggression	Dinh et al. (2014)
			S. histophorus		Aguilar (2000)
	Pectoral wave	Pectoral wave	*B. dussumieri*	Courtship	Clayton and Vaughan (1988)
	Tail twirl	Tail beat	*B. dussumieri*		Clayton and Vaughan (1988)
		Tail twirl	*B. boddarti*		Vichaidist et al. (2015)
		Tail undulation	*Ps. argentilineatus*	Courtship	Brillet (1980a)
			Ps. novaeguineaensis	Threat	Milward (1974)

(*Continued*)

Table 11.1 (*Continued*) Summary of Generalized Social Behaviors of Mudskippers

Body Postures	Proposed Name	Name	Species	Context	Authority
	Bulge	Bulge	*B. dussumieri*		Clayton and Vaughan (1988)
					Muhammad and Norma-Rashid (2005)
		Cheeks puffed	*Ps. modestus*	Courtship	Lee and Graham (2002)
		Bucco-pharyngeal extension			
	Yawn	Yawn	*B. dussumieri*		Clayton and Vaughan (1988)
			S. histophorus	While resting	Townsend and Tibbets (2005)
	Head up	Head Up	*B. dussumieri*		Clayton and Vaughan (1988)
			B. boddarti	Territorial	Swennen et al. (1995)
			Ps. modestus		Matoba and Dotsu (1977)
		Nod (Head up and down)	*Ps. argentilineatus*	Displacement	Brillet (1975)
	Gape (combination of "head up" and "yawn")	Gape	*B. dussumieri*		Clayton and Vaughan (1988)
			B. boddarti		Vichaidist et al. (2015)
		Gape	*Pn. schlosseri*	Mating/aggression	Mazlan et al. (2004)
			Ps. argentilineatus or		Brillet (1980b)
		Operculum display	*Ps. modestus*	Aggression	Kagawa et al. (2013)
	Quiver	Quiver	*B. dussumieri*		Clayton and Vaughan (1988)
		Quiver	*Ps. magnuspinnatus*	Pre-spawning courtship	Baeck et al. (2008)
		Body waving	*Ps. takita*		Takita et al. (2011)
		Snake	Species not specified		Muhammed and Norma Rashid (2005)
	Arching	Arched back	*Ps. modestus*		Matoba and Dotsu (1977)
		Lifting its back	*Ps. modestus*		Taru et al. (2006)
			Ps. magnuspinnatus		Baeck et al. (2008)
		Dorsal flexure of the body	*Ps. argentilineatus* or *Ps. kalolo*		Stebbins and Kalk (1961)

(*Continued*)

Table 11.1 (*Continued*) Summary of Generalized Social Behaviors of Mudskippers

Aerial Behaviour	Proposed Name	Name	Species	Context	Authority
	Flip (short distance maintaining contact with substrate)	Lateral flip	*B. dussumieri*		Clayton and Vaughan (1988)
		Flip into air	*Ps. novaeguineaensis*	Reproductive	Aguilar (2000)
	Jump (over long distance and partially through the air)	Jump	*B. dussumieri*		Clayton and Vaughan (1988)
		Jump	*Ps. argentilineatus*	Direct attack	Brillet (1975)
		Jump	*Ps. modestus*	Avoiding predators	Huang et al. (2013)
			Ps. gracilis	Jump	Milward (1974)
			Ps. takita		Takita et al. (2011)
			Pn. schlosseri		Mazlan et al. (2004)
		Skip	*S. histophorus*	Agonistic encounter	Townsend and Tibbetts (2005)
	Tail stand (fin erection should be described)	Leap	*B. dussumieri*		Clayton and Vaughan (1988)
			B. boddarti		Vichaidist et al. (2015)
		Vertical jump	*Ps. argentilineatus*		Brillet (1975)
			Ps. modestus	Courtship	Matoba and Dotsu (1977)
			Ps. magnuspinnatus	Courtship	Baeck et al. (2008)
		Tail stand	*S. histophorus*	Courtship	Townsend and Tibbets (2005)
			S. histophorus	Courtship or aggression	Takita et al. (2011)
		Stand	*S. histophorus*		Lee and Graham (2002)
		Standing on tail	*Pd. lanceolatus*	Could not be assigned	Swennen et al. (1995)

(*Continued*)

Table 11.1 (*Continued*) Summary of Generalized Social Behaviors of Mudskippers

Pair Orientation	Proposed Name	Name	Species	Context	Authority
	Reverse parallel formation (elaborate to "O" or "D" formation depending on body curvature of the pair)	Reverse parallel formation "O" posture (bodies curved) Antiparallel	*B. dussumieri* *B. boddarti* *Ps. argentilineatus* *Ps. magnuspinnatus* *Pn. schlosseri*	Aggressive Mating	Clayton and Vaughan (1988) Vichaidist et al. (2015) Brillet (1975) Baeck et al. (2008) Mazlan et al. (2004)
	Parallel (elaborate to "V" formation depending on pairs' positioning)	Parallel formation 'V' posture Show lateral body to each other	*B. dussumieri* *B. boddarti* *Ps. argentilineatus* *Ps. magnuspinnatus* *Ps. novaeguineaensis* *Pn. schlosseri*	Courtship Aggressive Courtship	Clayton and Vaughan (1988) Vichaidist et al. (2015) Brillet (1975) Baeck et al. (2007) Takita et al. (2011) Mazlan et al. (2004)
	T formation	T formation "T" posture "T" or "L" shape	*B. dussumieri* *Ps. argentilineatus* *Pn. schlosseri*	Aggressive Courtship	Clayton and Vaughan (1988) Brillet (1975) Mazlan et al. (2004)
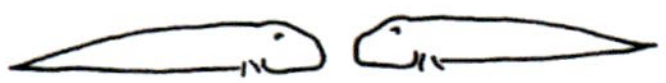	I formation	Face to face formation "I" posture	*B. birdsongi* *Ps. argentilineatus* *Ps. modestus*	Fighting Aggressive Courtship	Townsend and Tibbets (2005) Takita et al. (2011) Brillet (1975) Taru et al. (2007)
	Jostle	Jostle	*B. dussumieri*	Territorial	Clayton and Vaughan (1988)
	Jostle + specific dorsal fin terminology	Active pushing	*B. dussumieri*	Territorial	Muhammad and Norma-Rashid (2005)

activity "snake" (Table 11.1: Body Postures) and described it as "rapid lateral sinusoidal body movement" within shallow pools of water that displaced sediment. Takita et al. (2011) described the same activity as "twisting of the body" with the mud being pushed by the tail to the periphery of the territory. Additionally, although the description of "quivering" and its context suggest a straightforward behavior related to courtship (Clayton and Vaughan 1988), it remains to be confirmed whether these "shallow sinusoidal wave" movements have a different motivational basis compared with "snake" behavior (Table 11.1: Body Postures). The various descriptions of tail beats, if not their motivation (Table 11.1: Fin Displays), are in agreement, but for others there is insufficient precision. In aerial antics of mudskippers, there are at least three different actions (Table 11.1: Aerial Behavior), but it is not always clear which form is being described for a particular species. Hopefully, Table 11.1 will help to clarify and standardize observations, descriptions, naming, and interpretations of mudskipper behavior. Only with good descriptions of behavioral elements and their context can seemingly similar acts be distinguished.

11.2.1 Courtship–Agonistic Interplay

11.2.1.1 Courtship Displays

The sequence of events during courtship follows similar patterns in most observed species of mudskippers and is diagrammatically shown in a broad outline for *Periophthalmus modestus* (Matoba and Dotsu 1977: fig. 16) and in greater detail for *Ps. argentilineatus* (Brillet 1980b: fig. 3, 1970, 1969c; reported as *Ps. sobrinus*). Brillet (1980a, 1970, 1969a) described in detail the first and second phases of behavior of courting males in various categories: tail twirls, vertical jumps, back arching, mouth gaping, and bucco-pharyngeal expansion or contraction, dorsal-fin displays, and "snout" touch (touching females on their flanks). All these six categories are subsumed under a single category called "courtship displays" by Matoba and Dotsu (1977: fig. 16). Brillet's (1980a: fig. 3) Phase III also gives behavioral details as opposed to just "Orientation," "Lead-I," and "Lead-II," respectively, in Matoba and Dotsu (1977). Phase IV and "entering the nest" complete the sequence in both papers. Equally important in courtship displays (and aggressive interactions) is the relative position of the fishes, and Brillet (1975: fig. 11) provides an aerial view of their orientation for Phases I–III (see also Table 11.1: Pair Orientation).

The role of female mudskippers during courtship is a passive one, either remaining stationary or following the males. Courtship displays advance faster if females reduce their dorsal-fin displays. This is reflected in the Brillet diagram (Brillet 1980a: fig. 3) where the omission of steps shortens the display sequence once the female was ready to spawn. Conversely, the courtship displays could be prolonged if the female was not ready. The median duration for 16 courtship sequences (Phases I–IV) was 4 minutes, with extremes of 1 and 21 minutes (Brillet 1980a: tab. 1). Interruption of courtship displays by conspecifics or other organisms may lead to termination of the display. Some of the individual elements in these phases of courtship displays reflect the motivational conflict for males to allow females into their territories and for females to enter it as discussed in the next section.

11.2.1.2 Agonistic Displays

The agonistic behavior of *Ps. argentilineatus* was analyzed in a series of papers focusing on element frequency (Brillet 1980b), size and sex effects (Brillet 1981), and the role of the dorsal fins (Brillet 1983). Observations were made in both the field and laboratory; fishes captured for laboratory observation were reproductively inactive. Laboratory observations provided more data due to crowding in the observation tanks.

Sixteen agonistic behavioral units were divided into three groups: interactive behavior, "ritualized" postures, and avoidance behavior (Groups I, II, and III, respectively; see Table 11.2). The

Table 11.2 Frequency of the 16 Agonistic Units for Large (90–100 mm TL) and Small (43–55 mm TL) *Periophthalmus argentilineatus*

		Size Category	
	Agonistic Units	Large	Small
Group I	Erection of both dorsal fins	270	155
	Aggressive approach	179	327
	Tracking movement	153	142
	Direct attack	77	282
	Leap	24	116
	Chase	97	162
	Bite	8	7
	Tail undulation	40	1
	Total acts group I	848	1192
Group II	"O" posture	44	4
	"T" posture	24	2
	"V" posture	10	2
	"I" posture	28	
	Combat	18	26
	Total acts group II	124	34
Group III	Flight (usually by bounds)	590	936
	Departure (by crutching)	29	69
	Jump aside	7	15
	Total acts group III	626	1020
Overall total acts		1598	2246

Source: After Brillet C., *Biology of Behavior*, 5, 297–315, 1980b, Table II.

Table 11.3 Aggressive Initiatives between Groups of Large Males and Females of *Periophthalmus argentilineatus*

Groups	Victories	Defeats	No Result	Row Total
♂ x ♂	154	18	8	180
♂ x ♀	230	2	4	236
♀ x ♂	44	11	5	60
♀ x ♀	55	2	15	72
Column total	483	33	32	T = 548

Source: After Brillet C., *Biology of Behavior*, 6, 35–57, 1981, Table VIII.

result of each encounter was scored as a victory if the initiator allowed the opponent to escape and as a defeat if the initiator escaped from the opponent (Group III responses). Encounters where both fish escaped or in which one failed to respond to a threat were considered as "No Result." These were tabulated in Table 11.3 for sex differences in agonistic behavior.

Laboratory studies on groups of 10 individuals confined in a tank of 0.88 m^2 enabled comparisons to be made that were not possible in the field including differences in interactions between mature and immature individuals (Brillet 1980b), size and sex (Brillet 1981), and the role of the dorsal fins (1983). Throughout these papers, tables were presented for the number of agonistic units and the results of the encounters for these comparisons, but here only one example of each is discussed. Brillet made a distinction between the fish that initiated the interaction (the dominant fish) and the fish that responded (subordinate). For the first six agonistic behavioral units (Group I), over 85% were performed by the initiator (Brillet 1980b: tab. II). For simplicity, percentages are omitted here and only the total number of each of the 16 agonistic units given by the mature and immature fish groups is presented in Table 11.2.

Behavior of the immature fish was much less ritualized than that of adults, reflected by the fewer orientation postures by immature individuals and the higher incidences of bounds, fights, and direct attacks (Table 11.2: Group II Units; see also Table 11.1: Pair Orientation). The encounters between immature fish were resolved in 98% of cases, with the initiator being victorious in 96% of them. In encounters involving mature fish where there were more ritualized body postures, 12% were unresolved and the initiator was only victorious in 82% of these instances. To investigate the effects of size on the outcome of agonistic encounters, 10 fish were grouped into each of three size classes: 10 mm standard length (SL), 20 mm SL, or 45 mm SL (Brillet 1981: tabs. III, IV and V, respectively). In all groups, the larger, dominant fish initiated most encounters and were mostly victorious (94–100%). Conversely in the few encounters initiated by smaller subordinate fish, the percentage of their victories declined as the size discrepancy increased (68%, 36%, and 0%, respectively).

To investigate the effect of sexes, two groups of 10 fish were observed. Males (mean 93 mm SL) and females (mean 92 mm SL) of equal sizes made up one group, whereas in the other, the females were 20 mm larger than the males (mean 93 mm SL and 74 mm SL, respectively; Brillet 1981). The results of the encounters for the first group are shown in Table 11.3. As before, the largest dominant individuals (males in the first group and females in the second one) initiated most of the encounters and were mostly victorious (98% and 93% of the time, respectively). The percentage of victories initiated by the subordinate individual (females in the first group and males in the second one) was less (73% and 61%).

In summary, size, irrespective of sex, was the major determinant of dominance. However, in equally sized individuals, males dominated females. Dominance of females over males required a greater size disparity than that between males. Another way of representing these relationships is through the "dominance coefficient" (Table 11.4: Rows 1–6). The dominance coefficient is generated by adding victorious bouts as an initiator and responding opponent and dividing the resultant number by the total number of encounters (Brillet 1981). This allows the comparison of the behavior of one category of fish with that of its opponent (e.g., large vs. small or male vs. female). Thus in the 296 encounters between male and female fish, the male coefficient was 0.81 (Table 11.4) as calculated from Table 11.3 ([230 + 11]/[236 + 60]) (Brillet 1981).

Erection of the dorsal fins (Table 11.1: Group I Interactive Behavior) plays a major role in agonistic interactions; used more frequently by males than females and less frequently by dominants than by subordinates. Also, dorsal-fin erection displays occurred more frequently during the formation of hierarchies in the fish tanks, and these incidences declined once the hierarchy was established. Fin-erection displays would increase again if established social order was challenged either by introductions of conspecifics or by escalating territorial boundary disputes (Brillet 1983). The signal value of the dorsal fins to both sexes was investigated by the removal of either the first or both dorsal fins in *Ps. argentilineatus* (Brillet 1984).

Fish with both dorsal fins removed could, of course, not raise them and were perceived to be dominant individuals. The dominance coefficients (Table 11.4: Rows 7 and 8) reflect this, and this value is especially high for females without dorsal fins, approaching values similar to fish with a 20 mm size difference (Table 11.4: Row 2). The ritualized orientation postures also include erection of the two dorsal fins. The "O posture," in which two fish are next to each other, but face opposite directions, was the most frequently used orientation. Dorsal fins were erected throughout the territorial confrontation. Although leaning toward the opponent, the caudal fin may or may not be spread, but the two dorsal fins are always fully erect. The confrontation concludes when the dorsal fins are erect in the "V posture." However, they are only erected by the fish at the head of the "T posture" and are not erected by either fish in the "I posture" (Brillet 1975).

The dorsal fins of *S. histophorus* (Aguilar 2000) and *Ps. kalolo*, a sympatric species to *Ps. argentilineatus*, can be raised independently as well as together (Brillet 1984). Lacking this ability, the scope for signaling different levels of aggression is much reduced in *Ps. argentilineatus*. Additionally, males of *Ps. kalolo* do not flag their first dorsal fin when approaching females, which may be part of the mechanism of reproductive isolation between the two species (Brillet 1984).

Table 11.4 Brillet's Dominance Coefficients under Different Experimental Situations

Experimental Situation		Fish Category		Dominance Coefficient		Table No.
		A	B	A	B	Brillet (1981)
Size differences irrespective of sex (mean SL in mm)	1	94	86	0.81	0.14	III
	2	93	75	0.78	0.15	IV
	3	93	47	1.00	0.0	V
	4	65	50	0.78	0.15	VI
Sex differences (mean SL in mm)		Male	Female	Male	Female	
	5	93	92	0.81	0.16	VIII
	6	74	93	0.11	0.81	X
Removal of both dorsal fins. Same size (unspecified)		Intact	Modified	Intact	Modified	**Brillet (1983)**
	7	Male	Male	0.51	0.44	III
	8	Female	Female	0.20	0.75	

11.2.1.3 Common Elements and Changing Behavior

In males of *Ps. argentilineatus*, dorsal-fin erection, tail undulation, mouth gaping, shoving, crab-like approaches, and even short departures are common to both courtship and agonistic displays. However, during courtship displays the rate of fin erection is more rhythmic and undulation of the fin is slower compared with agonistic displays (Brillet 1980a). Females approach male territories for courtship in a similar stance to that of agonistic display and may elicit an aggressive response from the resident male (Phase I of courtship). The two fish meet in an "O posture" and the interaction could escalate into a full aggressive encounter unless the female, by remaining perfectly still without bending her body, does not respond resulting in a "D posture." Phases II and III of courtship could easily change to one of aggression. The "D posture" is maintained if the female remains stationary as the male circles her (Phase II) but can revert to an "O posture" if she moves. During successful courtship, a male never bites, but without sufficient sexual motivation, the "snout or nose touch" (Phase III) may become a bite and the female will flee. Alternately if the female strongly reacts to a snout touch by a short lateral jump away, this too will result in it becoming an agonistic encounter rather than a sexual one. Interruptions from conspecifics may also reduce the chance of a successful nuptial parade; chasing away these disruptive conspecifics may have changed the motivational state of the displaying pair (Brillet 1980a).

The detailed phases of agonistic and courtship behaviors in *Ps. argentilineatus* serve as a useful primer to the study of ethology in mudskippers (Brillet 1980b). However, future researchers of mudskipper behavior must bear in mind species differences in the phases of behavior. A case in point is observed in *Ps. kalolo*, a species that utilizes its dorsal fins differently from the sympatric *Ps. argentilineatus* (Brillet 1984). As in *Ps. argentilineatus*, *Ps. kalolo* performs the "O" orientation in courtship displays, but unlike in *Ps. argentilineatus*, males of *Ps. kalolo* do not circle the females, however, the head-to-tail formation remains. Additionally, high jumps with erect first and second dorsal fins, tail twirls, "crab-like approach," and "nose blow" that occur with *Ps. argentilineatus* are absent in *Ps. kalolo* (Brillet 1984).

11.2.2 Territoriality

The agonistic behavior described in the last section largely results from individuals trying to exclude con- or heterospecifics from a particular space. This space is where they spend most of their time and may be used for feeding or breeding. Where the space has no clear boundaries, it is called a home range; where it does, a territory (see Section 12.3.1.2).

The costs and benefits of such competition have been investigated under the central concept of evolutionary stable strategies with the hawk–dove–bourgeois models being used primarily to investigate agonistic contests (Lendrem 1986). Three forms of asymmetries have been identified as predicting which animal wins: its resource holding potential, resource value and the uncorrelated asymmetry: territorial ownership (Kokko et al. 2006). For example, this approach was used to demonstrate that the territorial contests of *Gobius cruentatus* were resolved using the bourgeois strategy (Picciulin et al. 2006).

11.2.2.1 Burrows, Saucers, Turrets, and Other Constructs

Burrows are important to mudskippers as spaces to seek refuge and protect developing eggs (Toba and Ishimatsu 2014; see Section 9.6.2). Burrows are constructed by all species with the exceptions of *Ps. modestus* (Baeck et al. 2008) and *Ps. argentilineatus* (reported as *vulgaris*, Nursall 1981). These species do not excavate their own burrows (except during overwintering) but appropriate burrows of other mudflat residents including crabs. Additionally, the surface structures at or near burrow openings that are composed of mud excavated from the burrow such as turrets and rimmed saucers are considered to function as vantage points for observation (Brillet 1975), territorial markers, and physical barriers against incoming tides (Bhatt et al. 2009). Heights of the two turrets of the burrows of *Ps. argentilineatus* (reported as *Ps. sobrinus*), for example, were found to be about three times taller in areas with weaker currents (Brillet 1976).

Egg or brood chambers, if present within burrows (see Chapter 9), are filled with air. When there are developing eggs, the air within the egg chamber is continually replenished and the nest is guarded (Toba and Ishimatsu 2014). Females of all species construct burrows at some stage of their life, even if only for overwintering (Baeck et al. 2007; Townsend and Tibbetts 1995), but their role in digging breeding burrows varies between species.

In a survey of 56 excavated burrows of *Ps. argentilineatus* (reported as *Ps. sobrinus*), 25 of 35 burrows occupied by males had a "diverticula" (=spawning/brood chamber) compared with three in 14 burrows occupied by females. These observations led to the conclusion that burrows were constructed by males and not females (Brillet 1976). In captive *Ps. argentilineatus* (reported as *Ps. vulgaris*), however, although solitary female burrows never contained egg chambers, paired females assisted males in the excavation of burrows for reproduction (Rupp 2015b). Using plaster casts, Rupp (2015b) confirmed Brillet's (1976) observation that there could be more than one breeding chamber in a burrow.

When males and females of *Boleophthalmus pectinirostris* were caged separately and allowed to build burrows, only burrows excavated by males included a brood chamber (Hong et al. 2007). Conversely, both males and females of *Periophthalmus magnuspinnatus* construct burrows used for reproduction together and remain as a pair for the duration of egg development (Baeck et al. 2008). Both males and females of *Pn. schlosseri*, *Periophthalmodon freycineti*, and *S. histophorus* partake in brood care as confirmed by the pair of fish gulping air before diving into the burrow, but it is unknown if females aid in burrow construction for reproduction (Takita et al. 2011; Lee et al. 2005). Females of *S. histophorus* are considered less likely to partake in the construction of reproduction burrows as the pair were never observed to jointly defend the shared territory (Takita et al. 2011) and males engaged in significantly more agonistic displays and snaking than did females that spent considerably more time feeding (Townsend and Tibbetts 2005).

Only males of *Ps. argentilineatus* and *B. pectinirostris* brood eggs, with *B. pectinirostris* additionally plugging exits to the burrow with mud balls after the departure of females (Hong et al. 2007). Given that air must be continually replenished to the brood chamber, Toba and Ishimatsu (2014) wondered if these plugs were ever removed, a question that has yet to be answered. Although Matoba and Dotsu (1977) caught pairs of *Ps. modestus* in burrows, these may only have been spawning as Kobayashi and Dotsu (1971) were of the opinion that no parental fish stayed to guard the eggs.

This would leave the eggs susceptible to predation. If burrows are above the water table and only inundated during spring tides, it is possible that air in egg chambers does not need to be actively replenished. However, for species such as *B. pectinirostris* that reside in the lower intertidal areas, this explanation will not stand.

Some species construct well-defined territories that are delimited by the deposition of mud some 2–4 cm high round the boundaries. In Queensland, at Brisbane (Townsend and Tibbetts 2005), Cardwell (Aguilar 2000), and Cairns (Takita et al. 2011), male *S. histophorus* occupy and largely construct these mud-walled territories; similar mud-walled territories are constructed by the same species in China (reported as *S. viridis*, Yang et al. 2003). Mud-walled territories of *B. dussumieri* (reported as *B. boddarti*) built and maintained by both sexes, in Khawr Mussa, Iran (Höpner and Maraschi 1999) and in Sulaibikhat Bay, Kuwait (Clayton and Vaughan 1982), create a contiguous polygonal mosaic. In the latter locality mud-walled territories were of similar area (≈1.6 m^2) and their occupants were of similar size and weight (≈240 mm and ≈64 g) irrespective of sex. Depending on the number of sides of each territorial polygon, individuals are adjacent to four to nine neighbors (Clayton and Vaughan 1982) that may be of the same or opposite sex (Clayton and Vaughan 1986). Occupants of territories actively maintained the mud walls and collaborative efforts were observed for shared walls. The construction of these walls by *B. dussumieri* is density dependent (Clayton 1987), which is also the case for *B. pectinirostris* (Chen et al. 2007). In Funing Bay, China, *B. pectinirostris* are held in ponds before being harvested (Chen et al. 2007). Within these ponds there are territories, where each territory has an area of ≈0.03 m^2, which is considerably smaller than territories of *B. dussumieri* (described earlier). Neither the size nor sex of the occupants was provided by Chen et al. (2007); however, in this artificial environment, territorial walls were only constructed during winter.

Without walls to delimit territorial boundaries, it is much harder to assess the area of influence of an individual mudskipper. In cases such as these, boundaries have to be plotted through the location of agonistic encounters. Beyond these boundaries, fish may wander further afield, thus extending the spatial location to become "home ranges" (Powell and Mitchell 2012; Börger et al. 2008).

11.2.2.2 Home Ranges

Mudskippers are known to move with the tide across mudflats and banks and channels of creeks (Clayton 1993). Tidal migration by mudskippers has been attributed to environmental or foraging factors (Takeda et al. 2011; Colombini et al. 1996, 1995; Ikebe and Oishi 1997, 1996, 1992). Five species of mudskippers are sympatric in the Pallarenda swamp in Townsville, Australia, where habitat partitioning occurs through interspecific aggressive interactions as investigated by Nursall (1981, 1974) and which is summarized in Table 11.5. Size and dominance hierarchy determine the outcomes of interactions between these five sympatric species. Through staged contests of paired species in the laboratory, Nursall (1974) determined that confrontations involved the erection of the dorsal fins, but the signaling varied between species (Nursall 1981; Table 11.5).

Two mudskipper species are reported to migrate with the tide: *Ps. argentilineatus* and *Periophthalmus gracilis* (Nursall 1981). The former migrates along the tideline and therefore encounters the dominant *Pn. freycineti* (reported as *Pn. schlosseri*). In contrast, *Ps. gracilis* moves perpendicular to the tideline, thus encountering both the dominant *Ps. argentilineatus* and *Periophthalmus minutus* with which it shows little antagonism (Nursall 1981, 1974). Segregation of four sympatric species in Morib, Peninsular Malaysia, however, was attributed to substrate preferences in relation to foraging, reproduction, and refuge (Muhammad and Norma-Rashid 2005).

In the same way that local topography affects burrow architecture (Brillet 1976, 1969b), it also affects the shape of territorial areas, and Brillet (1975) identified three types at two sites, Morondava and Tulear, in Madagascar. On level and non-vegetated areas, *Ps. argentilineatus* strongly defends its burrow and to a lesser extent, the surrounding foraging area; this represents the

Table 11.5 Partitioning of the Habitat and Dominance Relationships between Sympatric Species of Mudskippers from the Pallarenda Swamp, Townsville, Australia

Species (1974 Designation; 1981 Name)	Location	Movement	Size (mm) Mean Size/ Maximum Size	Behavior	Dominant To:	Subordinate To:	Dorsal Fins: D1 D2
Ps. argentilineatus (BB; as *Ps. vulgaris*)	Tidal and pool margins	Migrant	86/100	Aggressive and active	*Ps. novaeguineaensis* *Ps. minutus* *Ps. gracilis*		Colored strongly erected: D1 drop first
Ps. novaeguineaensis (RS; as *Ps. expeditionium*)	Steep creek banks	Non-migrant	72/90	Unaggressive and quiescent		*Ps. argentilineatus* *Ps. minutus* *Ps. gracilis*	Colorless weakly erected
Ps. minutus (RF; as *Periophthalmus* sp.)	Tidal flat with vegetation	Non-migrant fixed burrow	65/80	Less aggressive and less active	*Ps. novaeguineaensis*	*Ps. argentilineatus*	Colored
Ps. gracilis (SBB; *Ps. gracilis*)	Tidal margins	Migrant	57/60	Aggressive and active	*Ps. novaeguineaensis*	*Ps. argentilineatus*	Colored strongly erected D2 drop first
Pn. schlosseri	Tidal creek edge	Non-migrant fixed burrow	Not known/250	Not aggressive and less active	*Ps. argentilineatus* *Ps. novaeguineaensis* *Ps. minutus* *Ps. gracilis*	None	Black only D1 raised

Source: From Nursall J.R., *American Zoologist*, 14, 1099–1118, 1974; Nursall J.R., *Bulletin of Marine Science*, 31:730–735, 1981. (Table data are a summary of observations.)

first typical territorial area. On the steep banks of channels, however, the burrow and foraging areas of *Ps. argentilineatus* are separated, like two ends of a "dumbbell," and the corridor between these areas is undefended (Brillet 1975). In this second territorial-area type, efforts at defending smaller burrow areas are weaker than the larger foraging areas. The third type of territorial area is one on flat terrain but among pneumatophores. As with the first territorial-area type, *Ps. argentilineatus* forage around their burrows, but in this instance, the fish also move away from their burrows to forage, but exhibit little aggression toward intruding conspecifics (see Section 12.3).

Although the sex of these territorial *Ps. argentilineatus* in the dumbbell territories was not reported, because of a damaged tail, one of the four fish observed over 15 days was confirmed as using the same burrow and foraging area (Brillet 1975). Marked individuals were also central to the understanding of the role of aggressive behavior in access to limiting resources by the bay goby, *Lepidogobius lepidus* (Grossman 1980; also see Section 10.6).

At Inhaca Island where Stebbins and Kalk (1961) also described their activity, Gordon et al. (1968) also reported on the (type 1 above) strong burrow orientation of *Ps. argentilineatus* (reported as *Ps. sobrinus*). However, on Nosy Bé, Madagascar, where it inhabited rocky coasts, open mudflats and mangroves, the fish was not burrow oriented (Gordon et al. 1968). This suggests that territorial behavior within species may have greater flexibility than previously assumed.

The size of the territory influences the level of aggression shown by the occupant to intruders. In sparsely populated areas, *Ps. argentilineatus* foraged only about 60 cm away from their burrows (Brillet 1975), similar to the distance of 50–70 cm in higher density areas where border disputes between adjacent territorial individuals limited movement (Brillet 1975). In contrast, the distance fish moved away from their burrow in the second or third type of territory could be many meters. Over much of this distance, aggressive interactions were minimal or absent (Brillet 1975). Similarly, neutral zones between territories of individuals of *S. histophorus* allow conspecifics to pass without antagonistic interactions (Townsend and Tibbetts 2005). The fish in these neutral areas can be considered as errant individuals that are usually smaller but which may be sexually mature (Clayton and Vaughan 1988).

11.3 FACTORS AFFECTING BEHAVIOR

Behaviors displayed by mudskippers have important physiological and social functions. Rolling on the mud substrate and submersion in water, for example, aid in transpiration by maintaining sufficient moisture levels on the skin (Clayton 1993; Ip et al. 1991). Burrowing and territorial defense, on the other hand, are related to reproduction and resource competition (Clayton and Vaughan 1988; Brillet 1969b, c, d). Mudskippers are known to display aggression toward conspecifics (Brillet 1984; Nursall 1981, 1974) and other species found within the habitat (Mazlan et al. 2004). Despite the propensity of mudskippers to display aggressive behaviors, studies on heterospecific interactions within a guild of coral-feeding chaetodontid fishes found that phylogeny was not the best predictor of aggressive interaction (Blowes et al. 2013). Instead, dietary overlap and specialization were a better predictor and aggressive interspecific interactions where resource partitioning is promoted may contribute to increased biodiversity. For mudskippers, complementary evidence in support of this hypothesis may be obtained by including dietary information with already published studies of sympatric species of mudskippers in the Malacca Straits (Polgar and Crossa 2009) or Queensland (Nursall 1981). However, dietary overlap may not be the only factor; *B. boddarti* and *Pseudapocryptes elongatus* (reported as *Pseudapocryptes lanceolatus*), feeding predominantly on diatoms and algae, engage in interspecific aggressive displays while in cohabited pools but show no intraspecific aggression (Swennen et al. 1995).

It is likely that mate choice plays an important role in mudskipper courtship just as it does in other fishes (see Candolin and Wong 2008). Crucial to mudskippers' reproductive success is the ability to effectively brood the eggs, a parental quality that would be of interest to females in male nest-guarding species (e.g., *Ps. argentilineatus* and *B. pectinirostris*) and to both sexes where joint rearing is involved (e.g., *Ps. magnuspinnatus* and *S. histophorus*). Parental care is but one of several reasons for mates to be choosy and there are additional indirect benefits (Candolin and Wong 2008) of which condition and size are worth mentioning.

Nuptial coloration provides signals for mate selection and also supplies information on hormonal and reproductive status (Kodric-Brown 1998), the latter being influenced by the distribution of various pigments (see Johnson and Fuller 2014). Body and fin coloration of several species of mudskippers from northern Australia were described for field identification purposes (Takita et al. 2011) but additionally larger sizes and brighter colors could indicate higher fecundity (Candolin and Wong 2008) and greater mating success (Johnson and Fuller 2014), respectively.

Signaling behavior in the form of extended phenotypes (Schaedelin and Taborsky 2009) may also play a role in species recognition or mate selection. In the same way, sand pyramids are constructed to attract females to burrows of resident male ghost crabs (*Ocypode* spp.) that do not use acoustic signals (Clayton 2001), the turrets, saucers, and other constructions at or near burrows of mudskippers may also function as signals. For example, the burrows of *Ps. argentilineatus* (reported as *Ps. koelreuteri*) have a single turret, although those of *Ps. sobrinus* have two (Brillet 1984).

11.3.1 Neuro-Hormonal Control

The study of hormonal control in mudskippers has largely centered on physiological processes concerned with homeostasis (see Chapter 7; Clayton 1993), reproduction (Hong et al. 2014; Zhao et al. 2004, 2002; Zhang et al. 1989), and amphibious behavior (Sakamoto et al. 2011, 2015). Few studies address issues of social behavior in relation to hormonal control. The neuropeptide arginine vasotocin (VT) is implicated in regulating aggressive behavior in *Ps. modestus* (Kagawa 2014; Kagawa et al. 2013), as it does in other fish groups (e.g., Greenwood et al. 2008). Intracerebroventicular injections of a high dose (500 pg/g body weight of VT) were administered to three pairs of size-matched male *Ps. modestus*, followed by 10-minute staged encounters in an aquatic environment (Kagawa et al. 2013). Males injected with VT were more aggressive, with more attacks and higher instances of dorsal fin and opercular displays, than fishes in the control group. In a separate experiment there were encounters between six pairs of untreated fish where 15% of the experimental area was above the water level; dominant individuals moved significantly more frequently between water and land and spent more time on land than did the subordinates (Kagawa et al. 2013; Kagawa 2014). These fish were then sacrificed and contrary to expectation, the endogenous VT messenger RNA (mRNA) levels were higher in subordinate than in dominant individuals. As both subordinate and dominant individuals had similar plasma cortisol levels, this finding could not be explained as a stress response but it may be osmoregulatory in nature (Kagawa et al. 2013). The authors also suggested that mudskipper hierarchies were formed in the water, where most of the agonistic encounters took place. Although little is known of the behavior of mudskippers during submersion in the wild (Townsend and Tibbetts 2005), mudskipper hierarchies are unlikely to be formed in the water and the observations described earlier can be attributed to the small land area available within the experimental set-up.

Wai et al. (2006) were able to identify the distinct brain regions that were activated during behavior in *Ps. modestus* (reported as *Periophthalmus cantonensis*). They compared the percentage of all cells in five brain regions that were c-fos positive for three treatment groups against a control group. Two groups consisted of five fish each, one of which was subjected to glass rod agitation and the other of satiated fish. The third group was of five aggressive fish as established through being the dominant member in a 30-minute paired encounter (Wai et al. 2006). Dominance was assessed by dorsal-fin erection and chasing behavior. Unfortunately, the results were inappropriately analyzed by *t*-tests, but nevertheless suggested that when compared with the control individuals, the aggressive individuals of *Ps. modestus* showed significant percentage increases in the cells of the whole diencephalon but not the pituitary, and in the cells and fibers of medulla but not in the pons. Against expectation, there was little or no c-fos activation in the telencephalon, which is the area associated with avoidance and spatial memory (Mukuda 2010).

Agitated individuals also showed the same level of c-fos activity in the diencephalon, but its pattern of activation was very different to that of the aggressive fish. Although there was an unexpected absence of discrete nuclei activation in the aggressive fish, some fibers were activated. Their activation and the diffused nature of the diencephalon sites were considered as an "important equivalent of aggressive behavior" (Wai et al. 2006).

These fibers probably had their origin in the thalamus or hypothalamus and descended to the tectum and brain stem (Wai et al. 2006). Further work is needed to identify the point(s) of origin of these tracts, but until their origins are identified, these findings cannot be compared with the six brain nuclei, or the "social behavior network," described for other fish groups (Fernald and Maruska 2012).

11.3.2 Seasons

Seasonality of behavior is a result of varying temperature, food availability, and reproduction. Seasonality in reproductive cycles is, in turn, dependent on latitude and the rainy season. Populations farther from the equator are subject to colder temperatures than those in the tropics. A population of *S. histophorus* at 27° S, for example, has not been observed between August and mid-October, indicating that these fishes likely overwinter in burrows during that time (Townsend and Tibbetts 1995). Populations of *Ps. magnuspinnatus* and *Ps. modestus* at 34° N are reported to overwinter in burrows between November and March and reemerge in April to spawn between May and July (Baeck et al. 2008). Breeding and overwintering durations differ between these two species; *Ps. modestus* is more tolerant of lower temperatures than *Ps. magnuspinnatus*; so the overwintering duration for this species is shorter (Baeck et al. 2008). Northerly populations of *B. pectinirostris* (at 34° N, Jeong et al. 2004; Chung et al. 1991) also overwinter in their burrows.

Table 11.6 summarizes the duration of breeding seasons of mudskipper populations in localities arranged from north to south. Species such as *B. boddarti*, *Pn. schlosseri* (Dinh et al. 2015; Mazlan and Rohaya 2008; Sircar and Har 1975), and *Pd. elongatus* (Minh et al. 2010; Dinh et al. 2007) breed during the rainy season. Courtship displays, excavation of burrows for reproduction, and spawning will occur during this period, but observers must be aware of similarities between courtship and agonistic displays as discussed earlier. It is, therefore, necessary to also make observations outside the breeding season when the behavioral elements are not influenced by reproduction. Individual *Ps. magnuspinnatus*, for example, maintained space around them throughout the year but displayed territorial defense in the breeding season after courtship and pairing (Baeck et al. 2008). Similar seasonal observations are required for *Ps. modestus* (Matoba and Dotsu 1977) and *Ps. sobrinus* (Stebbins and Kalk 1961) to establish if they follow the same pattern of behavior.

11.3.3 Daily Cycles

In addition to seasonal changes, there are also immediate effects of diurnal and tidal influences. However, as these are discussed elsewhere in this book (see Section 10.2.3 and Section 12.3.4), they will not be discussed at length in this chapter. Brillet (1975) observed that the aggressive behavior in individuals of *Ps. argentilineatus* (reported as *Ps. sobrinus*) was the most intense during the falling tide and lessening with the rising tide. During high tide and toward the end of the low tide, aggressive behavior was absent.

Table 11.6 Summary of Reported Reproductive Seasons of Various Mudskipper Species, Presented from Northern to Southern Latitudes

Species	Location	Coordinates	Reproductive Season	Authority
B. pectinirostris	Korea: Obong-ri; Okgu-gun	(35°50′ N 126°30′ E)	Mid-June–late August	Ryu et al. (1995)
S. gigas	Korea: Jungdo Island	(35°30′ N 126°10′ E)	June–July	Kim et al. (2011)
B. pectinirostris	Korea: Wolyon-ri, Hoihyon-myon, Okku-gun; Chollabuk-do	(35°49′ N 126°38′ E)	June–July	Chung et al. (1991)
Ps. modestus	Korea: Suncheon Bay	34°52′ N 127°29′ E	April–September; peak in May	Kim et al. (2007)
B. pectinirostris	Korea: Songsanri; Kurori, Kangjin Bay	(34°36′ N 126°46′ E)	May–July	Jeong et al. (2004)
Ps. modestus	Japan: Shinhamamako Lagoon, Tokyo Bay	34°40′ N 139°55′ E	June–August	Taru et al. (2006)
Ps. modestus (as *Ps. cantonensis*)	Japan: Honmyo Estuary; Isahaya Bay; Ariake Sound; Kyushu	(32°51′ N 130°09′ E)	Late May–mid-August	Matoba and Dotsu (1977)
Ps. waltoni	Iraq: Khor Al-Zubair	(30°04′ N 47°57′ E)	March–April	Al Daham and Al Noor (2000)
Ps. waltoni	Kuwait: Sulaibikhat Bay	(29°19′ N 47°52′ E)	March–May	El Zaidy et al. (1975)
B. pectinirostris	China: Funing Bay, Fujian	26°53′ N 120°03′ E	May–June	Hong et al. (2007)
B. dussumieri	Pakistan: Korangi Creek, Karachi	24°47′ N 67°08′ E	April–May and July–September	Hoda and Akhtar (1985)
B. dussumieri (as *B. dentatus*)	India: Jodiya Coast, Saurashtra, Gujarat	(22°42′ N 70°16′ E)	January–February	Soni and George (1986)
Ps. modestus (as *Ps. cantonensis*)	China: Mai Po Marshes, Hong Kong	22°30′ N 114°02′ E	February–October; peaks March–May, September–October	Chan (1989)
B. pectinirostris	China: Shenzhen Bay	(22°29′ N 114°01′ E)	April–August; April–May	Cai (1996)
B. boddarti, *Pn. schlosseri*, *Ps. koelreuteri*	India: Ganges River Estuary, Midnapore, West Bengal	21°40′ N 87°37′ E	Peak spawning August–October	Sircar and Har (1975)
S. histophorus (as *S. viridis*)	China: Zhanjing Sea	(21°06′ N 110°20′ E)	April–June	Cao (2009)
B. dussumieri	India: Thani Creek, Mumbai	(19°21′ N 72°58′ E)	July–September	Mutsaddi and Bal (1970)
B. boddarti	Vietnam: Tran De District, Soc Trang Province	9°28′ N 106°12′ E	Peak spawning August–September	Dinh et al. (2015)
Pd. elongatus	Vietnam: Mekong Delta, Bac Lieu Province	09°14′ N 105°43′ E	Summer monsoon	Bucholtz et al. (2009)

(Continued)

Table 11.6 (*Continued*) Summary of Reported Reproductive Seasons of Various Mudskipper Species, Presented from Northern to Southern Latitudes

Species	Location	Coordinates	Reproductive Season	Authority
Pd. elongatus	Vietnam: Mekong Delta	8°5′ N 104°5′–106° E	June–November; peaks in July and October	Dinh et al. (2007)
Ps. barbarus	Sierra Leone: Freetown	(8°29′ N 13°15′ E)	May–June	Turay et al. (2006)
Ps. barbarus (as *Ps. papilio*)	Nigeria: Lagos Lagoon	6°24′ N–6°36′ N 3°20′ W–3°50′ W	February–March; October	Lawson (2010)
Ps. barbarus	Nigeria: Ikot Abasit, Imo River Estuary	4°34′ N 7°33′ E	February–May	Etim et al. (2002)
Ps. barbarus	Nigeria: New Calabar River	4°15′ N–4°45′ N 6°57′ E–7°04′ E	Year round; peaks in February and July	Chukwu et al. (2010)
Pn. schlosseri	Peninsular Malaysia: Sepang Mangroves	(2°36′ N 101°42′ E)	June–July and October–November	Mazlan and Rohaya (2008)

11.3.4 Sexual Dimorphism

Many mudskippers are not apparently sexually dimorphic and it is a challenge to distinguish if individuals being observed in field studies are either males or females and whether they are sexually mature. Sexual dimorphism does occur in some mudskippers, but these characters can only be readily observed upon closer inspection. For example, the first dorsal-fin bases are longer in males than in females of *Periophthalmus barbarus* (Udo 2002) and *Boleophthalmus poti* (Polgar et al. 2013) and are not easily observed in individuals in the field. Of potential use, however, are characters such as the color or the size disparity of the dorsal fins. In *Periophthalmodon septemradiatus*, for example, the two dorsal fins of the male are joined and the first fin ray on the large D1 is elongated. In females, however, the two fins are separate, and the first dorsal fin is extremely reduced (Khaironizam and Norma-Rashid 2003). In *Periophthalmus novaeguineaensis*, males have a larger first dorsal fin than do females. It is also differently colored (Takita et al. 2011) and has an elongated first spine (Polgar 2010). Color differences are also found in other species. For *Ps. gracilis*, there are gender differences in the color of the first dorsal fin and in *Ps. minutus* both the first and second dorsal fins differ in color and pattern (Takita et al. 2011). However, as the intraspecific differences in fin color between *Ps. modestus* and *Ps. magnuspinnatus* are only present in large individuals (Baeck et al. 2008), this implies that such color distinctions have a limited utility for gender identification in the field. Additionally, if the fins are only spread occasionally, as in these two species, it will further limit their usefulness.

The chances of correctly distinguishing between males and females in some mudskipper species can increase during the breeding season. Changes in color of mature individuals during courtship may aid in distinguishing the sexes. For example, two individuals of *Ps. minutus* found in breeding burrows were considered as a mating pair on the basis that their color patterns and behavior differed (Takeda et al. 2011). Other species reported to change body color during courtship displays are *Ps. kalolo* (Magnus 1972) and *Ps. modestus* (Matoba and Dotsu 1977). Similarly males of *Ps. argentilineatus* become paler, changing from brownish gray to a yellow or orange hue, whereas females remain gray but with a darkened area on the head below the eyes. Observers must again be cautioned that changes in body color also occur in agonistic encounters, for example, between individuals of *Ps. gracilis* (Nursall 1981). Additionally, in courtship and agonistic displays in males, there are reports of darkening of the eyes but not the head (Brillet 1980a). Color changes in males of *Ps. modestus* during courtship displays differ. Lee and Graham (2002) noted the change from brown to pale beige, whereas Huang (2013) reported the change from gray to orange or whitish gray. Regardless, because females do not change color, males and females of this species can be confidently discriminated.

Conversely, in paired territorial *S. histophorus*, it is the female that has black eye borders and the male that has white ones. Although both sexes are white ventrally, the whiteness extends onto the flank in females and is visible even in the field (Townsend and Tibbetts 2005). Whether these distinctions are true of unpaired individuals is unknown. Unpaired males can, however, be identified by their tail-standing behavior, the function of which is to attract a mate into his territory. Tail-standing jumps occur sporadically, in bursts of 3–10 consecutive jumps (mean burst duration 15 ± 9 seconds with jumps every 2.5 ± 2 seconds: N not reported; Aguilar 2000). It is a short-lived activity occurring only over a few days while establishing pairs. Aguilar (2000) observed 57 territories of *S. histophorus* at Cardwell, Queensland, where initially 34 of the territories had displaying males and 20 housed pairs. Two days later, the number of pairs found in the territories doubled, and only 11 males were displaying the tail-jumping behavior. Males of several species perform similar vertical leaps (Table 11.1; Aerial Behavior) as do females of *Ps. argentilineatus* when approaching a territory of another individual (Brillet 1980a). Based on size, Townsend and Tibbetts (2005) were also able to distinguish the sex of paired individuals of *S. histophorus*; males (≈150 mm total length [TL]) were always larger than females (≈120 mm TL). It is unlikely, however, that this can be generalized to other species. Matoba and Dotsu (1977) collected 18 mating pairs of *Ps. modestus* and in only six of the pairs were the males marginally larger. Consequently, a matched pairs *t*-test based on their data conversely showed that females were significantly larger than their mates (mean: male 55.2 mm TL, females 58.9 mm TL; $p = .01$ two-tailed test).

In *Ps. argentilineatus* (Brillet 1969c) and *Pd. elongatus* (Bucholtz et al. 2009; Swennen et al. 1995), there is no sexual dimorphism but males and females can be distinguished on the basis of their behavior during interactions in the breeding season. However, female *Ps. argentilineatus* have been reported to leap and tail twirl, and males are known to court other males (Brillet 1980a). Homosexual behavior in fishes has been known for a long time (Morris 1951) and occurs for a variety of reasons (Bailey and Zuk 2009), including increased potential for successful courtship (Bierbach et al. 2012). The most reliable method for gender identification is to mark and recapture individuals.

11.4 RECOMMENDATIONS FOR FUTURE STUDY

To better identify the behavioral elements concerned with courtship displays and mating, field observations should be conducted throughout the year so that baseline behavioral elements can be ascertained. Difficulties in distinguishing between males and females in many mudskipper species and harsh physical conditions faced by observers during in situ studies (see Chapter 15) remain as challenges to our comprehensive understanding of their behavior. Laboratory studies, where better recording facilities are available and the gender and social system can be manipulated, can be equally or more informative.

More complete ethograms, sequence analyses, and behavioral transition matrices should be prepared. Species signaling behavior, particularly in the use of the dorsal and caudal fins, needs to be revisited so that the fish's motivational states can be better identified or predicted.

Some immediate behavioral-related questions include, "Do fish with taller turrets or that leap more frequently attract more females?" "Is their coloration brighter than that of less active fish?" "Is the number of diverticula made by a male related to any of these characteristics?" "Is the size of the fish sufficient to override the bourgeois strategy?"

The study of mudskipper behavior should contribute more directly to the general theoretical framework of our understanding of teleost fish behavior.

REFERENCES

Abril-De-Abreu R., Cruz J. and Oliveira R.F. 2015. Social eavesdropping in zebrafish: Tuning of attention to social interactions. *Scientific Reports* 5:12678.

Aguilar N.M. 2000. Comparative physiology of air-breathing gobies. Section VI Natural History of Australian mudskippers. PhD dissertation, University of California, San Diego.

Al Daham N.K. and Al Noor S.S. 2000. Reproductive biology of the spotted mudskipper, *Periophthalmus waltoni* Koumans, 1941 (Pisces: Gobiidae) in Khor Al-Zubair, northwest Arabian Gulf. *Marina Mesopotamica* 15:425–437.

Baeck G.W., Takita T. and Yoon Y.H. 2008. Lifestyle of Korean mudskipper *Periophthalmus magnuspinnatus* with reference to a congeneric species *Periophthalmus modestus*. *Ichthyological Research* 55:43–52.

Bailey N.W. and Zuk M. 2009. Same-sex sexual behavior and evolution. *Trends in Ecology and Evolution* 24:349–446.

Bhatt N.Y., Patel S.J., Patel D.A. et al. 2009. Burrowing activities of goby fish in the recent intertidal mud flats along the Navinal coast, Kachchh, western India. *Journal of the Geological Society of India* 74:515–550.

Bierbach D., Jung C.T., Hornung S. et al. 2012. Homosexual behaviour increases male attractiveness to females. *Biology Letters* 9:1–4.

Blowes S.A., Pratchett M.P. and Connolly S.R. 2013. Heterospecific aggression and dominance in a guild of coral-feeding fishes: The roles of dietary ecology and phylogeny. *American Naturalist* 182:157–168.

Börger L., Dalziel B.D. and Fryxell J.M. 2008. Are there general mechanisms of animal home range behaviour? A review and prospects for future research. *Ecology Letters* 11:637–650.

Brillet C. 1969a. Premiere description de la parade nuptiale des poissons amphibies Periopthalmidae. *Comptes Rendus de l'Academie des Sciences Serie D* 269:1889–1892.

Brillet C. 1969b. Etude du comportement constructeur des poissons amphibies Periophthalmidae. *Terre et Vie* 23:496–520.

Brillet C. 1969c. Observations sur le comportement agressif des poissons amphibies Periophtalmes (*Periophthalmus koelreuteri* Pallas). *Comptes Rendus de l'Academie des Sciences Serie D* 269:1114–1117.

Brillet C. 1969d. Comportement territorial et hierarchie sociale chez les poissons amphibies du genre *Periophthalmus* (Teleostei Gobioidae). *Comptes Rendus de l'Academie des Sciences Serie D* 269:1322–1325.

Brillet C. 1970. Relations entre comportement sexual territoire et agressivitie chez les Periophtalmes. *Comptes Rendus de l'Academie des Sciences Serie D* 269:1507–1510.

Brillet C. 1975. Relations entre territoire et comportement agressif chez *Periophthalmus sobrinus* Eggert (Pisces Periophthalmidae) au laboratoire et en mileau naturel. *Zeitschrifte fur Tierpsychologie* 39:283–331.

Brillet C. 1976. Structure du terrier reproduction et comportement des jeunes chez le poisson amphibie *Periophthalmus sobrinus* Eggert. *Terre et Vie* 30:465–483.

Brillet C. 1980a. Comportement sexuel du poisson amphibie *Periophthalmus sobrinus* Eggert ses rapports avec le comportement agonistique. *Review Ecologie (Terre et Vie)* 34:427–468.

Brillet C. 1980b. Comportement agonistique du poisson amphibie *Periophthalmus sobrinus* Eggert. Analyse quantitative I. Frequence des divers elements des repertoire. *Biology of Behavior* 5:297–315.

Brillet C. 1981. Comportement agonistique du poisson amphibie *Periophthalmus sobrinus* Eggert. Analyse quantitative II Influence de la taile et du sexe sur l'establissement des relations des dominance/subordination et sur l'utilisation du repetoire agonistique. *Biology of Behavior* 6:35–57.

Brillet C. 1983. Agonistic behaviour in the amphibious fish *Periophthalmus sobrinus* Eggert. Quantitative analysis (III. Role of dorsal fins in dominance-subordination relations). *Biology of Behavior* 8:49–66.

Brillet C. 1984. Habitat et comportement du poisson amphibie *Periophthalmus koelreuteri africanus* Eggert. A Tulear comparaison avec le *P. sobrinus* Eggert sympatrique. *Terre et La Vie* 39:337–345.

Brillet C. 1986. Notes on the behaviour of the amphibious fish *Lophalticus kirkii* Gunther (Pisces Salariidae). Comparison with *Periophthalmus sobrinus* Eggert. *Review Ecolologie (Terre et Vie)* 41:361–375.

Bucholtz R.H., Meilvang A.S., Cedhagen T. et al. 2009. Biological observations on the mudskipper *Pseudapocryptes elongatus* in the Mekong delta, Vietnam. *Journal of the World Aquaculture Society* 40:711–723.

Cai Z. 1996. Population structure and reproductive characteristics of mudskipper *Boleophthalmus pectinirostris*, in Shenzhen Bay, China. *Acta ecologica sinica/Shengtai Xuebao. Beijing* 16:77–82.

Candolin U. and Wong B.B.M. 2008. Mate Choice. In *Fish Behavior* (eds. C. Magnhagen, V.A. Braithwaite, E. Forsgren et al.). Boca Raton: CRC Press, pp. 337–376.

Cao F.-J. 2009. Observation on the ovary development of *Scartelaos virids* [*viridis*]. *Journal of Tropical Oceanography* 28:123–129.
Chan K.-Y.P. 1989. The ecology of mudskippers (Pisces: Periophthalmidae) at the Mai Po Marshes Nature Reserve, Hong Kong. MSc Thesis, University of Hong Kong.
Chen S., Hong W., Zhang Q. et al. 2007. Why does the mudskipper *Boleophthalmus pectinirostris* form territories in farming ponds? *Journal of the Marine Biological Association of the United Kingdom* 87:615–619.
Chukwu K.O., Deekae S.N. and Gabriel U.U. 2010. Reproductive biology of *Periophthalmus barbarus* (Linneaus 1766) in new Calabar River, Nigeria. *Agriculture and Biology Journal of North America* 1:1158–1161.
Chung E.Y., An C.M. and Lee T.-Y. 1991. Sexual maturation of the bluespotted mud hopper, *Boleophthalmus pectinirostris* (Linnaeus). *Bulletin of the Korean Fisheries Society* 24:167–176.
Clayton D.A. 1987. Why mudskippers build walls. *Behaviour* 102:185–195.
Clayton D.A. 1993. Mudskippers. *Oceanography and Marine Biology An Annual Review* 31:507–577.
Clayton D. 2001. Acoustic calling in four species of ghost crabs: *Ocypode jousseaumei, O. platytarsus, O. rotundata* and *O. saratan* (Brachyura: Ocypodidae). *Bioacoustics* 12:37–55.
Clayton D.A. and Vaughan T.C. 1982. Pentagonal territories of the mudskipper *Boleophthalmus boddarti* (Pisces: Gobiidae). *Copeia* 1982:233–235.
Clayton D.A. and Vaughan T.C. 1986. Territorial acquisition in the mudskipper *Boleophthalmus boddarti* (Pisces: Gobiidae) on the mudflats of Kuwait. *Journal of Zoology London (A)* 209:501–519.
Clayton D.A. and Vaughan T.C. 1988. Ethogram of *Boleophthalmus boddarti* (Teleostei: Gobiidae) a mudskipper found on the mudflats of Kuwait. *Journal of the University of Kuwait (Science)* 15:115–138.
Colombini I., Berti R., Ercolini A., Nocita A. and Chelazzi L. 1995. Environmental factors influencing the zonation and activity patterns of a population of *Periophthalmus sobrinus* Eggert in a Kenyan mangrove. *Journal of Experimental Marine Biology and Ecology* 190:135–149.
Colombini I., Berti R., Nocita A. and Chelazzi L. 1996. Foraging strategy of the mudskipper *Periophthalmus sobrinus* Eggert in a Kenyan mangrove. *Journal of Experimental Marine Biology and Ecology* 197:219–235.
Dinh T.D., Ambak M.A., Hassan A. et al. 2007. Biology and population dynamics of the goby *Pseudapocryptes elongatus* in the coastal mud flat areas of the Mekong Delta, Vietnam. *Pakistan Journal of Biological Sciences* 10:3284–3294.
Dinh Q.M., Giang N.T.T., Duy N.N.L. et al. 2014. Burrow configuration and utilisation of the Blue-Spotted Mudskipper, *Boleophthalmus boddarti*, caught in Soc Trang, Vietnam. *Kasetsart University Fisheries Research Bulletin* 38:1–9.
Dinh M.Q., Nguyen T.T. and Nguyen T.K.T. 2015. Reproductive biology of the mudskipper *Boleophthalmus boddarti* in Soc Trang. *Journal of Biology* 37:362–369.
El Zaidy S., Eissa S.M. and Al Naqi Z. 1979. Ecological studies on the mud-skipper fish *Periophthalmus chrysospilos* (Bleeker) in Kuwait. *Bulletin of the Faculty of Science Cairo University* 48:113–134.
Etim L., King R.P. and Udo M.T. 2002. Breeding, growth, mortality and yield of the mudskipper *Periophthalmus barbarus* (Linneaus 1766) (Teleostei: Gobiidae) in the Imo River estuary, Nigeria. *Fisheries Research Amsterdam* 56:227–238.
Fernald R.D. and Maruska K.P. 2012. Social information changes the brain. *Proceedings of the National Academy of Science USA* 109:17194–17199.
Gordon M.S., Boetius J., Evans D.H. et al. 1968. Additional observations on the natural history of the mudskipper, *Periophthalmus sobrinus. Copeia* 1968:853–857.
Greenwood A.K., Wark A.R., Fernald R.D. et al. 2008. Expression of arginine vasotocin in distinct preoptic regions is associated with dominant and subordinate behaviour in an African cichlid fish. *Proceedings of the Royal Society B* 275:2393–2402.
Grossman G.D. 1980. Food, fights, and burrows: The adaptive significance of intraspecific aggression in the Bay Goby (Pisces: Gobiidae) *Ocecologia (Berlin)* 45:261–266.
Harris V.A. 1961. On the locomotion of the mudskipper *Periophthalmus koelreuteri* (Pallas) (Gobiidae). *Proceedings of the Zoological Society of London* 134:107–135.
Hoda S.M.S. and Akhtar Y. 1985. Maturation and fecundity of mudskipper *Boleophthalmus dentatus* in the northern Arabian Sea. *Indian Journal of Fisheries* 32:64–74.

Hong L.Y., Hong W.S., Zhu W.B. et al. 2014. Cloning and expression of melatonin receptors in the mudskipper *Boleophthalmus pectinirostris*: Their role in synchronizing its semilunar spawning rhythm. *General and Comparative Endocrinology* 195:138–150.

Hong W., Chen S., Zhang Q. et al. 2007. Reproductive ecology of the mudskipper *Boleophthalmus pectinirostris*. *Acta Oceanologica Sinica* (English Edition) 26:72–81.

Höpner T. and Maraschi S.M.K. 1999. Tidal flats in Iran: Intertidal treasure Khowr-e Mussa—Unraised. *Waddwn Sea Newsletter* 1:3–6.

Hu F., Nie L.-J. and Fu S.-J. 2015. Information dynamics in the interaction between a prey and a predator fish. *Entropy* 17:7230–7241.

Huang M.-C. 2013. Mudskippers in Tainan: The ecology and lifestyle of mudskipper *Periophthalmus modestus* in Tainan coastal wetland. *International Journal of Science and Engineering* 3:37–43.

Ikebe Y. and Oishi T. 1992. Diurnal change of behavior pattern in the Chinese mudskipper, *Periophthalmus cantonensis*. *Zoological Sciences* 9:1296.

Ikebe Y. and Oishi T. 1996. Correlation between environmental parameters and behaviour during high tides in *Periophthalmus modestus*. *Journal of Fish Biology* 49:139–147.

Ikebe Y. and Oishi T. 1997. Relationships between environmental factors and diel and annual changes of the behaviors during low tides in *Periophthalmus modestus*. *Zoological Science* 14:49–55.

Ip Y.K., Chew S.F. and Tang P.C. 1991. Evaporation and the turning behavior of the mudskipper *Boleophthalmus boddarti*. *Zoological Science (Tokyo)* 8:621–624.

Jeong S.-J., Han K.-H., Kim J.-K. et al. 2004. Age and growth of the blue spot mudskipper (*Boleophthalmus pectinirostris*) in the mud flat of Southwestern Korea. *Journal of the Korean Fisheries Society* 37:44–50.

Johnson A. and Fuller R.C. 2014. The meaning of melanin, carotenoid, and pterin pigments in the bluefin killifish, *Lucania goodei*. *Behavioral Ecology*. 26: 158-167.

Kagawa N. 2014. The role of arginine vasotocin in the agressive behavor of mudskippers. *Comparative Endocrinology* 40:99–101.

Kagawa N., Nishiyama Y., Kato K. et al. 2013. Potential roles of arginine-vasotocin in the regulation of aggressive behavior in the mudskipper (*Periophthalmus modestus*). *General and Comparative Endocrinology* 194:257–263.

Kanemori Y., Takegaki T. and Natsukari Y. 2006. Genetic population structure of the mudskipper *Boleophthalmus pectinirostris* inferred from mitochondrial DNA sequences. *Japanese Journal of Ichthyology* 53:133–141.

Khaironizam M.Z. and Norma-Rashid Y. 2003. First record of the mudskipper, *Periophthalmodon septemradiatus* (Hamilton) (Teleostei: Gobiidae) from Peninsular Malaysia. *Raffles Bulletin of Zoology* 51:97–100.

Kim J.K., Baek H.-J., Kim J.-W. et al. 2011. Sexual maturity and early life history of the mudskipper *Scartelaos gigas* (Pisces, Gobiidae): Implications for conservation. *Fisheries and Aquatic Sciences* 14:403–410.

Kim J.W., Yoon Y.H., Shin H.C. et al. 2007. Maturation and spawning of Shuttles hoppfish *Periophthalmus modestus* in the mud flat of Suncheon Bay, Korea. *Journal of the Korean Fisheries Society* 40:79–83.

Kobayashi T. and Dotsu Y. 1971. Nest and nesting behavior of the mudskipper *Periophthalmus cantonensis* in Ariake Sound. *Bulletin of the Faculty of Fisheries Nagasaki University* 32:27–39.

Kodric-Brown A. 1998. Sexual dichromatism and temporary color changes in the reproduction of fishes. *American Zoologist* 38:70–81.

Kokko H., Lopez-Sepulcre A. and Morrell L.J. 2006. From hawks and doves to self-consistent games of territorial behavior. *American Naturalist* 167:901–912.

Lawson E.O. 2010. Aspects of reproductive biology in mudskipper, *Periophthalmus papilio* from mangrove swamps of Lagos Lagoon, Lagos, Nigeria. *Journal of Fisheries International* 5:36–43.

Lee H.J. and Graham J.B. 2002. Their game is mud. *Natural History* 111:42.

Lee H.J., Martinez C.A., Hertzberg K.J. et al. 2005. Burrow air phase maintenance and respiration by the mudskipper *Scartelaos histophorus* (Gobiidae: Oxudercinae). *Journal of Experimental Biology* 208:169–177.

Lendrem D. 1986. Game theory models: Social behaviour. In *Modelling in Behavioural Ecology: An Introductory Text* (ed. D. Lendrem). London and Sydney: Croon Helm, pp. 162–192.

Magnus D.B.E. 1972. On the ecology of the reproduction behaviour of *Periophthalmus kalolo* Lesson at the east African coast. *Proceedings of the International Congress of Zoology* 17 Theme 6:1–4.

Matoba M. and Dotsu Y. 1977. Prespawning behaviour of the mudskipper *Periophthalmus cantonensis* in Ariake Sound. *Bulletin of the Faculty of Fisheries, Nagasaki University* 43:23–33.

Mazlan A.G., Faridaht Y., Shukort M-N. and Arshad A. 2004. Preliminary study on the behaviour of Giant Mudskipper (*Periophthalmodon schlosseri*) in Sepang Selangor. In *Proceedings of the First Joint Seminar on Coastal Oceanography* (ed. C. Nitithamyong) pp. 107–112. Bangkok: Department of Marine Science, Chulalongkom University.

Mazlan A.G. and Rohaya M. 2008. Size, growth and reproductive biology of the giant mudskipper, *Periophthalmodon schlosseri* (Pallas, 1770), in Malaysian waters. *Journal of Applied Ichthyology* 24:290–296.

Milward N.E. 1974. Studies on the taxonomy ecology and physiology of Queensland mudskippers. PhD dissertation, University of Queensland.

Minh T.H., Gallardo W.G. and Phuong N.T. 2010. Fishery and aquaculture of juvenile mudskipper *Pseudapocryptes elongatus* (Cuvier, 1816) in the coastal zone of Mekong Delta, Vietnam. *Asian Fisheries Science* 23:224–239.

Morris D. 1951. Homosexuality in the ten-spined stickleback (*Pygosteus pungitius* L.). *Behaviour* 4:233–261.

Muhammad A.A.S. and Norma-Rashid Y. 2005. Size class distribution and behaviour budgets of sympatric mudskippers utilising mangrove microhabitats in Morib. In *Ecology of Klang Strait. Faculty of Science, University of Malaya, Kuala Lumpur* (eds. A. Sasekumar and V.C. Chung), pp. 115–126. Kuala Lumpar: University of Malaya Press.

Mukuda T. 2010. Immunohistochemial identification of telencephalic neurons expressing c- Fos protein during learning and memory in the mudskipper. 15th Biennial Scientific Meeting of the International Society for Comparative Psychology. Awaji Island, May 19–21. Hiroshima University, Japan.

Mutsaddi K.B. and Bal D.V. 1970. Maturation and spawning of *Boleophthalmus dussumieri. Journal of the University of Bombay* 39:58–76.

Nursall J.R. 1974. Character displacement and fish behavior especially in coral reef communities. *American Zoologist* 14:1099–1118.

Nursall J.R. 1981. Behavior and habitat affecting the distribution of five species of sympatric mudskippers in Queensland. *Bulletin of Marine Science* 31:730–735.

Picciulin M., Sebastianutto L., Costantini M. et al. 2006. Aggressive territorial ethogram of the of the red-mouthed goby, *Gobius cruentatus* (Gmelin, 1789). *Electronic Journal of Ichthyology* 2:38–49.

Polgar G. 2010. Mudskippers: An Introduction For Aquarists. *Conscientious Aquarist Magazine*, http://www.wetwebmedia.com/ca/volume_7/volume_7_1/mudskippers.html

Polgar G. and Bartolino V. 2010. Size variation of six species of oxudercine gobies along the intertidal zone in a Malayan coastal swamp. *Marine Ecology Progress Series* 409:199–212.

Polgar G. and Crosa G. 2009. Multivariate characterisation of the habitats of seven species of Malayan mudskippers (Gobiidae: Oxudercinae). *Marine Biology* 156:1475–1486.

Polgar G., Jaafar Z. and Konstantinidis P. 2013. A new species of mudskipper, *Boleophthalmus poti* (Teleostei: Gobiidae: Oxudercinae) from the Gulf of Papua, Papua New Guinea, and a key to the genus. *Raffles Bulletin of Zoology* 61:311–321.

Polgar G., Malavasi S., Cipolato G. et al. 2011. Acoustic communication at the water's edge: Evolutionary insights from a mudskipper. *PLoS One* 6 e21434:1–9.

Powell R.A. and Mitchell M.S. 2012. What is a home range? *Journal of Mammalogy* 93:948–958.

Quan X., Jin X., Wang R. et al. 2012. Complete mitochondrial genome of the walking goby *Scartelaos histophorus* (Perciformes, Gobiidae). *Mitochondrial DNA* 23:298–300.

Rupp H -G. 2014. Mudskippers—Troublesome rearing. *Koralle* 15:58–67.

Rupp H -G. 2015a. Glorious mud *Practical Fishkeeping*, May 6, 2015, 64–69.

Rupp, H.-G. 2015b. Mudskipper breeding. A small sensation: successful breeding of mudskippers in the aquarium. *Aquaristik Fachmagazin and Aquarium Heute* 47:14–33.

Ryu B.S., Kim I.S. and Choi Y. 1995. Ecology and life history of *Boleophthalmus pectinirostris* in Korea. *Journal of the Korean Fisheries Society* 28:316–324.

Sakamoto T., Mori C., Minami S. et al. 2011. Corticosteroids stimulate the amphibious behavior in mudskipper: Potential role of mineralocorticoid receptors in teleost fish. *Physiology and Behavior* 104:923–928.

Sakamoto T., Nishiyama Y., Ikeda A. et al. 2015. Neurohypophysial hormones regulate amphibious behaviour in the mudskipper goby. *PloS One* 10(7):e0134605.

Schaedelin F.C. and Taborsky M. 2009. Extended phenotypes as signals. *Biological Reviews* 84:293–313.

Sircar A.K. and Har S.P. 1975. Ecology of an estuarine breeding ground of some gobioid fishes in west Bengal India. In *Pacific Science Association Special Symposium on Marine Sciences* (eds. B. Morton and J.R. Lee). Hong Kong: Government Printer, pp. 63–64.

Soni V.C. and George B. 1986. Age determination and length weight relationship in the mudskipper *Boleophthalmus dentatus*. *Indian Journal of Fisheries* 33:231–234.

Stebbins R.C. and Kalk M. 1961. Observations on the natural history of the mud-skipper *Periophthalmus sobrinus*. *Copeia* 1961:18–27.

Swennen C., Ruttanadakul N., Haver M. et al. 1995. The five sympatric mudskippers (Teleostei: Gobioidea) of Pattani area, southern Thailand. *Natural History Bulletin of the Siam Society* 43:109–129.

Takeda T., Hayashi M., Toba A. et al. 2011. Ecology of the Australian mudskipper *Periophthalmus minutus*, an amphibious fish inhabiting a mudflat in the highest intertidal zone. *Australian Journal of Zoology* 59:312–320.

Takita T., Larson H.K. and Ishimatsu A. 2011. The natural history of mudskippers in northern Australia, with field identification characters. *Beagle* 27:189–204.

Taru M., Sunobe T. and Uchino T. 2006. Reproductive ecology and distribution of juveniles of the mudskipper *Periophthalmus modestus* (Gobiidae) at Shinhamako Lagoon, recess of Tokyo Bay, Japan, with a note on its conservation. *Japanese Journal of Ichthyology* 53:159–165.

Toba A. and Ishimatsu A. 2014. Roles of air stored in burrows of the mudskipper *Boleophthalmus pectinirostris* for adult respiration and embryonic development. *Journal of Fish Biology* 84:774–793.

Townsend K.A. and Tibbetts I.R. 1995. Re-appearance of the blue mudhopper, *Scartelaos histophorus* (Pisces: Gobiidae) in the greater Brisbane area. *Memoirs of the Queensland Museum* 38:671–676.

Townsend K.A. and Tibbetts I.R. 2005. Behaviour and sexual dimorphism of the blue mudskipper, *Scartelaos histophorus* (Pisces: Gobiidae). *Proceedings of the Royal Society of Queensland* 112:53–62.

Turay I., Vakily J., Palomares M. et al. 2006. Growth, Reproduction and food of the mudskipper, *Periophthalmus barbarus* on mudflats of Freetown, Sierra Leone. In *Fishes in Databases and Ecosystems. Fisheries Centre Research Report* (eds. M. Palomares, K. Stergiou and D. Pauly) pp. 49–54. Vancouver, Canada: University of British Columbia, Fisheries Centre.

Tytler P. and Vaughan T. 1983. Thermal ecology of the mudskippers, *Periophthalmus koelreuteri* (Pallas) and *Boleophthalmus boddarti* (Pallas) of Kuwait Bay. *Journal of Fish Biology* 23:327–337.

Udo M.T. 2002. Intersexual plasticity in aspects of the biology of the mudskipper *Periophthalmus barbarus* (Gobiidae) in the mangrove swamps of IMO Estuary, Nigeria. *Journal of Environmental Sciences (China)* 14:95–101.

Vichaidist T., Sukhsangchan C. and Munthum Y. 2015. Life behaviors of the mudskippers (*Boleophthalmus boddarti* Pallas, 1770) in the mangrove nature education center, Samutsongkhram province. Proceedings of 53rd Kasetsart University Annual Conference, 3–6 February 2015. Kasetsart University, Thailand. *Smart Agriculture "The Future of Thailand." Plants, Animals, Veterinary Medicine, Fisheries, Agricultural Extension and Home Economics*, pp.1299–1306.

Wai M.S.M., Lorke D.E., Webb S.E. et al. 2006. The pattern of c-fos activation in the CNS is related to behavior in the mudskipper, *Periophthalmus cantonensis*. *Behavioural Brain Research* 167:318–327.

Yang K.Y., Lee S.Y. and Williams G.A. 2003. Selective feeding by the mudskipper (*Boleophthalmus pectinirostris*) on the microalgal assemblage of a tropical mudflat. *Marine Biology* 143:245–256.

Zhang Q.Y., Hong W.S., Dai Q.N. et al. 1989. Studies on induced ovulation, embryonic development and larval rearing of the mudskipper (*Boleophthalmus pectinirostris*). *Aquaculture* 83:375–385.

Zhao, W., Hong, W., Wu, D., Jiang, X. & Zhang, Q. 2002. Induced maturation and spawning by pheromones in female mudskipper (*Boleophthalmus pectinirostris*). *Journal of Oceanography in Taiwan Strait/ Taiwan Haixia. Xiamen*, 21, 284–291.

Zhao, W.-H., Hong, W.-S., Zhang, Q.-Y. & Ma, X.-L. 2004. Studies on 17[alpha]-P and PGs levels during maturation and spawning in fishes Bostrichthys sinensis and Boleophthalmus pectinirostris. *Oceanologia et Limnologia Sinica*, 35, 69–73.

CHAPTER 12

Emergent Patterns in Spatio-Temporal Ecology

Gianluca Polgar

CONTENTS

12.1 INTRODUCTION

Mudskippers of the genera *Boleophthalmus*, *Periophthalmodon*, *Periophthalmus*, *Scartelaos*, and *Zappa* (Teleostei, Gobiidae, Oxudercinae; Murdy 1989; "*Periophthalmus* lineage"; Agorreta et al. 2013; Agorreta and Rüber 2012) are amphibious gobies of aquatic ancestry that are "fully terrestrial for some portion of the daily cycle" (Polgar et al. 2010; Murdy 1989). These species form sympatric guilds at the interface between the terrestrial and marine domains (Webb 2012) and are distributed in tropical and subtropical coastal areas of the eastern Atlantic and Indo-Pacific regions. Mudskipper communities are spatially structured, being typically distributed along gently sloping intertidal gradients formed by soft terrigenous deposits, and can extend into supratidal areas (Takita et al. 2011; Polgar et al. 2010; Polgar and Bartolino 2010; Polgar and Crosa 2009). Above mean sea level (MSL; see Appendix A; Duke et al. 1998), in the upper intertidal and upper-mid intertidal zones, these deposits are often colonized by halophytic vascular plants such as mangroves or sedges, which drastically alter the intertidal ecological conditions.

Mudskippers perform key biological functions while exposed to air. Their habitats offer the "continuously humid" condition theorized by Inger (1957) as being one of the proximate selective factors that facilitated the water-to-land transition of Devonian prototetrapods, 385–365 million years ago (Niedzwiedzki et al. 2010; Polgar et al. 2010; Clack 2006; Liem 1987). The halophytic colonization of the intertidal zone shaping the habitat of several mudskipper species is also reminiscent of the vascular plant colonization of land that accompanied and prepared the arthropod and tetrapod invasions (Polgar et al. 2010; Algeo et al. 2001; Retallack 1997). For these reasons, irrespective of the controversial role of mudskippers as convergent models for the tetrapod land invasion (e.g., Graham and Lee 2004; Schultze 1999; Gordon 1998; Gordon and Olson 1995), mudskippers are one of the most studied living models of vertebrate water-to-land transitions (e.g., Standen et al. 2014; Gibb et al. 2013; Hsieh 2010; Polgar and Bartolino 2010; Polgar et al. 2010; Graham et al. 2007; Shimizu et al. 2006).

Mudskipper species present several physiological and anatomical adaptations to an amphibious lifestyle (e.g., Michel et al. 2015; Kuciel et al. 2013; Chew et al. 2007; Sayer 2005; Zhang et al. 2003; Aguilar et al. 2000; Graham 1997). They are also highly mobile throughout their life cycle, which includes a larval planktonic stage, an amphibious dispersalist stage ("errants," see Section 12.3.3) and an amphibious territorial stage (e.g., Ishimatsu and Graham 2011; Clayton 1993; Clayton and Vaughan 1986; Brillet 1975). Mudskippers live in highly dynamic and heterogeneous environments and behavior plays a range of adaptive roles in their autecology, allowing these species to (1) rapidly modify microhabitat conditions around their body (e.g., Ip et al. 1991; Clayton and Vaughan 1988); (2) select specific habitat conditions to maintain homeostasis (e.g., Taylor et al. 2005; Tytler and Vaughan 1983); and (3) build burrows that offer a safe habitat for the development of embryos and a refuge during high tide (HT) (e.g., Ishimatsu et al. 2007; Ishimatsu and Graham 2011; Baeck et al. 2008; Lee et al. 2005; Clayton and Vaughan 1986; Brillet 1969). Spatio-temporal patterns also regulate synecological behavioral interactions, such as foraging (e.g., Baeck et al. 2013; Ishimatsu et al. 2007; Townsend and Tibbetts 2005; Yang et al. 2003; Colombini et al. 1996) and antipredatory strategies (e.g., see Chapter 11).

The conundrum of vertebrate ecological and evolutionary water-to-land transitions is the identification of the selective factors that facilitated them (e.g., Clack 2006, 2002; McNamara and Selden 1993). In the intertidal zone, this transition is essentially a movement across the intertidal ecological gradient in a water-to-land direction, or a spatio-temporal change of the distribution of evolutionary lineages, at the interface between the marine and terrestrial domains (Webb 2012). Spatio-temporal patterns of behaviors with specific autecological and synecological adaptive values can be compared between different mudskipper species. Mapping such patterns on phylogenies, thus reconstructing evolutionary trajectories, would likely provide insights into the selective factors that facilitated the transition.

An exhaustive comparative analysis of the known spatio-temporal behavioral and ecological patterns of all mudskipper species is beyond the scope of this contribution. The published patterns of a case study species, the East African *Periophthalmus argentilineatus* Valenciennes, 1837, for which this knowledge is less fragmentary, are here detailed, integrated, and discussed. The choice of the genus *Periophthalmus* is motivated by the fact that (1) it is the most speciose mudskipper genus; (2) it occurs in a relatively large variety of habitat conditions; and (3) it is the sister taxon of oxudercine and amblyopine taxa in several molecular phylogenetic reconstructions (e.g., Tornabene et al. 2013; Agorreta et al. 2013; Agorreta and Rüber 2012; Chakrabarty et al. 2012; Thacker 2009). Some aspects of the biology of *Periophthalmus* species may, therefore, exemplify a range of plesiomorphic mudskipper traits that may be present in different species of the "*Periophthalmus* lineage" (Agorreta et al. 2013).

The objectives of this contribution are to (1) review, integrate, and describe the spatio-temporal autecological patterns of a *Periophthalmus* species at different scales and throughout its whole life cycle that could be used in future comparative analyses; and (2) use the East African *Periophthalmus argentilineatus* as a model to formulate testable hypotheses on the selective factors that facilitated the water-to-land transition in mudskippers.

12.2 METHODS

The autecology of the East African *Ps. argentilineatus* is reviewed. Several eco-ethological elements and their spatio-temporal patterns are described, that is, fertilization, hatching, and incubation, burrowing, courtship, feeding, and resting.

As the mudskippers' life cycle includes ecologically different ontogenetic stages, the eco-ethological elements must be separately analyzed and ecologically relevant spatio-temporal scales must be defined for each ontogenetic stage. Therefore, to achieve the first objective of this study, I subdivided the timeline in nested time intervals corresponding to the periodic variation of key abiotic factors such as illumination, tidal submersion, temperature, air relative humidity, and rainfall (e.g., Clayton and Snowden 2000; Ikebe and Oishi 1997, 1996; Colombini et al. 1995; Brillet 1980). Seasons, synodic phases (neap, spring tides), diel phases (night, day), and tidal phases (6-hour intervals around high and low tides [LT] describing a semi-diurnal cycle) were thus categorized. I used tidal data to describe the spatial distribution of this fish species by dividing the habitat into subtidal, lower, mid, and upper intertidal, and supratidal zones for each tidal phase (see Appendix A for abbreviations used in text). The spatio-temporal patterns of specific eco-ethological units were then explored and discussed to achieve the second objective of the study.

12.3 AUTECOLOGY OF THE EAST AFRICAN *PS. ARGENTILINEATUS*: AN ANNOTATED REVIEW

Molecular studies indicate that *Ps. argentilineatus* is a species complex in need of taxonomic revision (Polgar et al. 2014). The taxon described here corresponds to "clade I" in Polgar et al. (2014) and is genetically distinct from other Indo-Malayan populations, being distributed along the coasts of East Africa and Madagascar (e.g., Polgar et al. 2014; Colombini et al. 1995; Brillet 1975; Magnus 1972; Eggert 1935). The specimens examined in the studies cited here correspond to the morphological description of *Periophthalmus sobrinus* Eggert, 1935: a species Murdy (1989) considered to be a junior synonym of *Ps. argentilineatus* (Figure 12.1); I will provisionally refer to this taxon as *Ps.* cf. *sobrinus* in this review.

During his studies in western Madagascar, Brillet (1984a, 1980, 1976, 1975, 1970, 1969) behaviorally and morphologically discriminated two species: *Ps. sobrinus*, which in his papers before 1975 was reported as *Ps. koelreuteri* (Pallas 1770), a junior synonym of *Periophthalmus barbarus* (Linnaeus 1766) *sensu* Murdy (1989); and *Ps. koelreuteri africanus* Eggert, 1935, a junior synonym of *Periophthalmus kalolo* Lesson, 1831 *sensu* Murdy (1989), that co-occurred in only one of his study sites, Tuléar (=Toliara; Brillet 1984a, b, 1975). Magnus (1972, 1970), who conducted behavioral observations of this species in East Africa (near Mombasa, Kenya; Brillet 1976), misidentified *Ps. sobrinus* as *Ps. kalolo* (Brillet 1976; personal observation). Brillet (1975: figs 14–16) used four study sites: two flat (Morondava: station A, and Toliara—henceforth F1 and F2, respectively) and two sloped (Morondava: station B, and Toliara—henceforth S1 and S2, respectively) areas. Brillet (1975, 1969) provided one of the most detailed qualitative descriptions of the spatio-temporal habitat for a *Periophthalmus* species; he conducted most of his observations

(a)

(b)

Figure 12.1 *Periophthalmus sobrinus* Eggert, 1935 is a junior synonym of *Periophthalmus argentilineatus* Valenciennes *sensu* Murdy (1989). (a) In situ photograph of *Periophthalmus* cf. *sobrinus*, Cape Vidal, Sugar Loaf, St. Lucia, South Africa. (Courtesy of Luboš Mráz.) (b) Figure of *Periophthalmus sobrinus* in Eggert (1935).

in S1 and F1, using S2 and F2 as comparison sites, and utilized large artificial settings to observe mudskippers in "semi-natural" laboratory conditions, that is, reproducing tidal excursions, substrate conditions, and naturally occurring population densities. Further qualitative contributions to the eco-ethology of this species were provided by Gordon et al. (1968) and Stebbins and Kalk (1961), whereas quantitative contributions were provided by Colombini et al. (1995) and Berti et al. (1992, 1994). Age–length relationships and the effects of coastal pollution on growth were investigated by Kruitwagen et al. (2006), and the trophic ecology was studied by Colombini et al. (1996) and Kruitwagen et al. (2007).

12.3.1 Burrow and Habitat Distribution

12.3.1.1 Burrow Structure and Function

The morphology of the burrows of *Ps.* cf. *sobrinus* varies with different sediment composition and cohesiveness (Brillet 1969). Burrows are 18–80 cm deep (Brillet 1976, 1969; Magnus 1970), and deeper burrows are apparently distributed in more elevated areas, relative to mean high water (MHW, see Appendix A); the deepest ones were found on top of the banks of tidal channels (Brillet 1976). The lowest ones along the topographic profile, almost at mean low water (MLW) level, were found in muddy sand, where the less cohesive substrate prevents the building of true galleries (Brillet 1976). Burrows are typically "Y-shaped," that is, with two openings at a distance of 5–7 cm from each other. Walls or "turrets" of variable shape are built around the openings and are on average 2.5 cm high (0–10 cm); one of the two is always taller (Brillet 1976, 1969). Both openings lead symmetrically and merge into a main vertical shaft. This shaft enlarges in a lower chamber, and one or more lateral, upward-directed diagonal tunnels branch from the lower chamber or the vertical shaft, ending in ovoid chambers, where eggs are laid (Brillet 1969). Burrows always contain some water in the lower chamber during LT, suggesting that this chamber reaches the level of the water table; this hypothesis is supported by field and laboratory observations (Brillet 1975).

Mudskippers do not use burrows to avoid terrestrial predators (e.g., when chased by a human observer), but instead use them (1) during LT, probably as a source of water, as the substrate dries up and other sources of water become less available, or disappear entirely; (2) during HT, as an alternative to climbing emerged objects (e.g., mangrove roots and trunks) or moving to emerged areas; and (3) during the breeding season, to lay, fertilize, and guard eggs (Brillet 1969; see also below). Only the latter function is obligate.

12.3.1.2 Territorial and Burrowing Activities

Both sexes dig and use burrows in mud or sandy mud year round. Burrows are dug 20–80 cm below MHW, where they are submerged for 1–4 hours during mean HT (Brillet 1975). Individuals can dig and use more than one burrow both in the field and in the laboratory (Brillet 1975).

Territories are always established and defended around a burrow system (Magnus 1970; Brillet 1969) and their boundaries are better defined by interindividual interactions at higher burrow densities, as individuals are more aggressive with decreasing distance from their burrow (Brillet 1975). However, territories are not symmetrically arranged around the burrow, and in both semi-natural and natural conditions, a wider proportion of the defended area than that surrounding the burrow is wetted daily. Both in the field and in the laboratory, when tides are simulated, portions of the territory that are being covered by shallow moving water (shallower than the body depth of the fishes), that is, during flood tide (FT), are not defended by territorial individuals. However, portions of the territory that are being uncovered by shallow moving water, that is, during ebb tide (ET), or covered by shallow stagnant water, for example, in the field during LT, and in the laboratory when tides are not simulated, are actively defended (Brillet 1975; personal observation, in Indo-Malayan specimens of *Ps. argentilineatus*).

Territorial neighbors are able to recognize each other; the aggressive interactions between neighbors are ritualized, whereas attacks of territorial owners against non-neighbor intruders are direct and non-ritualized (Brillet 1975). However, territories are not defended from intruders smaller than 4 cm in total length (TL). Based on otoliths of 242 individuals from six localities in Tanzania, *Ps.* cf. *sobrinus* measures <6.5 cm TL at about 1 year of age (Kruitwagen et al. 2006); adults observed in Madagascar have been reported to reach approximately 10 cm TL (Brillet 1975). Therefore, the smaller (<4 cm TL) individuals observed by Brillet (1976, 1975) were probably young of the year. However, it is unclear how territorial individuals discriminate potential competitors, such as non-territorial "errants" (see Section 12.3.3) and non-neighbor territorial individuals performing intertidal movements ("promeneurs" or "shuttlers," Brillet 1975; Sections 12.3.4 and 12.3.5), from non-competitors, that is, post-settlement young of the year that probably do not burrow (Section 12.3.3). Distinctive coloration patterns or morphological traits are not apparent in smaller individuals, except for proportionally smaller dorsal fins. The recognition mechanism might be a short-term change of body coloration or a specific behavioral interaction, as in the case of females entering male territories during the breeding period that are not attacked if they do not react to the approaching male (Brillet 1980, 1975). Alternatively, acoustic signals may be involved (Polgar et al. 2011), or an aggressive territorial response might only be elicited by fishes larger than a minimum size, or erecting proportionally larger dorsal fins. No territory establishment events such as digging of new burrows or occupation of an abandoned burrow were documented in the field. In semi-natural conditions, territory establishment typically takes 1–2 days, if the fish density is below a critical density threshold (Brillet 1975). Above that threshold no burrowing occurs, apparently because the mudskippers are continuously engaged in aggressive interactions.

In the field (Morondava, Toliara, Soalala), burrow density was 0.6–1.7 m^{-2} in flatter areas ($n = 3$) and 0.3–0.8 m^{-2} in sloped areas ($n = 3$); the maximum size of a stable territory in semi-natural conditions was 1.4 m^2 (Brillet 1975). The lowest population density was observed at a site in Toliara, where two territorial individuals were the only ones in a range of 30 m (Toliara); one of these two individuals moved on average ~60 cm from its burrow (max = 3 m) during LT and showed very low levels of territorial interactions (Brillet 1975: fig. 17). In Morondava, the site F1 had the highest burrow density recorded, with 1.7 burrows m^{-2} (27 burrows in a quadrat of 16 m^2). Two territorial individuals in this area had similar average home ranges (50–70 cm from the burrow) during LT but were considerably more aggressive and their territorial boundaries were distinct (Brillet 1975). It, therefore, appears that the home range of *Ps.* cf. *sobrinus* is independent of density, whereas its territorial activity increases with density. In this respect, territorial *Ps.* cf. *sobrinus* can react to the presence of a conspecific from a distance of up to 4 m while out of water (Brillet 1980) and are able to react to 2–3 mm prey items at distances of >1 m (Brillet 1975). Assuming that territorial and courtship interactions are elicited by visual stimuli at distances of >0.5 m, this implies that territorial owners can see well beyond the average size of their territories and their home range is not determined by their visual field. High individual variation in the distance covered during intertidal movements and around the burrow during LT was also observed (Brillet 1975).

Burrows are found (1) on the banks of tidal channels; (2) in the *Avicennia* pneumatophore zone, that is, in the area seaward to the forest marine fringe, covered by *Avicennia* pneumatophores but not by plant stems (e.g., Polgar and Crosa 2009); and (3) in flat *Avicennia* pioneer forests, under the canopy (Brillet 1975; Magnus 1970).

In S1 (Figure 12.2a,b), territories were distributed on the steeper bank of a non-ephemeral tidal creek. Territorial individuals had a burrow system, 0–70 cm below MHW, on the upper portion of the erosive bank of the creek. Higher burrows at 0–20 cm below MHW were probably better protected from strong tidal currents and flotsam, and burrow density was higher here (Figure 12.2a,b: *r*). At the end of the wet season (March–April), storms created a mud deposit that increased the level of the bottom of the tidal creek. These mudskippers then built a separate burrow system on this deposit in May, ~70 cm below MHW (Figure 12.2a,b: *r*'; Brillet 1975: fig. 15a). During the first part of the dry season, fishes used both the lower (Figure 12.2a,b: *r*') and upper (Figure 12.2a,b: *r*) burrow systems during each tidal cycle and until the onset of the breeding season in August, when they started to use only the upper systems. The hot and wet season is from November to April in Madagascar (June and July are the coldest months of the dry season, Brillet 1976). In S2, territorial individuals occupied an upper burrow system on a steep slope, in

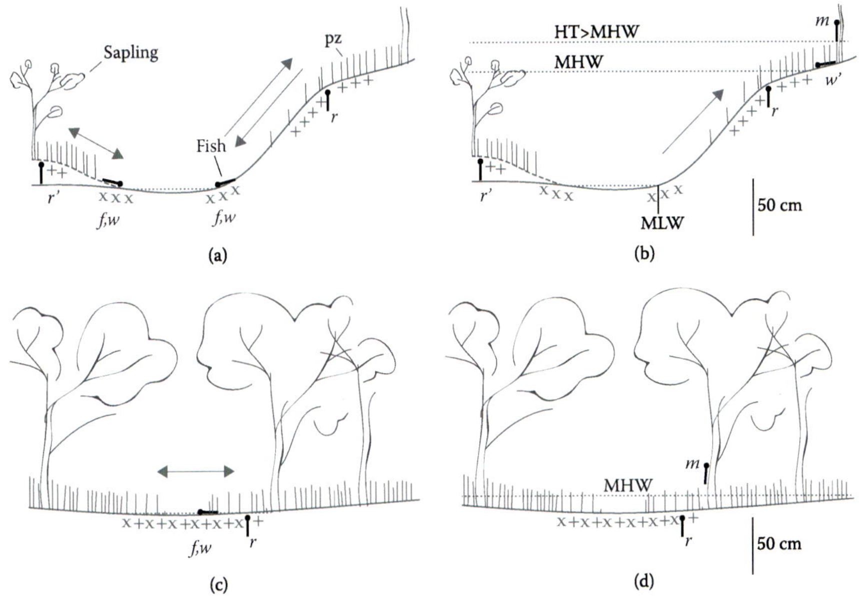

Figure 12.2 Diagrams of two types of habitat of metamorphosed individuals of *Periophthalmus* cf. *sobrinus*. (a) Site S1 during early ebb tide (ET) and low tide (LT): fish (*ball and stick symbols*) move (*downward arrow*) from their shelters on the upper bank of the tidal channel (+++; upper resting grounds) to lower areas on the banks of the channel (*xxx*; feeding grounds [FGs]), and feed here (*f*); during LT, as the substrate gets drier, the fish rest near the water's edge (*w*) or go back to their burrows (upward arrow) in the upper resting grounds (*r*); after the wet season, storms accumulated mud on the bottom of the tidal channel (*hatched profile*); some fish dug burrows here (*r'*, lower resting grounds) using them during LT (*double arrow*); (b) site S1 during FT and HT; *HT>MHW* = HT higher than MHW (*hatched lines*). Fish move (*upward arrow*) to the upper resting grounds during FT; territorial individuals either enter into their burrows during HT (*r*) or rest above the water's edge (*w'*) and climb vertical emerged substrates (*m*), like non-territorial individuals. Some fish may have stayed inside burrows in the lower resting ground (*r'*) during HT, although Brillet (1975) did not report it; (c) site F1 (bottom of an ephemeral tidal inlet) during ET and LT. Territorial fish feed (*f*) and rest (*w*) in the feeding territories around their burrows during both ET and LT, and enter into their burrows during LT in the same area (*r*) (*double arrow*); (d) site F1 during HT. Fish remain in the same area: some territorial individuals enter into their burrows (*r*), whereas other fish climb on vertical substrates (*m*) (Modified from Brillet C., *Zeitschrift für Tierpsychologie*, 39, 283–331, 1975.) See Appendix A for other abbreviations.

front of a dense *Avicennia* forest, and performed feeding activities on a lower mud bank, where they did not dig any burrows. These two areas were horizontally separated by a sand bank ~15 m long. In F1, burrows were on the forest floor, in an area with very low density of pneumatophores and plants, 10–15 cm below MHW (Figure 12.2c,d; Brillet 1975: fig. 14). In F2, burrows were dug on a forest floor densely covered by pneumatophores (Brillet 1975).

Under denser canopies, in site F1, burrows were found at greater distances from the water's edge of non-ephemeral tidal creeks (Brillet 1969: fig. 2b). This reflects the well-known positive correlation between mangrove canopy density and topographic height in the intertidal zone (Polgar and Crosa 2009). Brillet (1969) hypothesized that this may imply a lower dependence on water in more shaded habitats, where evaporation rates are probably lower. However, burrow density generally decreased in more densely vegetated areas (Brillet 1969). The drawings by Brillet (1975: fig. 14) suggest that F1 was on the bottom of an ephemeral tidal inlet (Polgar and Crosa 2009), thus being characterized by more humid conditions of the substrate.

Periophthalmus cf. *sobrinus* starts digging during ET, when the digging behavior is also more intense, and performs it irregularly during LT. Burrowing can also occur at night (Brillet 1980, 1975); in semi-natural conditions, fishes frequently start digging new burrows at night (Brillet 1969).

12.3.2 Reproduction

The breeding season for *Ps.* cf. *sobrinus* is reported to vary in different parts of southwestern Madagascar, from August to April in Morondava and from late August to early May in Toliara (Brillet 1980, 1969). The breeding season precedes the wet season by approximately 3 months, lasting through the whole wet season. This implies that larvae are released when higher rainfall and fluvial discharge trigger planktonic blooms in nearshore waters. According to Magnus (1970), males have a red margin on the second dorsal fin that is not present in females. Brillet (1980, 1975) did not observe this sexually dimorphic trait, but observed changes in coloration of the head and body of both sexes. During June and July, when mudskippers are not reproductively active, both sexes have a gray to brownish body coloration. During the breeding season, males acquire a yellowish to pale orange background color, while females remain grayish. During courtship, the male's eyes become dark brown, whereas the female's eyes become whitish and her snout dark brown. As a consequence, the coloration patterns of different sexes are reversed during courtship—yellowish to pale orange body plus dark brown eyes in the male versus dark gray body and dark brown snout and whitish eyes in the females (Brillet 1980). Courtship was observed at dusk and dawn, at temperatures <25°C, but never at night (Brillet 1980). Most courtship events were observed during ET (76%) and LT (23%) and none during HT (n = 79, Brillet 1980). During the breeding season, females leave their burrows and territory to enter male territories, triggering male courtship displays, although male solitary displays have been observed (Brillet 1980). Males have been reported to intercept receptive females within and outside their territories, at distances up to 4 m (Brillet 1980). During courtship, a male leads a female to his burrow, covering distances of up to 20 m ("promenade nuptial" or nuptial walk, Brillet 1980). Inside the male's burrow, the eggs are laid and fertilized (Brillet 1980, 1975; Magnus 1970), as suggested by the observation of a series of alternate dives of the partners (solitary stays), which have a duration of ~1 minute, the first stay of the male being significantly longer. Longer joint stays (~3 minutes) are then observed, followed by a final, longer joint stay (11–50 minutes, Brillet 1980). The duration of the cohabitation is variable and the female may leave shortly after she exits the burrow (4–72 minutes from the start of the courtship, Brillet 1980). However, Brillet (1980) never prolonged his observations for >100 minutes from the start of the courtship, and in several cases he failed to observe the partners exiting the burrow for the last time (Brillet 1980). It is unknown if the female returns to her own burrow after she leaves the male burrow; in some cases, the female visited the burrow of a neighboring male immediately after her exit from the first male burrow, suggesting that females can deposit more batches of eggs in a row (Brillet 1980). The fertilized eggs are guarded by the resident male throughout the incubation period (Brillet 1980, 1975; Magnus 1970). In a few cases, male egg guarders were observed to remain inside the burrow when it is flooded, during HT (Magnus 1970).

According to Magnus (1970), reproductive males dig burrows in 1–2 days, in a narrow unvegetated strip of mudflat in front of the mangrove forest, below mean high water neap (MHWN, i.e., at lower level than MHW). Brillet (1975) reported that males use reproductive burrows both in flat and sloped areas. In S1, females were always invited only into the upper burrows of the male owners of dumbbell territories (see Section 12.3.5), during the breeding season (Figure 12.2a,b: *r*). New burrows are built after the breeding season (Brillet 1980, 1975, 1969).

In Toliara, eggs were found inside burrows from November to March, and females with ripe ovaries were collected from November to May–June (Brillet 1980, 1976). Eggs of *Ps.* cf. *sobrinus* containing embryos with distinct heads and tails, at a similar developmental stage of embryos of *Ps. modestus* 52–63 hours after fertilization (reported as *Ps. cantonensis*, Kimura 1958: fig. 1F;

Kobayashi et al. 1972: fig. 1M–N), hatched after 72 hours at 28–30°C under laboratory conditions (Brillet 1976). A summary of the spatio-temporal distribution of *Ps.* cf. *sobrinus* embryos is in Table 12.1.

No information is available on the larvae of *Ps.* cf. *sobrinus*. Brillet (1976) unsuccessfully searched for the larvae with plankton nets in tide pools and mangrove tidal creeks throughout the year. Population genetics and biogeographical studies suggest that genetic connectivity among mudskipper populations is maintained by offshore larval dispersal (e.g., Chen et al. 2015; He et al. 2015; Kanemori et al. 2006; Mukai and Sugimoto 2006; Takeshima 1996; Chang and Lee 1994). However, direct spatio-temporal information on larval seasonal, synodic, tidal, and diel distribution is lacking. Early post-settlement young were reported in Toliara from the end of October to the end of June (Brillet 1976), that is, from 2 to 3 months after the start of the breeding period, and up to 2 months after the end of the breeding period (Brillet 1980).

12.3.3 Settlement and Post-Settlement

Newly metamorphosed young were found near the water's edge, often taking shelter in large aggregations in crevices of mangrove bark, under plant debris, and around tide pools (Brillet 1976). Post-settlement young rapidly become aggressive toward conspecifics, although they do not show ritualized aggressive behaviors and do not defend territories. Small individuals were also found in dense aggregations in locations with fewer larger and territorial individuals, suggesting the presence of spatial partitioning among different sizes and age classes (Brillet 1976).

In semi-natural conditions, non-territorial individuals were observed to enter unsupervised territories and even burrows of territorial individuals (Brillet 1975). This suggests that the typically smaller and less aggressive non-territorial individuals search either a suitable area to dig a new burrow, or colonize abandoned burrows during territory establishment ("individus errants" or "errants," Brillet 1975). However, Brillet (1976) reported that post-settlement young below 3 cm TL neither use nor are attracted to burrows of conspecifics or invertebrates. Brillet (1975) also reported that "young individuals" use burrows of crabs and *Amoya signata* (reported as *Ctenogobius pavidus*), presumably at a later ontogenetic stage. In an adaptive scenario, the observation that post-settlement young do not use burrows is consistent with the observation that territorial individuals do not attack smaller intruders (see Section 12.3.1.2; Brillet 1975).

12.3.4 Diel, Synodic, and Intertidal Movements of Metamorphosed Individuals

Brillet (1980, 1975) reported that *Ps.* cf. *sobrinus* is mainly diurnal, although he admitted that his nocturnal observations were probably unreliable (Brillet 1980). Gordon et al. (1968), who studied this species in Nosy Bé, a small island off northwestern Madagascar, and Colombini et al. (1995), who studied this species in East Africa (Tana River delta, Shekiko, Kenya), observed nocturnal intertidal migrations. The latter study found that these mudskippers are significantly more active on the mangrove forest floor and the adjacent mud bank of a tidal channel at night than during the day. This study concluded that *Ps.* cf. *sobrinus* is not influenced by the diel light cycle per se, but by the position of the water's edge, determined by synodic and tidal cycles, and probably by correlated environmental factors, such as temperature and relative humidity.

Colombini et al. (1995) carried out observations in a 50-m-wide mangrove forest fringing a tidal channel cut-off from the river delta. The forest had a gentle slope above MHWN, and an adjacent saltmarsh on its landward side, from which the forest was separated by a 10m-wide unvegetated land strip, and an erosive bank at the extreme high water spring (EHWS) level (Figure 12.3a). The authors described the spatial and temporal distribution patterns of fish "actively moving on the surface" and "resting on nests," where "nests" = burrows, as frequencies of sightings. They also measured time series of air temperature and relative humidity in the middle of the forest, 1 m above the

Table 12.1 Stages of Egg Fertilization, Development, and Hatching of *Periophthalmus* cf. *sobrinus* (Presence–Absence Matrix), at Different Temporal Scales

Fertilization								Development								Hatching								Incubation[c]	htw
Neaps				Springs												Neaps				Springs					
Night		Day		Night		Day		Wet Season				Dry Season				Night		Day		Night		Day			
HT	LT	HT	LT	HT	LT	HT	LT	S	U	M	L	S	U	M	L	HT	LT	HT	LT	HT	LT	HT	LT		
0	0	0	?[a]	0	0	0	1	0	(1)	1	0	0	0	0	0	?	?	?	?	?[b]	?	?	?	124–135 hrs	?

Source: Colombini I. et al., *Journal of Experimental Marine Biology and Ecology*, 190, 135–149, 1995; Brillet C., *Zeitschrift für Tierpsychologie*, 39, 283–331, 1975.

Note: Embryos (fertilized eggs) are demersal and occur in egg chambers of burrows dug in soft intertidal sediments. Timing of hatching and hatching temporal window is not known for this species. *?* = missing data; *1* = observed; *0* = not observed; *htw* = hatching temporal window; in parentheses, inferences and observations that need to be further investigated. HT and LT indicate intervals of ± 1 hour around the respective tidal phase. See Appendix A for abbreviations.

[a] Courtship and fertilization probably did not occur during low neap tides due to the fishes' low activity levels (Colombini et al. 1995).

[b] Hatching occurred at night during spring HT, in the congener *Periophthalmus modestus* (Ishimatsu et al. 2007).

[c] Incubation time was estimated from Brillet (1976), and the incubation times measured in the congener *Periophthalmus modestus* (Kobayashi et al. 1972; Kimura 1958).

ground. Observations were made along a 6-m-wide and 100-m-long belt transect, perpendicular to the creek's bank, which crossed the whole system. Three time scales were used: (1) the neap-spring tidal synodic cycle; (2) the semi-diurnal tidal cycle; and (3) the diel cycle. Two periods of 48 tidal hours were observed: 2 days of spring and 2 days of neap tides (adjacent synodic cycles). The study was conducted in November, in a wet season (the "short rains" of Kenya); therefore, the authors might have observed the courtship behavior described by Brillet (1980).

The authors reported that during spring tides all their sightings were of active individuals, whereas during neap tides almost half of the sightings were individuals resting on top of burrows. Total frequencies of sightings at night and during the day were comparable during spring tides, but significantly differed during neap tides: only 38% of all sightings during neap tides were made during the day; of this amount, 72% were mudskippers resting on burrows. Of the night sightings made during neap tides, 61% were active fish. Mudskippers followed the water's edge during both spring and neap FTs and ETs, and large aggregations were found at the erosive bank during extreme spring HT. In particular, fish were closer to the water's edge during neap tides than during spring tides. Accordingly, the mean position of the mudskipper during neap tides was ~36 m closer to the channel than during spring tides (Figure 12.3b,c).

During the night spring LT (LT ± 2 hours), individuals were spread over ~40 m along the transect, extending into the pneumatophore zone seaward of the forest (Figure 12.3b). During the day spring LT, they were distributed over a narrower space interval along the topographic gradient (~20 m), and their average position was shifted toward land by ~30 m relative to the night spring LT (Figure 12.3c). During the night neap LT (LT ± 1 hour), the fish were similarly distributed over a narrower space interval (~20 m), closer to the channel than during night spring LT. In both these cases, the mudskippers reached the upper portion of the non-vegetated mud bank of the creek (Figure 12.3a,b). During the day neap LT, when the water's edge moved onto the mud bank, no active fish were sighted here, while few individuals were sighted inside the forest (Figure 12.3c).

The frequency distribution of active mudskipper sightings was bimodal during spring tides (around HT and LT) and unimodal during neap tides (around HT). Overall, the difference in the frequency of sightings of active fish between spring tides (82%) and neap tides (18%) was highly significant during the day and still significant at night (58% and 42%, respectively). There was no significant difference in the frequency of sightings of active mudskippers between night and day during spring tides.

During spring LT, fish could actively feed in a wider area, likely because the substrate was still wet from the previous HT. The fact that the mudskippers were closer to the water's edge during LT at night than during the day might be the effect of a thermoregulatory behavior, as night temperatures in air were lower than in water. In fact, air temperature inside the forest correlated significantly to the position of the fish. However, this could also relate to the presence of visual predators such as waders in the pneumatophore zone and on the non-vegetated mud bank during the day. Diurnal visual predators may prompt the mudskipper to take shelter under the canopy and away from the bank. These hypotheses are not mutually exclusive.

During neap LT, because most of the forest floor was not wet from the previous HT, dry conditions likely inhibited fish activity during the day. Territorial individuals probably entered burrows or drastically decreased their activity and took shelter under plant debris, as in other *Periophthalmus* species (personal observation). The observation that humidity was correlated to the position of the mudskipper during neap tides supports this interpretation.

In the analysis of fish resting on burrows during neap tides, only eight burrows were found in the whole area. Burrows were positioned at three different levels from the channel—near the erosive bank, in an intermediate forested area, and in the pneumatophore zone (Figure 12.3a: *r*). The frequency of resting mudskippers on burrows was negatively correlated to humidity in the highest burrows, negatively correlated to humidity and temperature in the intermediate ones, and positively correlated to the tidal level, that is, to the distance of the water's edge from the eroding bank, in the

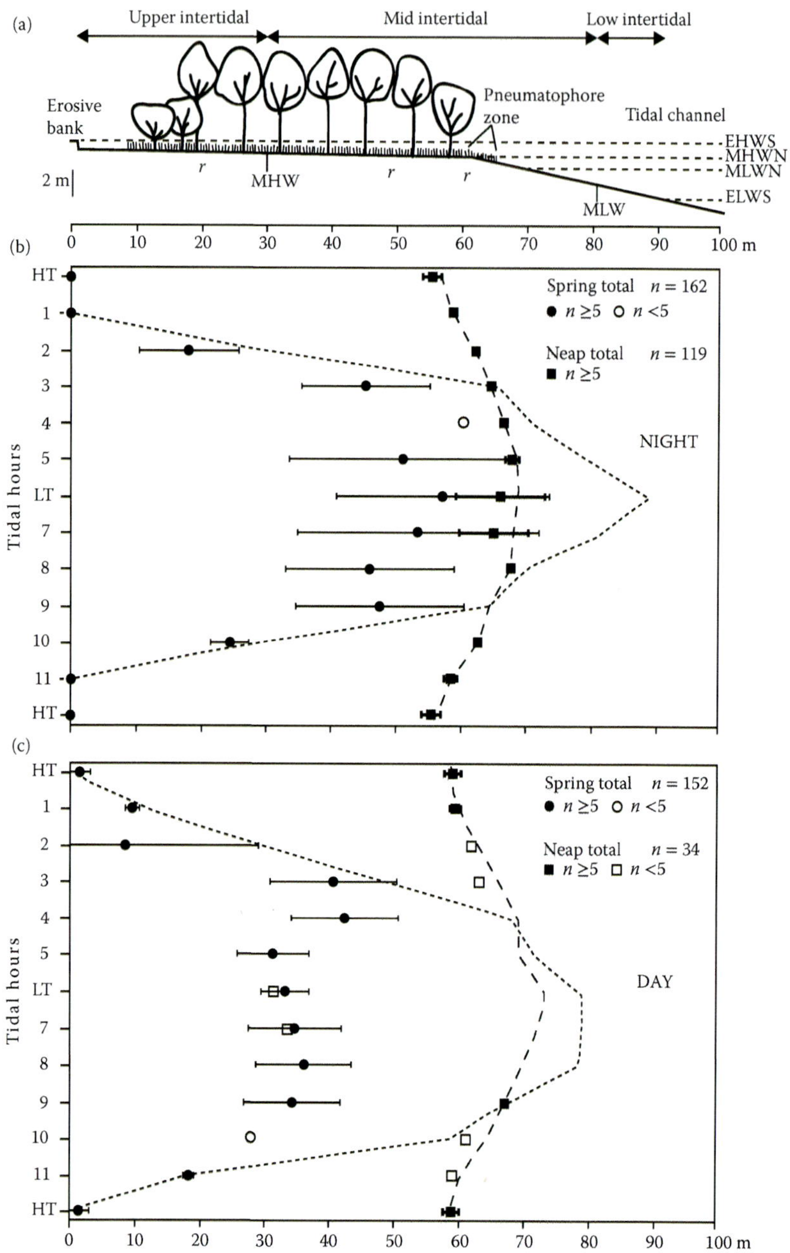

Figure 12.3 Diagrams showing the spatial and temporal distribution of metamorphosed individuals of *Periophthalmus* cf. *sobrinus* in a mangrove forest near Shekiko, Kenya. (a) Topographic profile of the belt transect made in the study site; MHW is assumed to be at the midpoint between EHWS and MHWN; MLW is assumed to be at the midpoint between ELWS and MLWN; r = position of burrows; (b) distribution of the individuals at night (mean positions with confidence limits), during spring (*circles*) and neap (*squares*) tides; *hatched* and *dotted lines* = positions of the neap and spring water's edge, respectively; n = number of sightings; (c) distribution of the individuals during the day, during spring and neap tides (Modified from Colombini I. et al., *Journal of Experimental Marine Biology and Ecology*, 190, 135–149, 1995.) See Appendix A for other abbreviations.

lowest ones. Overall, no significant difference between night and day was found, but significantly more fish were resting on burrows during LT, and significantly fewer during HT. These findings suggest that burrows were used to maintain physiological homeostasis during neap tides. No mudskipper could rest on burrows during spring HT, as the burrows were probably submerged. Although Colombini et al. (1995) reported that no fish remained inside burrows during HT, the design of this study suggests that they could have just missed the moment when the few territorial individuals (only eight burrows) entered the burrows during FTs.

Contrary to Brillet (1975; see Section 12.3.5), Colombini et al. (1995) described active individuals during HT, but considered mudskippers to be "resting or inactive" only when resting on burrows. Therefore, "active" individuals likely included fish both foraging and resting at the water's edge during HT (i.e., not on burrows). It is also unclear if they focused only on mudskippers on the surface of the forest floor or also counted those perching and resting on mangrove roots. Further, these authors did not discriminate between territorial and non-territorial individuals.

12.3.5 Daily Activity of Territorial Individuals

Brillet (1975) focused on burrowing territorial individuals, without discriminating sexes, and only conducted observations during the day. He observed for several hours a group of individuals in different tidal phases, in habitats with different topographies (S1, S2, F1, F2; see Section 12.3); however, due to the lack of tagging techniques, he could follow distinct individuals only in a few cases (four to five fishes), using distinctive morphological features. In all of his study sites, peak diurnal activity took place during the first part of ET, when mudskippers moved from their HT shelters and positioned themselves immediately above the FGs, still submerged, and during the last part of ET and LT, when the FGs were exposed to air. Activity decreased as the fish waited for the ET to expose the FGs; and after LT, as the exposed substrate gradually became drier. Activity increased briefly during FT, when intertidal movements to HT shelters occurred; and again decreased during HT, as the mudskipper either took shelter inside their burrows, or perched on mangrove roots and trunks, or rested on the exposed substrate at the water's edge. Peaks in feeding activity during spring LT and neap HT were also demonstrated by Colombini et al. (1996), consistent with the activity described by Colombini et al. (1995) (see Section 12.3.4). Colombini et al. (1996) also documented reduced feeding activity at night.

In S1, fish never fed near the burrows in the upper portion of the sloped creek bank, where water was not retained during ET (Brillet 1975). During the first part of ET, territorial mudskippers, including reproductive males (Brillet 1980, 1975), came out of the upper burrows and crossed the channel, perching on the emerged substrate above and around an area positioned lower on the banks of the tidal channel, still underwater. During the last part of ET, when this area was exposed, they rapidly moved to the water's edge (Figure 12.2a: *downward arrow*) and performed intense competitive feeding here (Figure 12.2a: *f*; Brillet 1975; Magnus 1970). This behavioral sequence implies the capacity to orientate toward an area that is not directly visible, demonstrating the directional orientation ability of this species (Berti et al. 1994, 1992). Brillet (1975) further observed that territorial individuals defending a specific area near the water's edge while feeding, also defended a corresponding area around their upper burrow on the bank slope during FT, and that the two areas were connected by scarcely defended corridors ("couloirs"); the feeding area was larger and more actively defended than the upper area around the burrow. He called these aggressively defended areas "dumbbell territories" ("haltére territoires," Brillet 1975).

In S2, during ET, most of the mudskippers migrated from their burrows on top of a steep bank in front of a dense *Avicennia* forest to a lower feeding area analogous to the feeding area in S1. However, the fish did not move directly to the foot of the steep bank where the burrows were located, where no flat and wet area was exposed during LT. Instead, they covered a distance of 15–30 m along the bank and then across a sand bank, rapidly proceeding in a single line and without stopping to feed, to reach a flat muddy area near the water's edge, without burrows, where they fed actively and competitively.

In F1, the mudskippers fed in the proximity of their burrows during ET, without moving to tide pools or tidal channels. This shows that suitable conditions for feeding were available in the unvegetated and wet mud immediately around the burrows (Figure 12.2c).

In F2, although the area around the burrow was wet and small tide pools were abundant during LT, territorial individuals often came out of their burrows during ET and moved for several meters to reach the edge of large tide pools to feed. Brillet (1975) speculated that the high density of pneumatophores (up to 350 m^{-2}) in this area inhibited burrowing, thus accounting for lower fish density and less frequent aggressive interactions. Feeding activity around the burrows was also apparently inhibited by the high pneumatophore density, possibly due to either lower prey abundance or inappropriate conditions for movement and predation. Brillet (1975) did not report if any burrows occurred near these large tide pools.

In both S1 and F1, mudskippers fed and interacted in the FGs during early LT, where territorial activities and competitive feeding occurred (Figure 12.2a,c: *f*; Brillet 1975). As the exposed substrate became drier during LT, feeding activity decreased; the mudskippers either rested near the residual tide pools (Figure 12.2a,c: *w*) or moved back to their resting grounds, taking shelter in their burrows (Figure 12.2a: *upward arrow*, *r*; Figure 12.2c: *r*; Brillet 1975, 1969).

On sloped banks (S1, S2), the mudskippers moved back (Magnus 1970) to their upper territories during FT, following the water's edge and feeding opportunistically as water covered the exposed substrate (Brillet 1975; Magnus 1970). Some fish in S1 may have entered the burrows in the lower resting grounds and stayed there until the next ET (see Section 12.3.1.1; Figure 12.2b: *r'*); however, Brillet (1975) did not report this. As FT flooded the channel, if the upper burrows were not submerged (e.g., neap FT), territorial individuals stayed at the water's edge, near their burrows, and showed a high level of tolerance toward intruders (Figure 12.2b: *w'*). If water reached the upper burrows, some individuals entered the burrows before they were completely submerged, and possibly stayed inside until the next ET (Figure 12.2b: *r*). Other mudskippers followed the water's edge to a more elevated position, resting on the exposed substrate near the water's edge during HT (Brillet 1975; Magnus 1970). If the water covered the flatter area above the banks of the channel, the fish that did not enter their burrows climbed on emerged mangrove roots and trunks (Figure 12.2b: *m*; Brillet 1975).

In flat areas (F1, F2), the water floods the inlet network and the forest floor from multiple directions during FT (Polgar and Crosa 2009). In F2, the mudskippers that earlier moved to feeding areas near tide pools, returned to their resting areas and rested on the vegetation or entered their burrows, possibly staying inside them during HT (Brillet 1975). In F1, the territorial individuals that fed around their burrows during ET moved for <1 m toward the water's edge to feed there, as the water approached their burrows. At this time, they also defended their territories from neighbors coming out of their burrows and from non-territorial migrants (contrary to territorial owners in S1). Then, as the water reached their burrows, they either entered them (Figure 12.2d: *r*) or moved away for several meters, to rest on the emerged substrate. Eventually, these latter individuals climbed on the emerged vegetation (Figure 12.2d: *m*). Brillet (1975) noted fewer mudskippers perching on the vegetation in Morondava (F1) than in Toliara (F2), where the vegetation was denser.

Territorial individuals perform burrow maintenance typically during ET before intertidal migrations, and during LT (Brillet 1969). During ET, tidal migrations typically elicited numerous aggressive interactions (Brillet 1975), as territories were often trespassed both by territorial and non-territorial individuals. Migrations during FT also elicited a brief increase in aggressive interactions as water reduced the available area of the feeding areas, but aggressiveness was inhibited when fish rested on emerged substrates during HT and distance between individuals was reduced (Brillet 1975; Magnus 1970). Courtship events frequently occurred during ET and FT (Brillet 1980, 1975; Magnus 1970) and elicited aggressive interactions as courting males and females trespassed neighboring male territories (Brillet 1980). During FT and ET in flat areas and during FT in sloped areas, the burrows were not visible from the point of view of the mudskippers, another example of the adaptive value of their directional orientation ability (Berti et al. 1994, 1992).

Comparing the results of Colombini et al. (1995) to those of Brillet (1980, 1975, 1976, 1969), it appears that the more recent analysis mostly describes the movements ("active fish") of non-territorial individuals that do not use burrows (errants; see Section 12.3.3), in a relatively flat forested area. Errants were also probably numerically dominant in this study, as up to 50–60 sightings per tidal hour during HT were recorded, whereas only eight burrows were found (Colombini et al. 1995). On the other hand, the individuals that "rested on nests" during neap tides (Colombini et al. 1995) likely were territorial individuals, the only ones that defend and regularly use burrows (Brillet 1975). Although he did not measure the synodic cycle, the "irregular" use of burrows by territorial individuals as described by Brillet (1975) is consistent with the synodic periodicity in the use of the "occasional" burrows described by Colombini et al. (1995). A synthesis of these spatio-temporal patterns is summarized in Table 12.2.

Gordon et al. (1968) reported the presence of *Ps.* cf. *sobrinus* in rocky and cobblestone beaches; the use of these intertidal habitats is not infrequent in other species of *Periophthalmus* (Clayton 1993). However, it is unlikely that such areas are suitable for egg laying and building of burrows, unless the rocky substrate presents cracks and crevices filled with mud, where fishes can dig burrows

Table 12.2 Metamorphosed *Periophthalmus* cf. *sobrinus* (Presence–Absence Matrix), at Different Spatio-Temporal Scales

	Night										Day									
	HT					LT					HT					LT				
	S	U	Mf	Mu	L	S	U	Mf	Mu	L	S	U	Mf	Mu	L	S	U	Mf	Mu	L
Wet Season																				
Neap Tides																				
Slope	?	?	?	?	?	?	?	?	?	?	0	0	b(f)r	0	0	0	0	bc[a]	0	c[a]fr
Flat	0	0	bfr	0	0	0	0	b(f)r	fr	0	0	0	bfr	0	0	0	0	bc[a](f)r	0	0
Spring Tides																				
Slope	?	?	?	?	?	?	?	?	?	?	0	(fr)	(bfr)	0	0	0	0	bc	0	cfr
Flat	0	fr	(b)fr	0	0	0	0	(b)fr	fr	0	0	fr	(b)fr	0	0	0	0	(b)cfr	0	0
Dry Season																				
Neap Tides																				
Slope	?	?	?	?	?	?	?	?	?	?	0	0	(bfr)	0	(b)	0	0	b	0	bfr
Flat	?	?	?	?	?	?	?	?	?	?	0	0	(bfr)	0	0	0	0	b(f)r	0	0
Spring Tides																				
Slope	?	?	?	?	?	?	?	?	?	?	0	(fr)	(bfr)	0	(b)	0	0	b	0	bfr
Flat	?	?	?	?	?	?	?	?	?	?	0	0	(b)fr	0	0	0	0	(b)fr	0	0

Source: Colombini I. et al., *Journal of Experimental Marine Biology and Ecology*, 190, 135–149, 1995; Brillet C., *La Terre et la Vie*, 4, 496–520, 1969; Brillet C., *Zeitschrift für Tierpsychologie*, 39, 283–331, 1975; Brillet C., *Revue d'Ecologie (La Terre et la Vie)*, 34, 427–468, 1980.

Note: Selected eco-ethological elements are reported: use of burrow (*b*), courtship (*c*), feeding (*f*), and resting, either perching on roots and trunks or at the water's edge (*r*). *?* = missing data; *0* = not observed; in parentheses, inferences and observations that need to be further investigated. Both courtship and feeding are typically associated with territorial behaviors (Brillet 1975); *Slope* = sloped habitats; *Flat* = flat habitats (Brillet 1975). In Colombini et al. (1995), MHW is assumed to be at intermediate horizontal distance between MHWN and EHWS, and MLW at intermediate horizontal distance between ELWS and MLWN (see also Figures 12.2 and 12.3a). Night observations are not available for sloped habitats (Brillet 1980, 1975, 1969) and during the dry season (Colombini et al. 1995). *Mf* = mid-intertidal, forested area (under the canopy or in the pneumatophore zone); *Mu* = mid-intertidal, unvegetated area; see Table 12.1 and Appendix A for other abbreviations.

[a] Colombini et al. (1995) reported a lower activity level during neap LT than during spring LT; the breeding period was assumed to approximately correspond to the wet season.

(northern Sulawesi, *Ps. argentilineatus*, pers. obs.). If no soft substrates are available in the area or can be reached by the fishes with intertidal movements, these would possibly be sink populations, which do not produce larvae.

12.4 SPATIO-TEMPORAL PATTERNS IN THE WATER-TO-LAND TRANSITION

12.4.1 Spatio-Temporal Variation of Eco-Ethological Elements in *Ps.* cf. *sobrinus*

The spatio-temporal distribution of fertilization and embryonic development is relatively invariant in *Ps.* cf. *sobrinus*, although some variation was observed in the spatial distribution of reproductive burrows (Table 12.1). However, the distribution of the behaviors of metamorphosed individuals, including territorial owners, is extremely variable, both in space and time (Table 12.2). For example, there are four possible combinations of the spatio-temporal distribution of the use of burrow: (1) no variation of occurrence in space and time: each fish regularly uses a burrow system in the same area; (2) variation of space in time: each fish uses a burrow in a portion of its habitat in a time interval, and another burrow in another portion of its habitat in a different time interval; (3) variation in time but not in space: each fish remains in the same area, but uses the burrow only during specific intervals of time; and (4) variation in space and time: each fish uses a burrow only in a specific portion of its habitat, in a specific time interval.

Periophthalmus cf. *sobrinus* exhibits variation both in space and time in the use of burrow, courtship, feeding, and resting (Table 12.2). In particular, the spatio-temporal variation of burrowing and feeding in this species appears to convey relevant information with respect to its adaptation to semi-terrestrial habitats.

12.4.2 Burrowing and Feeding Grounds

Late Ordovician burrows dug in soil are among the earliest animal fossil tracks on dry land, suggesting that burrows dug in soft substrates have always played a key role in the colonization of land by semi-aquatic animals (Retallack 2001; Retallack and Feakes 1987). Several aquatic species of air-breathing vertebrates, including sarcopterygian dipnoi (e.g., Fishman et al. 1992) and some mudskippers, are able to withstand prolonged periods of emersion while buried in fine sediments, or inside burrows, via aestivation, as in the case of *Pseudapocryptes elongatus* (reported as *Pseudapocryptes lanceolatus*, Hora 1935); or via overwintering, as in the case of *Ps. modestus* (Kobayashi et al. 1971) and *Scartelaos histophorus* (Tamura et al. 1976). In arid and temperate climates, burrowing appears to be the key adaptation of tropical intertidal species. For example, harsh climatic conditions are mitigated by tides in the northern Persian Gulf, as prevalent HT conditions occur during the day in the hotter months and prevalent LT conditions during the day in the colder months (Jones and Clayton 1983). However, tropical intertidal macrofauna, including mudskippers, occur only in tidal mudflats in this region, whereas rocky shores, which offer unsuitable conditions to intertidal burrowers, host mostly warm temperate communities that are tolerant of variable climate (Clayton 1985; Jones and Clayton 1983). In humid tropical climates, burrowing and tidal mudflat habitats appear to be less related to avoidance of environmental extremes in mudskippers, as evidenced by the occasional occurrence of mudskippers on rocky, cobblestone, and sandy shores (Clayton 1993), also at distances of >1 km from tidal mud deposits (personal observation). The highly porous, three-dimensional medium offered by silt and clay can contain large amounts of interstitial water when water logged, subsequently attaining plastic and cohesive to semi-liquid mechanical properties, with increasing amounts of water. Terrigenous organic matter enhances these mechanical and hygroscopic properties. It, therefore, appears that burying and burrowing in organic mud has been one of the key adaptations facilitating land invasions in several aquatic animal lineages, protecting these organisms from more extreme and variable subaerial environmental conditions, and reducing water loss.

In *Ps.* cf. *sobrinus*, tidal movements and the distribution of territories (Brillet 1975) demonstrate the presence of two functionally different areas that can be spatially and temporally separated, hereby defined as "feeding grounds" (FGs) and "burrowing grounds" (BGs). FGs are typically larger and more actively defended during spring ET and LT, and neap HT (competitive feeding), whereas BGs are smaller and defended only during the first part of spring FT. The defended areas immediately around the burrows in the upper portion of the slopes in S1 and S2 (and after the wet season, in the lower area near the FGs) are BGs, analogous to those in F1 and F2 (Figure 12.2; Brillet 1975). All the burrows observed by Brillet (1975) were at 0–70 cm below MHW, and burrow density was highest at 0–20 cm below MHW (Figure 12.2b: *r*); in F1, burrows were found at 10–15 cm below MHW (Figure 12.2d: *r*). This suggests preferred microhabitat conditions for the BGs, as also supported by the following observations: (1) burrows were not present in other flat forested areas >70 cm below MHW (Brillet 1969: 498); and (2) in S1, the burrows dug on the bottom of the tidal channel, ~70 cm below MHW (Figure 12.2a,b: *r'*; Brillet 1975: fig. 15a), were never used for breeding at the onset of the breeding season in August, and were abandoned the next month, before the start of the wet season. Reproductive males always led females inside the upper burrows of their dumbbell territories (Figure 12.2a: *r*). Similarly, (1) the lower grounds of S1 and S2 where the fish fed during ET and LT; (2) the area immediately around the burrow in F1; and (3) the edge of the large tide pools of F2, are all functionally analogous FGs.

Only in F1, the two types of areas (FG, BG) were spatially overlapped. Consistently, this is the only case in which the area around the burrows was actively defended by territorial individuals both during FT and ET; F1 was also the area with the highest burrow density, suggesting it offered conditions preferred by the local mudskipper population, for territory establishment (Brillet 1975). At HT, fewer fish were perching on the trees in F1 than in F2, despite the higher density of burrows, suggesting that most of the individuals entered the burrows, and that fewer non-territorial individuals (errants) inhabited this area. The burrows dug in S1 after the wet season (Figure 12.2a: *r'*) were also near the FGs; but for the mentioned reasons, these were not functionally analogous to the BGs in F1. The fact that these lower burrows were used in S1 only during the dry season by mudskippers that also owned upper resting grounds and burrows, and never for breeding, suggests that they were dug to tolerate drier conditions during LT without climbing to the upper resting grounds (Figure 12.2a: *upward arrow*, *r*). Due to the accumulated mud on the bottom of the creek and its reduced frequency of submersion, the bank probably dried up more rapidly during LT. Therefore, these S1 areas should be considered lower, non-reproductive BGs.

The lower level of territorial aggression observed in BGs than in FGs in S1, S2, and F2, and the higher level of aggression observed in the BGs of F1 is also consistent with this interpretation, because BGs correspond to FGs only in F1. The relatively low level of aggression in F2 around the FGs is an exception. Several non-mutually exclusive hypotheses can explain this observation. First, as suggested by Brillet (1975), the high density of pneumatophores in this area might have reduced visual aggressive interactions. A similar effect created by mud walls was observed in *Boleophthalmus dussumieri* (Clayton 1987). The presence of vertical visual barriers reduced the frequency and intensity of aggressive interactions between territorial individuals of *Ps. variabilis* in semi-natural conditions (Polgar 2004). Second, aggression in *Ps.* cf. *sobrinus* is proportional to population density; F2 burrow density was lower than in other areas. Last, most of the individuals inhabiting this suboptimal area were probably less aggressive, non-territorial errants, as suggested by the many individuals perching on emerged spots during HT.

Peaks of feeding activity in *Ps.* cf. *sobrinus* always occur on wet substrates, during ET and early LT during spring tides, and during HT during neap tides. Non-mutually exclusive explanations can be (1) higher abundance of potential prey; (2) food manipulation in water, for example, small tide pools (personal observation); and (3) smaller time needed to find pockets of water to recharge the hydraulic tongue (Michel et al. 2015). The frequent observation that after a successful hunt *Periophthalmus* species return to shallow-water pools or to the burrow, if pools are unavailable supports these two latter hypotheses. Clayton and Snowden (2000) observed this in *Periophthalmus waltoni*, and I observed it in *Periophthalmus gracilis* and *Periophthalmus novaeguineaensis*.

Consistent with the need for water in order to effectively forage, territorial *Ps.* cf. *sobrinus* perform intertidal migrations to wet FGs, at lower intertidal level, if these are not available around their BGs (S1, S2, F2). Similar intertidal migrations to lower intertidal areas associated with feeding activity have also been observed in other mudskipper species, such as *Periophthalmus chrysospilos* (Polgar and Crosa 2009), *Periophthalmodon schlosseri* (Chua 2009), and *Boleophthalmus boddarti* (personal observation). It was also observed in the rockskipper blenny *Alticus kirkii* (reported as *Lophalticus kirkii*, Brillet 1986). Where FG and BG are overlapped like in flat-forested areas on the bottom of ephemeral tidal inlets (F1), territorial mudskippers likely have more time to feed during ET than in other areas. In fact, in these cases (1) they do not perform intertidal movements; and (2) the exposed substrate likely remains wet for a longer time under the canopy. A higher burrow density was measured in F1 than in S1, and in general, in flat than in sloped areas (Brillet 1975), suggesting that the former areas are preferred for territory establishment. On the other hand, territorial *Ps.* cf. *sobrinus* in F2 and S2 probably had less time to feed, due to their intertidal migrations across areas such as dense root networks, sand banks, and steep banks of tidal creeks, in which they did not feed.

The more frequent use of the burrows by fish living in F1 relative to S1 during HT (Brillet 1975) suggests that both climbing and staying inside burrows during HT are more energetically expensive than resting on flat, exposed, and wet substrates, and the latter option is preferred if available (S1; Figure 12.2b: *w*'). In the intertidal areas of Kukup town in the southern tip of Peninsular Malaysia, *Ps. chrysospilos* and *Pn. schlosseri* often rest on floating wooden planks during HT, instead of perching on the available mangrove saplings and trunks. In the same area, inside the pioneer mangrove forests of Kukup Island, where no horizontal emerged surfaces are available, these species perch vertically on the vegetation during HT (personal observation; see also Chua 2009).

12.4.3 The "Pull" or "Push" Dilemma

Hypotheses on the ecological factors driving land invasion in aquatic organisms living in shallow waters historically revolved around two alternatives. In the first scenario, organisms that first colonize semi-terrestrial habitats would have access to unexploited trophic resources (e.g., Inger 1952). This "pull hypothesis" predicts that vascular plants should have colonized land systems before animals, accumulating large amounts of unexploited subaerial trophic resources unavailable underwater. However, the first animals that came out of water were probably not herbivores, but decomposers, predators, and scavengers, that fed on organic detritus and other decomposers in semi-terrestrial habitats, before the vascular plant invasion and the accumulation of large trophic resources on land (Hasiotis et al. 2003; MacNaughton et al. 2002; McNamara and Selden 1993). Therefore, it appears unlikely that access to resources similar in quality and quantity to those already available in water compensated for the considerable amount of energy needed to cope with the harsh subaerial conditions (Retallack 2001; McNamara and Selden 1993; Retallack and Feakes 1987). In the second scenario, intense direct competition in shallow aquatic ecosystems would have driven some animal lineages to take shelter in shallower waters and on emerged substrates. There is fossil evidence of such dynamics during the water-to-land transition of different animal groups, such as late Silurian eurypterids predating semi-aquatic euthycarcinoids, and Devonian placoderms predating tetrapodomorphs (Clack 2002; McNamara and Selden 1993). For any water-to-land transition, this "push hypothesis" predicts the presence of higher levels of direct competition in water than on land, and the evolution of amphibious behaviors as antipredatory adaptations.

In *Ps.* cf. *sobrinus*, as in other mudskippers and rock skippers, FG is never at higher topographic level than the shelter area (BG in mudskippers), along the intertidal gradient. In other words, if these fish migrate to feed, they always move to the water's edge, never landwards. Therefore, at least in these cases, the hypothesis of an evolutionary "pull" from subaerial unexploited trophic resources (i.e., at higher topographic level) is falsified. Future studies and more inclusive comparative analyses may test if this is a general pattern in mudskippers.

The pattern observed in these species depicts a transition from lower intertidal areas where the food is available (FG) to higher intertidal areas where there is lower aquatic predation risk, and is consistent with the "push" hypothesis. In particular, above MSL, the high spatial complexity of the intertidal vegetation offers shelter from both terrestrial predators like wading birds (Clayton 1993) and aquatic predators such as fishes, swimming crabs, crocodiles, and semi-aquatic snakes (Kamal et al. 2014; Nagelkerken et al. 2010; Laegdsgaard and Johnson 2001; Clayton 1993; personal observation). Higher intertidal habitats and emersion also offer protection from aquatic parasites (Polgar et al. 2009; Pearse 1933, 1932, 1929).

12.4.4 Predation Risk and Reproduction

The only obligate function played by burrows in *Ps.* cf. *sobrinus* is reproduction (see Section 12.3.1.1). Several mudskipper species use burrows to maintain physiological homeostasis during LT and avoid the water column during HT, but these functions are variable within and between species (see Clayton 1993). However, the burrow is always used to lay and fertilize eggs in all the studied mudskipper species. The "push" hypothesis suggests that the landward displacement of the area around the burrow (BG) relative to the FG (Brillet 1975; S1, S2, F2; Sections 12.3.2, 12.3.5, and 12.4.2) may be a specific antipredatory strategy. Investigations may be conducted on potential aquatic egg predators. In the tree frog *Endropsophus ebraccatus*, oviposition site selection in transitional habitats has been demonstrated to be strongly driven by the risk of aquatic predation (Touchon and Worley 2015).

Along this line of reasoning, during the evolution of amphibious traits in the most recent common ancestor of mudskippers, such ecological "push" would have positively selected lineages that moved their BGs to higher intertidal levels. The hypothetical aquatic ancestors likely lived at lower intertidal or subtidal levels and utilized FGs around their burrows (FG = BG). Burrowing is reported in several aquatic gobies that likely have a similar behavioral ecology, including *Croilia mossambica* (Blaber and Whitfield 1977), *Gobioides peruanus* (Van Tassell 2010), *Microgobius gulosus* (Gaisner 2005), and several coral reef species of the genera *Amblygobius*, *Asterropteryx*, *Echinogobius*, *Oxyurichthys*, *Signigobius*, and *Valenciennea* (Brandl and Bellwood 2014). The network of ephemeral inlets on the forest floor, where substrate conditions are favorable for both FGs and BGs (Polgar and Crosa 2009; F1 in *Ps.* cf. *sobrinus*), may have acted as an ecological and evolutionary route from the subtidal zone and tidal channels to the drier, vegetated areas. Consistently, early post-settlement young of *Ps.* cf. *sobrinus*, more vulnerable to aquatic predators, were only found in forested areas (Brillet 1976).

Brillet (1969) did not describe the construction of the two openings of the Y-shaped burrow in *Ps.* cf. *sobrinus*. However, Kobayashi et al. (1971) observed in the laboratory that *Ps. modestus* digs a Y-shaped burrow by first making a hole digging downward and then changing the digging direction and creating a second opening by digging upward. Then the fish digs a vertical shaft starting from the lowest point of the "U," completing the "Y" shape. This digging technique is probably conserved among congeners digging Y-shaped burrows, as I observed the same sequence in *Ps. variabilis* (unpublished data). This sequence suggests a hypothetical evolutionary sequence from an intertidal aquatic ancestor living in U-shaped burrows. Brillet (1975) observed in semi-natural conditions a vigorous undulation behavior performed by *Ps.* cf. *sobrinus*, while resting upside down inside the burrow, in the apparent attempt to flush the burrow. This suggests that the function of an ancestral "U-shaped" burrow dug in muddy intertidal sediments was to easily create a unidirectional current, thus rapidly flushing the oxygen-depleted stagnant water, especially after a prolonged period of emersion. During LT, the hypothetical intertidal ancestor presumably rested in the deeper and wetter portion of the burrow, when exposed to air, and possibly buried itself in the more humid bottom, as water evaporated. A variant of this behavior might have been to dig downward, thus reaching deeper and more humid conditions. Brillet (1969) observed that in semi-natural conditions, *Ps.* cf. *sobrinus* selected more humid patches of mud to start digging. Digging vertically in a drying U-shaped burrow would have allowed ancestors of mudskippers to reach the water table, thus being able to colonize higher intertidal grounds. In this view, the presence of deeper burrows of *Ps.* cf. *sobrinus* on the banks of tidal channels (Brillet 1976) appears to be related to the deeper position of the water table in the proximity of tidal channels (Montalto et al. 2007).

12.5 LIMITS, DEVELOPMENTS, AND CONCLUDING REMARKS

Although *Ps.* cf. *sobrinus* is one of the better-studied mudskipper species from an eco-ethological standpoint, this review highlights (1) the need for more studies including both qualitative descriptions and quantitative analyses; (2) several gaps in the spatio-temporal distribution (Tables 12.1 and 12.2); and (3) the absence of information on the larval stages. The first point, particularly important to better define the post-settlement ontogenetic stages and the patterns of territory establishment, could be addressed by marking specimens subcutaneously, to readily identify individuals in the field (e.g., Gaisner 2005). The second point can be addressed by quantitative field studies conducted in different spatio-temporal conditions (e.g., in different seasons; Colombini et al. 1995). The third point can be solved by conducting taxonomic and ecological studies of the ichthyoplankton communities using barcoding, morphological taxonomy, and otolith elemental composition methods in coastal areas where mudskippers are abundant, and by conducting developmental studies in the laboratory (e.g., Rolls 2014; Ratnasingham and Hebert 2007; Richards 2006; Kobayashi et al. 1972).

The choice of the spatio-temporal scales naturally affects the scope of the analysis. Larger spatial scales than those adopted here, at the geographic or topographic level, would elucidate geomorphological features indicative of different ecosystems, such as estuaries, lagoons protected by coastal ridges, lower tracts of rivers, and the respective mudskipper communities (e.g., Polgar et al. 2017). The interaction of tidal cycles and climate at a seasonal scale may highlight ecological patterns that affect reproductive cycles, being associated with mudskipper community composition and diversity (e.g. Jones and Clayton 1983). Smaller spatial scales should also be explored, since intertidal and semi-terrestrial subaerial and infaunal habitats are considerably heterogeneous relative to size and movements in mudskippers (e.g., Polgar et al. 2010; Polgar and Crosa 2009; Ishimatsu et al. 2000). Environmental parameters such as light intensity, temperature, air humidity, and soil evaporation can be investigated at smaller spatio-temporal scales using small data loggers (e.g., iButtons® Maxim Integrated). Such studies could better investigate rapid patterns of variation, for example, around ETs and FTs, and during specific phases of the diel cycle, such as sunset, dawn, and midday, that might go unnoticed at larger spatio-temporal scales.

I would like to conclude by stressing the need for studies describing mudskippers' behavioral patterns and life cycles. Although quantitative analyses allow broader and more objective comparisons and are the ultimate goal of any scientific investigation, qualitative descriptions, presently considered old fashioned and inferior to quantitative approaches, are the very philosophical foundation of any sound scientific question. It is indeed surprising how this trend in the scientific community has not changed direction since the 1970s (Lorenz 1973). Only after carefully describing the details of different mudskipper lifestyles can we conduct comparative "analyses on a broad front" (Lorenz 1948). Only at a later stage can such analytical processes guide sound hypothesis testing and quantitative experimental designs in a phylogenetic scenario. The need for descriptive data unavailable in recent international scientific literature is clearly demonstrated by the number of old works cited in this study, several ones written in different languages. The diverse, complex, and dynamic nature of the mudskippers' processes, mechanisms, and environments implies that there is little chance that any naively reductionist approach significantly improves our understanding of their ecology and evolution.

Appendix A

Abbreviations and Definitions

EHWS: extreme high water spring
ELWS: extreme low water spring
ET: ebb tide
HT: high tide (semi-diurnal cycle)
FT: flood tide
LT: low tide (semi-diurnal cycle)
MHW: mean high water
MHWN: mean high water neap
MLW: mean low water
MLWN: mean low water neap
MSL: mean sea level
S: supratidal zone = above EHWS
U: upper intertidal zone = above MHW and below EHWS
M: mid-intertidal zone = above MLW and below MHW
L: low-intertidal zone = below MLW and above ELWS

ACKNOWLEDGMENTS

The authors would like to thank E.O. Murdy and Z. Jaafar for their useful comments and criticisms.

REFERENCES

Agorreta A. and Rüber L. 2012. A standardized reanalysis of molecular phylogenetic hypotheses of Gobioidei. *Systematics and Biodiversity* 10:375–390.

Agorreta A., San Mauro D., Schliewen U. et al. 2013. Molecular phylogenetics of Gobioidei and phylogenetic placement of European gobies. *Molecular Phylogenetics and Evolution* 69(3):619–633.

Aguilar N.M., Ishimatsu A., Ogawa K. and Huat K.K. 2000. Aerial ventilatory responses of the mudskipper, *Periophthalmodon schlosseri*, to altered aerial and aquatic respiratory gas concentrations. *Comparative Biochemistry and Physiology* 127:285–292.

Algeo T.J., Scheckler S.E. and Maynard J.B. 2001. Effects of the middle to late Devonian spread of vascular land plants on weathering regimes, marine biotas, and global climate. In *Plants Invade the Land—Evolutionary and Environmental Perspectives* (eds. P.G. Gensel and D. Edwards). New York, NY: Columbia University Press, pp. 213–236.

Baeck G.W., Takita T. and Yoon Y.H. 2008. Lifestyle of Korean mudskipper *Periophthalmus magnuspinnatus* with reference to a congeneric species *Periophthalmus modestus*. *Ichthyological Research* 55:43–52.

Baeck G.W., Yoon Y.H. and Park J.M. 2013. Ontogenetic and diel changes in diets of two sympatric mudskippers *Periophthalmus modestus* and *Periophthalmus magnuspinnatus* on the tidal flats of Suncheon Bay, Korea. *Fisheries Science* 79(4):629–637.

Berti R., Chelazzi L., Colombini I. and Ercolini A. 1992. Direction-finding ability in a mudskipper from the delta of the Tana River (Kenya). *Tropical Zoology* 5:219–228.

Berti R., Colombini I., Chelazzi L. and Ercolini A. 1994. Directional orientation in Kenyan population of *Periophthalmus sobrinus* Eggert: Experimental analysis of the operating mechanism. *Journal of Experimental Marine Biology and Ecology* 181:135–141.

Blaber S. and Whitfield A.K. 1977. The biology of the burrowing goby *Croilia mossambica* Smith (Teleostei, Gobiidae). *Environmental Biology of Fishes* 1(2):197–204.

Brandl S.J. and Bellwood D.R. 2014. Pair-formation in coral reef fishes: An ecological perspective. *Oceanography and Marine Biology: An Annual Review* 52:1–80.

Brillet C. 1969. Etude du comportement constructeur des poissons amphibies Periophtalmidae. *La Terre et la Vie* 4:496–520.

Brillet C. 1970. Relations entre comportement sexuel, territoire et agressivité chez les Periophtalmes. *Comptes Rendus Hebdomadaires des Scéances de l'Académie des Sciences Paris* 269:1507–1510.

Brillet C. 1975. Relations entre territoire et comportement agressif chez *Periophthalmus sobrinus* Eggert (Pisces, Periophthalmidae) au laboratoire et en milieu naturel. *Zeitschrift für Tierpsychologie* 39:283–331.

Brillet C. 1976. Structure du terrier, reproduction et comportement des jeunes chez le poisson amphibie *Periophthalmus sobrinus* Eggert. *La Terre et la Vie* 3:465–483.

Brillet C. 1980. Comportement sexuel du poisson amphibie *Periophthalmus sobrinus* Eggert ses rapports avec le comportement agonistique. *Revue d'Ecologie (La Terre et la Vie)* 34:427–468.

Brillet C. 1984a. Étude comparative de la parade nuptiale chez deux espèces de Poissons amphibies sympatriques (Pisces, Periophthalmidae). *Comptes Rendus de l'Académie des Sciences Série III, Paris* t. 298 (12):347–350.

Brillet C. 1984b. Habitat et comportement du poisson amphibie *Periophthalmus koelreuteri* africanus Eggert. A Tuléar comparaison avec le *P. sobrinus* Eggert sympatrique. *Revue d'Ecologie (La Terre et la Vie)* 39(3):337–345.

Brillet C. 1986. Notes sur le comportement du poisson amphibie *Lophalticus kirkii* Gunther (Pisces—Salariidae). Comparaison avec *Periophtalmus sobrinus* Eggert. *Revue d'Ecologie (La Terre et la Vie)* 41(4):361–376.

Chakrabarty P., Davis M.P. and Sparks J.S. 2012. The first record of a trans-oceanic sister-group relationship between obligate vertebrate troglobites. *PLoS One* 7(8):e44083. doi:10.1371/journal.pone.0044083.

Chang J.-T. and Lee S.-C. 1994. Genetic variation of the Chinese mudskipper, *Periophthalmus cantonesis* (Osbeck, 1762) (Pisces; Perciformes, Gobiidae) from Taiwan. *Zoological Studies* 33(1):34–43.

Chen W., Hong W., Chen S., Wang Q. and Zhang Q. 2015. Population genetic structure and demographic history of the mudskipper *Boleophthalmus pectinirostris* on the northwestern pacific coast. *Environmental Biology of Fishes* 98:845–856.

Chew S.F., Sim M.Y., Phua Z.C., Wong W.P. and Ip Y.K. 2007. Active ammonia excretion in the giant mudskipper, *Periophthalmodon schlosseri* (Pallas) during emersion. *Journal of Experimental Zoology* 307(6):357–369.

Chua J.L.T. 2009. Tidal migration of the giant mudskipper, *Periophthalmodon schlosseri* (Pallas, 1770) in a Singapore mangrove. Unpublished Undergraduate Research Opportunities in Science Report. National University of Singapore, Singapore, 30 pp.

Clack J.A. 2002. *Gaining Ground: The Origin and Evolution of Tetrapods.* Bloomington, IN: Indiana University Press, 369 pp.

Clack J.A. 2006. The emergence of early tetrapods. *Palaeogeography Palaeoclimatology Palaeoecology* 232:167–189.

Clayton D.A. 1993. Mudskippers. *Oceanography and Marine Bulletin: An Annual Review* 31:507–577.

Clayton D.A. 1985. Ecology of mudflats with particular reference to those of the northern Arabian Gulf. In *Marine Environment and Pollution. Conference on the Marine Environment and Pollution* (eds. R. Halwagy, D. Clayton and M. Behbehani). Kuwait City: Kuwait Foundation for the Advancement of Science and Environmental Protection Council, Kuwait University, pp. 83–96.

Clayton D.A. 1987. Why mudskippers build walls. *Behaviour*, 102:185–195.

Clayton D.A. and Snowden R. 2000. Surface activity in the mudskipper, *Periophthalmus waltoni* Koumans 1941 in relation to prey activity and environmental factors. *Tropical Zoology* 13:239–249.

Clayton D.A. and Vaughan T.C. 1986. Territorial acquisition in the mudskipper *Boleophthalmus boddarti* (Teleostei, Gobiidae) on the mudflats of Kuwait. *Journal of Zoology, London(A)* 209:501–509.

Clayton D.A. and Vaughan T.C. 1988. Ethogram of *Boleophthalmus boddarti* (Pallas) (Teleostei, Gobiidae), a mudskipper found on the mudflats of Kuwait. *Journal of the University of Kuwait (Sciences)* 15:115–138.

Colombini I., Berti R., Ercolini A., Nocita A. and Chelazzi L. 1995. Environmental factors influencing the zonation and activity patterns of a population of *Periophthalmus sobrinus* Eggert in a Kenyan mangrove. *Journal of Experimental Marine Biology and Ecology* 190:135–149.

Colombini I., Berti R., Nocita A. and Chelazzi L. 1996. Foraging strategy of the mudskipper *Periophthalmus sobrinus* Eggert in a Kenyan mangrove. *Journal of Experimental Marine Biology and Ecology* 197:219–235.

Duke N.C., Ball M.C. and Ellison J.C. 1998. Factors influencing biodiversity and distributional gradients in mangroves. *Global Ecology and Biogeography Letters* 7:27–47.

Eggert B. 1935. Beitrag zur Systematik, Biologie und geographischen Verbreitung der Periophthalminae. Ergebnisse einer durch die Notgemeinschaft der Deutschen Wissenschaft ermöglichten Reise nach Niederländisch-Indien 1926–1927 und der Sundaexpedition der Notgemeinschaft der Deutschen Wissenschaft 1929–1930. *Zoologische Jahrbücher, Abteilung für Systematik, Ökologie und Geographie der Tiere (Jena)* 67(1–2):29–116.

Fishman A.P., Galante R.J., Winokur A. and Pack A.I. 1992. Estivation in the African lungfish. *Proceedings of the American Philosophical Society* 136(1):61–72.

Gaisner A. 2005. Parental care and reproductive behavior of the clown goby, *Microgobius gulosus*, with observations on predator interactions. *Environmental Biology of Fishes* 73:341–356.

Gibb A.C., Ashley-Ross M.A. and Hsieh S.T. 2013. Thrash, flip, or jump: The behavioral and functional continuum of terrestrial locomotion in teleost fishes. *Integrative and Comparative Biology* 53(2):295–306.

Gordon M.S. 1998. African amphibious fishes and the invasion of the land by tetrapods. *South African Journal of Zoology* 33(2):115–118.

Gordon M.S., Boëtius J., Evans D.H. and Oglesby L.C. 1968. Additional observations on the natural history of the mudskipper, *Periophthalmus sobrinus*. *Copeia* 4:853–857.

Gordon M.S. and Olson E.C. 1995. *Invasions of the Land: The Transitions of Organisms from Aquatic to Terrestrial Life*. New York, NY: Columbia University Press, 312 pp.

Graham J.B. 1997. *Air-Breathing Fishes. Evolution, Diversity and Adaptation*. San Diego, CA: Academic Press, 299 pp.

Graham J.B. and Lee H.J. 2004. Breathing air in air: In what ways might extant amphibious fish biology relate to prevailing concepts about early tetrapods, the evolution of vertebrate air breathing, and the vertebrate land transition? *Physiological and Biochemical Zoology* 77(5):720–731.

Graham J.B., Lee H.J. and Wegner N.C. 2007. Transition from water to land in an extant group of fishes: Air breathing and the acquisition sequence of adaptations for amphibious life in oxudercine gobies. In *Fish Respiration and Environment* (eds. M.N. Fernandes, F.T. Rantin, M.L. Glass and B.G. Kapoor). Enfield, NH: Science Publishers, pp. 255–288.

Hasiotis S.T., Lieberman B.S., Dilliard K.A. and Pope M.C. 2003. Earliest traces of life on land: Arthropod trackways and other ichna in Early Cambrian (*Nevadella* zone) intertidal to supratidal deposits, Mackenzie Mountains, Northwest Territories, Canada (abs.): Geological Society of America Annual Meeting, Abstracts with Programs, Seattle, Washington, 472 pp.

He L., Mukai T., Chu K.H., Ma Q. and Zhang J. 2015. Biogeographical role of the Kuroshio current in the amphibious mudskipper *Periophthalmus modestus* indicated by mitochondrial DNA data. *Scientific Reports* 5:15645. doi:10.1038/srep15645.

Hora S.L. 1935. Ecology and bionomics of the gobioid fishes of the Gangetic Delta. Paper presented at the XII International Congress of Zoology, Lisbon, pp. 841–865.

Hsieh S.T. 2010. A locomotor innovation enables water-land transition in a marine fish. *PLoS One* 5:e11197. doi:10.1371/journal.pone.0011197.

Ikebe Y. and Oishi T. 1996. Correlation between environmental parameters and behaviour during high tides in *Periophthalmus modestus*. *Journal of Fish Biology* 49:139–147.

Ikebe Y. and Oishi T. 1997. Relationships between environmental factors and diel and annual changes of the behaviors during low tides in *Periophthalmus modestus*. *Zoological Science* 14:49–55.

Inger R.F. 1952. Walking fishes of southeastern Asia travel on land. *Chicago Natural History Museum Bulletin* 23:4–7.

Inger R.F. 1957. Ecological aspects of the origin of the tetrapods. *Evolution* 11:373–376.

Ip Y.K., Chew S.F. and Ping C.T. 1991. Evaporation and the turning behavior of the mudskipper *Boleophthalmus boddaerti*. *Zoological Science, Tokyo* 8(3):621–624.

Ishimatsu A. and Graham J.B. 2011. Roles of environmental cues for embryonic incubation and hatching in mudskippers. *Integrative and Comparative Biology* 51(1):38–48.

Ishimatsu A., Takeda T., Kanda T., Oikawa S. and Khoo K.H. 2000. Burrow environment of mudskippers in Malaysia. *Journal of Bioscience* 11:17–28.

Ishimatsu A., Yoshida Y., Itoki N., Takeda T., Lee H.J. and Graham J.B. 2007. Mudskippers brood their eggs in air but submerge them for hatching. *Journal of Experimental Biology* 210:3946–3954.

Jones D.A. and Clayton D.A. 1983. The systematics and ecology of crabs belonging to the genera *Cleistostoma* de Haan and *Paracleistostoma* de Man on Kuwait mudflats. *Crustaceana* 45(2):183–199.

Kamal S., Lee S.Y. and Warnken J. 2014. Investigating three-dimensional mesoscale habitat complexity and its ecological implications using low-cost RGB-D sensor technology. *Methods in Ecology and Evolution* 5:845–853.

Kanemori Y., Takegaki T. and Natsukari Y. 2006. Genetic population structure of the mudskipper *Boleophthalmus pectinirostris* inferred from mitochondrial DNA sequences. *Japanese Journal of Ichthyology* 53(2):133–141 (in Japanese with English summary).

Kimura M. 1958. Early development of the mud-skipper, *Periophthalmus cantonensis* (Osbeck). *Bulletin of the Japanese Society of Scientific Fisheries* 23(12):754–757.

Kobayashi T., Dotsu Y. and Miura N. 1971. Nest and nesting behavior of the mudskipper *Periophthalmus cantonensis*, in Ariake Sound. *Bulletin of the Faculty of Fisheries, Nagasaki University* 32:27–39 (in Japanese with English abstract).

Kobayashi T., Dotsu Y. and Miura N. 1972. Egg development and rearing experiments of the larvae of the mud skipper, *Periophthalmus cantonensis. Bulletin of the Faculty of Fisheries, Nagasaki University* 35:49–62 (in Japanese with English abstract).

Kruitwagen G., Hecht T., Pratap H.B. and Bonga S.E.W. 2006. Changes in morphology and growth of the mudskipper (*Periophthalmus argentilineatus*) associated with coastal pollution. *Marine Biology* 149:201–211.

Kruitwagen G., Nagelkerken I., Lugendo B.R., Pratap H.B. and Bonga S.E.W. 2007. Influence of morphology and amphibious life-style on the feeding ecology of the mudskipper *Periophthalmus argentilineatus. Journal of Fish Biology* 71:39–52.

Kuciel M., Żuwała K. and Satapoomin U. 2013. Comparative morphology (SEM) of the peripheral olfactory organ in the Oxudercinae subfamily (Gobiidae, Perciformes). *Zoologischer Anzeiger* 252(4):424–430.

Laegdsgaard P. and Johnson C. 2001. Why do juvenile fish utilise mangrove habitats? *Journal of Experimental Marine Biology and Ecology* 257:229–253.

Lee H.J., Martinez C.A., Hertzberg K.J., Hamilton A.L. and Graham J.B. 2005. Burrow air phase maintenance and respiration by the mudskipper *Scartelaos histophorus* (Gobiidae: Oxudercinae). *Journal of Experimental Biology* 208:169–177.

Liem K.F. 1987. Functional design of the air ventilation apparatus and overland excursions by teleosts. *Fieldiana Zoology* 37:i–iii, 1–29.

Lorenz K. 1948. The natural science of the human species: An introduction to comparative research. *The "Russian Manuscript"* (1944–1948). Edited from the author's posthumous works by A. von Cranach. Cambridge: MIT Press, 1996.

Lorenz K. 1973. The Fashionable fallacy of dispensing with description. *Die Naturwissenshaften* 60:1–9.

MacNaughton R.B., Cole J.M., Dalrymple R.W., Braddy S.J., Briggs D.E.G. and Lukie T.D. 2002. First steps on land: Arthropod trackways in Cambrian-Ordovician eolian sandstone, southeastern Ontario, Canada. *Geology* 30(5):391–394.

Magnus D.B.E. 1970. Bewegungsweisen, Nahrungserwerb und Fortpflanzungverhalten des Schlammspringers *Periophthalmus kalolo* (Freilandaufnahmen). Publikationen zu Wissenschaftlichen Filmen, Sektion Biologie, Serie 14(7): Film D 1282. Institut für den Wissenschaftlichen Film (IWF), Göttingen.

Magnus D.B.E. 1972. On the ecology of the reproduction-behavior of *Periophthalmus kalolo* Lesson at the East-African coast. Paper presented at the XVII International Congress of Zoology, Monte-Carlo, S. 1 (S.d.) n. 6.

McNamara K. and Selden P. 1993. Strangers on the shore. *New Scientist* 139:23–37.

Michel K.B., Heiss E., Aerts P. and Van Wassenbergh S. 2015. A fish that uses its hydrodynamic tongue to feed on land. *Proceedings of the Royal Society, Series B* 282:20150057. doi:10.1098/rspb.2015.0057.

Montalto F., Parlange J.-Y. and Steenhuis T.S. 2007. A simple model for predicting water table fluctuations in a tidal marsh. *Water Resources Research* 43(3):W03439. doi:10.1029/2004WR003913.

Mukai T. and Sugimoto M. 2006. Genetic population structure of the mudskipper, *Periophthalmus modestus*, in Japan inferred from mitochondrial DNA sequence variations. *Japanese Journal of Ichthyology* 53(2):151–158.

Murdy E.O. 1989. A taxonomic revision and cladistic analysis of the oxudercine gobies (Gobiidae: Oxudercinae). *Records of the Australian Museum, Supplement* 11:1–93.

Nagelkerken I., De Schryver A.M., Verweij M.C., Dahdouh-Guebas F., van der Velde G. and Koedam N. 2010. Differences in root architecture influence attraction of fishes to mangroves: A field experiment mimicking roots of different length, orientation, and complexity. *Journal of Experimental Marine Biology and Ecology* 396:27–34.

Niedzwiedzki G., Szrek P., Narkeiwicz K., Narkeiwicz M. and Ahlberg P.E. 2010. Tetrapod trackways from the early Middle Devonian period of Poland. *Nature* 463:43–48.

Pearse A.S. 1929. Observations on certain littoral and terrestrial animals at Tortugas, Florida, with special reference to migrations from marine to terrestrial habitats. *Papers from the Tortugas Station of the Carnegie Institute of Washington* 391:205–223.

Pearse A.S. 1932. Observations on the ecology of certain fishes and crustaceans along the banks of the Matla River at Port Canning. *Records of the Indian Museum, Calcutta* 34:289–298.

Pearse A.S. 1933. The gobies at Paknam. *Journal of the Siamese Natural History Society, Supplement* 9:173–178.

Polgar G. 2004. Adattamenti all'ambiente peritidale di alcune specie malesi dei generi *Periophthalmus*, *Periophthalmodon* e *Boleophthalmus* (Pisces, Gobiidae, Oxudercinae, Periophthalmini): Ricerche ecologiche e messa a punto di un apparato per l'allevamento in condizioni semi-naturali. Laurea degree dissertation, University Sapienza of Rome.

Polgar G. and Bartolino V. 2010. Size variation of six species of oxudercine gobies along the intertidal zone in a Malayan coastal swamp. *Marine Ecology and Progress Series* 409:199–212.

Polgar G., Burreson E.M., Stefani F. and Kamrani E. 2009. Leeches on mudskippers: Host-parasite interaction at the water's edge. *Journal of Parasitology* 95(4):1021–1025.

Polgar G. and Crosa G. 2009. Multivariate characterization of the habitats of seven species of Malayan mudskippers (Gobiidae: Oxudercinae). *Marine Biology* 156(7):1475–1486.

Polgar G., Ghanbarifardi M., Milli S., Agorreta A., Aliabadian M., Esmaeili H.R. and Khang T.F. (2017) Ecomorphological adaptation in three mudskippers (Teleostei: Gobioidei: Gobiidae) from the Persian Gulf and the Gulf of Oman. *Hydrobiologia*. doi:10.1007/s10750-017-3120-8

Polgar G., Malavasi S., Cipolato G., Georgalas V., Clack J.A. and Torricelli P. 2011. Acoustic communication at the water's edge: Evolutionary insights from a mudskipper. *PLoS One* 6:e21434.

Polgar G., Sacchetti A. and Galli P. 2010. Differentiation and adaptive radiation of amphibious gobies (Gobiidae: Oxudercinae) in semi-terrestrial habitats. *Journal of Fish Biology* 77:1645–1664.

Ratnasingham S. and Hebert P.D.N. 2007. BOLD: The barcode of life data system (www.barcodinglife.org). *Molecular Ecology Notes* 7:355–364.

Retallack G.J. 1997. Early forest soils and their role in Devonian global change. *Science* 276:583–585.

Retallack G.J. 2001. *Scoyenia* burrows from Ordovician palaeosols of the Juniata formation in Pennsylvania. *Palaeontology* 44(2):209–235.

Retallack G.J. and Feakes C.R. 1987. Trace fossils evidence for Late Ordovician animals on land. *Science* 235:61–63.

Richards W.J. 2006. *Early Life Stages of Atlantic Fishes: An Identification Guide for the Western Central North Atlantic*. Boca Raton, FL: CRC.

Rolls H.J. 2014. Using otolith elemental composition to track the habitat use, movements, and life history patterns of common snook (*Centropomus undecimalis*) and red drum (*Sciaenops ocellatus*) in the Tampa Bay estuary. PhD dissertation, University of South Florida.

Sayer M.D.J. 2005. Adaptations of amphibious fish for surviving life out of water. *Fish and Fisheries* 6:186–211.

Schultze H.-P. 1999. The fossil record of the intertidal zone. In *Intertidal Fishes: Life in Two Worlds* (eds. M.H. Horn, K.L.M. Martin and M.A. Chotkowski). San Diego, CA: Academic Press, pp. 373–392.

Shimizu N., Sakai Y., Hashimoto H. and Gushima K. 2006. Terrestrial reproduction by the air-breathing fish *Andamia tetradactyla* (Pisces; Blenniidae) on supralittoral reefs. *Journal of Zoology* 269:357–364.

Standen E.M., Du T.Y. and Larsson H.C.E. 2014. Developmental plasticity and the origin of tetrapods. *Nature* 513(7516):54–58.

Stebbins R.C. and Kalk M. 1961. Observations of the natural history of the mudskipper. *Copeia* 1:18–26.

Takeshima M. 1996. Genetic divergence among populations of the mudskipper, *Periophthalmus modestus*, in East Asia. MSc dissertation, University of the Ryukyus.

Takita T., Larson H.K. and Ishimatsu A. 2011. The natural history of mudskippers in northern Australia, with field identification characters. *The Beagle, Records of the Museum and Art Galleries of the Northern Territory* 27:189–204.

Tamura S.O., Morii H. and Yuzuriha M. 1976. Respiration of the amphibious fishes *Periophthalmus cantonensis* and *Boleophthalmus chinensis* in water and on land. *Journal of Experimental Biology* 65:97–107.

Taylor J.R., Cook M.M., Kirkpatrick A.L., Galleher S.N., Eme J. and Bennet W.A. 2005. Thermal tactics of air-breathing and non air-breathing gobiids inhabiting mangrove tidepools on Pulau Hoga, Indonesia. *Copeia* 4:886–893.

Thacker C.E. 2009. Phylogeny of Gobioidei and placement within Acanthomorpha, with a new classification and investigation of diversification and character evolution. *Copeia* 1:93–104.

Tornabene L., Chen Y. and Pezold F. 2013. Gobies are deeply divided: Phylogenetic evidence from nuclear DNA (Teleostei: Gobioidei: Gobiidae). *Systematics and Biodiversity* 11(3):345–361.

Touchon J.C. and Worley J.L. 2015. Oviposition site choice under conflicting risks demonstrates that aquatic predators drive terrestrial egg-laying. *Proceedings of the Royal Society Series B* 282(1808):20150376. doi:10.1098/rspb.2015.0376.

Townsend K.A. and Tibbetts I.R. 2005. Behaviour and sexual dimorphism of the blue mudskipper, *Scartelaos histophorus* (Pisces: Gobiidae). *Proceedings of the Royal Society of Queensland* 112:53–62.

Tytler P. and Vaughan T. 1983. Thermal ecology of the mudskippers *Periophthalmus koelreuteri* (Pallas) and *Boleophthalmus boddarti* (Pallas) of Kuwait Bay. *Journal of Fish Biology* 23:327–337.

Van Tassell J. 2010. *Gobioides peruanus*. The IUCN Red List of Threatened Species 2010: e.T183981A8210729. http://dx.doi.org/10.2305/IUCN.UK.2010-3.RLTS.T183981A8210729.en (accessed April 2016).

Webb T.J. 2012. Marine and terrestrial ecology: Unifying concepts, revealing differences. *Trends in Ecology and Evolution* 27(10):535–541.

Yang K.Y., Lee S.Y. and Williams G.A. 2003. Selective feeding by the mudskipper (*Boleophthalmus pectinirostris*) on the microalgal assemblage of a tropical mudflat. *Marine Biology* 143:245–256.

Zhang J., Taniguchi T., Takita T. and Ali A.B. 2003. A study on the epidermal structure of *Periophthalmodon* and *Periophthalmus* mudskippers with reference to their terrestrial adaptation. *Ichthyological Research* 50:310–317.

CHAPTER 13

Care and Management in Captivity

Richard Mleczko and Hans-Georg Rupp

CONTENTS

13.1 INTRODUCTION

Mudskippers are a source of fascination for aquarists due to their atypical life history. There are 43 species of mudskippers (subfamily Oxudercinae) distributed in 10 genera, of which four genera are highly amphibious: *Boleophthalmus*, *Scartelaos*, *Periophthalmodon*, and *Periophthalmus*. The latter two genera are most frequently encountered in the aquarium trade. Common names used when referring to mudskippers are varied and include the following: mudhoppers, Johnny jumpers, kangaroo fish, google-eyed mangrove fish, and climbing fish. In Japan, where mudskippers are popular pets, they are known as "mutsugoro" (*Boleophthalmus* spp.) and "tobihaze" (*Periophthalmus* spp.). In Germany, they are known as "Schlammspringer" and "Glotzaugen," translated to "mud-hoppers" and "goggle-eyes," respectively.

Two of the earliest aquarium articles on mudskippers are by Simpson (1955) and Klausewitz (1968). Many early articles had errors, such as incorrect species names and distribution reports (Stratton 2001; Lucanus 1998; Editor 1976; Simpson 1955). Two of the first factually accurate articles are by Polunin (1972) and Murdy (1986). Hobbyists seeking further information on mudskippers turned to scientific articles by Murdy (1989) and Clayton (1993). Despite the longstanding interest in mudskippers, scientists and aquarists alike lament the paucity of data on the autecology of these fishes. Clayton (1993) even went as far as commenting that, "It is doubtful if there is any other group of fishes in which so much general interest is based on so little scientific knowledge." The advent of the Internet allowed aquarists to share their experiences in mudskipper husbandry and observations of mudskippers behavior. In 1995, the first website devoted to mudskippers—Richard's Mudskipper (first author of this chapter) and Brackish Goby Homepage—went live. Today, there are many websites and hobby groups maintained by both aquarists and scientists that act as hubs for information exchange.

13.2 MUDSKIPPER SPECIES IN THE AQUARIUM TRADE

Numerous species of mudskippers are imported for the aquarium trade each year (see Table 13.1). However, the identities of these fishes are questionable due to misidentifications by aquarist magazines and aquarium shops. In aquarium shops in Europe, *Periophthalmus barbarus* and *Parapocryptes serperaster* are readily available. The former species is imported from West African countries such as Gambia and Nigeria, whereas the latter is likely imported from India (Unidentified Fish Importer, personal communication). Aquarium shops in North America also import *Ps. barbarus* (Figure 13.1) from West Africa in addition to another commonly encountered species, *Periophthalmus argentilineatus*, (Chapter 12, Figure 12.1a) which is imported from Thailand. In both European and North American aquarium shops, *Periophthalmodon septemradiatus* (Figure 13.2) is at times available for purchase and imported from Vietnam (Gianluca Polgar, 2015, personal communication). Exported primarily from India, *Periophthalmus novemradiatus* (Figure 13.3) is now established in American, Japanese, and German markets (Monks 2006; personal observation). Although many mudskipper species are native to Australia, few are available in the aquarium trade. The most commonly sold species are *Ps. argentilineatus*, *Periophthalmus gracilis,* and *Periophthalmus minutus*. Occasionally, *Periophthalmus novaeguineaensis* and *Periophthalmodon freycineti* are sold but the latter is less popular due to its larger size of up to 30 cm in length (Mleczko 1998). *Periophthalmus modestus* (Figure 13.4) is native to Japan and kept by mudskipper enthusiasts there (personal observation).

Species that belongs to two mudskipper genera, *Periophthalmodon* and *Periophthalmus* are commonly available to hobbyists (Monks 2006; Schaefer 2005) and can accommodate well to aquaria. In their natural habitat, these fishes are vigorous hunters that prey on a wide variety of organisms, including, crabs, shrimps, small fishes, and terrestrial arthropods. *Periophthalmus argentilineatus* have also been reported to consume algae in aquaria (R. Mleczko, personal observation). Species of *Periophthalmus* are variable in size, but those sold in aquarium shops are typically 3–10 cm in length (R. Mleczko, personal observation). The body color of these two genera is generally brown, with bright spots and stripes on the head and dorsal fins. The first dorsal fin is usually more colorful than the second. In several species, the first dorsal fins of the males are slightly larger with elongate spines. Species of *Periophthalmodon* attain sizes larger than species of *Periophthalmus*. *Periophthalmodon schlosseri* (Figure 13.5) and *Pn. freycineti* may reach up to 30 cm in the wild (R. Mleczko, personal observation).

Mudskippers other than the two genera above are traded with varying regularity. *Apocryptes bato* (Figure 13.6), *Pa. serperaster,* and *Pseudapocryptes elongatus* (Figure 13.7) are encountered in aquarium shops occasionally. Species of *Apocryptodon*, *Boleophthalmus* (Figure 13.8), and *Scartelaos* (Figure 13.9) are rarely traded. These fishes feed on algae, which in the wild is skimmed

Table 13.1 **Summary of Available Information for Species of Oxudercine Gobies Available in the Aquarium Trade**

Species	Common Name(s)	Trade Status/ Availability	Ease of Maintenance	Water Conditions	Diet	Temperament	Remarks
Apocryptes bato	Ghost goby	Less Common	Relatively easy; should be given a ledge and deep area for swimming	Brackish, SG >1.005, T: 26–32°C	Live insects, worms, brine shrimp, commercial fish feed, dried shrimp, and bloodworms	Less aggressive	
Boleophthalmus boddarti	Google-eyed goby, comb-toothed mudskipper	Rare	Difficult	Brackish, SG >1.005, T: 26–32°C	Algae and *Spirulina* flakes	Unknown	Not suitable for the home aquarium due to its specific diet of diatoms
Parapocryptes serperaster	Mud-rill skipper	Rare	Relatively easy; should be given a ledge and deep area for swimming	Brackish, SG >1.005, T: 26–32°C	Live insects, worms, brine shrimp, commercial fish feed, dried shrimp, and bloodworms	Not aggressive	
Periophthalmus argentilineatus	Barred mudskipper, silver-lined mudskipper, and big-fin mudskipper	Common	Relatively easy to feed and maintain	Brackish, SG >1.005, T: 26–32°C	Live insects, worms, brine shrimp, commercial fish feed, dried shrimp, and bloodworms	Territorial and aggressive	Avoid overfeeding, tendency to be bullied by other larger fishes
Periophthalmus barbarus	Atlantic mudskipper, butterfly mudskipper	Common	Relatively easy to feed and maintain but individuals can be aggressive	Brackish, SG >1.005, T: 26–32°C	Live insects, worms, brine shrimp, commercial fish feed, dried shrimp, and bloodworms	Aggressive to conspecifics	
Periophthalmus chrysospilos	Gold-spotted mudskipper	Common	Relatively easy to feed and maintain	Brackish, SG >1.005, T: 26–32°C	Live insects, worms, brine shrimp, commercial fish feed, dried shrimp, and bloodworms	Territorial and aggressive	Not a hardy species
Periophthalmus gracilis	Pygmy mudskipper	Common	Relatively easy to feed and maintain	Brackish, SG >1.005, T: 26–32°C	Live insects, worms, brine shrimp, commercial fish feed, dried shrimp, and bloodworms	Territorial and aggressive	Avoid overfeeding, tendency to be bullied by other larger fishes
Periophthalmus kalolo	Common mudskipper	Common	Relatively easy to feed and maintain	Brackish, SG >1.005, T: 26–32°C	Live insects, worms, brine shrimp, commercial fish feed, dried shrimp, and bloodworms	Territorial and aggressive	Avoid overfeeding, tendency to be bullied by other larger fishes
Periophthalmus malaccensis	None	Common	Relatively easy to feed and maintain	Brackish, SG >1.005, T: 26–32°C	Live insects, worms, brine shrimp, commercial fish feed, dried shrimp, and bloodworms	Territorial and aggressive	Avoid overfeeding, tendency to be bullied by other larger fishes

(*Continued*)

Table 13.1 (*Continued*) Summary of Available Information for Species of Oxudercine Gobies Available in the Aquarium Trade

Species	Common Name(s)	Trade Status/ Availability	Ease of Maintenance	Water Conditions	Diet	Temperament	Remarks
Periophthalmus minutus	Red-finned mudskipper	Common	Relatively easy to feed and maintain	Brackish, SG >1.005, T: 26–32°C	Live insects, worms, brine shrimp, commercial fish feed, dried shrimp, and bloodworms	Territorial and aggressive	Avoid overfeeding, tendency to be bullied by other larger fishes
Periophthalmus modestus	Tobihaze (Japanese), Lai hau (Cantonese), and Malddug-mangdungo (Korean)	Common	Relatively easy to feed and maintain	Brackish, SG >1.005, T: 26–32°C	Live insects, worms, brine shrimp, commercial fish feed, dried shrimp, and bloodworms	Territorial and aggressive	Avoid overfeeding, tendency to be bullied by other larger fishes
Periophthalmus novaeguineaensis	New Guinea mudskipper	Rare	Relatively easy to feed and maintain	Brackish, SG >1.005, T: 26–32°C	Live insects, worms, brine shrimp, commercial fish feed, dried shrimp, and bloodworms	Less aggressive	Not a hardy species
Periophthalmus novemradiatus	Indian mudskipper, Indian dwarf mudskipper	Common	Relatively easy to feed and maintain	Brackish, SG >1.005, T: 26–32°C	Live insects, worms, brine shrimp, commercial fish feed, dried shrimp, and bloodworms	Not aggressive	Recommended for beginners given its ease of maintenance
Periophthalmodon freycineti	Giant mudskipper	Rare	Difficult	Brackish, SG >1.005, T: 26–32°C	Live insects, worms, brine shrimp, commercial fish feed, dried shrimp, and bloodworms	Territorial and aggressive	Too large for most home aquaria
Periophthalmodon schlosseri	Giant mudskipper	Rare	Difficult	Brackish, SG >1.005, T: 26–32°C	Live insects, worms, brine shrimp, commercial fish feed, dried shrimp, and bloodworms	Territorial and aggressive	Too large for most home aquaria
Periophthalmodon septemradiatus	None	Common	Relatively easy to feed and maintain	Brackish, SG >1.005, T: 26–32°C	Live insects, worms, brine shrimp, commercial fish feed, dried shrimp, and bloodworms	Territorial and aggressive	Males are moderately aggressive toward one another
Pseudapocryptes elongatus	Asian dragon goby	Rare	Relatively easy; should be given a ledge and deep area for swimming	Brackish, SG >1.005, T: 26–32°C	*Spirulina* flakes, tubifex worms, and sinking pellets	Not aggressive	Shells and PVC pipes can be used as hiding places
Scartelaos histophorus	Blue mudhopper, eel-skipper, and bearded goby	Rare	Relatively easy to feed and maintain	Brackish, SG >1.005, T: 26–32°C	Live insects, worms, brine shrimp, commercial fish feed, dried shrimp, and bloodworms; *Spirulina* and algae flakes	Territorial and aggressive	

Sources: From information augmented by personal observations and Schaefer, F., *Brackish-Water Fishes: All About Species, Care and Breeding*, AQUALOG Special, Rodgau, 2005.

Figure 13.1 Lateral view of *Periophthalmus barbarus* in an aquarium in Germany. (Courtesy of Frank Schaefer.)

Figure 13.2 Lateral view of *Periophthalmodon septemradiatus* in an aquarium. (Courtesy of Gianluca Polgar.)

Figure 13.3 A pair of *Periophthalmus novemradiatus* in the midst of dorsal fin display or "flagging," in an aquarium in Germany. (Courtesy of Hans-Georg Rupp.)

Figure 13.4 Lateral view of *Periophthalmus modestus* in an aquarium in Japan. (Courtesy of James Fatherree.)

Figure 13.5 Dorsal view of *Periophthalmodon schlosseri* in the wild from Malaysia. (Courtesy of Bernard Dupont.)

Figure 13.6 Lateral view of *Apocryptes bato* in an aquarium in Germany. (Courtesy of Frank Schaefer.)

Figure 13.7 Lateral view of *Pseudapocryptes elongatus* in an aquarium in Germany (Courtesy of Frank Schaefer.)

Figure 13.8 Lateral view of *Boleophthalmus boddarti* in an aquarium in Germany. (Courtesy of Frank Schaefer.)

Figure 13.9 Lateral view of *Scartelaos histophorus* in an aquarium in Germany. (Courtesy of Frank Schaefer.)

off the mud surface, thus making their husbandry impractical. *Zappa confluentus* has a very narrow distributional range and is not traded. Each of these eight genera is either uncommon in the wild or has a small distributional range. For these reasons, it is unlikely that they will ever become readily available in the aquarium trade.

13.3 MUDSKIPPERS IN CAPTIVITY

Mudskippers easily adapt to an aquarium setup that resembles their natural environment. The ideal setup is a paludarium that enables the mudskipper to spend time both in and out of the water. A spacious tank must be provided for these fishes, especially for species that consume prey out of water. Mudskippers are found in tropical and subtropical areas, in hot and humid habitats. Replicating these conditions in the paludarium is essential for their long-term health. In temperate environs or during colder months, paludariums must be kept heated. Below 15°C mudskippers become lethargic, at 10°C the fish become completely inactive, at 5°C they stop breathing (Tamura et al. 1976), and if kept this cold for too long they will die. Neither will they do well in excessively hot conditions; protracted exposure to water temperatures above 34°C will lead to death (Gordon et al. 1985). A lid on the aquarium is absolutely essential to maintain the warm, humid air these fishes need. A water temperature range of 26–32°C is optimal. They do well in water that has a specific gravity (SG) of 1.005–1.015 as this mimics the range found in nature (Monks 2006; Mleczko 1998). The water should be hard (not less than about 120 mg/L of calcium carbonate) and have a pH of about 8 (Mleczko 1998). Unfortunately, many aquarium stores sell mudskippers as freshwater fishes because fishes that do not require special care are more attractive to buyers. However, there are reports of mudskippers such as *Periophthalmus weberi* (Larson 2008) being found in freshwater and *Ps. barbarus* seeking out refuge in a chlorinated swimming pool in Benin, West Africa (Richard Markham, 2001, personal communication). In an ideal captive environment, mudskippers can be expected to live anywhere from 2 to 7 years (R. Mleczko, personal observation).

Mudskippers are best kept in their own tank, without any other non-amphibious fish species. This avoids the problem of providing sufficient water for companion fishes while maintaining dry areas for the mudskippers. Possible tank mates include other fishes able to tolerate brackish water such as *Poecilia* spp. and *Stigmatogobius* spp. for smaller paludariums, and *Toxotes* spp. and *Scatophagus* spp. for larger ones. When including larger fish species that are carnivorous, it is important to consider the sizes of other tank mates. Because they require saline conditions, mudskippers do not make good tank mates for newts, frogs, or turtles despite their similar need for a mix of aquatic and terrestrial environments.

13.3.1 Food and Feeding

Mudskippers readily consume live food such as small fishes, brine shrimps, small shrimps and crabs, worms, mosquito larvae, and small insects like flies. They will also eat dry fish food such as freeze-dried shrimp or krill, freeze-dried worms, frozen brine shrimp, sinking shrimp pellets, and flake food (Monks 2006). As with many other captive fishes, it is better to be cautious when feeding and provide them smaller amounts of food. Overfeeding leads to problems with digestion, in particular, acute constipation. When constipated, the anal region appears reddish and stretched and the mudskipper may experience a painful death. If overfed with dry food, expansion of food in the stomach may result in the fishes experiencing bloated guts. To avoid digestive issues, mudskippers should be fed every second day with a staple diet of live or frozen foods that they would encounter in nature. Prepared foods, such as flake and pellets, should be used to augment the diet only once or twice a week.

13.3.2 Diseases and Health Problems

Little is known of health problems in mudskippers compared to other aquarium fishes. As with other fishes maintained in brackish water, they appear less prone to bacterial infections when compared to those in freshwater. However, as with any captive fishes, wounds and ulcers must be promptly treated. Bacterial gill infections, in particular, can cause the gill filaments to fuse making gaseous exchange less efficient. Newly purchased mudskippers may carry a variety of parasites. In the wild, mudskippers have been found with copepods (Zeehan Jaafar, 2016, personal communication) and marine piscicolid leeches (Polgar et al. 2009). Among the more common parasites are protozoans in the gall bladder and gills (Kalavati et al. 2002) and copepods and trematodes adhering to the buccal cavity, gill arches, and gill filaments (Mhaisena et al. 1996). In addition, parasitic worms known as acanthocephalans can make their homes in the intestines (Mhaisena et al. 1996).

An indication of unhealthy mudskippers is when they chronically fail to keep their pectoral and caudal fins open. Normally, these fins are partially or fully spread out; fins that are "pinched" may mean that these fishes are experiencing discomfort (R. Mleczko, personal observation). This serves as a crude assessment method, and if found in mudskippers, detailed investigations into the fish and aquarium conditions must ensue. Altercations over territory and resources may lead to physical attacks that may in turn lead to stress and other health-related issues, especially if mudskippers are of disparate sizes (R. Mleczko, personal observation).

13.3.3 Behavior

Mudskippers are responsive and interactive pets and can be trained to approach the aquarist to obtain food. They will even rest on your finger or climb onto your hand (Figure 13.10). When the paludarium includes an extensive dry area, resident mudskippers appear more territorial. These fishes are aggressive toward each other as they vie for the coveted sections of the paludarium. As in the wild, mudskippers in aquaria will develop a hierarchy with the dominant fish, usually the largest one getting the best position and being first to the food. Mudskippers rarely cause serious injuries to each other during antagonistic displays. However, an individual was observed with one eye hanging onto the orbital socket by a thread of tissue, presumably having been injured during an altercation with its tank mates. The eye eventually fell off and the scar healed, and the fish remained healthy and lived on as a one-eyed mudskipper (Figure 13.11). On another occasion, a mudskipper grabbed another by its caudal fin and dragged it underwater. In smaller paludariums or those with higher stock densities, mudskippers appear more tolerant of one another, even sharing the same branch or piece of driftwood to rest on. Mudskippers are extremely interactive and their behavioral repertoire includes

Figure 13.10 Pet mudskipper *Periophthalmus* sp. on hand, Australia. (Courtesy of Tom Martin.)

Figure 13.11 Injured *Periophthalmus variabilis* in an aquarium in Germany. (Courtesy of Hans-Georg Rupp.)

nudging their tank mates, leaping into the air, flagging (Figure 13.3), attaching to the glass wall of the aquarium, and puffing up their gill chambers with air so as to float in the water. Digging of burrows and courtship displays can also be observed when the proper materials are provided in the tank. Feeding time is one of the most entertaining aspects to mudskipper husbandry as they are skilled and ferocious hunters with excellent eyesight. The smallest worm or fly will be found and a live moth will provide endless fun as the mudskippers battle and try to outsmart each other and their prey.

Mudskippers often raise and lower the first and second dorsal fins, a term known as "flagging" (Figure 13.3). Mudskippers flag as a threat display or to attract mates, thus, a mudskipper kept on its own will rarely flag. The first dorsal fin of many species is brightly colored, ranging from hues of reds to blues.

13.3.4 Reproduction

For mudskippers to successfully breed in captivity, the tank must include a tidal regime (see Section 13.4.2.6) and an adequate quantity of suitable substrate for burrows. Without these, most breeding attempts will have a high probability of failure (H-G. Rupp, personal observation). The Tokyo Sea Life Park achieved the first successful breeding of mudskippers in aquaria (Engelmann 2005). Rupp was successful in breeding mudskippers using a home setup (Rupp 2014, 2015a,b). Artificial tidal regimes ensure that the male can dig the burrow during low tide, when the mud flat dries out. For 10 days after spawning, during low tide, the male takes fresh air into the egg chamber where the eggs are attached to its ceiling and upper parts of the walls (Figure 13.12). During the simulated high tide, air-adding behavior is very rarely seen. Once fully developed, the male removes the air out of the egg chamber during the nocturnal high tide to trigger hatching by immersing the eggs in water after which the larvae leave the burrow.

Rupp experimented with artificial substrates (mixtures of sand and clay) but was only successful when mud from the tidal flats of the North Sea was used. Providing 15–25 cm of substrate within the aquarium allowed mudskippers to dig stable burrows that did not collapse when submerged. If cardboard is used to shield the tank walls at the height of the substrate, it sometimes encourages the mudskippers to build their egg chambers against the glass. In these instances, the spawning as well

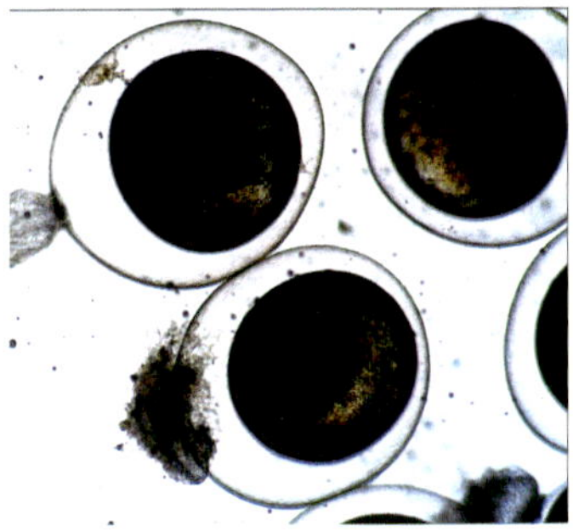

Figure 13.12 Eggs of *Periophthalmus variabilis*, obtained from an egg chamber within an aquarium in Germany. (Courtesy of Hans-Georg Rupp.)

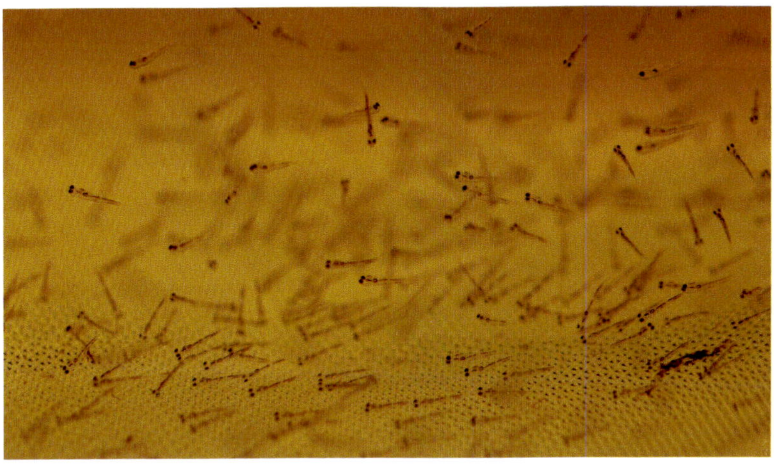

Figure 13.13 Larvae of *Periophthalmus variabilis* in an aquarium in Germany. (Courtesy of Hans-Georg Rupp.)

as the care of the brood and the hatching can be observed once the cardboard is removed, provided the water is not too turbid.

After leaving the burrow, the larvae are positively phototactic. Using light, these larvae must be carefully caught and transferred into a separate breeding tank, easily achieved if a tidal system is in place. For this purpose, it is necessary that the outflow from the basin be directed to a plankton net that is upstream from the sump in a separate container with a constant water level. The mesh for the net has to be chosen so that, while the larvae are contained (Figure 13.13), the mesh would not clog too quickly with residue from the turbid water. The larvae of *Ps. variabilis* and *Ps. novemradiatus* are about 3 mm upon hatching, with a head diameter of about 0.35 mm and 0.2 mm at the chest. A major difficulty is ensuring the correct rearing parameters, as one has difficulty observing the tiny larvae even with a magnifying glass. If the larvae are not observable under the water surface, one can only speculate about their state of health. After multiple rearing attempts under different conditions, the greatest success was achieved using a Kreisel tank (Figure 13.14) with a permanent light source from above the tank, combined with the green-water method (Kobayashi et al. 1972).

Once the yolk sac is consumed, a good starter food for mudskipper larvae is the rotifer *Brachionus plicatilis*. But two difficulties have to be managed: If the rotifers are artificially enriched in a separate vessel and later placed in the breeding tank, they will consume the available algae and will loose nutritional quality. Furthermore the water quality will decrease. To prevent the accumulation of high concentrations of NH_4^+ and NO_2 concentrations, a homemade trickle filter (Figure 13.15) with a

Figure 13.14 Kreisel tank to house larvae of *Periophthalmus variabilis* in an aquarium in Germany, here shown without green water. (Courtesy of Hans-Georg Rupp.)

plankton net to prevent the larvae from escaping is recommended. The larvae grow very slowly, measuring 5 mm and 8 mm in the third and fourth weeks respectively. When switching to *Artemia* nauplii for feeding the larger larvae offer both *Brachionus* and *Artemia* so that the larvae can adjust. Because it is unclear when the larvae start feeding on *Artemia* nauplii due to the poor water visibility and different growth rate between individuals (see Figure 13.16), a broad food range is recommended.

After approximately 6 weeks, metamorphosis begins for *Ps. variabilis*. At first, the mudskipper larvae are often attached to the wall of the breeding tank. They then detach from the tank wall and began leaving the water and move to exposed surfaces (Figure 13.17). About 3 months post-hatching, the juveniles will begin feeding on frozen bloodworms (Figure 13.18).

13.4 MUDSKIPPER HUSBANDRY

13.4.1 Paludarium Design

The environmental and physiological requirements of mudskippers call for specifically designed paludariums. The natural habitat of mangrove forests cannot be replicated in the paludarium but similar conditions can be achieved. Unlike other aquarium fishes, mudskippers can be viewed from

Figure 13.15 Kreisel tank with trickle filter, Germany. (Courtesy of Hans-Georg Rupp.)

the sides, and above the tank. Any large rectangular or square receptacle with one side open is suitable as a tank. Furthermore, these shapes maximize the separation between mudskippers and reduce territorial aggression. A tank with 2–5 cm of water is acceptable as a holding, quarantine, or transport tank. A permanent design that is quick to set up with minimum effort is a tank half filled with brackish water. Mudskippers naturally burrow so a soft substrate such as fine sand or mud is advised. While mud is most preferred by these fishes, it reduces visibility due to the fine sediments that remain suspended in the water column. Out-of-water areas are essential; so flat stones, driftwood, and a beach or mudflat area are highly recommended. These perch areas will allow mudskippers to rest while out of water. If mud or a mix of fine sand and mud is used as substrate, mangrove plants or those that can tolerate brackish water can be introduced into the tank.

Figure 13.16 Two individuals of *Periophthalmus novemradiatus* from the same clutch of eggs grow at different rates, Germany. (Courtesy of Hans-Georg Rupp.)

Figure 13.17 Newly-metamorphosed juvenile of *Periophthalmus variabilis* emerging from the water in an aquarium, Germany. (Courtesy of Hans-Georg Rupp.)

Figure 13.18 Thirteen week-old juvenile of *Periophthalmus novemradiatus* feeding on bloodworms in an aquarium in Germany. (Courtesy of Hans-Georg Rupp.)

Figure 13.19 A mudskipper, *Periophthalmus novemradiatus*, using an artificial burrow made from PVC pipes in an aquarium in Germany. (Courtesy of Hans-Georg Rupp.)

The sloping aquarium design is popular in paludariums with mudskippers. This design has deeper water at one end and the water depth gets progressively shallower toward the other, followed by a dry area. The deeper end can accommodate other fishes. Mudskippers will mostly utilize the shallower and dry end of the tank. The slope is achieved by increasing the height of substrate, from lowest at the water end to highest at the land end. Over time, the substrate, especially fine sand and mud, will move resulting in a slope of very shallow gradient. The point at which the slope deteriorates most is at the water–substrate interface. Other than periodic maintenance, a retaining wall or vertical partition can be installed to combat deterioration of the slope. At the water–land interface, a piece of plastic or glass can be glued in place either as a vertical wall or at an angle to retain the substrate. Alternatively, large rocks or driftwood can be wedged at the area to prevent the deterioration.

In the wild, mudskippers excavate burrows in the mud substrate. In the paludarium, these burrows or tunnels are viable only if the mud is of the correct consistency, which is difficult to achieve. The introduction of artificial burrows to the substrate can make good hiding places for mudskippers (Figure 13.19). This can be achieved through the use of plastic or polyvinyl chloride (PVC) pipes. Hollow ceramic objects or large shells partially buried in the substrate are also suitable. The diameter of the PVC pipes has to be considered so that the fish can turn around while inside the pipe.

13.4.1.1 Building a Paludarium

The following paludarium design is suggested specifically for mudskippers. Partition the tank such that one-third is filled with water while the remaining two-thirds is dry. Use a silicone sealer to mount in place a piece of glass or plastic as a partition separating these two areas. Ideally, the height of the partition is about a third of the tank height. Before mounting the partition consider transforming the smooth surface to one that is more realistic. This can be achieved by applying two-part epoxy resin or silicone rubber over the surface and sprinkling some sand over it. The sand will remain on the surface when dried. When the partition is ready, mount at the desired location. While the silicone is still wet, sprinkle with sand to attain better aesthetics as dried silicone is obvious under aquarium lightning. While both silicone and epoxy resin are safe for fishes, for good measure, flush with water before populating the aquarium.

Choose an appropriate substrate for the paludarium. Fine sand, mud, or a mix of both is recommended. If using sand, wash and flush until the water runs clear. Place the tank in the desired final location—once the substrate is added, the tank may be too heavy to move. Fill the water area of the paludarium with substrate to about a depth of about 2 cm. In the dry section, add the substrate until it reaches the top of the partition. Decorate the land area with driftwood, dried mangrove roots or rocks. Insert into the sand artificial burrows, if desired. Into the substrate, plant mangrove seedlings or other plants tolerant of brackish water (see Monks 2006). Add brackish water to the water area. Finally, ensure that the paludarium is fitted with a tight lid and is free of gaps. Any gaps for power leads and air hoses must be sealed. Mudskippers will escape from the aquarium given a chance, as they are able to scale wires, pipes, and the sides of glass tanks. Figure 13.20 shows the completed paludarium.

13.4.2 Ideal Environmental Conditions

13.4.2.1 Water Parameters

Water chemistry is more important when keeping other brackish fishes with mudskippers and less important in a mudskipper only setup (Monks 2006). This opinion is based on our many years

Figure 13.20 Top view of a completed paludarium made from Perspex with mangrove plants and driftwood added, while a Perspex partition separates the "land" and "water" components of the tank, Australia. (Courtesy of Richard Mleczko.)

of brackish water fishkeeping experience as well as numerous field trips to mudskipper habitats (Monks 2006: 195; Mleczko 2001, 1998). The following average set of parameters is recommended for a typical brackish water setup: SG between 1.005 and 1.015, water hardness 120–200 ppm calcium carbonate, water pH 8.0–8.5. In the wild, mudskippers occur in a wide salinity range; from freshwater (SG ≈ 0; Larson 2008) to hypersaline conditions (SG >1.027; Mleczko 1998) although laboratory experiments showed mudskippers generally favored water with some salt content (Gordon et al. 1985).

When growing brackish water and mangrove plants, additional parameters must be considered (Monks 2006: 202). In addition to waste products from the mudskippers and other fish that contribute as nutrients, plants also require iron and magnesium. Trace amounts of magnesium are naturally present in estuarine water, natural seawater, and artificial seawater. Epsom salts, essentially magnesium sulfate, can be added to the water should more magnesium be required. The addition of aquarium-safe plant fertilizer may also be beneficial. It is recommended to test the water regularly for nitrates and ammonia but this becomes less important if mangrove plants are used as they can efficiently remove nitrates and ammonia.

13.4.2.2 Heating

Heating is essential for paludariums in regions where the air temperature falls below 25°C. Any type of submersible aquarium heater with a thermostat is suitable as long as the power rating is high enough to maintain the water temperature within the range of 26–32°C. In the wild, mudskippers can be found in extreme temperature regimes. For example, in Kuwait, the temperature can vary from 8°C in winter to 45°C in summer (Clayton and Vaughan 1988), although the mudskippers themselves are usually only active when the air temperatures are between 13°C and 40°C. Through laboratory experiments, Gordon et al. (1985) demonstrated that mudskippers voluntarily seek out water temperatures over 34°C. When the water temperature is kept within the range stated above, the air temperature in the tank will usually be a few degrees lower than the water temperature, and this will result in a relative humidity of between 80% and 90%. Condensed water droplets will collect on the glass of the tank, and eventually, run down the glass. In our experience, mold or fungus rarely grows in the tanks, whereas algae will grow, which is eaten by mudskippers or other fishes.

13.4.2.3 Lighting

Lighting is less critical for the mudskippers, but important if plants are present in the paludarium. Whereas standard fluorescent aquarium lighting is acceptable for mudskippers, a fluorescent light with an output spectrum optimized for plant growth should be considered if plants are present. Getting enough light is critical for these plants. If the output of the fluorescent light is not sufficient, the use of metal-halide lamps may aid the situation.

13.4.2.4 Filtration

Water filtration in a paludarium housing only mudskippers may not be necessary. However, if the tank is stocked with other aquatic fishes, a filtration system should be considered. Almost any filter system can be used, except for an under-gravel filter, which would not be suitable if the above design is adopted. Some commonly used filtration systems are corner or box filter, a trickle filter system, or an external filter. A filter system will have the added benefit of circulating the water and thereby increasing heat transfer, and distributing waste products making the nitrogen cycle more efficient. Partial water changes are still recommended on a fortnightly basis whereby one-quarter of the water is replenished. If no filtration system is in place, partial water changes are essential. The presence of brackish water and mangroves plants will also act a chemical filtration system (Monks 2006). Aquarists who include mangrove plants in their paludarium report dramatic decreases in nitrates and phosphates. If using mud obtained from a mudflat for an attempt at breeding mudskippers (as discussed in Section 13.3.3), a sponge as part of the filtration system, cleaned regularly, will maximize the visibility of the water.

13.4.2.5 Aeration

Aeration is not critical as mudskippers are able to utilize atmospheric air. In the absence of a filter pump, aeration can circulate the water. An air pump is already a part of the corner or box filter system. Air pumps aid in maintaining a positive air pressure within the paludarium and also flush out any lingering odors.

13.4.2.6 Tide Simulation

Tide simulation to replicate the natural tidal conditions is necessary for breeding mudskippers in aquaria (see Section 13.3.3). The most specialized tidal simulation system consists of one or more paludariums stacked above a sump (Rupp 2013). The core of the system is a canister filter, which pumps the water via an external heater into the top tank. Except for the sump, each basin is connected with two bulkheads to the tank below it. The water flows from the sump into the canister filter and is then pumped into the top tank. Tides result from the arrangement of the pairs of bulkheads. The first is constructed as an overflow and is always open. It marks the high-tide level. Low tide happens every 12 hours when the second-lowest outflow is opened. The water flows for a period of 6 hours through an electrically operated ball valve, which is controlled by a timer (Figure 13.21). By changing the settings on the computer, alternating tidal cycles can be established for the tanks. Low tide in one or more tanks automatically creates high tide in the reservoir basin. Depending on the layout of each paludarium, and the resulting released or retained water volumes, two or more basins can be synchronized. However, the sump must never be allowed to overflow or completely drain. Additionally, the water that evaporates from the reservoir tank must be replaced regularly, as its level drops relatively quickly. This problem can be significantly reduced if the sump is designed as an aquarium, which permits a much larger store of water than what is needed for high tide conditions in the upper tanks. This design guarantees that the water is constantly filtered and heated while tidal cycles are simulated. The system can be optimized through the extraction of the biofilm in the reservoir at high tide, which can be achieved with a surface skimmer. Depending on the thickness and slope of the substrate layer, very different zones of wet and dry can be created, and the terrestrial part undergoes a major expansion at low tide.

An alternative design is a system consisting of one or more paludariums and a sump (Figure 13.22), but using two pumps controlled by timers. These are placed in the reservoir tank and run alternately for 6 hours each. Their flow rate is differentially adjusted by ball valves or performance. They pump water via a common hose and an external heater into the top tank; from there, it passes into the respective lower tank and, finally, via a trickle filter back into the sump.

Tides result from the two outlets, which are placed one above the other. If the low-capacity pump is operating, all the water exits from the smaller, lower placed overflow outlet (low tide). However, if the higher performance pump is operating, that outlet is too small to drain all the supplied water, so the water level rises gradually until it overflows over the top and creates high tide. Both pumps must have check valves to prevent the water from one pump returning through the other into the reservoir tank. A simplification of this design is to use a single powerful pump, which is switched on and off every 6 hours. Although the sump can be filtered around the clock with an additional canister filter, heating the individual paludariums is not possible at low tide unless using a cable heater inside the substrate or thermal radiation.

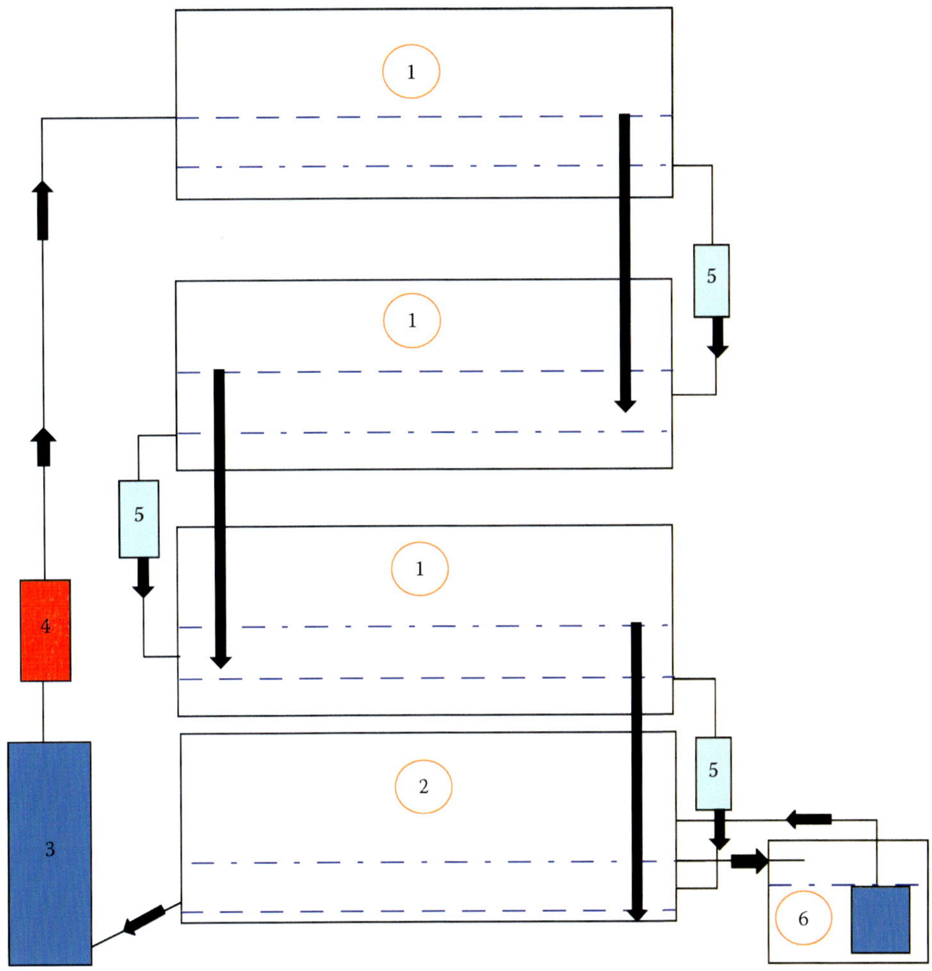

Figure 13.21 Typical three-tank setup with one sump. (1, Paludarium; 2, Sump; 3, Canister filter; 4, External heater; 5, Water timer; 6, Surface skimmer.)

13.4.2.7 Other Organisms in the Paludarium

When considering other animals to introduce to paludariums with mudskippers, start with those able to tolerate brackish water. Many estuarine gobies such as *Brachygobius* spp., *Redigobius* spp., or *Stigmatogobius* spp. are excellent candidates. For a complete list of compatible brackish water fishes, refer to Monks (2006). Crustaceans from a brackish environment, such as *Uca* spp. or *Pseudosesarma* spp. are also suitable, if they are large enough in relation to the mudskippers. Schaefer (2005) recommends against keeping crabs and mudskippers together as crabs have been reported as prey items for larger mudskippers. Indeed, especially during molting, crabs are very vulnerable and can lose extremities to the aggression of the mudskippers. However, Mleczko has kept them together and found that the mudskippers and crabs avoided one another with only infrequent interactions. On the other hand, when Mleczko introduced hermit crabs (*Pagurus sinuatus*) into the paludarium, they were predated on by the mudskippers. Mangrove snails are also excellent paludarium mates as they confer the added benefit of feeding and controlling algal growth. Due to the brackish water, terrapins, frogs, newts, salamanders, and axolotls are not suitable as paludarium mates for mudskippers.

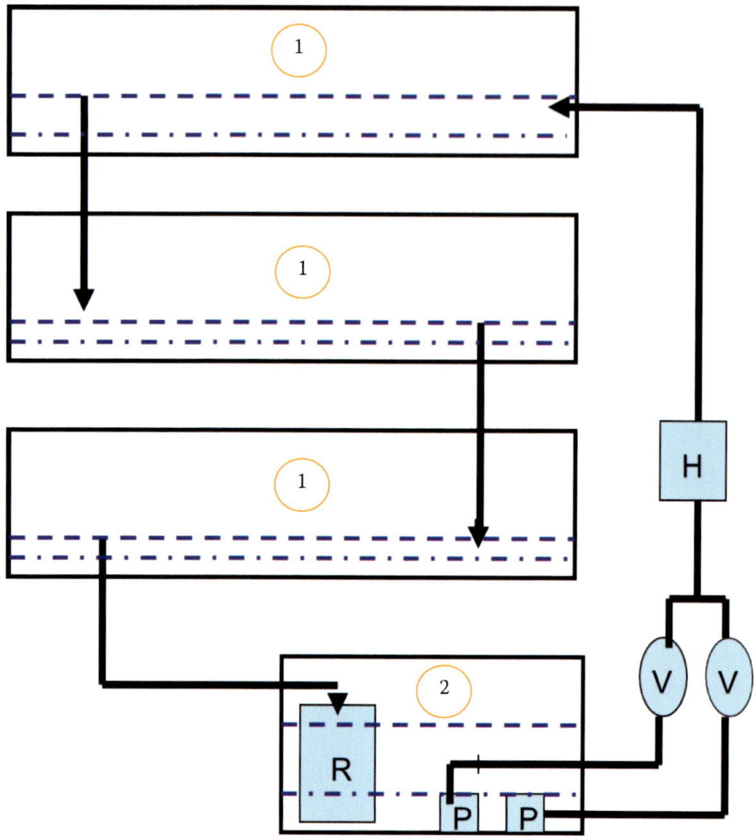

Figure 13.22 Simplified three-tank setup with one sump. (1, Paludarium; 2, Sump; H, Heater; V, Valve; P, Pump; R, Trickle filter.)

13.5 CLOSING REMARKS

Mudskippers are atypical fishes and require specialized attention that is significantly different from other aquarium fishes. Paludariums, with wet and dry areas, are more suitable for these fishes. Driftwood or large rocks make excellent out-of-water perches. Maintaining ideal paludarium conditions require effort but maintenance can be easy once a working system is achieved. How one creates the paludarium is only limited by the imagination. A trip to the coast, a mangrove forest or a mudflat will provide much inspiration for paludarium design. Breeding is possible if mud and a tidal system are implemented.

ACKNOWLEDGMENTS

The authors would like to thank most of the other chapter authors for their continued support over the last 20 years as information was very scarce in the early days of mudskipper keeping. We would also like to thank Frank Schaefer for the use of his photos and acknowledge his contributions to mudskipper keeping and the aquarium hobby.

REFERENCES

Clayton D.A. 1993. Mudskippers. *Oceanography and Marine Biology: An Annual Review* 31:507–577.

Clayton D.A. and Vaughan T.C. 1988. Ethogram of *Boleophthalmus boddarti* (Pallas), a mudskipper found on the mudflats of Kuwait. *Journal of the University of Kuwait (Science)* 15(1):115–138.

Editor. 1976. *Boleophthalmus pectinirostris. Tropical Fish Hobbyist* 25(4):83–84.

Engelmann W.E. 2005. *Zootierhaltung. Fische*. Frankfurt: Verlag Harri Deutsch, 890 pp.

Gordon M.S., Gabaldon D.J. and Yip A.Y. 1985. Exploratory observations on microhabitat selection within the intertidal zone by the Chinese mudskipper fish *Periophthalmus cantonensis. Marine Biology* 85:209–215.

Kalavati C., Dorothy K.P. and Pandian P.P. 2002. Gall bladder inhabiting Myxosporidia from fishes of Andaman Sea. *Journal of Parasitic Diseases* 26(2):89–100.

Klausewitz W. 1968. Some observations on East African mudskippers. *Tropical Fish Hobbyist* 16:52–65.

Kobayashi T., Dotsu Y. and Miura N. 1972. Egg development and rearing experiments of the larvae of the mud skipper, *Periophthalmus cantonensis. Bulletin of the Faculty of Fisheries, Nagasaki University* 35:49–62.

Larson H.K. 2008. Weber's mudskipper *Periophthalmus weberi*: New record for the Daly river. *Northern Territory Naturalist* 20:19–21.

Lucanus O. 1998. A fish out of water—Mudskippers. *Tropical Fish Hobbyist XLVI* 7:20–22.

Mhaisena F.T. and Al-Malikib N.S. 1996. Parasites, diseases and food of the dark-blotched mudskipper *Periophthalmus waltoni* (Perciformes: Gobiidae) in the Khor Al-Zubair estuary (Iraq). *Zoology in the Middle East* 13:85–88.

Mleczko R. 1998. The mudskippers and mangroves of Cairns. *Practical Aquariums and Water Gardens* 21:20–21.

Mleczko R. 2001. Mudskippers, part 2: Aquaria. *Practical Aquariums and Water Gardens* 36:8–10.

Monks N. (ed.) 2006. *Brackish Water Fishes: An Aquarist's Guide to Identification, Care & Husbandry*. New Jersey, NJ: TFH Publications, 384 pp.

Murdy E.O. 1986. Mudskippers of Malaysia. The lords of the mudflats. *Freshwater and Marine Aquarium* 9:20–23.

Murdy E.O. 1989. A taxonomic revision and cladistic analysis of the oxudercine gobies (Gobiidae: Oxudercinae). *Records of the Australian Museum*. (Suppl. 11):1–93.

Polgar G., Burreson E.M., Stefani F. and Kamrani E. 2009. Leeches on mudskippers: Host-parasite interaction at the water's edge. *Journal of Parasitology* 95(4):1021–1025.

Polunin I. 1972. Who says fish can't climb trees? *The National Geographic Magazine* 141:85–91.

Rupp H-G. 2013. Fiddler crabs in captivity: A real challenge. *Coral* 10(3):88–96.

Rupp H-G. 2014. Schlammspringer—Der mühsame Weg zur Nachzucht. *Koralle* 87, 15(3):58–67.

Rupp H-G. 2015a. Glorious mud. *Practical Fishkeeping* May 06, 2015:64–69.

Rupp H-G. 2015b. Schlammspringerzucht. *Aquaristik Fachmagazin* 243, 47(3):14–33.

Schaefer F. 2005. *Brackish-Water Fishes: All about Species, Care and Breeding*. Rodgau: AQUALOG Special, 80 pp.

Simpson D.A. 1955. Fish out of water. *Aquarium Journal* 26:181–183.

Stratton R.F. 2001. From fish to philosopher: The mudskipper. *Tropical Fish Hobbyist* XLIX 7:134–138.

Tamura S.O., Morii H. and Yuzuriha M. 1976. Respiration of the amphibious fishes *Periophthalmus cantonensis* and *Boleophthalmus chinensis* in water and on land. *Journal of Experimental Biology* 65(1):97–107.

CHAPTER 14

Taxa and Habitat Conservation

Beth Polidoro

CONTENTS

14.1 INTRODUCTION

Mudskippers (subfamily Oxudercinae) are a group of 43 species adapted to a variety of intertidal habitats with changing tides and variation in salinities, including mangrove areas, tidal pools, tidal mudflats, and freshwater coastal swamps. Mudskippers are distributed throughout the tropical and subtropical Indo-West Pacific, with only one species, *Periophthalmus barbarus,* present in the eastern central Atlantic. Throughout these regions, mangroves are dominant coastal features that provide important ecosystem services, including primary habitat for mudskippers.

Mangroves are woody plants, comprising 70 species globally, that are uniquely situated at the interface between land and sea in tropical and subtropical latitudes (Polidoro et al. 2010). Mangrove plant species have several anatomical and physiological adaptations to saline, water-saturated soils, disperse their seeds by water, and have specialized structures called pneumatophores or aerial roots that allow for aeration of roots (Figure 14.1a through c). The importance of mangrove forests to coastal ecosystems and human livelihoods are well documented. Mangrove forests provide critical wildlife habitats, support coastal fisheries, filter sediments and pollution, protect coastal zones from flooding and erosion, and sequester carbon (Bouillon 2011; Donato et al. 2011; Spalding et al. 2010; Mumby et al. 2004). Ecosystem services provided by mangrove forests are estimated to be worth at least US$1.6 billion per year worldwide (Field et al. 1998; Costanza et al. 1997).

(a)

(b)

Figure 14.1 (a) Inundated mangrove community at high tide, mouth of Sungai [=River] Api-Api, Singapore. (Courtesy of Zeehan Jaafar.) (b) Exposed mangrove community and mud flat at low tide, Waingapu, Sumba Island, Indonesia. (Courtesy of Zeehan Jaafar.)

(Continued)

(c)

Figure 14.1 (*Continued*) (c) Specialized aerial roots, or pneumatophores, facilitate gaseous exchange in hypoxic environments, Berlayar Creek, Singapore. (Courtesy of Zeehan Jaafar.)

Like mudskippers, mangrove plants can exist in varying conditions of high or low salinities, extreme tides, strong winds, high temperatures and muddy, anaerobic or hypoxic soils (Kathiresan and Bingham 2001). These unique environments allow mangroves to support rich assemblages of other species. The muddy or sandy sediments in mangrove forests support a variety of epibenthic, infaunal, and meiofaunal invertebrates, while inundated channel areas are home to diverse communities of phytoplankton, zooplankton, and fishes (Kathiresan and Bingham 2001). In some mangrove forests, mudskippers form mixed colonies with digging crabs (Sayer and Davenport 1991), and may exhibit habitat zonation (Colombini et al. 1995; Clayton 1993). Mangrove plants may also serve as a direct food source for some mudskipper species. For example, *Ps. barbarus* has been documented to feed on detritus from *Avicennia nitida* (Irvine 1947).

The burrowing activity of many species found in mangrove forests, including mudskippers, leads to fine-scale habitat heterogeneity, localized oxygenation of soil, and more dynamic flows of dissolved nutrients from the soil matrix (Hogarth 2007). For example, some mudskippers dig extensive burrows into mangrove sediments, and experience extremely hypoxic conditions during periods of low tide. To survive, *Periophthalmodon schlosseri* creates burrows as deep as 125 cm with special excavated chambers that hold pockets of air. *Pn. schlosseri* fills the pockets by gulping air at the surface and releasing the air in the subterranean chambers, thus escaping, rather than tolerating, hypoxic conditions (Ishimatsu et al. 1998). Such burrowing and air-trapping activities, particularly where mudskipper populations are dense, can significantly affect oxygenation and chemistry of mangrove sediments (Kristensen et al. 1995).

Like mudskippers, mangrove plants are adapted to different environments in the coastal zone, and display distinct community zonation from low to high water, or topography, based on changes in

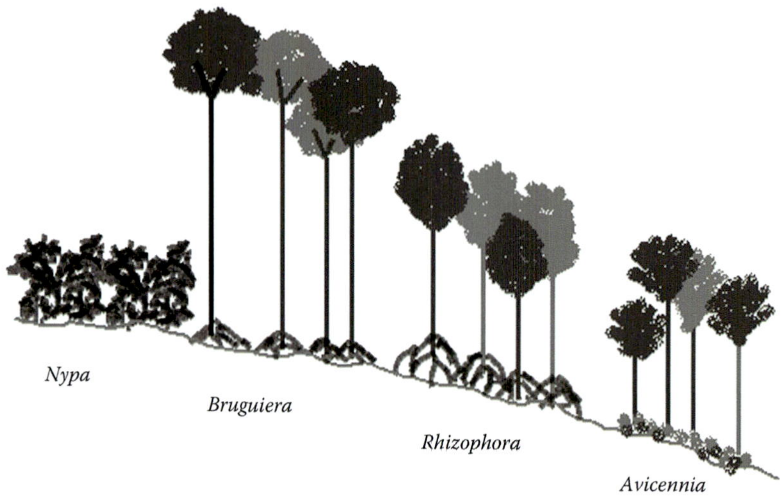

Figure 14.2 Simplified conceptual zonation model of floral mangrove community, landward to the left and seaward to the right. (Courtesy of Beth Polidoro.)

inundation and salinity (Figure 14.2). Some mangrove plants, such as *Bruguiera* spp, are more common in landward areas dominated by freshwater and upstream estuarine habitats. Species in these upstream estuarine or high intertidal regions often have specific freshwater habitat requirements and are usually patchily distributed (Polidoro et al. 2014; Duke et al. 1998). Other mangrove plants, such as *Avicennia* spp. and *Sonneratia* spp., have higher salinity and inundation tolerance, and are thus more common in the low intertidal zones and along coastlines (Polidoro et al. 2014; Dahdouh-Guebas et al. 2002). Many of the dominant mangrove plant species in these downstream estuarine and low intertidal regions have high salinity requirements and benefit from regular, if not constant, tidal flooding (Duke et al. 1998).

14.2 GLOBAL THREATS TO MANGROVE HABITAT

Mangrove forests around the globe are threatened by a number of impacts, including climate change, coastal development, pollution, aquaculture, and harvest of wildlife, fuelwood, and timber (Ellison 2015; Polidoro et al. 2010; Primavera et al. 2004; Valiela et al. 2001). In recent decades, mangrove habitats have been rapidly reduced and degraded (Giri et al. 2011; Spalding et al. 2010; FAO 2007), to current global estimates of less than 138,000 km^2 of mangrove forests remaining (Giri et al. 2011). Globally, more than 35% of mangrove area is estimated to have been lost since 1980, with mangrove deforestation continuing to occur at average rates of at least 1–2% per year (Alongi 2015, 2002). Of the 70 species of mangrove plants, 11 species (16%) are at elevated risk of extinction, primarily due to widespread clearing of mangrove habitats within their limited distributions, in combination with slow growth and relatively low regeneration rates (Polidoro et al. 2010). In general, tropical ecosystems are relatively slow to recover, with conservative estimates of at least 20–50 years for mangrove areas to show recovery from damage and degradation (Thorhaug 1989).

14.2.1 Removal of Mangrove Areas

Almost half (44%) of the world's population lives within 150 km of a coastline (Cohen et al. 1997). Heavily populated coastal zones have led to the widespread clearing of mangrove areas for coastal development or resource use. Around the globe, mangrove areas are routinely cleared to construct shrimp and fish farms, salt pans, urban and industrial areas, roads, ports, airports, tourist resorts, and for the development of rice, coconut, and other agricultural productions. Approximately 38% of global mangrove loss has been attributed to the clearing of mangrove areas solely for aquaculture (Ellison 2008). Additionally, approximately 26% of mangrove forests worldwide are degraded due to overexploitation for fuelwood and timber production (Valiela et al. 2001). Removal of mangrove areas directly or indirectly contributes to a number of coastal problems including: alteration of natural tidal flows, increased coastal damage during storms, alteration of the groundwater table, reduced water quality, release of toxic wastes, introduction of excess nutrients, and alteration of natural food chains (Alongi 2002).

Mangrove forests, as the epitome of the terrestrial–marine ecotone, effectively connect terrestrial and marine ecosystems. In many regions, the hydrology of terrestrial forests, including groundwater, stream, and overland flow, is directly connected to mangrove assemblages, which in turn are connected to near-shore seagrass beds and coral reefs. These hydrological connections are critical in sustaining the functionality of all of these ecosystems. Presently, many of these hydrologic connections are fragmented by deforestation of terrestrial and mangrove forests, urban encroachment, and other infrastructure, but are rarely functionally managed as interdependent ecosystems.

The majority of mangrove plant species that are at elevated risk of extinction have small distribution ranges in areas that are dominated by mangrove removal. For example, *Heritiera fomes* and *Heritiera globosa*, classified as Endangered on the IUCN Red List of Threatened Species, are associated with freshwater, upstream estuarine areas in south and Southeast Asia. Where they occur, an estimated 50–80% of mangrove areas were removed for coastal development, creation of fishponds, and the expansion of palm and timber plantations (Ong 2003). Of special concern is *Bruguiera hainesii*, another mangrove plant found in the high and intermediate intertidal estuarine zones. Although likely a hybrid between *Bruguiera cylindrica* and *Bruguiera gymnorhiza* (Ono et al. 2016), it has a current known population of fewer than 200 trees (75 in Malaysia, 120 in Papua New Guinea, and three in Singapore), due to extensive mangrove forest removal for coastal development within its distributional range. Similar to *Heritiera* spp., the recovery of *B. hainesii* is limited by its slow growth and low regeneration rate (Polidoro et al. 2010).

14.2.2 Coastal Pollution

Mangrove forests are also at the intersection of the flow of both terrestrial and marine pollution (Figure 14.3). Land pollutants (including pesticides, petrochemicals, industrial chemicals, heavy metals, and sewage) primarily move seaward, and into mangrove ecosystems through overland flow, coastal rivers, and streams. Marine pollutants (including marine debris, plastics, discarded and relict fishing gear, oil spills, and other associated contaminants) are washed ashore into mangrove and other coastal areas through wave, wind, and current action (Eriksen et al. 2014; Debrot et al. 2013; Burns et al. 1993). However, the distribution of the pollutants within the sediments, uptake and/or bioaccumulation by mangrove plants and other organisms, as well as the subsequent degradation and/or storage of these pollutants in the mangrove ecosystem are influenced by a number of different factors (Figure 14.4). These variables include the hydrology of the mangrove ecosystems (which can influence salinity, pH, the frequency and duration of inundation, and concentrations of dissolved oxygen), the properties of the sediments or soils, the class of pollutants (e.g., heavy metals, pesticides, petrochemicals), and climate variations in rainfall, temperature, erosion, and storm activity.

Despite filtering and removing contaminants through mangrove root uptake (Yang et al. 2008; Tam and Wong 1996; Clough et al. 1983), there is growing evidence that exposure to terrestrial and marine chemicals may reduce the survival and health of mangrove ecosystems (Lee 2009; Duke et al. 2005; Peters et al. 1997). However, the effects for the majority of pollutants on mangrove plants are largely unknown (Lacerda 1998); there are likely interspecific differences in chemical and concentration

Figure 14.3 Marine and terrestrial debris deposited in mangrove areas from wind, wave and tidal actions, Lim Chu Kang, Singapore. (Courtesy of Zeehan Jaafar.)

tolerances, and variations in natural conditions (such as salinity and inundation/oxidation) that influence chemical bioavailability and environmental behavior. Although mangrove habitats serve as a sink for trapping heavy metals and other pollutants, mangrove areas can rerelease heavy metals and other pollutants into the environment when disturbed by anthropogenic or natural threats.

Mangrove ecosystems are thought to be the most sensitive and vulnerable of tropical habitats to petrochemical or oil contamination (e.g., Lewis et al. 2011). The first casualties of oil in a mangrove forest are benthic and burrowing fauna, which can die on the day of oiling (Duke et al. 2000). Oil contamination from petroleum storage, refining facilities, near-shore oil exploration, and shipping, can occur in areas directly adjacent to mangrove habitats. In the low energy and anaerobic mangrove ecosystem, oil accumulation can be difficult to remove and slow to biodegrade (Lewis et al. 2011; Lewis 1979). There are several reported impacts of oil on mangrove plants, and include stunted growth, limb loss, defoliation, reduced germination, increased mutation, and lethality (Tam et al. 2005; Levings et al. 1997; Burns et al. 1994; Garrity et al. 1980a, 1994; Getter et al. 1980b). Recovery of mangrove habitats from a single catastrophic oil exposure may take 20 years or longer due to the long-term persistence of oil trapped in anoxic sediments (Burns et al. 1993).

Many studies have documented the ability of the fine-grained particles and clays in mangrove soils and sediments to sequester cadmium, chromium, mercury, copper, lead, zinc, and other trace metals (Lewis et al. 2011; Tam and Wong 1996; Silva et al. 1990; Harbison 1986). This uptake can lead to bioaccumulation of metals in mangrove leaves, stems, bark flowers and roots, and impact other

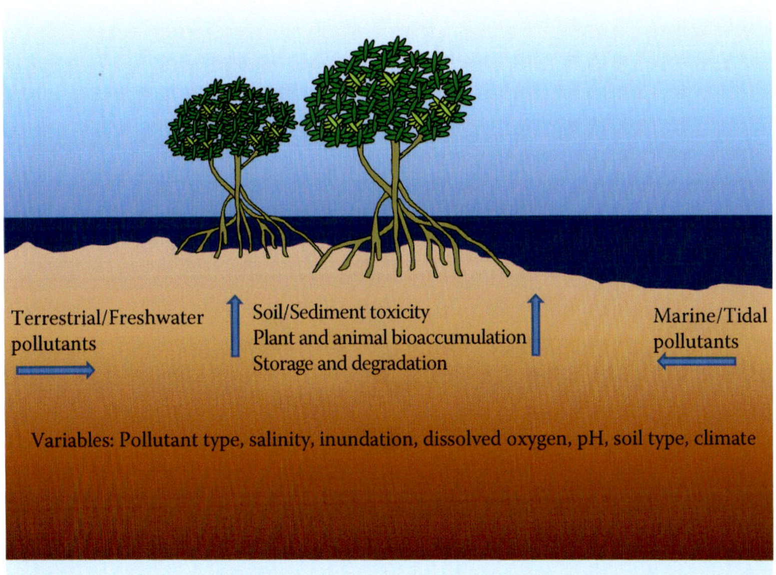

Figure 14.4 Pathways for pollutant and toxin accumulation in mangrove ecosystems. (Modified from Lewis, M. et al., *Environmental Pollution*, 159, 2328–2346, 2011.)

biota that interact with mangrove ecosystems, including fishes, invertebrates, and algae (Sandilyan and Kathiresan 2014; Lewis et al. 2011). However, there are very few studies on the impacts of land or marine-based pollutants on mangrove-associated biota. In the Mtoni estuary in Tanzania, a mangrove area with high concentrations of a variety of pollutants (e.g., heavy metals, organophosphate pesticides, petrochemicals, and polychlorinated biphenyls), a population of *Periophthalmus argentilineatus* showed suppressed growth and disturbance of eye development compared to other non-polluted mangrove areas (Kruitwagen et al. 2006). In controlled laboratory studies, Bob-Manuel (2012) showed that *Ps. barbarus* experienced high mortality when exposed to low concentrations of water-soluble diesel fuel, potentially similar to concentrations found around the Niger Delta.

Mudskippers have been proposed as potential bioindicators of pollutants in mangrove and other coastal areas (Ansari et al. 2014). Several studies have documented physiological changes in mudskippers when exposed to various pollutants, particularly heavy metals (e.g., Ansari et al. 2014; Ahmed et al. 2011; Bu-Olayan and Thomas 2008). Although mudskippers may be able to bioaccumulate relatively high concentrations of some heavy metals in their tissues (Polgar et al. 2010), they may show more sensitivity to other classes of pollutants, such as oil and petrochemicals, pesticides, and polychlorinated biphenyls (Feng et al. 2003; Nakata et al. 2003, 2002; Parmar and Patel 1993). With additional research, mudskippers may be able to serve as bioindicators for levels of selected contaminants in mangrove and other coastal ecosystems by measuring changes in their morphological, anatomical, physiological, and genetic parameters.

14.2.3 Climate Change and Sea-Level Rise

Climate change, including the impacts of sea-level rise, is a major threat to mangrove and coastal areas (Alongi 2015; Ellison 2015), primarily because it can disrupt the specific habitat requirements of mangrove biota (Gilman et al. 2008; Ellison 2005). Sea-level rise may cause the mortality of mangrove plants in present coastal zone positions, and result in their reestablishment at higher elevations. However, landward migration of mangroves can be blocked by topography, coastal development, or decreased recruitment (seedling growth) of species that normally occur lower in the tidal spectrum (Li et al. 2014; Ksiksi et al. 2012; Faraco et al. 2010; Di Nitto et al. 2008). Mangrove areas in low-relief islands in carbonate settings without freshwater input may be the most sensitive to sea-level rise, due to lack of sediment accretion (Ellison 2015). Direct removal of mangrove vegetation can also lead to erosion that can increase land submersion and loss, further exacerbating the impacts of sea-level rise (Hayden and Granek 2015).

Increased air and sea temperatures associated with climate change can alter respiration and photosynthesis of mangrove plants, resulting in reduced productivity at low latitudes and increased winter productivity at high latitudes (Ellison 2015; Alongi 2008). Other impacts of climate change include increased wave, wind, precipitation, and storm action leading to mangrove forest damage, erosion, and sedimentation (Alongi 2015; Ellison 2015; Faraco et al. 2010). Flooding, caused by increased precipitation, storms, or relative sea-level rise can also result in decreased productivity, photosynthesis, and survival (Ellison 2000). Large storms have resulted in mass mortality in more than 10 Caribbean mangrove forests in the last 50 years (Armentano et al. 1995; Jimenez et al. 1985). As with many ecosystems and species, the comprehensive impacts of climate change are not yet fully understood. Increased carbon dioxide concentrations in the atmosphere may also increase mangrove productivity by changing flowering and reproductive cycles, and expanding ranges of mangroves into higher latitudes (Ellison 2015).

14.3 REGIONAL MANGROVE HABITAT DECLINES

Mangrove habitats are found in 123 countries and territories globally (Spalding et al. 2010), with variable rates of loss and severities of local threats. The largest expanses of mangrove forests are found in four areas: the Sundarbans (Bangladesh and India), the Niger Delta (Nigeria), northern Brazil, and southern Papua New Guinea. However, more than 75% of mangroves are concentrated in just 15 countries with variable rates of mangrove area decline since 1980 (Table 14.1). Approximately

Table 14.1 Fifteen Countries with the Largest Cumulative Mangrove Areas in the Year 2010, and Percentage Loss of Mangrove Areas Since 1980

	Country	Mangrove Area in 2010 (km^2)	Percent of Global Total (%)	Approximate Decline Since 1980 (%)
1	Indonesia	31,129.9	22.6	31.0
2	Australia	9,779.8	7.1	0.5
3	Brazil	9,626.8	7.0	90.5
4	Mexico	7,419.2	5.4	27.0
5	Nigeria	6,536.7	4.7	0.2
6	Malaysia	5,053.9	3.7	16.2
7	Myanmar	4,945.8	3.6	8.7
8	Papua New Guinea	4,801.2	3.5	30.3
9	Bangladesh	4,365.7	3.2	0
10	Cuba	4,215.4	3.1	0
11	India	3,682.8	2.7	11.6
12	Guinea Bissau	3,386.5	2.5	23.9
13	Mozambique	3,188.5	2.3	3.1
14	Madagascar	2,780.8	2.0	9.1
15	Philippines	2,631.4	1.9	18.6

Source: Modified from Giri, C. et al., *Global Ecology and Biogeography*, 20, 154–159, 2011; FAO, *The World's Mangroves 1980–2005, FAO Forestry Paper 153*, Forest Resources Division, Rome, 2007.

42% of the world mangroves occur in Asia, followed by 20% in Africa, 15% in North and Central America, 12% in Oceania, and 11% in South America. Although there are likely several centers of origin for the evolution of mangrove plant species (Ricklefs and Latham 1993), the highest numbers of mangrove plant species occur in Southeast Asia (Polidoro et al. 2010). Mudskippers are found throughout the world's mangrove areas, but are absent from the mangrove areas around the islands of the Eastern Pacific, the Caribbean, and along the coasts of North and South America.

14.3.1 Mangroves of Asia

The Indo-Malay-Philippine archipelago, or Coral Triangle region, hosts the highest diversity of mangroves, with more than 35 species co-occurring in some areas (Abantao et al. 2015; Polidoro et al. 2010). This region has also experienced the highest proportion of mangrove area loss globally, with an average of more than 30% of mangrove area reduced across the region since 1980 (Polidoro et al. 2010; Spalding et al. 2010; FAO 2007; Figure 14.5). Coastal development and the conversion of tidal wetlands to fish ponds are the primary causes of mangrove removal in many areas of Southeast Asia (Primavera et al. 2004; Figure 14.6). Mangroves in this region are also particularly vulnerable to fuelwood and timber harvesting, as well as increasing coastal pollution and heavy metal contamination from increasing industrialization (Sandilyan and Kathiresan 2014).

Per country, some of the highest rates of loss have occurred in Vietnam (42%), China (34%), and Indonesia (31%) over the past 30 years (FAO 2007; Figure 14.5). Localized loss is nowhere greater than in central Luzon, Manila Bay, Philippines where 99.8% of mangroves present in 1951 were completely extirpated by 1994 (Primavera 2000). Bangladesh stands out in the region, as the extent of mangrove forests in the Sundarbans has remained relatively constant since the 1970s (Giri et al. 2007), although large areas of mangroves have been degraded due to timber harvesting and reductions in freshwater input (Hoque et al. 2006). Although India has lost only approximately 12% of its mangrove area since 1980, remaining habitat is severely degraded (Kathiresan and Bingham 2001). For this reason, 67% of the mangrove plants, 52% of the macroalgae, 10% of the invertebrates, and 4% of the vertebrates found in mangrove ecosystems in India are considered nationally endangered due to the decline and degradation of mangrove habitat (Rao et al. 1998).

Of particular concern in this region may be the mudskipper, *Periophthalmus modestus*. This species occurs in intertidal and mudflat areas of China, Korea, Taiwan, and southern Japan, where some of the highest mangrove declines and rates of coastal pollution in the region have been recorded (Wang et al. 2013; Spalding et al. 2010; FAO 2007). Within its relatively restricted range, *Ps. modestus* is adversely impacted by shrimp farming and development of marshes, swamps, estuaries, and tidal mud flats (Wang et al. 2013; Biao and Kaijin 2007). It is also reportedly harvested for the aquarium trade and for traditional Chinese medicine (Tang 1987 in Ming-Chih 2013), although the impacts of these activities are unknown.

14.3.2 Mangroves of West Africa

Mangrove habitats across Africa have on average, declined by approximately 14% since 1980 (FAO 2007). The highest losses occurred in western central Africa, where more than 60% of mangrove habitats have been lost along the coasts of the Congo, Cote d'Ivoire, and Liberia (Figure 14.5). Approximately 12% of the world's mangrove areas occur along the coast of western and central Africa, with 5% of the world's mangrove forests found in and around Nigeria (Giri et al. 2011; Spalding et al. 2010), with the Niger River Delta representing one of the largest continuous tracts of mangroves in the world. In western central Africa, mangrove areas are important for fisheries and as a source of timber and fuelwood (Feka 2015; Feka and Ajonina 2011). The mangrove wood industry is a major source of income to many coastal communities, and a direct source of fuel for 30% of the coastal population (Feka 2015).

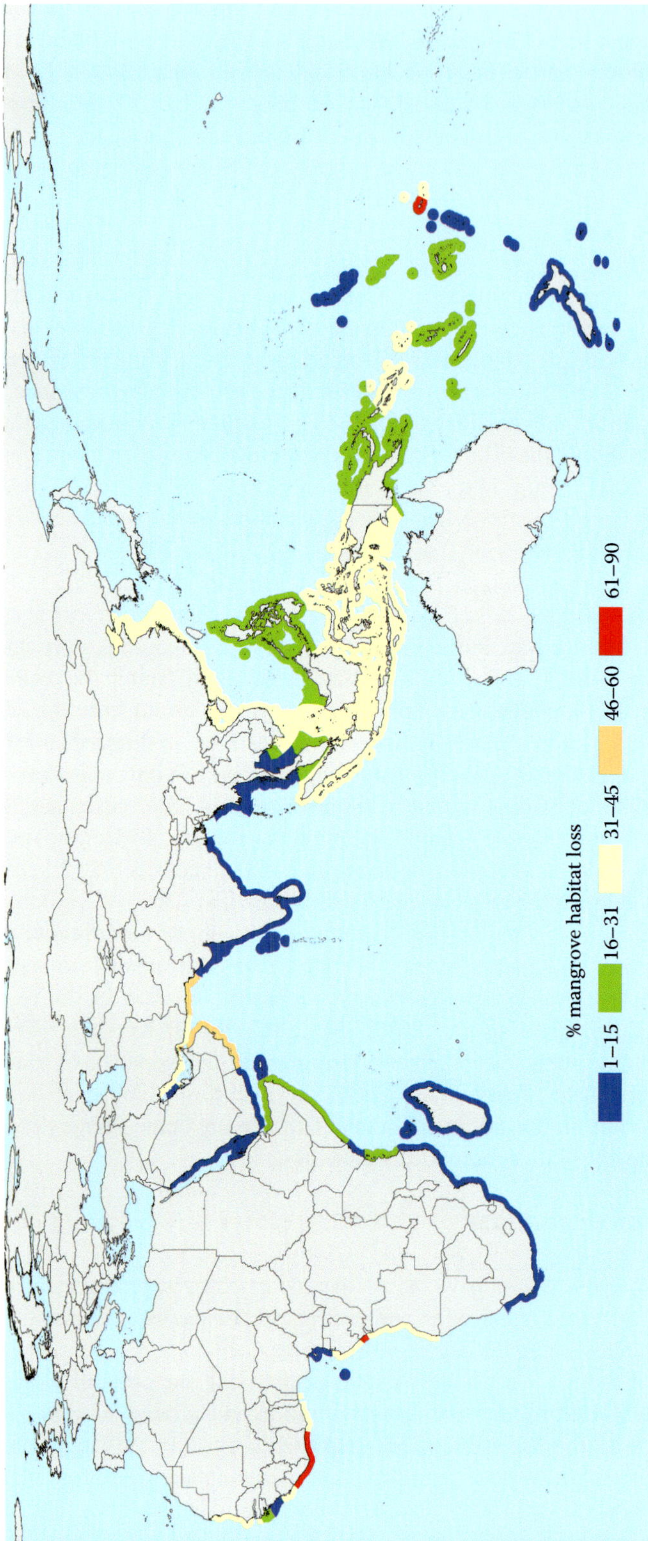

Figure 14.5 Percent of mangrove habitat loss since 1980 in countries with mudskipper populations. (Courtesy of Beth Polidoro.)

Figure 14.6 Estuarine mangrove area in Kuala Langat, Peninsular Malaysia, cleared for fish and shrimp ponds. (Courtesy of Zeehan Jaafar.)

Widespread losses have occurred from conversion of mangrove areas to urban or agricultural lands. Oil and gas extraction in the Bight of Biafra, which extends from the Niger River Delta to Gabon, has also severely degraded mangrove areas, drinking water, and fish stocks. Oil exploration and extraction activities in this area, including dredging of channels, digging new canals, clearance for platforms, pipelines, and seismic surveys have severally changed hydrology, salinities, and siltation rates (Spalding et al. 2010). Direct pollution from oil spills is a regular occurrence, with more than 2,000 incidents recorded from 1997 to 2001 (Nwilo and Badejo 2006). In the Niger Delta region, total hydrocarbons measured in sediment, surface water, and mudskippers exceeded the national permissible limit for inland waters and seafood (Clinton et al. 2009).

Periophthalmus barbarus, the only mudskipper species in this region, is relatively common, and occurs in fresh, marine, and brackish waters on exposed intertidal mudflats of mangrove forests, coastal lagoons, and estuaries from southern Morocco to Angola, including Sao Tome and Principe Islands (Stiassny et al. 2007). In the Niger Delta region, *Ps. barbarus* is harvested in subsistence fisheries and targeted for the commercial aquarium trade. Although little data are available on the volume and value of ornamental fishes traded in Nigeria, *Ps. barbarus* was exported at approximately US$1 per fish in 2006–2007 (Ukaonu et al. 2011). In the Imo River Estuary in southeast Nigeria, *Ps. barbarus* is overexploited, with 79% of the available stock fished annually (Udoh et al. 2013). In general, mudskipper fisheries, along with other tropical artisanal fisheries, are considered poorly managed (Udoh et al. 2013).

14.3.3 Mangroves of East Africa and Arabian (Persian) Gulf

In East Africa, mangrove forests are found along the entire coastline, including the western shores of Madagascar, and have suffered relatively fewer losses than other regions. The highest mangrove habitat declines along the east African coastline since 1980 within mudskipper distribution are 24% in Somalia and 18% in Tanzania (FAO 2007). Although variable, mangrove protection in this region is theoretically quite good, with all mangrove forests in Kenya and Tanzania under some degree of protection (Spalding et al. 2010). In reality, permits for mangrove plant extraction are regularly granted (Kairo 2003). As in West Africa, one of the most important threats to mangroves in East Africa is the selective harvesting of timber and fuelwood. Although entire forests are not cleared, selective logging has caused many mangrove areas to be severely degraded (Dahdouh-Guebas et al. 2004, 2000; Obade et al. 2004; Kairo 2003). Along the coasts of East Africa, two mudskipper species are present, *Ps. argentilineatus* and *Periophthalmus kalolo*, both widely distributed across the Indo-Pacific to Tonga and the Samoa Islands (Murdy 1989). These species may be found in the aquarium trade, but given their widespread distributions, commercial collections have local impacts or cause extirpations.

To the north, in the Arabian (Persian) Gulf and Gulf of Oman, arid climates and lack of freshwater input restricts mangrove areas to coastal lagoons and large alluvial fans associated with intermittent rivers. As they are some of the only trees in the region, mangrove plants have high ecological significance for a variety of organisms, including fishes, birds, and livestock (Spalding et al. 2010). Significant mangrove habitat losses have occurred primarily along the coasts of Iran and Pakistan, with 31% and 54% of mangrove areas lost respectively since 1980 (FAO 2007). In the Gulf region, mangrove areas are also degraded by wood extraction and aquaculture development. The risk of oil spills is also large, as the Strait of Hormuz receives heavy oil tanker traffic carrying 20–40% of global oil supply each year (Spalding et al. 2010).

The Gulf region is home to at least four species of mudskippers, two of which are known only from the Arabian (Persian) Gulf to the coasts of Pakistan: *Periophthalmus waltoni* and *Scartelaos tenuis*. In addition to mangrove areas, both species are also dependent on mud and sand flats, habitats which have experienced substantial deterioration in the region due to sea bottom dredging and coastal development for tourism and industry, both of which have resulted in changes of coastal water flow and sedimentation rates (Sheppard et al. 2010). Of special concern may be *Ps. waltoni*, which is the most common mudskipper in the region, but is reported to be one of the main fish products with relatively high market value in Khouzestan, Iran (Sary and Mohammadi 2012). Several studies in the Gulf region have documented mercury, heavy metal, and polycyclic aromatic hydrocarbon concentrations in *Ps. waltoni*, and proposed its use as a bioindicator due to its sensitivity to pollution (Sary and Mohammadi 2012; Shirani et al. 2012; Bu-Olayan and Thomas 2008).

14.3.4 Mangroves of Australia and Oceania

Mangrove area losses in Australia and the islands of Oceania have been relatively low over the past 30–40 years, with less than 1% loss in Australia reported since 1980 (FAO 2007). The most extensive mangrove areas are found in the wetter north and northeast of the country, which supports about 6% of the global mangrove area (Spalding et al. 2010). Sparsely human populated, the primary threats to mangrove areas in this region are destruction and deterioration from severe tropical cyclones (Spalding et al. 2010). However, many of the mangrove areas in the region are protected from anthropogenic threats through conservation zoning and establishment of protected areas such as Kakadu National Park and Cobour Peninsula Ramsar Site. Australia is home to one of the highest diversity of mudskippers globally, where 11–15 mudskipper species occur. A number of mudskipper species are endemic to this region including *Apocryptodon wirzi*, *Boleophthalmus birdsongi*, *Boleophthalmus caeruleomaculatus*, and *Periophthalmus darwini*.

Threats to, and protection of, mangroves in the island nations of Oceania are variable, as conservation is often a mix of formal legal measures and traditional management practices (Spalding et al. 2010). However, mangrove losses continue to occur due to harvesting for firewood, charcoal and timber, and in some regions, for large-scale commercial fuelwood enterprises (Senibulu 2000). Mangrove habitats are removed or deteriorated by conversion to agriculture, urban, and tourism areas, and because land is limited, sites for solid waste disposal (FAO 2007). In Oceania, mangrove-associated fisheries are of considerable importance, as many of the region's national economies and households are dependent on the marine resources that utilize mangroves. It is estimated that 60% of Fiji's commercial fisheries and 83% of subsistence species (fishes, crabs, prawns, molluscs, etc.) depend on mangrove areas at a point in their life history (Ellison 2004). Only two species of mudskippers occur in the islands of Oceania, *Ps. argentilineatus* and *Ps. kalolo*, both of which are widespread across the Indo-Pacific. With relatively few species of mudskippers and mangrove plants, the islands of the Western Pacific represent a terminal biogeographic margin for global mudskipper and mangrove plant distribution. Throughout the central Pacific, both mangrove plants and mudskippers are naturally absent. Although mudskippers are absent, areas in the Eastern Tropical Pacific and along the coasts of North and South America host different communities and mangrove plant species (Polidoro et al. 2010).

14.4 RECOMMENDATIONS FOR MUDSKIPPER AND MANGROVE CONSERVATION

Although the ranges of many mangrove plant and mudskipper species may overlap with marine and coastal protected areas, the level of protection and enforcement of these areas is widely varied. There are approximately 1,200 protected areas worldwide, which include mangrove areas, with 7–25% of global mangrove areas protected (Giri et al. 2011; Spalding et al. 2010). Additionally, there are 13 international treaties and instruments that afford some protection to mangrove areas in general, some of which have been in force for over 70 years (Table 14.2). However, these international treaties and instruments can also suffer from lack of implementation and enforcement, and do not necessarily confer legal protection to mangrove ecosystems. For these reasons, local and regional conservation actions, in combination with protective legislation and improved enforcement in protected areas, are critical to ensuring the maintenance and restoration of healthy mangrove ecosystems and mudskipper populations.

Table 14.2 International Conventions and Treaties That Include Protection of Mangrove Areas

International Convention or Treaty	Year
United Nations Convention on Biological Diversity	1992
United Nations Convention on Climate Change	1992
Convention on the Control of Transboundary Movements of Hazardous Wastes and their Disposal	1989
Convention for Protection and Development of the Marine Environment of the Wider Caribbean Region	1983
International Tropical Timber Agreement	1983
United Nations Convention on the Law of the Sea	1982
International Convention for the Prevention of Pollution from Ships	1973
Convention on International Trade in Endangered Species of Wild Fauna and Flora	1973
Convention on Prevention of Marine Pollution	1972
Convention Concerning the Protection of World Cultural and Natural Heritage	1972
Ramsar Convention on Wetlands of International Importance Especially as Waterfowl Habitat	1971
UNESCO Man and the Biosphere Program[a]	1971
Convention on Nature Protection and Wildlife Preservation in the Western Hemisphere	1940

Source: Modified from Ellison and Farnsworth, *Biotropica*, 28: 549–565,1996

[a] Not a legally binding convention, but a platform for international cooperation.

In some countries, such as in Pakistan, Cuba, and Bangladesh, the restoration and reforestation of mangroves has slightly increased overall mangrove areas (Alongi 2002). However, successful regeneration of mangrove areas is generally achieved by planting monocultures of fast-growing species, such as *Rhizophora* or *Avicenna* species. Many rare and slow-growing species, such as those found in upstream estuarine areas or with specific environmental requirements, are neither replaced, nor successfully replanted. In some cases, cleared mangrove forests fail to recover even after decades of abandonment, likely due to the changed hydrodynamics, salinity, acidity, and low nutrient levels associated with the initial removal or disturbance of mangroves (Farley et al. 2010). In addition, the long-term effects of mangrove reforestation on the accumulation, dissipation, and toxicity of heavy metals in mangrove sediments and mangrove organisms are largely unknown (Zhou et al. 2010). In sum, mangrove areas may be able to be rehabilitated in some regions through reforestation, but specific mangrove species and prior states of ecosystem health cannot likely be effectively restored (Polidoro et al. 2014).

In addition to increasing protection and restoration efforts of current mangrove areas, specific management strategies may help protect and preserve mangrove areas from the impacts of sea-level rise, increased severe storm events, and other anthropogenic causes of sediment loss and erosion. Natural factors that promote sediment accretion in mangroves include healthy adjacent seagrass and coral areas to protect mangroves, which reduce wave energy and supply organic matter and sand (Ellison 2015). Other management strategies to promote vertical accretion in mangroves include rehabilitation of degraded mangroves to promote growth of root mats, design of shore structures to allow for longshore sediment drift, design of dams to allow sediments to pass through, and control of boat wakes to reduce erosion (Ellison 2015).

Finally, with 16% of the world's mangrove plant species at the risk of extinction, and continued average decline of 1–2% in global mangrove area every year (Alongi 2015, 2002; Polidoro et al. 2010), more research is needed to determine the impact of mangrove forest loss and decline on the multitude of organisms that depend on these areas for all or part of their life history. For example, in a study of 850 terrestrial vertebrate species found in mangrove areas and with International Union for Conservation of Nature (IUCN) Red List of Threatened Species evaluations for extinction risk, approximately 40% were assessed as having an elevated risk of extinction due to extensive mangrove habitat loss (Luther and Greenberg 2009). However, as of 2016, only five of the 43 species of mudskippers have been evaluated for extinction risk under the IUCN Red List Categories and Criteria (IUCN 2016). As these unique fishes are largely restricted to mangrove and mangrove-associated habitats, assessments are needed to determine potential risk for each species. To proceed with sound management and conservation efforts, information is needed on the impacts of known threats (e.g., habitat loss and deterioration, pollution, harvesting for commercial fisheries and the aquarium trade, and the impacts of climate change) on local, regional, and global mudskipper populations.

REFERENCES

Abantao S.C., Apacible T.C., Cortez S.P., Pereda L.T. and Yllano O.B. 2015. Mangrove species diversity and on-site impact assessment of mangal coastal areas. *Expert Opinion on Environmental Biology* 4:3.

Ahmed K., Mehedi Y., Haque R. and Mondol P. 2011. Heavy metal concentrations in some macrobenthic fauna of the Sundarbans mangrove forest, southwest coast of Bangladesh. *Environmental Monitoring and Assessment* 177:505–514.

Alongi D.M. 2002. Present state and future of the world's mangrove forests. *Environmental Conservation* 29:331–349.

Alongi D.M. 2008. Mangrove forests: Resilience, protection from tsunamis, and responses to global climate change. *Estuarine, Coastal and Shelf Science* 76:1–13.

Alongi D.M. 2015. The impact of climate change on mangrove forests. *Current Climate Change Reports* 1:30–39.

Ansari A.A., Trivedi S., Saggu S. and Rehman H. 2014. Mudskipper: A biological indicator for environmental monitoring and assessment of coastal waters. *Journal of Entomology and Zoology Studies* 2:22–33.
Armentano T.V., Doren R.F., Platt W.J. and Mullins T. 1995. Effects of hurricane Andrew on coastal and interior forests of southern Florida: Overview and synthesis. *Journal of Coastal Research Special Issue* 2:111–114.
Biao X. and Kaijin Y. 2007. Shrimp farming in China: Operating characteristics, environmental impact and perspectives. *Ocean and Coastal Management* 50:538–550.
Bob-Manuel F.G. 2012. Acute toxicity tests of different concentrations of diesel fuel on the mudskipper, *Periophthalmus koelreuteri* (Pallas 1770): (Gobiidae). *Journal of Human Ecology* 39:171–174.
Bouillon S. 2011. Storage beneath mangroves. *Nature Geoscience* 4:1–2.
Bu-Olayan A.H. and Thomas B.V. 2008. Trace metals toxicity and bioaccumulation in mudskipper *Periophthalmus waltoni* Koumans 1941 (Gobiidae: Perciformes). *Turkish Journal of Fisheries and Aquatic Sciences* 8:215–218.
Burns K.A., Garrity S.D. and Levings S. 1993. How many years until mangrove ecosystems recover from catastrophic oil spills? *Marine Pollution Bulletin* 26:239–248.
Burns K.A., Garrity S.D., Jorissen F. et al. 1994. The Galeta oil spill. II. Unexpected persistence of oil trapped in mangrove sediments. *Estuarine, Coastal and Shelf Science* 38:349–364.
Clayton D.A. 1993. Mudskippers. *Oceanography and Marine Biology: An Annual Review* 31:507–577.
Clinton H.I., Ujagwung G.U. and Horsfall M. 2009. Evaluation of total hydrocarbon levels in some aquatic media in an oil polluted mangrove wetland in the Niger Delta. *Applied Ecology and Environmental Research* 7:111–120.
Clough B.F., Boto K.G. and Attiwell P.M. 1983. Mangroves and sewage: A re-evaluation. In *Biology and Ecology of Mangroves. Tasks for Vegetation Science* (ed. H.J. Teas) pp. 151–161. The Hague: Kluwer Academic Publishers.
Cohen J.E., Small C., Mellinger A., Gallup J. and Sachs J. 1997. Estimates of coastal populations. *Science* 278:1209–1213.
Colombini I., Berti R., Ercolini A., Nocita A. and Chelazzi L. 1995. Environmental factors influencing the zonation and activity patterns of a population of *Periophthalmus sobrinus* Eggert in a Kenyan mangrove. *Journal of Experimental Marine Biology and Ecology* 190:135–149.
Costanza R., d'Arge R., de Groot R. et al. 1997. The value of the world's ecosystem services and natural capital. *Nature* 387:253–260.
Dahdouh-Guebas F., Mathenge C., Kairo J.G. and Koedam N. 2000. Utilization of mangrove wood products around Mida Creek (Kenya) amongst subsistence and commercial users. *Economic Botany* 54:513–527.
Dahdouh-Guebas F., Van Pottelbergh I., Kairo J.G., Cannicci S. and Koedam N. 2004. Human-impacted mangroves in Gazi (Kenya): Predicting future vegetation based on retrospective remote sensing, social surveys, and tree distribution. *Marine Ecology Progress Series* 272:77–92.
Dahdouh-Guebas F., Verneirt M., Cannicci S., Kairo J.G., Tack J.F. and Koedam N. 2002. An exploratory study on grapsid crab zonation in Kenyan mangroves. *Wetlands Ecology and Management* 10:179–187.
Debrot, A.O., Bron, P.S., Leon, R. and Meesters, H.W.G. 2013. Marine debris in mangroves and on the seabed: largely-neglected litter problems: editorial. *Marine Pollution Bulletin* 72: 1–1.
Di Nitto D., Dahdouh-Guebas F., Kairo J.G., Decleir H. and Koedam N. 2008. Digital terrain modelling to investigate the effects of sea level rise on mangrove propagule establishment. *Marine Ecology Progress Series* 356:175–188.
Donato D.C., Kauffman J.B., Murdiyarso D., Kurnianto S., Stidham M. and Kanninen M. 2011. Mangroves among the most carbon-rich forests in the tropics. *Nature Geoscience* 4:293–297.
Duke N.C., Ball M.C. and Ellison J.C. 1998. Factors influencing biodiversity and distributional gradients in mangroves. *Global Ecology and Biogeography Letters* 7:27–47.
Duke N.C., Bell A.M., Pederson D.K., Roelfsoma C.M. and Nash S.B. 2005. Herbicides implicated as the cause of severe mangrove dieback in the Mackay region, NE Australia: consequences for marine plant habitats of the TBR World Heritage Area. *Marine Pollution Bulletin* 51:308–324.
Duke N.C., Burns K.A., Swannell R.P.J., Dalhaus O. and Rupp R.J. 2000. Dispersant use and a bioremediation strategy as alternate means of reducing the impact of large oil spills on mangrove biota in Australia: The Gladstone field trials. *Marine Pollution Bulletin* 41:403–412.
Ellison A.M. 2008. Managing mangroves with benthic biodiversity in mind: Moving beyond roving banditry. *Journal of Sea Research* 59:2–15.

Ellison A.M., Bank M.S., Clinton B.D. et al. 2005. Loss of foundation species: Consequences for the structure and dynamics of forested ecosystems. *Frontiers in Ecology and the Environment* 3:479–486.

Ellison J.C. 2000. How South Pacific mangroves may respond to predicted climate change and sea-level rise. In *Climate Change in the South Pacific: Impacts and Responses in Australia, New Zealand, and Small Island States* (eds. A. Gillepsie and W.C.G. Burns) pp. 289–301. Netherlands: Springer.

Ellison J.C. 2004. *Vulnerability of Fiji's Mangroves and Associated Coral Reefs to Climate Change: Review for the World Wildlife Fund.* Australia: World Wildlife Fund.

Ellison J.C. 2005. Holocene palynology and sea-level change in two estuaries in Southern Irian Jaya. *Palaeogeography, Palaeoclimatology, Palaeoecology* 220:291–309.

Ellison J.C. 2015. Vulnerability assessment of mangroves to climate change and sea-level rise impacts. *Wetlands Ecology and Management* 23:115–137.

Ellison A.M. and Farnsworth, E.J. 1996. Anthropogenic disturbance of Caribbean mangrove ecosystems: Past impacts, present trends, and future predictions. *Biotropica* 28:549–565.

Eriksen, M., Lebreton, L.C., Carson, H.S. et. al. 2014. Plastic pollution in the world's oceans: more than 5 trillion plastic pieces weighing over 250,000 tons afloat at sea. *PloS One* 9: e111913.

FAO. 2007. *The World's Mangroves 1980–2005, FAO Forestry Paper 153.* Rome: Forest Resources Division, FAO, 77 pp.

Faraco L.F.D., Andriguetto-Filho J.M. and Lana P.C. 2010. A methodology for assessing the vulnerability of mangroves and fisherfolk to climate change. *Pan-American Journal of Aquatic Sciences* 5:205–223.

Farley J., Batker D., De la Torre I. and Hudspeth T. 2010. Conserving mangrove ecosystems in the Philippines: Transcending disciplinary and institutional borders. *Environmental Management* 45:39–51.

Feka N.Z. 2015. Sustainable management of mangrove forests in West Africa: A new policy perspective? *Ocean and Coastal Management* 116:341–352.

Feka N.Z. and Ajonina G.N. 2011. Drivers causing decline of mangrove in West-Central Africa: A review. *International Journal of Biodiversity Science, Ecosystem Services and Management* 7:217–230.

Feng T., Guo X.Q., Zheng W.Y. and Yuan D.X. 2003. Detection of DNA strand breaks in the liver of *Boleophthalmus pectinirostris* treated with benzo (a) pyrene. *Bulletin of Environmental Contamination and Toxicology* 71:263–269.

Field C.B., Osborn J.G., Hoffman L.L. et al. 1998. Mangrove biodiversity and ecosystem function. *Global Ecology and Biogeography Letters* 7:3–14.

Garrity S.D., Levings S.C. and Burns K.A. 1994. The Galeta Oil Spill.1.The long-term effects on physical structure of the mangrove fringe. *Estuarine, Coastal and Shelf Science* 38:327–348.

Getter C.D., Snedaker S.C. and Brown M.S. 1980a. Assessment of biological damages at the Howard Star oil spill site, Hillsborough Bay and Tampa Bay, Florida. Report to Florida Department of Natural Resources, Tallahassee, FL, 65 pp.

Getter C.D., Nussman J.M., Gundlach E.R. and Scott G.I. 1980b. Biological changes of mangroves and sand beach communities at the Peck Slip oil spill site, Eastern Puerto Rico. RPI Report to NOAA. RPI/R/80/2/18-6. Office of Marine Pollution Assessment, Boulder, CO. Research Planning Institute, Inc, Columbia, South Carolina, 63 pp.

Gilman E., Ellison J., Duke N.C. and Field F. 2008. Threats to mangroves from climate change and adaptation options: A review. *Aquatic Botany* 89:237–250.

Giri C., Ochieng E., Tieszen L.L. et al. 2011. Status and distribution of mangrove forests of the world using earth observation satellite data. *Global Ecology and Biogeography* 20:154–159.

Giri C., Pengra B., Zhu Z., Singh A. and Tieszen L.L. 2007. Monitoring mangrove forest dynamics of the Sundarbans in Bangladesh and India using multi-temporal satellite data from 1973 to 2000. *Estuarine, Coastal and Shelf Science* 73:91–100.

Harbison, P.A.T. 1986. Mangrove muds—A sink and a source for trace metals. *Marine Pollution Bulletin* 17:246–250.

Hayden H.L. and Granek E.F. 2015. Coastal sediment elevation change following anthropogenic mangrove clearing. *Estuarine, Coastal and Shelf Science* 165:7–74.

Hogarth P.J. 2007. *The Biology of Mangroves and Seagrasses.* New York, NY: Oxford University Press, 273 pp.

Hoque M.A., Sarkar M.S.K.A., Khan S.A.K.U., Moral M.A.H. and Khurram A.K.M. 2006. Present status of salinity rise in Sundarbans area and its effect on Sundari (*Heritiera fomes*) species. *Research Journal of Agriculture and Biological Sciences* 2:115–121.

Irvine F.R. 1947. *The Fishes and Fisheries of Gold Coast.* London: Crown Agent, 352 pp.
Ishimatsu A., Hishida Y., Takita T. et al. 1998. Mudskippers store air in their burrows. *Nature* 391:237–238.
IUCN. 2016. IUCN Red List of Threatened Species. http://www.iucnredlist.org. (accessed January 1, 2016).
Jimenez J.A., Lugo A.E. and Cintron G. 1985. Tree mortality in mangrove forests. *Biotropica* 17:177–185.
Kairo J.G. 2003. Kenya. In *Report on the Africa Regional Workshop on the Sustainable Management of Mangrove Forest Ecosystems* (eds. D.J. Macintosh and E.C. Ashton) pp. 20–25. Accra: Center for African Wetlands.
Kathiresan K. and Bingham B.L. 2001. Biology of mangroves and mangrove ecosystems. *Advances in Marine Biology* 40:81–251.
Kristensen E., Holmer M., Banta G.T., Jensen M.H. and Hansen K. 1995. Carbon, nitrogen and sulphur cycling in sediments of the AO NAM BOR Mangrove forest, Phuket, Thailand: A review. *Phuket Marine Biological Centre Research Bulletin* 60:37–64.
Kruitwagen G., Hecht T., Pratap H.B. and Bonga S.W. 2006. Changes in morphology and growth of the mudskipper (*Periophthalmus argentilineatus*) associated with coastal pollution. *Marine Biology* 149:201–211.
Ksiksi T.S., Youssef T. and Abdelmawla E. 2012. Sea level rise and Abu Dhabi coastlines: An initial assessment of the impact on land and mangrove areas. *Journal of Ecosystem Ecography* 2:115.
Lacerda L.D. 1998. *Biogeochemistry of Trace Metals and Diffuse Pollution in Mangrove Ecosystems.* Okinawa: International Society for Mangrove Ecosystems, 61 pp.
Lee S.Y. 2009. Tropical mangrove ecology: Physical and biotic factors influencing ecosystem structure and function. *Australian Journal of Eco*logy 24:355–366.
Levings S.C., Garrity S.D., Van Vleet E.S. and Wetzel D.S. 1997. Sublethal injury to red mangroves two years after oiling. Proceedings of the 1997 International Oil Spill Conference. Washington DC: American Petroleum Institute, pp. 1040–1041.
Lewis M., Pryor R. and Wilking L. 2011. Fate and effects of anthropogenic chemicals in mangrove ecosystems: A review. *Environmental Pollution* 159:2328–2346.
Lewis R.R. 1979. Oil and mangrove forests: The aftermath of the Howard Star oil spill. *Florida Scientist* 42:26.
Li S., Meng X., Ge Z. and Zhang L. 2014. Vulnerability assessment of the coastal mangrove ecosystems in Guangxi, China, to sea-level rise. *Regional Environmental Change.* 15:265–275. doi:10.1007/s10113-014-0639-3
Luther D. and Greenberg R. 2009. Mangroves: A global perspective on the evolution and conservation of their terrestrial vertebrates. *BioScience* 59:602–612.
Ming-Chih H. 2013. Mudskippers in Tainan: The ecology and lifestyle of mudskipper *Periophthalmus modestus* in Tainan coastal wetland. *International Journal of Science and Engineering* 3:37–43.
Mumby P.J., Edwards A.J., Arias-Gonzalez J.E. et al. 2004. Mangroves enhance the biomass of coral reef fish communities in the Caribbean. *Nature* 427:533–536.
Murdy E.O. 1989. A taxonomic revision and cladistic analysis of the oxudercine gobies (*Gobiidae: Oxudercinae*). *Records of the Australian Museum* 11:1–93.
Nakata H., Sakai Y., Miyawaki T. and Takemura A. 2003. Bioaccumulation and toxic potencies of polychlorinated biphenyls and polycyclic aromatic hydrocarbons in tidal flat and coastal ecosystems of the Ariake Sea, Japan. *Environmental Science and Technology* 37:3513–3521.
Nakata H., Sakai Y. and Miyawaki T. 2002. Growth-dependent and species-specific accumulation of polychlorinated biphenyls (PCBs) in tidal flat organisms collected from the Ariake Sea, Japan. *Archives of Environmental Contamination and Toxicology* 42:222–228.
Nwilo P.C. and Badejo O.T. 2006. Impacts and management of oil spill pollution along the Nigerian coastal areas. *Administering Marine Spaces: International Issues* 119:1–15.
Obade P.T., Dahdouh-Guebas F., Koedam N., de Wulf R. and Tack J. 2004 GIS-based integration of interdisciplinary ecological data to detect land-cover changes in creek mangroves at Gazi Bay, Kenya. *Western Indian Ocean Journal of Marine Science* 3:11–27.
Ong J.E. 2003. Plants of the Merbok mangrove, Kedah, Malaysia and the urgent need for their conservation. *Folia Malaysiana* 4:1–18.
Ono J.O., Yong J.W.H., Takayama K. et al. 2016. *Bruguiera hainesii,* a critically endangered mangrove species, is a hybrid between *B. cylindrica* and *B. gymnorhiza* (Rhizophoraceae). *Conservation Genetics* 17:1–8.

Parmar P.G. and Patel S.K. 1993. Studies on acute toxicity of pesticide endosulfan on glycogen content of *Boleophthalmus dussumieri* (Cuv. and Val.) from Bhavnagar Coast. *Proceedings of the Academy of Environmental Biology* 2:59–62.

Peters E.C., Gassman N.J., Firman J.C., Richmond R.H. and Power E.A. 1997. Ecotoxicology of tropical marine ecosystems. *Environmental Toxicology and Chemistry* 161:12–40.

Polgar G., Sacchetti A. and Galli P. 2010. Differentiation and adaptive radiation of amphibious gobies (Gobiidae: Oxudercinae) in semi-terrestrial habitats. *Journal of Fish Biology* 77:1645–1664.

Polidoro B.A., Carpenter K.E., Collins L. et al. 2010. The loss of species: Mangrove extinction risk and geographic areas of global concern. *PLoS One* 5:1–10.

Polidoro B.A., Carpenter K.E., Dahdouh-Guebas F., Ellison J.C., Koedam N.E. and Yong J.W.H. 2014. Global patterns of mangrove extinctions and implications for coastal biodiversity loss. In *Coastal Conservation* (eds. J.L. Lockwood and B. Maslo). Rutgers University, NJ: Cambridge University Press.

Primavera J.H. 2000. Development and conservation of Philippine mangroves: Institutional issues. *Ecological Economics* 35:91–106.

Primavera J.H., Sadaba R.B., Lebata M.J.H.L. and Altamirano J.P. 2004. *Handbook of Mangroves in the Philippines—Panay*. Philippines: SEAFDEC Aquaculture Department and UNESCO Man and the Biosphere ASPACO Project, 106 pp.

Rao A.T., Molur S. and Walker S. 1998. *Report of the Workshop on Conservation Assessment and Management Plan for Mangroves of India, July 21–25, 1997*. Coimbatore, India: Zoo Outreach Organization, 106 pp.

Ricklefs R.E. and Latham R.E. 1993. Global patterns of diversity in mangrove floras. In *Species Diversity in Ecological Communities: Historical and Geographical Perspectives* (eds. R.E. Ricklef and D. Schluter) pp. 215–229. Chicago, IL: University of Chicago Press.

Sandilyan S. and Kathiresan K. 2014. Decline of mangroves–A threat of heavy metal poisoning in Asia. *Ocean and Coastal Management* 102:161–168.

Sary A.A. and Mohammadi M. 2012. Mercury concentrations in commercial fish from freshwater and saltwater. *Bulletin of Environmental Contamination and Toxicology* 88:162–165.

Sayer M.D.J. and Davenport J. 1991. Amphibious fish: Why do they leave water? *Reviews in Fish Biology and Fisheries* 1:159–181.

Senibulu M. 2000. The role of the national government in the economic development of mangroves in Fiji, presentation to the International Workshop Asia-Pacific Cooperation Research for Conservation of Mangroves, March 26–30, 2000. Okinawa, Japan.

Sheppard C., Al-Husiani M., Al-Jamali F. et al. 2010. The Gulf: A young sea in decline. *Marine Pollution Bulletin* 60:13–38.

Shirani M., Mirvaghefia A., Farahmanda H. and Abdollah M. 2012. Biomarker responses in mudskipper (*Periophthalmus waltoni*) from the coastal areas of the Persian Gulf with oil pollution. *Environmental Toxicology and Pharmacology* 34:705–713.

Silva, C.A.R., Lacerda, L.D. and Rezende, C.E. 1990. Metals reservoir in a red mangrove forest. *Biotropica* 22:339–345.

Spalding M., Kainuma M. and Collins L. 2010. *World Atlas of Mangroves*. London, UK: Earthscan, 319 pp.

Stiassny M.L.J., Teugels G.G. and Hopkins C.D. 2007. *The Fresh and Brackish Water Fishes of Lower Guinea, West-Central Africa*. Vol 1/II. Paris: IRD Éditions, 800 pp.

Tam N.F.Y. and Wong Y.S. 1996. Retention and distribution of heavy metals in mangrove soils receiving wastewater. *Environmental Pollution* 94:283–291.

Tam N.F.Y., Wong T.W.Y. and Wong Y.S. 2005. A case study on fuel oil contamination in a mangrove swamp in Hong Kong. *Marine Pollution Bulletin* 51:1092–1100.

Tang W.C. 1987. Chinese medicinal materials from the sea. *Abstracts in Chinese Medicine* 1:571–600.

Thorhaug A. 1989. Dispersed oil effects on tropical near shore ecosystems. *Oil Dispersants: New Ecological Approaches. ASTM STP* 1018:257–273.

Udoh J.P., Brownson I., Udo M.T. and Ofor C. 2013. Population dynamics of mudskipper, *Periophthalmus barbarus* (Linneaus 1766) (Teleostei: Gobiidae) in the Artisanal Fishery of Imo River Estuary, Southeast Nigeria. *Journal of Fisheries and Aquaculture* 4:148–153.

Ukaonu S.U., Mbawuike B.C., Oluwajoba E.O. and Williams A.B. 2011. Volume and value of ornamental fishes in the Nigerian export trade. *Agriculture and Biology Journal of North America* 2:661–664.

Valiela I., Bowen J.L. and York J.K. 2001. Mangrove forests: One of the world's threatened major tropical environments. *BioScience* 51:807–815.

Wang S.L., Xu X.R., Sun Y.X., Liu J.L. and Li H.B. 2013. Heavy metal pollution in coastal areas of South China: A review. *Marine Pollution Bulletin* 76:7–15.
Yang Q., Tam N.F.Y., Wong Y.S. et al. 2008. Potential use of mangroves as constructed wetland for municipal sewage treatment in Futian, Shenzhen, China. *Marine Pollution Bulletin* 57:735–743.
Zhou Y.W., Zhao B., Peng Y.S. and Chen G.Z. 2010. Influence of mangrove reforestation on heavy metal accumulation and speciation in intertidal sediments. *Marine Pollution Bulletin* 60:1319–1324.

CHAPTER 15

Challenges and Future Research: An Australian Perspective

Helen K. Larson

CONTENTS

15.1 MUDSKIPPERS IN AUSTRALIA

Mudskippers have been observed and recorded in Australia since 1770, but surprisingly little work has been carried out on their natural history and ecology. The presence of estuarine crocodiles and often difficult access to suitable habitats in remote regions may have something to do with this. Australian mudskipper species occur in mangroves and on mudflats from Shark Bay, Western Australia, to Brisbane, Queensland (Hoese and Larson 2006). Some have relatively small distributions (e.g., *Periophthalmus darwini* from Broome, Western Australia, to Melville Island and Darwin, Northern Territory) while others range widely (e.g., *Periophthalmus argentilineatus* from Shark Bay to Townsville). Of the six genera found in Australia, only *Periophthalmus* species are regularly observed. These are the most terrestrial species and tend to live over firmer substrates than the others, and thus come into contact with humans more readily.

The first published record of a mudskipper in Australia is that of Joseph Banks, who observed them in May 1770, at Thirsty Sound, northeast of Rockhampton, Queensland (Whitley 1970). The fish were probably *Ps. argentilineatus*, one of the most common and abundant species in Australia. Aboriginal rock art (which predates western records by thousands of years) portrays many identifiable fish species, but not mudskippers, probably because they are not desired as food nor linked to any stories (see Chaloupka 1993).

Of the 43 recognized oxudercine goby species known worldwide, 13 species are present in Australia, six of which are endemic to Australia and New Guinea: *Apocryptodon madurensis* of the Indo-west Pacific, *Apocryptodon wirzi* endemic to Australia and New Guinea, *Boleophthalmus birdsongi* endemic to Australia, *Boleophthalmus caeruleomaculatus* endemic to Australia, *Periophthalmodon freycineti* of the western Pacific, *Ps. argentilineatus* of the Indo-west Pacific, *Ps. darwini* endemic to Australia, *Periophthalmus gracilis* and *Periophthalmus minutus* of the

Figure 15.1 *Scartelaos histophorus* displaying on mudflat at low tide, Cairns Esplanade, Queensland, Australia. (Courtesy of Jeff Larson.)

western Pacific, *Periophthalmus novaeguineaensis* endemic to Australia and New Guinea, *Periophthalmus takita* endemic to Australia, *Periophthalmus weberi* of the western Pacific, and *Scartelaos histophorus* of the Indo-west Pacific (Figure 15.1).

15.2 MUDSKIPPER STUDIES

Of the oxudercine species now known, those from Australia, Indonesia, Japan, Kuwait, Malaysia, and Thailand that have been most studied. Limited work has been published on species from the African continent, India, Korea, New Guinea, and Singapore. A number of studies have been carried out using fish sourced through the aquarium trade or identified only as *Periophthalmus* sp. (e.g., Berti et al. 1992).

From an Australian viewpoint, we do not know much about the biology, reproduction, and population sizes of our oxudercine gobies. The amphibious mudskippers—*Boleophthalmus*, *Periophthalmodon*, *Periophthalmus*, and *Scartelaos*—are the best studied, as they can be observed with ease while on mudflats close to human populations. The most-cited (and possibly the least-often actually read) work is Norm Milward's PhD thesis (Milward 1974), in which he documents the ecology, distribution, food and feeding, and physiology of five species of Queensland mudskippers (that he identified as *Periophthalmodon schlosseri*, *Periophthalmus expeditionum*, *Ps. gracilis*, *Periophthalmus vulgaris*, and *S. histophorus*). Milward compared his Australian specimens to those held at the Australian Museum (Sydney), Natural History Museum (London), and Naturalis (Leiden). All the specimens he collected appear to be lost. However, one can determine what species he had from his descriptions. His *Pn. schlosseri* is now *Pn. freycineti*, *Ps. expeditionum* is *Ps. takita*, *Ps. vulgaris* is *Ps. argentilineatus*, whereas *Ps. gracilis* and *S. histophorus* remain the same.

Many of the sampling localities mentioned in Milward's work have since been modified. For example, the Cardwell foreshore in north Queensland no longer supports mudflats, especially after the serious cyclones of 2006 and 2011 (cyclones Larry and Yasi, respectively). The shore now consists mainly of granite boulders brought in to protect the nearby road, which was destroyed by Cyclone Yasi in 2011. Milward did note some erosion of Cardwell mudflats by weather and cyclones in 1971, 1973, and 1974 but indicates that *Scartelaos* populations always recovered, albeit slowly (1974:76–77). Some of the river mouths and other locations he visited and sampled are now modified by rock armor, boat ramps, marinas, and other tourist infrastructure, none of which assist the survival of mudflats. One of the most conspicuous of these modifications is at the Cairns foreshore along the Esplanade. The wide mudflats here not only support large populations of oxudercines but are also acknowledged as being valuable wintering/feeding habitats for migratory shorebirds (e.g., Ports North 2014; Birdlife Northern Queensland 2013). The council's signs along the Esplanade describe the activities and importance of these birds and fishes, yet it is presently destroying the site for them by bringing in more sand in an attempt to make a beach. Mudskippers are now almost completely absent from areas near the Esplanade pathway, which used to be the easiest site in Cairns to see *Pn. freycineti* and other mudskippers up close.

The most recent review of northern Australian amphibious mudskippers is that of Takita et al. (2011), which summarizes the natural history of nine of the 13 known species. However, there is a whole suite of "mud-slitherer" oxudercines that are poorly known, in Australia and elsewhere: *Apocryptes bato*, *Apocryptodon* spp., *Oxuderces* spp., *Parapocryptes* spp., *Pseudapocryptes* spp., and *Zappa confluentus*. These fishes are rarely truly terrestrial and are difficult to observe as they move through mud and muddy pools in turbid waters, often with head and eyes barely showing (or not at all). The only way to interact with them is to get into the mud with them and collect by hand or nets.

So what is actually known of Australian oxudercines? A survey of literature produced eight research papers specifically referring to Australian species. All of these concern the amphibious taxa (*Boleophthalmus*, *Periophthalmodon*, *Periophthalmus*, and *Scartelaos*), with the greatest amount of information given in Nursall (1981), Takita et al. (2011), and Townsend and Tibbetts (2005). Other publications list oxudercine species: various popular books, checklists, and reports (e.g., Larson et al. 2013; Hoese and Larson 2006; Allen et al. 2002), but almost nothing is recorded about the basic biology, distribution, and population of the other taxa, *Apocryptodon madurensis* (Australian populations may consist of two species) and *An. wirzi*, other than indicating that the species exist. These two species are classic mud-slitherers, usually only seen when the observer is sitting on very shallow mudflats, watching holes and turbid pools for movement. A New Guinea relative, *Zappa* from the Fly River estuary, resembles *S. histophorus* in its activity (Polgar et al. 2010; Murdy 1989).

Australia has three endemic species (*B. birdsongi*, *Ps. darwini*, and *Ps. takita*), whereas its immediate neighbor, New Guinea, has two (*Boleophthalmus poti* and *Z. confluentus*). Two species occur only in Australia and New Guinea, *B. caeruleomaculatus* and *An. wirzi*. In contrast, India has only two endemics (*Parapocryptes rictuosus* and the poorly known *Scartelaos cantoris*). No other countries have an endemic mudskipper.

Of the seven Australian–New Guinea endemic species, only the three Australian ones have been studied in any detail as to habitat requirements and life history. This work is summarized in Takita et al. 2011; the paper discusses nine of the mudskipper species known to occur in Australia (*B. birdsongi*, *B. caeruleomaculatus*, *Pn. freycineti*, *Ps. argentilineatus*, *Ps. darwini*, *Ps. minutus*, *Ps. novaeguineaensis*, *Ps. takita*, and *S. histophorus*). Polgar et al. (2010) examined the habitat differentiation of nine mudskipper species found in the Fly River delta region of New Guinea (*An. wirzi*, *B. caeruleomaculatus*, *B. poti* [reported as *Boleophthalmus* sp.]), *Pn. freycineti*, *Ps. darwini*, *Ps. novaeguineaensis*, *Ps. takita*, *Ps. weberi*, *S. histophorus*, and *Z. confluentus*). This study found some similarities (and differences) to that documented by Takita et al. (2011); the species assemblages differ somewhat although there are six shared species. The most terrestrial (upriver) New Guinea species (*Ps. weberi*) was tolerant of the widest range of conditions in habitat form and salinity (Polgar et al. 2010). In Australia, *Ps. weberi* has been less often recorded (and observed), with *Ps. minutus* being reported as the most terrestrial (inhabiting the highest tidal zone), although it does not travel as far upstream as does *Ps. weberi*. These two species in particular provide an interesting contrast in lifestyles in apparently hostile environments for a fish.

15.3 MUDSKIPPER CHALLENGES

Why is so little known of Australian mudskippers? A quick look at the various publications' authors and their affiliations gives a clue. Several people worked on the group during study for university degrees (and then worked elsewhere), the late Toru Takita of Nagasaki University regularly brought groups of students to work at a range of locations from Broome to Cairns, while I spent many years collecting gobioid fishes throughout the Northern Territory but with insufficient time and funding to work on mudskipper ecology. Another clue to the limited data on mudskippers is location. Mudskippers are mostly tropical, and in tropical Australia, there are several disincentives for researchers to enter mangroves and mudflats. The mud itself is often seen a major problem, as are crocodiles (Figure 15.2), biting insects and the difficulty of getting around in a substrate that may consist almost completely of mangrove pneumatophores, *Nypa* branches, or deep mud (Figures 15.3 and 15.4). Chief among these disincentives are estuarine crocodiles, which create considerable logistical problems even when researchers are in a small boat (Problem: How to do anything in a mangrove without becoming a prey item?). Spending a day sitting on an exposed mudflat in northern Australia, trying to observe "mud-slitherers" such as *Apocryptodon* is a similar logistical problem—one needs an assistant as crocodile-spotter. Another problem is the climate. Northern Australia has a tropical monsoonal climate, with intense "wet season" rains and cyclones making fieldwork difficult to quite impossible at times (Figure 15.5). Heavy seas can erode beaches and mudflats and even "normal" conditions can move sediments around, making long-term studies difficult.

Another issue is that of "health and safety," which seems straightforward, but Australia's Commonwealth and state governments have a range of rules, overseen by Safe Work Australia, which must be tracked. There are also animal ethics policies, legislation, and committees that must be followed. Each state has its own animal welfare or animal research Act that must fit within the Commonwealth's "Australian code for the care and use of animals for scientific purposes 8th edition" (as of 2015). These Acts variously apply to fishes, and there are exemptions under most Acts to permit recreational and commercial fishing to continue (in a manner that would not be permitted for a terrestrial vertebrate). After all the paperwork is done, one may still need a range of persons to assist (e.g., a watcher for crocodiles, a first-aider, and a heat stress expert).

Figure 15.2 Mudskipper research disincentive #1—*Crocodylus porosus*, Daintree River, north Queensland, Australia. (Courtesy of Jeff Larson.)

Figure 15.3 Mudskipper research disincentive #2—navigating mud, mangrove creek in Merbok mangroves, Kedah, Peninsular Malaysia. (Courtesy of Lynne Parenti.)

Figure 15.4 Mudskipper research disincentive #3—difficult habitat to move about in, Kuala Muda, Perak, Peninsular Malaysia. (Courtesy of Zeehan Jaafar.)

Destruction of mangrove areas hinders all work on mudskippers and all other estuarine organisms, although there is less destruction occurring in Australia when compared to many other countries such as Vietnam (its natural ecosystems severely damaged by wars). Australia is lucky in having the third largest extent of mangroves in the world, after Indonesia and Brazil (Duke 2006). However, as over 85% of the population lives near or on the coast (we are the world's driest inhabited

Figure 15.5 Mudskipper research disincentive #4—cyclones can wreck one's study area, Clump Point, Mission Beach, Queensland, Australia. (Courtesy of Helen Larson.)

continent), there is considerable pressure to develop more coastal land, despite some increase in awareness of mangrove systems. There is still a lack of understanding of the significance and value of mangrove systems in terms of "environmental services" (as environmental functions are called in an attempt to make them valuable in dollar terms to developers and politicians).

But the overriding problem is basically money, or lack of access to it. Funding for basic biological research has been slowly declining in Australia. Parliamentary Library estimates are that science research and development spending would be 0.56% of gross domestic product (GDP) in 2014–2015 (Parliamentary Library 2014). This is the equal lowest since records began in 1978–1979. The Organization for Economic Cooperation and Development has data for several countries, with the most recent for 2013. This shows Australia ranked only 18th out of 20 countries, with Australia's spending at 0.441% of GDP ahead only of Greece at 0.391% and the Slovak Republic at 0.369% (ABC Fact Check 2014). At the time of writing, at least AU$ 420 million has been cut from five key Australian science and research agencies, with the competitive grant system in serious slow decline. There are probably similar accounts of funding cuts and shifts in political attitude toward basic research in many other countries.

Another challenge to mudskipper research, less pronounced in Australia, but more severe throughout Southeast Asia, is that of finding unmodified and non-degraded mangroves and mudflat habitats. Mangrove forests are regularly destroyed for marinas, prawn farms, or housing estates, with mangrove creeks being put to a variety of uses (Figure 15.6).

A final challenge for mudskipper research, which is common to all fish studies, is that fishes are not perceived as "sexy" and thus less worthy of spending money, time, and effort on their study. Warm-blooded taxa continue to attract funding, especially so in Australia where cute furry animals such as koalas and kangaroos attract considerable media attention, and can influence political decisions as to where funds should be spent (Figure 15.7a,b). Non-furry animals, such as birds, frogs, snakes, butterflies, and fishes, usually get less attention and funding. Fishes have the additional disadvantage as being perceived as cold-blooded and incapable of feeling pain or possessing much intelligence, to their detriment (Brown 2015).

Figure 15.6 Relatively benign use of a mangrove stream as a boat landing site, Sungai Langat, Selangor, Malaysia. (Courtesy of Zeehan Jaafar.)

Funds for fish research are greatly oriented toward fisheries (easily perceived as being useful for humans) such as grants provided by the Fisheries Research and Development Council, while funding provided for exploratory and basic research (less easily perceived as being useful) is less. For example, the Australian Research Council provided AU$ 86.9 million in 2015 for wide-ranging research projects; only four involved fishes (on baitfish, inshore ecosystems, manta

Figure 15.7 (a) Marsupials are regarded as cute and loveable; Carol Walker with a juvenile Lumholtz tree kangaroo. (Courtesy of Jeff Larson.) (b) Mudskippers such as *Periophthalmus weberi* from Daly River, northern Australia, are not regarded in the same way even though they just as willingly perch on you. (Courtesy of Helen Larson.)

rays, and freshwater fishes). The Australian Biological Resources Study (ABRS) states that its aims are to discover what plants, animals, and other organisms occur in Australia; where are they found; and how can that information be provided to the public. ABRS does provide support and funding for research on Australia's flora and fauna, although the government has recently reduced this funding. "Nature discovery" projects such as Bush Blitz (http://www.bushblitz.org.au/) are focused on terrestrial organisms in existing reserves and national parks, with a few freshwater surveys being done—the usual challenges to mudskipper research apply (and it is unfortunate that few intertidal areas are included within Australia's National Reserve System, which Bush Blitz operates in).

Given all these challenges facing mudskipper research in Australia in particular, a few individuals manage to overcome them. The best example is that of the late Toru Takita. He noted (in 2011) that biological data on mudskippers were scarce and that his work was underpinned by his intention to "aid further investigations of mudskippers" and the need for further study (Takita et al. 2011). His collaborative approach meant that he encouraged groups of students and his colleagues Tatsusuke Takeda, Kiyoshi Soyano, and Sadao Kawaguchi to come to northern Australia, resulting in publications on ecology, physiology, natural history, and taxonomy (e.g., Takeda et al. 2012; Takita et al. 2011; Larson and Takita 2004; Zhang et al. 2000). All were generous with the information they gathered and the hypotheses they were putting together—I have a collection of handwritten notes and sketches from Takita, Takeda, and Kawaguchi, with my favorite being a conceptual model of the relationships between mangrove plants, mudskippers, and soil dynamics (Figure 15.8).

15.4 CONSERVATION ISSUES

Lack of funding means that up-to-date knowledge of the population status of known species is not always captured. Only five of the 43 known oxudercine species have been assessed (at the time of writing) under the International Union for Conservation of Nature (IUCN) Red List criteria for threatened species. They are *Apocryptodon punctatus*, *Boleophthalmus boddarti*, *Periophthalmus barbarus*, *Pseudapocryptes elongatus*, and *S. cantoris*. None are in a threatened category; four were listed as Least Concern. *Scartelaos cantoris*, apparently endemic to the Andaman Islands, is one of the most poorly known oxudercine species, with only the three type specimens known (Murdy 1989) and thus was listed as Data Deficient. These Red List assessments were done in 2009 (*Pseudapocryptes*), 2010 (*Apocryptodon*, *Periophthalmus*, *Scartelaos*), and 2012 (*Boleophthalmus*). More attention needs to be paid to the conservation status of these fishes, due to their dependence upon mangroves and intertidal soft substrate habitats, all of which are often modified or destroyed by human activity.

Although oxudercines are able to tolerate changes in temperature and salinity, some species are limited to particular habitats or zones within habitats, with the genus *Periophthalmus* being facultatively associated with mangroves. This dependence on specific habitats will acquire conservation significance if these habitats are threatened. And the question of climate change globally remains to be addressed.

Available mangrove habitats are declining, disappearing worldwide by 1–2% per year, with mangrove losses during the last quarter-century ranging between 35% and 86% (MacKinnon et al. 2012; Duke et al. 2007). In Australia, mangrove destruction is mostly due to clearing associated with coastal development (Duke et al. 2007; Duke 2006). A species-specific dieback of the widespread grey mangrove *Avicennia marina* was observed in Queensland, and affected more than 30 km^2 of mangrove forests in five separate estuaries, beginning in 1998 and apparently due to high concentrations of the herbicide diuron and excessive nutrients in runoff from agriculture (Duke et al. 2005).

Extensive intertidal mudflats, essential habitat for many oxudercines such as *Boleophthalmus* and *Scartelaos* (as well as for migratory shorebirds and uncountable numbers of invertebrates) are even less valued than mangroves. Intertidal flats are often heavily affected by human influence, but there is less certainty of their present distribution, status, and trends. However, current estimates of the rate of intertidal habitat loss in Asia are equal to or greater than recorded losses for mangroves, tropical forest, and seagrasses (MacKinnon et al. 2012). These losses have not slowed and their continued occurrence will eventually negatively affect all oxudercine populations depending upon them.

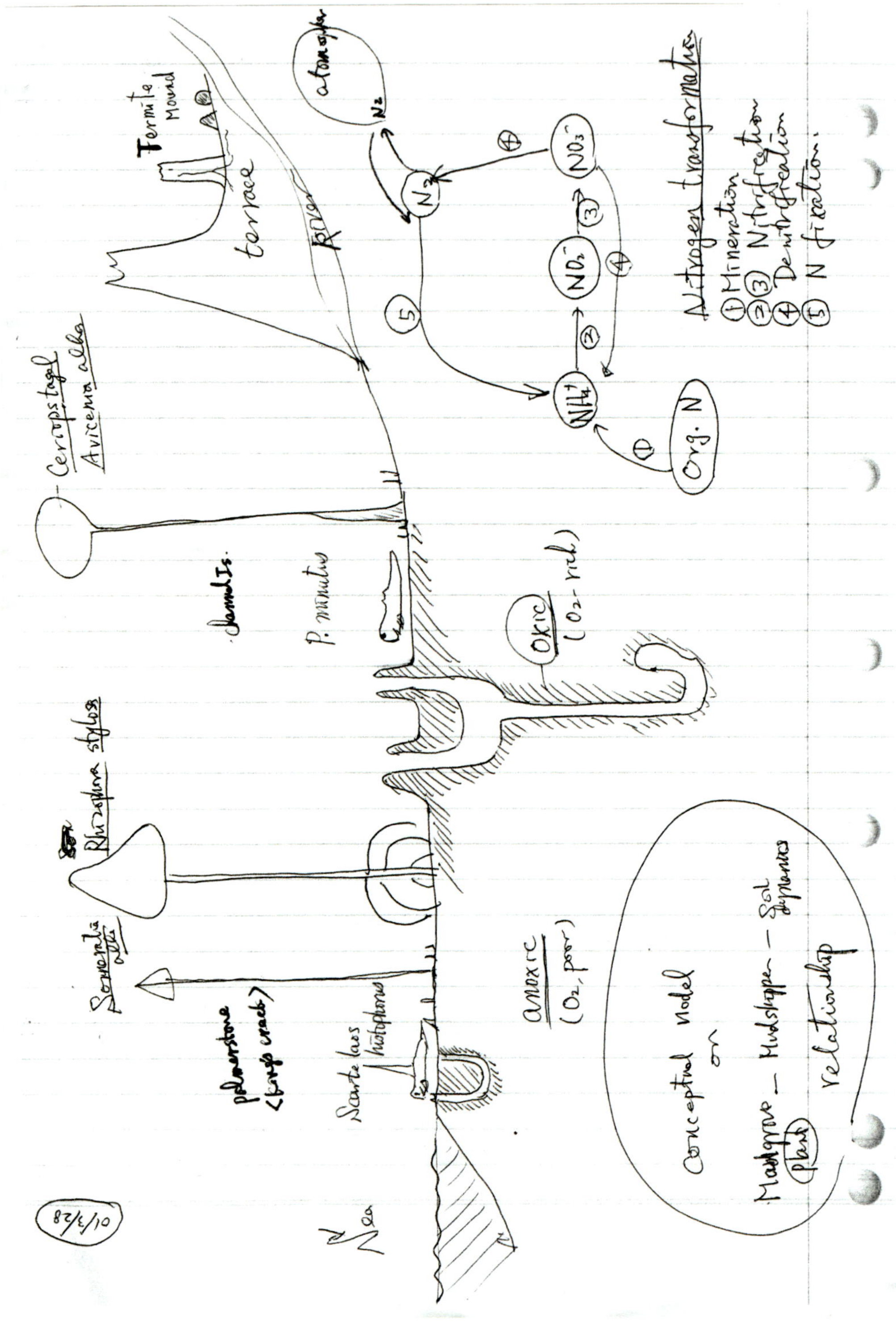

Figure 15.8 Conceptual model of relationships between mangrove plants, mudskippers, and soil dynamics, based on Northern Territory sites. (Courtesy of Sadao Kawaguchi.)

15.5 MUDSKIPPERS FOR THE FUTURE

How can mudskippers help us? What is their future? Several studies have shown (e.g., Polgar 2008; Kruitwagen et al. 2006) that mudskippers can be used as biomonitors for habitat degradation of mangroves and mudflat ecosystems, such as increase in abnormalities (e.g., anophthalmia) and changes in the species–area relationship. On the other hand, mudskippers make charismatic and appealing flagship species, highlighting the ecological value of threatened tropical coastal estuarine and mangrove ecosystems (Polgar and Sasekumar 2010). And of course, the same disincentives for research also hampers such use of these fishes—it is difficult to engage people with an animal that lives in an environment perceived to be hostile.

So there is much to be done. Ishimatsu and Gonzales (2011) listed seven "future research themes" in their chapter on mudskipper adaptations to terrestrial life. These were listed as follows:

1. Behavior of non-amphibious oxudercines should be investigated as some of these species exhibit partial amphibious activity. *Apocryptes*, *Oxuderces*, *Pseudapocryptes*, and *Zappa* have been observed by various workers to exhibit some amphibious behavior although that of *Apocryptes* has not been confirmed (Ishimatsu and Gonzales 2011). It is not clear from the few published works whether all species in each genus are amphibious. For example, I have never seen the northern Australian *An. wirzi* do anything other than burrow or swim though sloppy mud with only the eyes protruding (even when pursued with hands or net).
2. Burrow spawning and air storage needs further study, as not many species have been investigated. Ishimatsu and his colleagues first documented the storage of air in *Pn. schlosseri* burrows (Ishimatsu et al. 1998). Further investigations have shown that *Boleophthalmus pectinirostris* (Toba and Ishimatsu 2014) also store air in burrows, as do *Boleophthalmus dussumieri* (Ishimatsu et al. 1998), *Oxuderces nexipinnis* (reported as *Oxuderces dentatus*, Ishimatsu et al. 1998), *Ps. argentilineatus* (Brillet 1976), *Periophthalmus chrysospilos* (Ishimatsu et al. 1998), *Periophthalmnus magnuspinnatus* (Baeck et al. 2008), *Periophthalmnus modestus* (Ishimatsu et al. 2007), and *S. histophorus* (Lee et al. 2005). Is this behavior shared among all oxudercines? And are the details of these behaviors the same?
3. Potential for ovovivipary should be investigated, although it appears to be unlikely in mudskippers. It has been reported in *Ps. argentilineatus* (reported as *Ps. vulgaris*) by Harms (1934) although this was questioned by Ishimatsu and Gonzales (2011).
4. The potential for ammonia detoxification and tolerance should be studied in mudskipper embryos, as they develop in air, with no way to dispose of ammonia into the surrounding water as in most developing fishes. Ishimatsu et al. (2007) determined that *Ps. modestus* embryos took a week to develop before they hatched—what do they do with excess ammonia? Adults do have a high tolerance to excess ammonia and various ways of dealing with it (see Ishimatsu and Gonzales 2011).
5. How do newly hatched larvae escape safely from their hypoxic burrows; that is, do adults assist? Is the hatching time linked to tides? Etou et al. (2007) determined that *Ps. modestus* egg chambers contained hypoxic water (as do most intertidal burrows in soft sediment). This is a most interesting problem to be solved.
6. How do temperate climate mudskippers manage to overwinter in their burrows? It is known that some mudskippers use air storage in their burrows as a refuge during high tide (Toba and Ishimatsu 2014; Baeck et al. 2008; Lee et al. 2005). But what behavioral and physiological features do these temperate habitat mudskippers have that enable them to overwinter in burrows, such as *Ps. magnuspinnatus*, *Ps. modestus*, and *Periophthalmnus waltoni*?
7. What is the function of the elongate anterior narial tube that hangs down from a triangular fleshy part of the snout over the upper lip? This feature is an oxudercine synapomorphy (Murdy 1989). Possession of an elongate, downward-turned anterior narial tube is shared by a number of other gobioids, especially many species of amblyopine gobies. Many gobionelline gobies (*Brachygobius*, *Eugnathogobius*, *Chlamydogobius*, *Hemigobius*, *Luciogobius*, *Mugilogobius*, *Pseudogobius*, *Stigmatogobius*, and *Tamanka*) and most amblyopine gobies (*Amblyotrypauchen*, *Caragobius*, *Gymnoamblyopus*, *Odontamblyopus*, *Paratrypauchen*, *Pseudotrypauchen*, *Trypauchen*, and *Trypauchenichthys*) have similar overhanging anterior narial tubes with varying amounts of fleshy

tissue supporting the tube, although none have the distinct triangular fleshy flap surrounding the tube (Murdy 2003, 2006, 2008a,b; Murdy and Ferraris 2003; Murdy and Shibukawa 2001, 2002, 2003; Larson 2001). Several eleotrid fish genera also have very long anterior narial tubes that point downward, such as *Bostrychus, Eleotris*, and *Ophiocara* (Larson 2008; Hoese and Kottelat 2005; Akihito et al. 1988), while others do not, having shorter, upwardly or forwardly oriented narial tubes such as in *Butis*, *Hypseleotris*, *Odontobutis*, and *Ophieleotris* (Akihito et al. 1988). Why these structures are different is unknown, and there does not appear to have been any broad survey of them undertaken. Some of the gobionelline gobies and eleotrid fishes share the same mangrove habitats as mudskippers. None are amphibious, with the possible exception of *Eugnathogobius polylepis*, one specimen of which had been collected alive from the surface of dry mud among mangroves and other specimens obtained partly buried in exposed mud or in minuscule puddles at low tide (Larson 2009). Sniffing out the function (and reason) for the downturned narial tube would be an interesting project. To this list of Ishimatsu and Gonzales, I would add the following:

8. The basic biology of most mudskipper species still needs investigation, as the main focus has been on mudskippers of the east coast of Australia, the Arabian Gulf, Japan, and Malaysia. Most work has been done on the eastern tropical coast of Australia, Arabian Gulf, Japan, and Malaysia, with the late Toru Takita leading the way. Work is still required at these locations. Researchers in developing countries need assistance in their work, whether it is in the form of money, advice, mentoring, providing (often basic) references, or collaboration. The politics of the workplace and the government of the day unfortunately often prevent researchers providing such assistance. This works both ways, with researchers in developing countries often prevented from sending material overseas to colleagues and collaborating institutions by a range of laws and lack of funds. And the ongoing destruction of mangroves and mudflats hinders all work on mudskippers and indeed, a multitude of other estuarine species.
9. The taxonomy of mudskippers is still to be resolved. Murdy (1989) provided the first review of the subfamily, its genera, and species, recognizing 10 genera and 34 species; now there are 43 species known. However, with continuing work, it has become clear that a number of the presently accepted species of mudskippers are actually species complexes. Work on these is underway (see this volume). Additionally, recent work by several authors, using several different genes and a range of taxa, has shown that the Oxudercinae and Amblyopinae (Agorreta et al. 2013; Tornabene et al. 2013; Agorreta and Rüber 2012; Thacker 2003) are closest to the "Stenogobius group" of the Gobionellinae and the Sicydiinae, as hinted in Murdy's morphological work (Murdy 1989). Further work using a greater number of oxudercine and amblyopine species is needed, as there has been some congruence between the different studies, but the use of different genes and taxon selection has made clear comparisons difficult.
10. Further work to be done on the "extreme" species, such as the highly terrestrial *Ps. minutus*, which is able to withstand desiccation, high temperatures, and high salinities (Takeda et al. 2012) and on other species inhabiting hostile conditions, such as *Ps. waltoni* (Clayton and Snowden 2000). *Periophthalmus weberi*, while not so terrestrial, inhabits fresh to almost fresh water (unlike other mudskippers) and seems to have a narrow set of environmental conditions that it needs (Larson 2008; Khaironizam and Norma-Rashid 2003). It would be worthwhile to corroborate this and determine more aspects of its biology. The exposed shallow mudflat pools inhabited by taxa such as *Apocryptodon* can reach 40°C or more in the monsoonal summer—do the fish burrow down into the substrate to keep cool?
11. Terrestrial sound production has been recently demonstrated to occur in *Periophthalmodon septemradiatus* (Polgar et al. 2011). Acoustic communication (both in and out of water) should be investigated in more oxudercine species. Sound production (in water) has been known in gobioid fishes for some time, with several species known to vocalize during courtship, such as the gobiids *Bathygobius soporator, Gobiosoma bosc*, *Gobius niger*, and *Neogobius melanostomus* (Mok 1981), *Padogobius martensii*, *Padogobius nigricans*, *Knipowitschia punctatissima*, and *Zosterissesor ophiocephalus* (Lugli et al. 1996), and the freshwater gudgeon *Odontobutis obscura* (Takemura 1984). Polgar et al. (2011) provided a recent overview of gobioid vocalization. Amorim and Neves (2008) demonstrated that *Pomatoschistus pictus* vocalized to defend its territory; and defense appears to be the impetus for the *Periophthalmodon* vocalization reported by Polgar.
12. Bubble blowing in *S. histophorus* as reported by Townsend and Tibbetts (2005)—do other oxudercines perform this? This behavior was "frequently" observed, in which the fish emerged from

its burrow, opened its mouth, and literally blew a bubble, either to be drawn back into the mouth or spitting it out onto the mud. Townsend and Tibbetts commented that this particular action had not been recorded for any other mudskipper. Capturing an air bubble in the oral chamber is known to occur in gobiids as a result of hypoxic stress and is linked to aquatic surface respiration (Gee and Gee 1995, 1991). Bubble intake is known to occur in the terrestrial mudskippers *Periophthalmodon* and *Periophthalmus*, but the air is held within the oral and opercular chambers (Sponder and Lauder 1981). Air from the oral chamber is lost when these terrestrial species open the jaws to capture terrestrial prey, but air is maintained in the expanded gill chamber by diverticula and the ventromedial valve (Clayton 1993; Milward 1974). Milward (1974) describes captive *Periophthalmus* releasing air bubbles via the mouth and opercular opening, in a similar manner to that described by Townsend and Tibbetts (2005). Less terrestrial species, including *Scartelaos*, do not have the diverticula and ventromedial valve (Kok et al. 1998; Milward 1974).

13. Parasites. There are not many published records of parasites of mudskippers, whether external or internal. Recently, Polgar et al. (2009) described leeches on *S. tenuis* from Iran; this was the first record of an external parasite on a mudskipper. Bray and Justine (2013), while describing a new species of digenean parasite from *Ps. argentilineatus* from New Caledonia, documented some of the internal parasites of mudskippers, noting that few helminths, cestodes, copepods, isopods, leeches, monogeneans, or adult nematodes were found in their specimens of *Ps. argentilineatus*. A study of mudskippers from Kuwait Bay apparently found no parasites, internal or external; it was unclear which species of *Periophthalmus* (probably *Ps. waltoni*) were sampled (Al-Behbehani and Ebrahim 2010). A full survey of oxudercine internal parasites remains to be done. External parasites of mudskippers are clearly less well known, with Polgar et al. (2009) commenting that "... mudskippers' ectoparasites should be specifically adapted to sub-aerial conditions and peculiar host behaviors, such as air gulping (Graham 1997), air-phase maintenance in burrows (Ishimatsu et al. 2000, 1998), prolonged exposure to air (Sayer and Davenport 1991), rapid and drastic temperature changes out of water (Tytler and Vaughan 1983), and intense illumination." At least one instance of a cymothoid isopod parasite is known (Figure 15.9) from a *Ps. argentilineatus* in a north Queensland mangroves; the photographer was unable to catch the fish. The parasites of mudskippers, external and internal, would make a fascinating study topic.

Figure 15.9 Parasitic isopod posterior to right pectoral-fin base of *Periophthalmus argentilineatus*, Cowley Bay, Queensland, Australia. (Courtesy of Russell Constable.)

So, there are at least 13 project themes to be worked on, along with ongoing studies. How to progress such studies further? I have yet to meet a Fairy Godmother and mudskippers need one.

REFERENCES

ABC Fact Check. 2014. *Fact check: Science, research and innovation spending cut to 'historic low'*. http://www.abc.net.au/news/2014-10-07/adam-bandt-research-development-spending-claim-checks-out/5789134 (accessed September 2015).

Agorreta A. and Ruber L. 2012. A standardized reanalysis of molecular phylogenetic hypothese of Gobioidei. *Systematics and Biodiversity* 10(3): 375–390.

Agorreta A., San Mauro D., Schliewen U., et al. 2013. Molecular phylogenetics of Gobioidei and phylogenetic placement of European gobies. *Molecular Phylogenetics and Evolution* 69(3): 619–633.

Akihito P., Hayashi M. and Yoshino T. 1988. Suborder Gobioidei. In *The Fishes of the Japanese Archipelago* (eds. H. Masuda, K. Amaoka, C. Araga, T. Uyeno and T. Yoshino) pp. 236–289. Tokyo: Tokai University Press.

Al-Behbehani B.E. and Ebrahim H.M.A. 2010. Environmental studies on the mudskippers in the intertidal zone of Kuwait Bay. *Nature and Science* 8(5):79–89.

Allen G.R., Midgley S.H. and Allen M. 2002. *Field Guide to the Freshwater Fishes of Australia*. Perth: Western Australian Museum, 394 pp.

Amorim M.C.P. and Neves A.S.M. 2008. Male painted gobies (*Pomatoschistus pictus*) vocalise to defend territories. *Behaviour* 145:1065–1083.

Baeck G.W., Takita T. and Yoon Y.H. 2008. Lifestyle of Korean mudskipper *Periophthalmodon magnuspinnatus* with reference to a congeneric species *Periophthalmus modestus*. *Ichthyological Research* 55:43–55.

Berti R., Chelazzi L., Colombini I. and Ercolini A. 1992. Direction-finding ability in a mudskipper from the delta of the Tana river (Kenya). *Tropical Zoology* 5: 219–228.

Birdlife Northern Queensland. 2013. *Migratory Shorebirds*. Cairns Esplanade: Birding Information No. 11.

Bray R.A. and Justine J-L. 2013. A digenean parasite in a mudskipper: *Opegaster ouemoensis* sp. n. (Digenea: Opecoelidae) in *Periophthalmus argentilineatus* Valenciennes (Perciformes: Gobiidae) in the mangroves of New Caledonia. *Folia Parasitologica* 60(1):7–16.

Brillet C. 1976. Structure du terrier, reproduction et comportement des jeunes chez le poisson amphibie *Periophthalmus sobrinus* Eggert. *Revue d'Ecologie, (la Terre et la Vie)* 30:465–483. [paper not seen]

Brown C. 2015. Fish intelligence, sentience and ethics. *Animal Cognition* 18(1):1–17.

Chaloupka G. 1993. *Journey in Time: The World's Longest Continuing Art Tradition*. Chatswood: Reed, 256 pp.

Clayton D.A. 1993. Mudskippers. *Oceanography and Marine Biology: An Annual Review* 31: 507–577.

Clayton D.A. and Snowden R. 2000. Surface activity in the mudskipper, *Periophthalmus waltoni* Koumans 1941 in relation to prey activity and environmental factors. *Tropical Zoology* 13:239–249.

Duke N.C., Bell A.M., Pederson, D.K., Roelfsema C.M. and Nash S.B. 2005. Herbicides implicated as the cause of severe mangrove dieback in the Mackay region, NE Australia: Consequences for marine plant habitats of the GBR World Heritage Area. *Marine Pollution Bulletin* 51:308–324.

Duke N. 2006. *Australia's Mangroves. The Authoritative Guide to Australia's Mangrove Plants*. Brisbane: University of Queensland, 200 pp.

Duke N., Meynecke J-O., Dittmann S. et al. 2007. A world without mangroves. *Science* 317:41–42.

Etou A., Takeda T., Yoshida Y. and Ishimatsu A. 2007. Oxygen consumption during embryonic development of the mudskipper (*Periophthalmus modestus*): Implication for the aerial development in burrows. In *Fish Respiration and Environment* (eds. M.N. Fernanes, F.T. Rantin, M.L. Glass and B.G. Kapoor) pp. 83–91. Enfield, CT: Science Publishers.

Gee J.H. and Gee P.A. 1991. Reactions of gobioid fishes to hypoxia: Buoyancy control and aquatic surface respiration. *Copeia* 1:17–28.

Gee J.H. and Gee P.A. 1995. Aquatic surface respiration, buoyancy control and the evolution of air breathing in gobies (Gobiidae: Pisces). *Journal of Experimental Biology* 198:79–89.

Harms W.J.H. 1934. *Wandlungen des Artgefüges Unter Natürlichen und Künstlichen Umweltbedingungen: Beobachtungen an Tropischen Verlandungszonen und am Verlandenden Federsee*. Leipzig: Johan Ambrosius Barth, 212 pp. [paper not seen]

Hoese D.F. and Kottelat M. 2005. *Bostrychus microphthalmus*, a new microphthalmic cavefish from Sulawesi (Teleostei: Gobiidae). *Ichthyological Exploration of Freshwaters* 16(2):183–191.

Hoese D.F. and Larson H.K. 2006. Gobiidae. Gobies. In *Zoological catalogue of Australia* (eds. P.L. Beesley and A. Wells) Vol. 35, Parts 1–3, pp. 1612–1697. Canberra: ABRS and CSIRO Publishing.

Ishimatsu A., Hishida Y., Takita T. et al. 1998. Mudskippers store air in their burrows. *Nature* 391: 237–238.

Ishimatsu A., Yoshida Y., Itoki N., Takeda T., Lee H.J. and Graham J.B. 2007. Mudskippers brood their eggs in air but submerge them for hatching. *The Journal of Experimental Biology* 210:3946–3954.

Ishimatsu A. and Gonzales T.T. 2011. Mudskippers: Front runners in the modern invasion of land. In *The Biology of Gobies* (eds. R.A. Patzner, J.L. Van Tassell, M. Kovačić and B.G. Kapoor) pp. 609–638. Enfield, CT: Science Publishers.

Khaironizam M.Z. and Norma-Rashid Y. 2003. First record of the mudskipper, *Periophthalmodon septemradiatus* (Hamilton) (Teleostei: Gobiidae) from Peninsular Malaysia. *Raffles Bulletin of Zoology* 51:97–100.

Kok W.K., Lim C.B., Lam T.J. and Ip Y.K., 1998. The mudskipper *Periophthalmodon schlosseri* respires more efficiently on land than in water and vice versa for *Boleophthalmus boddaerti*. *Journal of Experimental Zoology* 280:86–90.

Kruitwagen, G., Hecht T., Pratap H.B. and Wendelaar Bonga, S.E. 2006. Changes in morphology and growth of the mudskipper (*Periophthalmus argentilineatus*) associated with coastal pollution. *Marine Biology* 149:201–211.

Larson H. 2001. A revision of the gobiid fish genus *Mugilogobius* (Teleostei: Gobioidei), and its systematic placement. *Records of the Western Australian Museum Supplement* 62:1–233.

Larson H.K. 2008. A new species of the gudgeon *Bostrychus* (Teleostei: Gobioidei: Eleotridae), from peninsular Malaysia. *The Beagle, Records of the Museums and Art Galleries of the Northern Territory* 24:147–150.

Larson H.K. 2009. Review of the gobiid fish genera *Eugnathogobius* and *Pseudogobiopsis* (Gobioidei: Gobiidae: Gobionellinae), with descriptions of three new species. *Raffles Bulletin of Zoology* 57(1):127–181.

Larson H.K. and Takita T. 2004. Two new species of *Periophthalmus* (Teleostei: Gobiidae: Oxudercinae) from Northern Australia, and a re-diagnosis of *Periophthalmus novaeguineaensis*. *The Beagle, Records of the Museums and Art Galleries of the Northern Territory* 20:175–185.

Larson H.K., Williams R.S. and Hammer M.P. 2013. The fish fauna of the Northern Territory: An annotated checklist. *Zootaxa* 3696:1–293.

Lee H.J., Martinez C.A., Hertzberg. K.J., Hamilton A.L. and Graham J.B. 2005. Burrow air phase maintenance and respiration by the mudskipper Scartelaos histrophorus (Gobiidae: Oxudercinae). *The Journal of Experimental Biology* 208: 169–177.

Lugli M., Torricelli P., Pavan G. and Miller P.J. 1996. Breeding sounds of male *Padogobius nigricans* with suggestions for further evolutionary study of vocal behaviour in gobioid fishes. *Journal of Fish Biology* 49:648–657.

MacKinnon J., Verkuil Y.I. and Murray N. 2012. IUCN situation analysis on East and Southeast Asian intertidal habitats, with particular reference to the Yellow Sea (including the Bohai Sea). *Occasional Paper of the IUCN Species Survival Commission* 47:70.

Milward N.E. 1974. Studies on the taxonomy, ecology and physiology of Queensland mudskippers. PhD thesis, University of Queensland.

Mok H.K. 1981. Sound production in the naked goby, *Gobiosoma bosci* (Pisces, Gobiidae)—A preliminary study. In *Hearing and Sound Communication in Fishes* (eds. W.N. Tavolga, A.N. Popper and R.R. Fay) pp. 447–456. New York, NY: Springer-Verlag.

Murdy E.O. 1989. A taxonomic revision and cladistic analysis of the oxudercine gobies (Gobiidae: Oxudercinae). *Records of the Australian Museum Supplement* 11:1–93.

Murdy E.O. 2003. A review of *Amblyotrypauchen* (Teleostei: Gobiidae), a genus of blind amblyopine gobies. *Proceedings of the Biological Society of Washington* 116(2):330–336.

Murdy E.O. 2006. A revision of the gobiid fish genus *Trypauchen* (Gobiidae: Amblyopinae). *Zootaxa* 1343:55–68.

Murdy E.O. 2008a. *Paratrypauchen*, a new genus for *Trypauchen microcephalus* Bleeker, 1860, (Perciformes: Gobiidae: Amblyopinae) with a redescription of *Ctenotrypauchen chinensis* Steindachner, 1867, and a key to '*Trypauchen*' group of genera. *aqua, International Journal of Ichthyology* 14(3):115–128.

Murdy E.O. 2008b. *Trypauchenichthys larsonae*, a new species of amblyopine goby from Australia (Gobiidae: Amblyopinae) with a key to the species in the genus. *aqua, International Journal of Ichthyology* 14(2):59–68.

Murdy E.O. and Shibukawa K. 2001. A revision of the gobiid fish genus *Odontamblyopus* (Gobiidae: Amblyopinae). *Ichthyological Research* 48:31–43.

Murdy E.O. and Shibukawa K. 2002. Redescription of the gobiid fish genus *Pseudotrypauchen* (Gobiidae: Amblyopinae) and its significance in amblyopine phylogeny. *Marine and Freshwater Research* 53(2):253–258.

Murdy E.O. and Shibukawa K. 2003. A revision of the Indo-Pacific fish genus *Caragobius* (Gobiidae: Amblyopinae). *Zootaxa* 301:1–12.

Murdy E.O. and Ferraris C.J. 2003. *Gymnoamblyopus novaeguineae*, a new genus and species of worm goby from Papua New Guinea (Gobiidae: Amblyopinae). *Zootaxa* 150:1–6.

Nursall J. 1981. Behavior and habitat affecting the distribution of five species of mudskipper in Queensland. *Bulletin of Marine Science* 31(3): 730–735.

Parliamentary Library. 2014. Australian government support for R&D, 1978–79 to 2014–15. Client advice.

Polgar G., Burreson E.M., Stefani F. and Kamrani E. 2009. Leeches on mudskippers: Host–parasite interaction at the water's edge. *Journal of Parasitology* 95(4):1021–1025.

Polgar G. 2008. Species-area relationship and potential role as a biomonitor of mangrove communities of Malayan mudskippers. *Wetlands Ecology and Management* 17(2):157–164.

Polgar G. and Sasekumar A. 2010. The potential of ecotourism and flagship species, and the urgent need for more holistic approaches and higher levels of connectivity for endangered tropical mudflats and coastal swamps. In *Ecotourism: Management, Development and Impact* (eds. A. Krause and E. Weir) pp. 43–87. Hauppauge, NY: Nova Science Publishers.

Polgar G., Malavasi S., Cipolato G., Georgalas V., Clack J.A. and Torricelli P. 2011. Acoustic communication at the water's edge: Evolutionary insights from a mudskipper. *PLoSOne* 6(6):1–9.

Polgar G., Sacchetti A. and Galli P. 2010. Differentiation and adaptive radiation of amphibious gobies (Gobiidae: Oxudercinae) in semi-terrestrial habitats. *Journal of Fish Biology* 77:1645–1664.

Ports North. 2014. Cairns shipping development project. *Draft Environmental Impact Statement* 2:38.

Sponder D.L, and Lauder G.V. 1981. Terrestrial feeding in the mudskipper Periophthalmus (Pisces: Teleostei): a cineradiograhic analysis. *Journal of Zoology, London* 193: 517–530.

Takeda T., Hayashi M., Toba A., Soyano K. and Ishimatsu A. 2012. Ecology of the Australian mudskipper *Periophthalmus minutus*, an amphibious fish inhabiting a mudflat in the highest intertidal zone. *Australian Journal of Zoology* 59(5):312–320.

Takemura A. 1984. Acoustical behaviour of the freshwater goby *Odontobutis obscura*. *Bulletin of the Japanese Society of Scientific Fisheries* 59:561–564. [paper not seen]

Takita T., Larson H.K. and Ishimatsu A. 2011. The natural history of mudskippers in northern Australia, with field identification characters. *The Beagle, Records of the Museums and Art Galleries of the Northern Territory* 27:189–204.

Thacker C.E. 2003. Molecular phylogeny of the gobioid fishes (Teleostei: Perciformes: Gobioidei). *Molecular Phylogenetics and Evolution* 26: 354–368.

Toba A. and Ishimatsu A. 2014. Roles of air stored in burrows of the mudskipper *Boleophthalmus pectinirostris* for adult respiration and embryonic development. *Journal of Fish Biology* 84:774–793.

Tornabene L., Chen Y. and Pezold F. 2013. Gobies are deeply divided: phylogenetic evidence from nuclear DNA (Teleostei: Gobioidei: Gobiidae). *Systematics and Biodiversity* 11(3): 345–361.

Townsend K.A. and Tibbetts I.R. 2005. Behaviour and sexual dimorphism of the blue mudskipper, *Scartelaos histophorus* (Pisces: Gobiidae). *Proceedings of the Royal Society of Queensland* 112:53–62.

Whitley G.P. 1970. *Early History of Australian Zoology*. Mosman: Royal Zoological Society of New South Wales, 75 pp.

Zhang J., Taniguchi T., Takita T. and Ali A.B. 2000. On the epidermal structure of *Boleophthalmus* and *Scartelaos* mudskippers with reference to their adaptation to terrestrial life. *Ichthyological Research* 47(4):359–366.

Index

For Product Safety Concerns and Information please contact our EU
representative GPSR@taylorandfrancis.com Taylor & Francis Verlag GmbH,
Kaufingerstraße 24, 80331 München, Germany

Batch number: 10722098

Printed by Printforce, the Netherlands